前言 Preface

再版說明

初版至今已近 3 年，其間 AI 發展迅速，各項套件 / 演算法均有重大變革，例如大語言模型（LLM）、生成式 AI（Generative AI）、Keras v3、YOLO v11、Gymnasium，全書實作的範例程式幾乎全部翻修，並新增更多應用程式，也包括更多的演算法說明，例如 Transformer、Diffusion Model…等。

為何撰寫本書

從事機器學習教育訓練已近 6 年，其間也在『IT 邦幫忙』撰寫 100 多篇的文章（https://ithelp.ithome.com.tw/users/20001976/articles），從學員及讀者的回饋獲得許多寶貴意見，期望能將整個歷程集結成冊，同時，相關領域的進展也在飛速變化，過往的文章內容需要翻新，因此藉機再重整思緒，想一想如何能將演算法的原理解釋得更簡易清晰，協助讀者跨入 AI 的門檻，另外，也避免流於空談，盡量增加應用範例，希望能達到即學即用，不要有過多理論的探討。

AI 是一個將資料轉化為知識的過程，演算法就是過程中的生產設備，最後產出物是模型，再將模型植入各種硬體裝置，例如電腦、手機、智慧音箱、自駕車、醫療診斷儀器、…等，這些裝置就擁有特殊專長的智慧，再進一步整合各項技術就構建出智慧製造、智慧金融、智慧交通、智慧醫療、智慧城市、智慧家庭、…等應用系統。AI 的應用領域如此的廣闊，個人精力有限，當然不可能具備十八般武藝，樣樣精通，惟有從基礎紮根，再擴及有興趣的領域，因此，筆者撰寫這本書的初衷，非常單純，就是希望讀者在紮根的過程中，貢獻一點微薄的力量。

本書主要的特點

由於筆者身為統計人，希望能『以統計/數學為出發點』，介紹深度學習必備的數理基礎，但又不希望內文有太多數學公式的推導，讓離開校園已久的在職者看到一堆數學符號就心生恐懼，因此，嘗試以『程式設計取代定理證明』，縮短學習歷程，增進學習樂趣。

TensorFlow 2.x 版有巨大的變動，預設模式改為 Eager Execution，並以 Keras 為主力，整合 TensorFlow 其他模組，形成完整的架構，本書期望對 TensorFlow/Keras 架構作完整性的介紹。

演算法介紹以理解為主，輔以大量圖表說明，摒棄長篇大論。

完整的範例程式及各種演算法的延伸應用，以實用為要，希望能觸發讀者靈感，能在專案或產品內應用。

介紹日益普及的演算法與相關套件的使用，例如 YOLO（物件偵測）、GAN（生成對抗網路）/DeepFake（深度偽造）、OCR（辨識圖像中的文字）、臉部辨識、BERT/Transformer、ChatBot、強化學習、自動語音辨識（ASR）等。

目標對象

深度學習的入門者：必須熟悉 Python 程式語言及機器學習基本概念。

資料工程師：以應用系統開發為職志，希望能應用各種演算法，進行實作。

資訊工作者：希望能擴展深度學習知識領域。

從事其他領域的工作，希望能一窺深度學習奧秘者。

閱讀重點

第一章介紹 AI 的發展趨勢，鑑古知今，瞭解前兩波 AI 失敗的原因，比較第三波發展的差異性。

第二章介紹深度學習必備的統計/數學基礎，不僅要理解相關知識，也力求能撰寫程式解題。

第三章介紹 TensorFlow/Keras 基本功能，包括張量（Tensor）運算、自動微分及神經網路模型的組成，並說明梯度下降法求解的過程。

第四章開始實作，依照機器學習 10 項流程，撰寫完整的範例，包括 Web、桌面程式。

第五章介紹 TensorFlow 進階功能，包括各種工具，如 TensorBoard、TensorFlow Serving、Callbacks。

第六～八章介紹圖像/視訊的演算法及各式應用。

第九～十一章介紹生成式 AI，包括 Encoder-decoder、生成對抗網路（GAN）及擴散模型（Diffusion Model），也說明 Stable Diffusion 及 Dall-E 實作。

第十二章介紹各種影像應用，包括臉部辨識、OCR、車牌辨識及圖像去背。

第十三～十四章介紹自然語言處理及各式應用。

第十五章介紹大型語言模型（Large Language Model, LLM），包括 Transformer 演算法、ChatGPT 實作及企業導入實務。

第十六章介紹語音辨識的原理/演算法/實作/應用，包括 OpenAI Whisper 實作及應用範例。

第十七章介紹強化學習（Reinforcement Learning, RL）的概念/演算法/實作/應用，包括 Gymnasium、Stable Baselines3 套件實作及應用範例。

本書範例程式碼全部收錄在 https://github.com/mc6666/DL_Book2，還包括各章的參考資料及超連結，方便使用者複製。

致謝

原本只想整理過往文章集結成書，沒想到相關技術發展太快，幾乎全部重寫，結果比預計時程多花了三倍的時間才完成，但因個人能力有限，還是有許多議題成為遺珠之憾，仍待後續的努力，過程中要感謝家人的默默支持，也感謝深智出版社的大力支援，使本書得以順利出版。

由於個人能力有限，內容如有疏漏、謬誤或有其他建議，
歡迎來信指教（mkclearn@gmail.com）。

目錄 Contents

第一篇　深度學習導論

第 1 章　深度學習（Deep Learning）導論

1-1　人工智慧的三波浪潮 1-1
1-2　人工智慧的未來趨勢 1-4
1-3　AI 的學習地圖 1-5
1-4　機器學習應用領域 1-7
1-5　機器學習開發流程 1-8
1-6　開發環境安裝 1-10
　　　參考資料（References）... 1-16

第 2 章　神經網路（Neural Network）原理

2-1　必備的數學與統計知識 2-2
2-2　線性代數（Linear Algebra）... 2-4
　　2-2-1　向量（Vector）... 2-5
　　2-2-2　矩陣（Matrix）... 2-10
　　2-2-3　聯立方程式求解 2-14
2-3　微積分（Calculus）... 2-17

	2-3-1	微分（Differentiation）	2-18
	2-3-2	微分定理	2-24
	2-3-3	偏微分（Partial Differentiation）	2-27
	2-3-4	線性迴歸求解	2-32
	2-3-5	積分（Integration）	2-39
2-4	機率（Probability）與統計（Statistics）		2-44
	2-4-1	資料型態	2-45
	2-4-2	抽樣（Sampling）	2-47
	2-4-3	基礎統計（Statistics Fundamentals）	2-51
	2-4-4	機率（Probability）	2-61
	2-4-5	機率分配（Distribution）	2-70
	2-4-6	假設檢定（Hypothesis Testing）	2-81
	2-4-7	小結	2-96
2-5	線性規劃（Linear Programming）		2-97
2-6	最大概似法（MLE）		2-102
2-7	神經網路（Neural Network）求解		2-107
	2-7-1	神經網路（Neural Network）	2-107
	2-7-2	梯度下降法（Gradient Descent）	2-111
	2-7-3	神經網路求解	2-116
	參考資料（References）		2-118

第二篇　TensorFlow 基礎篇

第 3 章　TensorFlow 架構與主要功能

3-1	常用的深度學習套件	3-1
3-2	TensorFlow 架構	3-3

3-3	張量（Tensor）運算	3-5
3-4	自動微分（Automatic Differentiation）	3-11
3-5	神經層（Neural Network Layer）	3-17
	參考資料（References）	3-23

第 4 章　神經網路實作

4-1	撰寫第一支神經網路程式	4-2
	4-1-1　最簡短的程式	4-2
	4-1-2　程式強化	4-4
	4-1-3　實驗	4-16
4-2	Keras 模型種類	4-24
	4-2-1　順序型模型（Sequential Model）	4-24
	4-2-2　Functional API	4-28
4-3	神經層（Layer）	4-30
	4-3-1　完全連接神經層（Dense Layer）	4-31
	4-3-2　Dropout Layer	4-34
4-4	激勵函數（Activation Function）	4-34
4-5	損失函數（Loss Functions）	4-40
4-6	優化器（Optimizer）	4-43
4-7	效能衡量指標（Performance Metrics）	4-49
4-8	超參數調校（Hyperparameter Tuning）	4-55
	參考資料（References）	4-59

第 5 章　TensorFlow 常用指令與功能

5-1	特徵轉換	5-2
5-2	模型存檔與載入（Model Saving and Loading）	5-4

5-3	模型彙總與結構圖（Summary and Plotting）		5-11
5-4	回呼函數（Callbacks）		5-14
	5-4-1	EarlyStopping Callback	5-15
	5-4-2	ModelCheckpoint Callback	5-16
	5-4-3	TensorBoard Callback	5-17
	5-4-4	自訂 Callback	5-19
	5-4-5	自訂 Callback	5-22
	5-4-6	取得優化器的學習率變化	5-23
	5-4-7	小結	5-26
5-5	TensorBoard		5-26
	5-5-1	TensorBoard 功能	5-26
	5-5-2	測試	5-28
	5-5-3	寫入圖片	5-31
	5-5-4	效能調校（Performance Tuning）	5-32
	5-5-5	敏感度分析（What-If Tool, WIT）	5-34
	5-5-6	小結	5-35
5-6	模型佈署（Deploy）		5-35
	5-6-1	網頁開發	5-36
	5-6-2	桌面程式開發	5-37
5-7	TensorFlow Dataset		5-40
	5-7-1	產生 Dataset	5-41
	5-7-2	圖像 Dataset	5-46
	5-7-3	TFRecord 與 Dataset	5-48
	5-7-4	TextLineDataset	5-52
	5-7-5	Dataset 效能提升	5-55
	參考資料（References）		5-56

第 6 章 卷積神經網路（Convolutional Neural Network）

6-1	卷積神經網路簡介	6-2
6-2	卷積（Convolution）	6-4
6-3	濾波器（Filter）	6-7
6-4	池化層（Pooling Layer）	6-13
6-5	CNN 模型實作	6-14
6-6	資料增補（Data Augmentation）	6-22
6-7	可解釋的 AI（eXplainable AI, XAI）	6-33
6-8	卷積神經網路的缺點	6-41
	參考資料（References）	6-42

第 7 章 預先訓練的模型（Pre-trained Model）

7-1	預先訓練模型的簡介	7-2
7-2	採用完整模型	7-5
7-3	採用部分模型	7-11
7-4	轉移學習（Transfer Learning）	7-16
7-5	Batch Normalization 說明	7-21
	參考資料（References）	7-26

第三篇　進階的影像應用

第 8 章 物件偵測（Object Detection）

8-1	圖像辨識模型的發展	8-2
8-2	影像金字塔與滑動視窗	8-4
8-3	方向梯度直方圖（HOG）	8-7

8-3	R-CNN 系列演算法	8-20
8-4	R-CNN 改良	8-34
8-5	YOLO 演算法簡介	8-39
8-6	YOLO 訓練與推論	8-43
8-7	YOLO 各項功能	8-43
8-8	圖像分類（Image Classification）	8-45
8-9	物件偵測（Object Detection）	8-49
	8-9-1　YOLO 物件偵測（Object Detection）	8-50
	8-9-2　TensorFlow Object Detection API	8-54
8-10	資料標記（Data Annotation）	8-56
8-11	物件偵測的效能衡量指標	8-58
8-12	實例分割（Instance Segmentation）	8-59
8-13	姿態辨識（Pose Estimation）	8-63
8-14	旋轉邊界框物件偵測（Oriented Bounding Boxes Object Detection）	8-67
8-15	物件追蹤（Object Tracking）	8-71
8-16	YOLO 測試心得	8-76
8-17	總結	8-76
	參考資料（References）	8-77

第 9 章　生成式 AI（Generative AI）

9-1	編碼器與解碼器（Encoder-decoder）	9-3
9-2	自動編碼器（AutoEncoder）	9-3
9-3	變分自編碼器（Variational AutoEncoder）	9-9
9-4	Conditional VAE	9-16
9-5	U-Net	9-19
9-6	風格轉換（Style Transfer）-- 人人都可以是畢卡索	9-26
9-7	快速風格轉換（Fast Style Transfer）	9-38

9-8	本章小結	9-42
	參考資料（References）	9-43

第 10 章 生成對抗網路（GAN）

10-1	生成對抗網路介紹	10-2
10-2	生成對抗網路種類	10-5
10-3	DCGAN	10-8
10-4	Progressive GAN	10-22
10-5	Conditional GAN	10-30
10-6	Pix2Pix	10-35
10-7	CycleGAN	10-49
10-8	CartoonGAN	10-60
10-9	GAN 挑戰	10-62
10-10	深度偽造（Deepfake）	10-63
	參考資料（References）	10-67

第 11 章 擴散模型（Diffusion Model）

11-1	擴散模型（Diffusion Model）原理	11-2
11-2	擴散模型（Diffusion Model）實作	11-7
11-3	MidJourney 簡介	11-11
11-4	Stable Diffusion 簡介	11-13
11-5	DreamStudio 使用	11-15
11-6	Stable Diffusion 本機安裝	11-16
11-7	Stable Diffusion API	11-22
11-8	Stable Diffusion Extension	11-26
11-9	ControlNet in Diffusers	11-29

11-10	NitroFusion	11-37
11-11	OpenAI DALL‧E	11-39
11-12	本章小結	11-42
	參考資料（References）	11-43

第 12 章 其他影像應用

12-1	臉部辨識（Facial Recognition）	12-1
12-2	臉部偵測（Face Detection）	12-3
12-3	MTCNN 演算法	12-8
12-4	臉部追蹤（Face Tracking）	12-11
12-5	臉部特徵點偵測	12-18
12-6	臉部驗證（Face Verification）	12-26
12-7	光學文字辨識（OCR）	12-29
12-8	EasyOCR	12-33
12-9	車牌辨識（ANPR）	12-39
12-10	影像去背（Background Removing）	12-44
12-11	本章小結	12-48
	參考資料（References）	12-49

第一篇
深度學習導論

近幾年（2022~）生成式 AI（Generative AI）已引爆另一波 AI 狂潮，從 2022/07/12 發表的 Midjourney 到 2022/11/30 上線的 ChatGPT，顛覆世人的想像，只要輸入一段咒語（正式名稱為提示, Prompt），就可以生成一幅令人驚豔的圖像或是一篇論文，之後，各種大語言模型（LLM）及圖像生成模型（Diffusion Model）不斷被訓練出來，科技大廠 Google、Microsoft、Meta、Telsa 皆投入巨資，追逐此一巨大的商機，也因此造就硬體廠商的蓬勃發展，包括 NVidia 的市值碾壓 CPU 霸主 Intel，台灣協力廠商也因而受惠，股票價格一飛衝天，到底這波熱潮會如何發展，讓我們一起跟上這波大潮，盡情衝浪吧。

在進入深度學習的殿堂前，學生常會想問以下的問題：

1. 人工智慧已歷經三波浪潮，這一波是否又將進入寒冬？
2. 人工智慧、資料科學、資料探勘、機器學習、深度學習到底有何關聯？
3. 機器學習開發流程與一般應用系統開發有何差異？
4. 深度學習的學習路徑為何？建議從哪裡開始？
5. 為什麼要先學習數學與統計，才能學好深度學習？
6. 先學哪一種深度學習框架比較好？TensorFlow 或 PyTorch？
7. 如何準備開發環境，進行模型建構及推論？

第一章就先針對以上的問題作一簡單的介紹，第二章則介紹深度學習必備的數學與統計基礎知識，有別於學校的教學，我們會偏重在程式的實作。

第 1 章

深度學習
（Deep Learning）導論

1-1 人工智慧的三波浪潮

人工智慧（Artificial Intelligence, AI）並不是最近幾年才興起的，其實它已經是第三波熱潮了，前兩波都經歷 10 餘年，就邁入寒冬，這一波熱潮至今也已經十多年了（2010~），是否又將邁入寒冬呢？我們就先來重點回顧一下過往的歷史吧。

第 1 章 深度學習（Deep Learning）導論

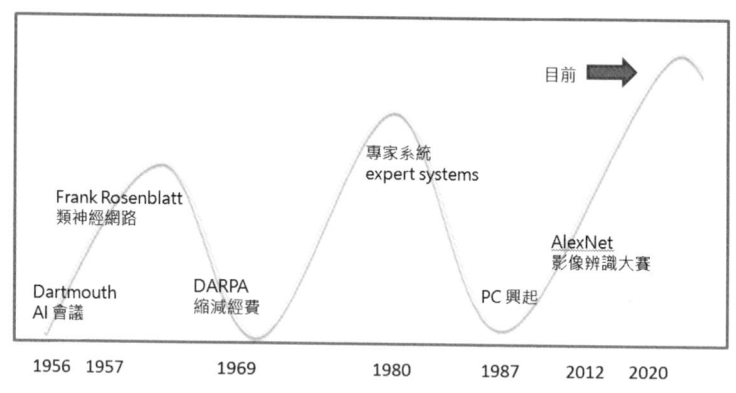

▲ 圖 1.1 人工智慧的三波浪潮

1. 1956 年在達特茅斯（Dartmouth）學院舉辦 AI 會議，確立第一波浪潮的開始。

2. 1957 年 Frank Rosenblatt 創建感知器（Perceptron），即簡易的神經網路，然而當時並無法解決複雜多層的神經網路，直至 1980 年代才被想出解決辦法。

3. 1969 年美國國防部（DARPA）基於投資報酬率過低，決定縮減 AI 研究經費，AI 邁入第一波寒冬。

4. 1980 年專家系統（Expert System）興起，企圖將各行各業專家的內隱知識，外顯為一條條的規則，從而建立起專家系統，不過因陳義過高，且需要使用大型且昂貴的電腦設備，才能夠建構相關系統，又不巧適逢個人電腦（PC）的流行，相較之下，鋒頭就被掩蓋下去了，至此，AI 邁入第二波寒冬。

5. 2012 年多倫多大學 Geoffrey Hinton 研發團隊利用分散式計算環境及大量影像資料，結合過往的神經網路知識，開發了 AlexNet 神經網路，參加 ImageNet 影像辨識大賽，大放異彩，一下子把錯誤率降低了 10% 以上，就此興起 AI 第三波浪潮，至今仍方興未艾。

第三波熱潮至今已逾十餘年（2010~），會不會又將邁入寒冬呢？觀察這一波熱潮，相較於過去前兩波，具備了以下幾項的優勢：

1-1 人工智慧的三波浪潮

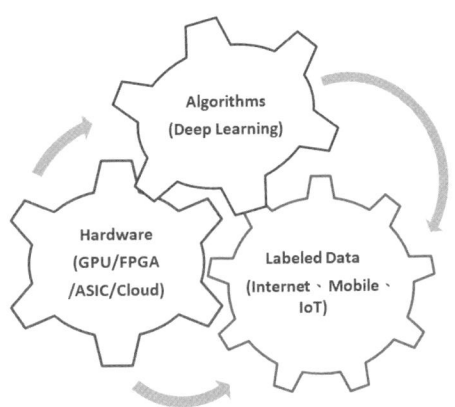

▲ 圖 1.2 第三波 AI 浪潮的觸媒

1. 先架構基礎功能，從影像、語音、文字辨識開始，再逐步往上構建各式的應用，例如自駕車（Self Driving）、對話機器人（ChatBot）、機器人（Robot）、智慧醫療、智慧城市…等，這種由下往上的發展方式比較扎實。

2. 硬體的快速發展：

    2.1 摩爾定律的發展速度：IC 上可容納的電晶體數目，約每隔 18 個月至兩年便會增加一倍，簡單來說，就是 CPU 每隔兩年便會增快一倍，過去 50 年均循此軌跡發展，此定律在未來十年也應該會繼續適用，之後也許量子電腦（Quantum Computing）等新科技會繼續接棒，它號稱目前電腦要計算幾百年的工作，量子電腦只要 30 分鐘就搞定了，如果成真，那時可能又是另一番光景了。

    2.2 雲端資料中心的建立：大型 IT 公司在世界各地興建大型的資料中心，採取『用多少付多少』（Pay as you go）。由於模型訓練時，通常需要大量運算，採用雲端方案，一般企業就不須在前期購買昂貴設備，僅需支付必要運算的費用，也不需冗長的採購流程，只要幾分鐘就能開通（Provisioning）所需設備，省錢省時。2025 年的 AI 伺服器也將引爆各 IT 大廠的搶購，為資料中心增添強盛軍備。

    2.3 GPU/NPU/TPU 的開發：深度學習主要是以張量運算為基礎，GPU 這方面的計算能力比 CPU 快上很多倍，專門生產 GPU 的美商輝達（NVidia）公司因而大放異彩，市值超越 Intel[1]，當然其他硬體及系統廠商也不會錯失如此龐大的商機，因此各式的 NPU（Neural-network Processing Unit）或 xPU 紛紛出籠，積極搶食這塊大餅，另外，今年（2025）應該會出現許多 AI PC，透過大型

第 1 章　深度學習（Deep Learning）導論

模型量化（Quantization），不管是手機或單晶片電腦（Raspberry Pi）都有望安裝影像/語言模型進行推論，『邊緣運算』（Edge Computing）將愈來愈盛行，例如手機、PC、機器人、無人機…等。

3. 演算法推陳出新：過去由於計算能力受限，許多無法在短時間訓練完成的演算法紛紛出爐，尤其是神經網路，現在已經可以建構上百層的模型，運算的參數也可以高達上兆個，都能在短時間內調校出最佳模型，因此，模型設計就可以更複雜，推理能力也能夠更完備。

4. 大量資料的蒐集及標記（Labeling）：人工智慧必須仰賴大量資料，讓電腦學習，從中挖掘知識，近年來網際網路（Internet）及手機（Mobile）盛行，企業可以透過社群媒體蒐集到大量資料，再加上物聯網（IoT）也可藉由感測器產生源源不斷的資料，作為深度學習的養份（訓練資料），科技大廠銀彈充足，可以雇用大量的人力，進行資料標記，提升資料品質，使得訓練出來的模型越趨精準。

因此，綜合以上的趨勢發展，第三波的 AI 熱潮會邁入寒冬短期內還看不到吧。

1-2 人工智慧的未來趨勢

OpenAI 認為 AI 未來的發展應朝著『人工通用智慧』（Artificial General Intelligence, AGI）[2] 方向前進，到達那個境界時，電腦將比人類還要聰明，OpenAI CEO Sam Altman 將 AI 進展分為 5 個層次（Levels），如下：

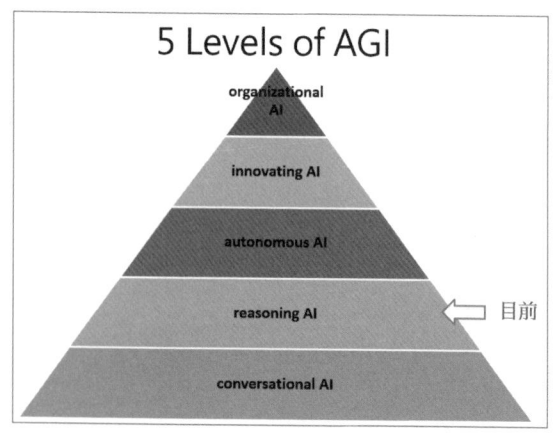

▲ 圖 1.3　AGI 5 個層次

1. Conversational AI：電腦能夠與人類溝通，例如 ChatGPT、CoPilot…等。
2. Reasoning AI：電腦能夠解決基礎的推理問題，包括數學求解及邏輯判斷，例如 ARC 2024 競賽題目[3]，類似 IQ 測驗的題目。

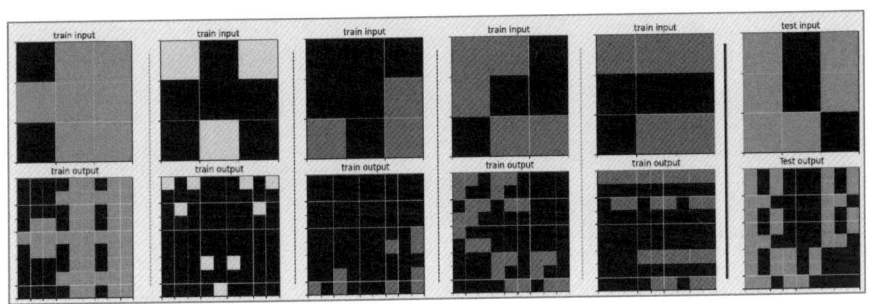

3. Autonomous AI：電腦可擔任『代理人』（Agent），能獨立下決策解決問題。
4. Innovating AI：不只能執行業務，更能改善流程，思考如何作的更好。
5. Organizational AI：電腦間能協同合作，互相支援。

目前已推進至 Reasoning AI 階段，包括 OpenAI GPT-o3、Claude Sonnet、DeepSeek R1…等模型，相信未來發展的速度會越來越快，我們要拉緊韁繩，隨著狂奔的 AI 前進。

1-3 AI 的學習地圖

AI 發展分為三個階段，分別為『人工智慧』（Artificial Intelligence）、『機器學習』（Machine Learning）、『深度學習』（Deep Learning），每一階段都更聚焦在特定的演算法，機器學習是人工智慧的部份領域，而深度學習屬於機器學習中的一環。

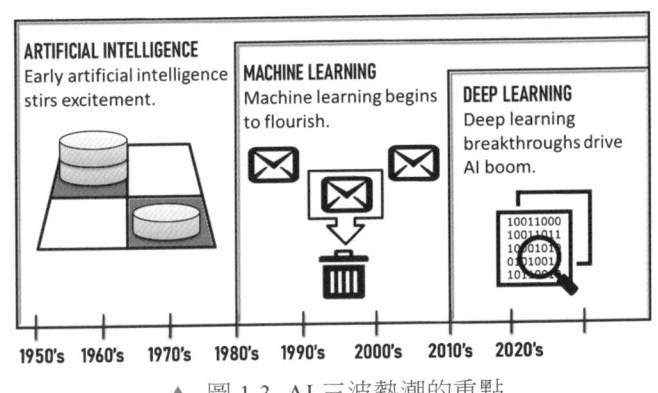

▲ 圖 1.3 AI 三波熱潮的重點

第 1 章　深度學習（Deep Learning）導論

而一般教育機構規劃 AI 的學習地圖即依照這個軌跡，逐步深入各項技術，通常分為四個階段：

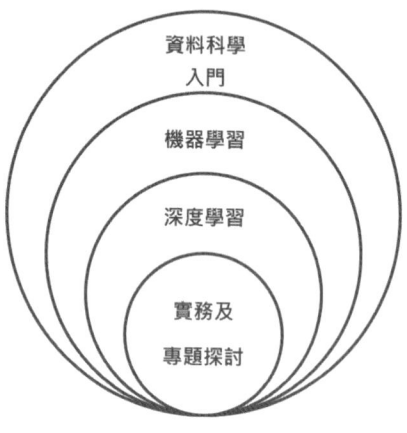

▲ 圖 1.4　AI 學習地圖

1. 資料科學（Data Science）入門：內容包括 Python/R 程式語言、數據分析（Data Analysis）、大數據平台（Hadoop、Spark）…等。

2. 機器學習（Machine Learning）：包含一些經典的演算法，如迴歸、羅吉斯迴歸、支援向量機（SVM）、集群（Clustering）…等，這些演算法雖然簡單，卻非常實用，在一般企業內被普遍使用。機器學習的大分類如下：

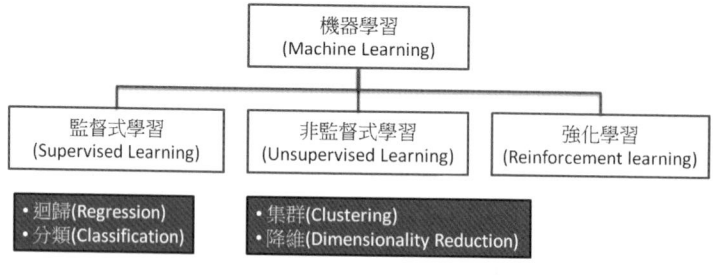

▲ 圖 1.5　機器學習的分類

- 監督式學習（Supervised Learning）：訓練資料有答案（Y），可以誘導（監督）演算法優化的方向，若 Y 為連續型變數，則稱為迴歸（Regression）演算法，反之，Y 為離散型變數（有限類別），則稱為分類演算法（Classification）。

- 非監督式學習（Unsupervised Learning）：訓練資料沒有答案（Y），藉由特徵資料 (x) 的相似性，將資料分群，稱為集群（Clustering）演算法，另外也可以用於降維（Dimensionality Reduction）。
- 強化學習（Reinforcement learning，簡稱 RL）：可以不提供訓練資料，藉由錯誤中學習（Try and Error），已獲得最大報酬為目標，尋求最佳行動策略。
- 除了以上分類，還有半監督式學習（Semi-supervised Learning）、自我學習（Self Learning）、聯合學習（Federated Learning）…等，百花齊放，千萬不要被既有的分類限制了你的想像。
- 另外，資料探勘（Data Mining）與機器學習的演算法大量重疊，其間的差異，有一說是資料探勘是著重挖掘資料的隱藏樣態（Pattern），而機器學習則著重於預測。

3. 深度學習（Deep Learning）：深度學習屬於機器學習中的一環，所謂深度（Deep）是指多層式架構的模型，例如各種神經網路（Neural Network）模型、強化學習（Reinforcement learning,RL）演算法…等。
4. 實務及專題探討（Capstone Project）：將各種演算法應用於各類領域 / 行業，強調專題探討及產業應用實作。

1-4 機器學習應用領域

機器學習的應用其實早已不知不覺間融入到我們的生活中了，例如：

1. 社群軟體大量運用 AI，預測使用者的行為模式、過濾垃圾信件。
2. 電商運用 AI，依據每位消費者的喜好推薦合適的商品。
3. 還有各式的 3C 產品，包括手機、智慧音箱，以人臉辨識取代登入，語音辨識代替鍵盤輸入，結合智慧家庭 / 保全。
4. 聊天機器人（ChatBot）取代客服人員，提供行銷服務。
5. 製造機器人（Robot）提供智慧製造、老人照護、兒童陪伴，凡此種種，不及備載。

第 1 章　深度學習（Deep Learning）導論

AI 相關的應用領域如下圖：

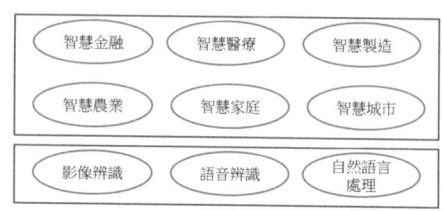

▲ 圖 1.6　AI 應用領域

目前相對熱門的研發領域包括：

1. 醫療：各種疾病的診斷（Medical Diagnostics）及新藥的開發。
2. 聊天機器人（ChatBot）：包括行銷、銷售及售後服務的支援。
3. 自駕車（Self Driving）：包括無人計程車。
4. 製造業：機器手臂、機器人（Robot）、人形機器人…等。
5. 軍事用途：無人機、飛彈、無人艇、戰鬥機器狗…等。

1-5 機器學習開發流程

機器學習開發流程（Machine learning workflow）與一般軟體開發略有不同，它著重在資料的整理與模型的訓練，目前有許多種建議的流程，例如 CRISP-DM（Cross-industry Standard Process for Data Mining）、Google Cloud 建議的流程 [4]…等，個人較偏好的流程如下：

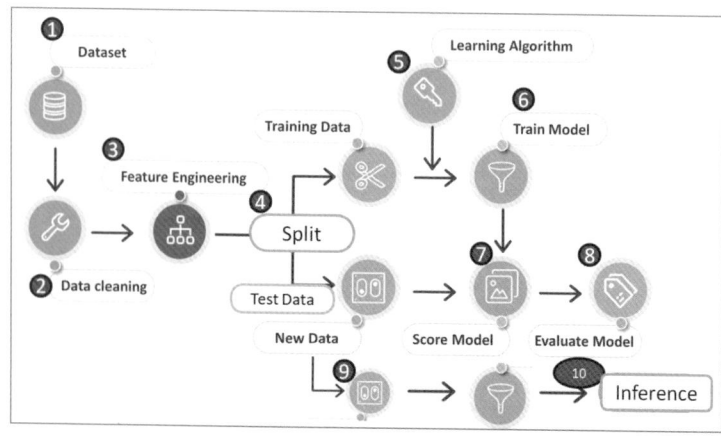

▲ 圖 1.7　機器學習開發流程（Machine Learning Workflow）

1-5 機器學習開發流程

流程概分為 10 個步驟，不含初期的企業需求瞭解（Business Understanding），只包括實際開發的步驟：

1. 蒐集模型訓練所需的資料，彙整為資料集（Dataset）。

2. 資料清理（Data Cleaning）及資料探索與分析（Exploratory Data Analysis, EDA）：EDA 通常是以描述統計量（Descriptive Statistics）及統計圖觀察資料的特性與分佈，瞭解資料的平均值/標準差、極端值（Outlier）、變數之間的關聯性…等。資料清理則包括欄位格式、編碼、長度…等屬性的統一、遺失值（Missing value）、重複值…等處理。

3. 特徵工程（Feature Engineering）：原始蒐集的資料未必是影響預測目標的關鍵因素，有時候需要進行資料轉換，以找到關鍵的影響因子，例如後續章節將影像/文字/語音轉換為向量。

4. 資料切割（Data Split）：將全部資料切割為訓練資料（Training Data）及測試資料（Test Data），前者提供模型訓練之用，後者用在衡量模型效能，例如準確度，切割的主要原因是確保測試資料不會參與訓練，以維持評估的公正性，即 Out-of-Sample Test。

5. 選擇演算法（Learning Algorithms）：依據問題的類型選擇適合的演算法。

6. 模型訓練（Model Training）：將訓練資料餵入演算法，進行模型訓練。

7. 模型計分（Score Model）：以測試資料進行模型推論，計算準確度或其他效能指標，評估模型效能。

8. 模型評估（Evaluate Model）：以不同的參數組合或各種演算法訓練多個模型，比較效能指標，找到最佳參數與模型。

9. 佈署（Deploy）：複製最佳模型至正式環境（Production Environment），開發使用介面或 API，提供服務。

10. 推論（Inference）：用戶端將新資料或提示輸入至模型內，進行預測（Predict）、生成（Generation）、分類（Classification）、集群（Clustering）…等。

機器學習開發流程與一般應用系統開發有何差異？最大的差別如下圖：

1. 一般應用系統則利用輸入資料與轉換邏輯產生輸出，例如撰寫報表，根據轉換規則將輸入欄位轉換為輸出欄位，但機器學習會先產生模型，再根據模型進行預測，重用性（Reuse）高。

2. 機器學習除了輸入資料外，還會蒐集大量歷史資料或在網際網路中爬一堆資料，作為建模的養分。

3. 後續產生的資料可再餵入演算法，重新訓練，不斷自我學習，使模型更聰明。

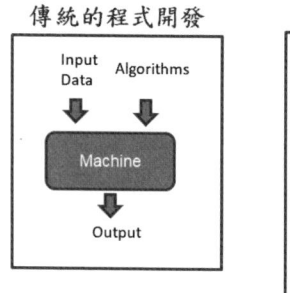
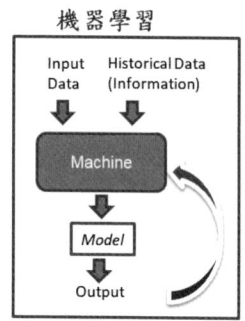

▲ 圖 1.8 機器學習與一般應用系統開發流程的差異

1-6 開發環境安裝

Python 是目前機器學習主流的程式語言，可以直接在本機安裝開發環境，亦能使用雲端環境，首先介紹本機安裝的程序，建議依照以下順序安裝：

1. 安裝 Anaconda：建議直接安裝 Anaconda，它內含 Python 及上百個常用套件，不需逐一安裝各個套件。先至 Anaconda 官網 [5] 下載安裝檔，依作業系統選擇安裝檔，例如在 Windows 作業系統安裝時，依指示點選下一步，惟執行到下列畫面時，全部都勾選，將安裝路徑加入至環境變數 Path 內，這樣就能在任何目錄下均可執行 Python 程式。Mac/Linux 安裝時會自動加入 Anaconda 安裝路徑。

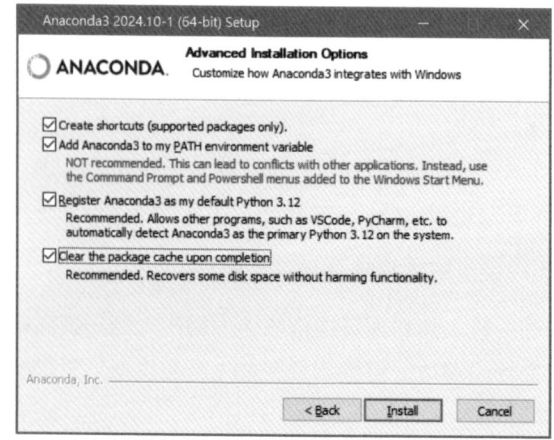

▲ 圖 1.9 勾選第 2 項，將安裝路徑加入至環境變數 Path 內

1-6 開發環境安裝

2. 安裝 TensorFlow 最新版：Windows 作業系統安裝時，在檔案總管的路徑輸入 cmd，開啟 DOS 視窗，Mac/Linux 則須開啟終端機，輸入：

   pip install tensorFlow -U

   ** 此指令同時支援 CPU 及 GPU，但要支援 **GPU 須另外安裝驅動程式及 SDK**。

   ** 依作業系統開啟 DOS 視窗或終端機，本書以下統一稱為開啟終端機。

3. 安裝測試：安裝完在終端機內輸入 Python，進入互動式環境，再輸入以下指令測試：

   > import tensorFlow

   > exit()

   - 若出現下列錯誤：

     from tensorflow.python._pywrap_tensorflow_internal import*

     ImportError:DLL load failed:The specified module could not be found.

     可能是 Windows 作業系統較舊，須安裝 MSVC 2022 runtime[6]。

   - 如果還是出現 Dll cannot be loaded. 也可使用 Anaconda 提供的 conda 指令來安裝，或安裝舊版的 TensorFlow，例如 2.10。conda install tensorflow

   - 建議還是安裝 TensorFlow 最新版（v2.18 或以上），才能搭配 Keras v3.x，安裝 TensorFlow 時會同時安裝 Keras，兩者在版本上必須匹配，否則匯出現許多錯誤，因為 Keras v2.x 與 v3.x 有顯著的差異。

   - Keras 原是獨立的套件，TensorFlow 已將它納入麾下，目前是 TensorFlow 開發的主流方式，因為它提供的語法較簡潔。

4. 若要支援 GPU，還需安裝 CUDA Toolkit、cuDNN SDK，可直接使用 conda 指令，也可至 NVidia 官網下載/安裝，conda 指令如下，注意，目前 Windows 作業系統下只有 v2.10 及更舊的版本支援 GPU：

   conda install-c conda-forge cudatoolkit=12.3 cudnn=8.9.7

5. CUDA Toolkit、cuDNN SDK 相關版本會隨著 TensorFlow 更新會有所不同，請參考 TensorFlow 官網『Install TensorFlow with pip』說明[7]，如下：

第 1 章　深度學習（Deep Learning）導論

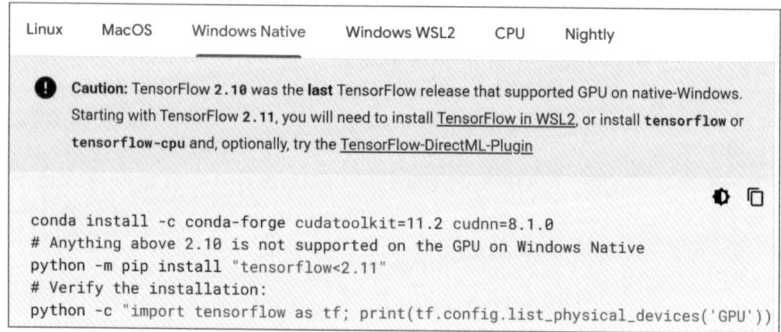

▲ 圖 1.10　TensorFlow 支援的 CUDA Toolkit 版本

注意，TensorFlow 目前只支援 **NVidia** 獨立顯卡，若是較舊型的顯卡必須查閱驅動程式搭配的版本資訊請參考 NVIDIA 官網說明 [8]，例如下表：

▲ 圖 1.11　CUDA Toolkit 版本與驅動程式的搭配

6. 奉勸各位讀者，太舊型的顯卡，若不能安裝 TensorFlow 支援的版本，就不用安裝 CUDA Toolkit、cuDNN SDK 了，因為顯卡記憶體過小，執行 TensorFlow 常常會發生記憶體不足（OOM）的錯誤，徒增困擾，GPU 記憶體至少要有 6GB 以上，使用會比較順暢。

7. CUDA Toolkit、cuDNN SDK 安裝成功後，檢查環境變數 Path 中是否有 CUDA Toolkit、cuDNN SDK 安裝路徑下的 bin、libnvvp。

1-6 開發環境安裝

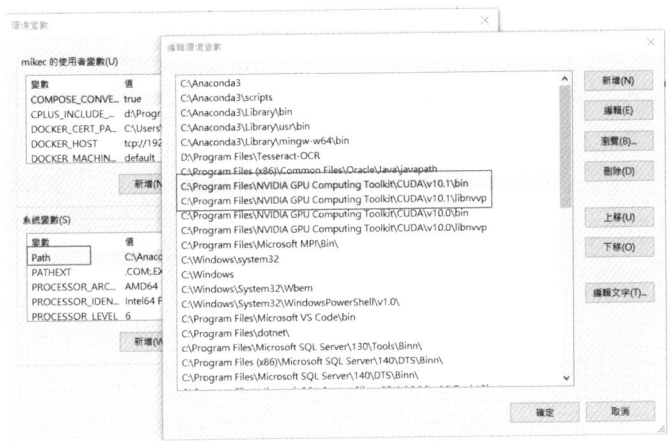

8. 安裝若有其他問題，可參考筆者的部落文『Day 01：輕鬆掌握 Keras』[9]。

接下來我們談談雲端環境的開通程序，Google、Kaggle、Hugging Face、AWS、Azure 都有提供深度學習的雲端開發環境，本文僅介紹 Google 雲端環境 Colaboratory，簡稱 Colab，只要有 Gmail 帳號就能免費使用，具備以下特點：

1. 常用的套件均已預先安裝，包括 TensorFlow、PyTorch。
2. 免費的 GPU：為 NVIDIA Tesla T4 GPU 顯卡，含 15GB 記憶體，真是佛心啊。
3. 開啟 Notebook 檔案時，會即時開通虛擬機，限連續使用 12 小時，逾時的話虛擬環境會被回收，所有資料會被刪除。
4. 若需要更多的 GPU 資源，可升級至 Colaboratory Pro，請參閱『選擇合適的 Colab 方案』[10]。

開通程序：

1. 使用瀏覽器，登入 Google 雲端硬碟（Google drive）。
2. 建立一個目錄，例如『0』，並切換至該目錄。

1-13

第 1 章 深度學習（Deep Learning）導論

3. 在螢幕中間按滑鼠右鍵，點選『更多』>『連結更多應用程式』。

4. 在搜尋欄輸入『Colaboratory』，找到後點擊該 App，按『Connect』按鈕即可開通。

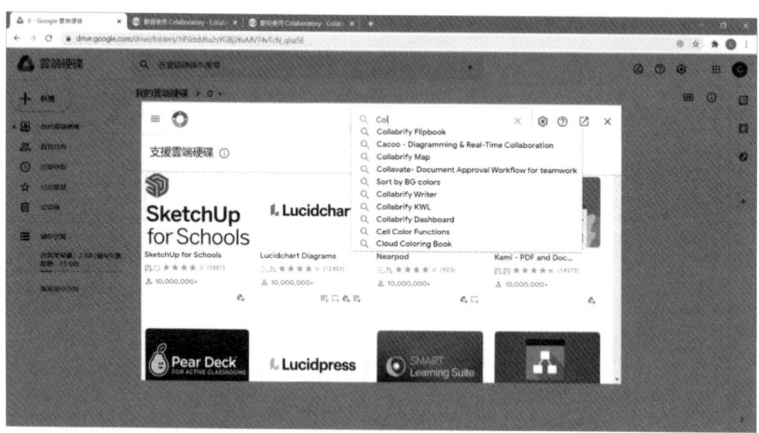

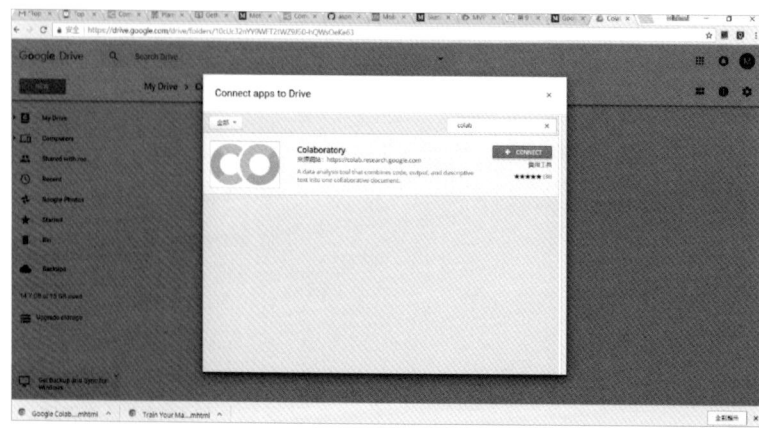

1-6 開發環境安裝

5. 開通後，即可開始使用，在左上方，點選『New』新增一個 Colaboratory 檔案。

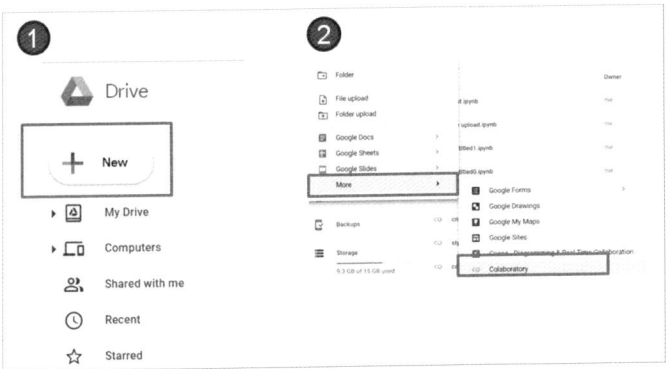

6. Colaboratory 會自動開啟虛擬環境，建立一個空白的 Notebook 檔案，副檔名為 ipynb，幾乎所有的雲端環境及大數據平台 Databricks 都以 Notebook 為主要使用介面。

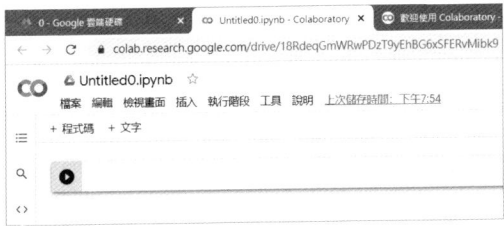

7. 要開啟既有的 Notebook 檔案，在 Google 雲端硬碟內滑鼠雙擊（Double click）檔案名稱即可。本機的 Notebook 檔案也可上傳至雲端硬碟，直接使用，完全不用轉換，非常方便。

8. 操作時若要支援 GPU 或 TPU，可設定運行環境如下，其中 TPU 為 Google 發明的 xPU。

1-15

第 1 章　深度學習（Deep Learning）導論

9. 更多的 Colaboratory 相關操作說明，可參考 Colaboratory 官網 [11]。

本書所附的範例程式，大部份為 Notebook 檔案，因為 Notebook 可以分格執行，反覆修改，不須重新啟動程式，也可使用 Markdown 語法撰寫美觀的說明，包括數學式。

在本機開啟 Notebook 檔案的方式很簡單，只要在檔案總管，切換至範例程式所在的目錄，在路徑列輸入 jupyter notebook，即可啟動網站並自動帶出網頁，再點選 Notebook 檔案，進行編輯及執行，相關的操作可以參考『Jupyter Notebook:An Introduction』[12] 或『Jupyter Notebook 完整介紹、安裝及使用說明』[13]。如果讀者不習慣使用 Jupyter Notebook，可以在 Jupyter Notebook 內點選【File】>【Save and Export Notebook As】>【Executable Script】，轉換為 .py 檔案。

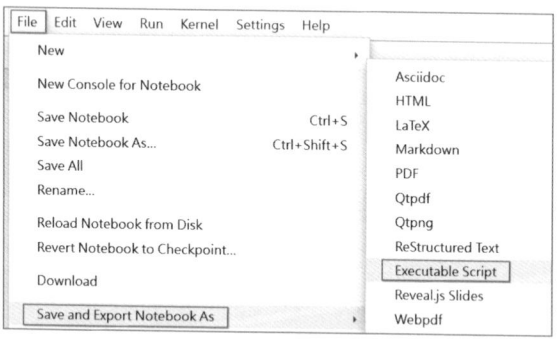

Jupyter 還有另外一個 Lab 套件，類似 Notebook，也可以編輯 Notebook 檔案，讀者也可以採用，指令為 jupyter lab。

▎參考資料（References）

[1] Dylan Yeh、陳建鈞，《市值首度超越 Intel！NVIDIA 贏在哪裡？》，2020
(https://www.bnext.com.tw/article/58410/nvidia-valuation-soars-past-intel-on-graphics-chip-boom)

[2] OpenAI's 5 Levels Of 'Super AI
(https://www.forbes.com/sites/jodiecook/2024/07/16/openais-5-levels-of-super-ai-agi-to-outperform-human-capability/)

參考資料（References）

[3]　ARC 2024 競賽題目

(https://www.kaggle.com/competitions/arc-prize-2024)

[4]　 Google Cloud 官網指南

(https://cloud.google.com/ai-platform/docs/ml-solutions-overview)

[5]　Anaconda 官網

(https://www.anaconda.com/download/success)

[6]　MSVC 2022 runtime

(https://aka.ms/vs/16/release/vc_redist.x64.exe)

[7]　TensorFlow 官網『Install TensorFlow with pip』說明

(https://www.tensorflow.org/install/pip)

[8]　NVIDIA 官網說明

(https://developer.nvidia.com/cuda-toolkit-archive)

[9]　陳昭明,《Day 01：輕鬆掌握 Keras》, 2020

(https://ithelp.ithome.com.tw/articles/10233272)

[10]　選擇合適的 Colab 方案

(https://colab.research.google.com/signup?utm_source=notebook_settings& utm_medium=link&utm_campaign=premium_gpu_selector)

[11]　Colaboratory 官網說明

(https://colab.research.google.com/notebooks/intro.ipynb)

[12]　Mike Driscoll, Jupyter Notebook: An Introduction

(https://realpython.com/jupyter-notebook-introduction/)

[13]　Jupyter Notebook 完整介紹、安裝及使用說明

(https://medium.com/ai-for-k12/jupyter-notebook-完整介紹-安裝及使用說明-846b5432f044)

MEMO

第 2 章

神經網路（Neural Network）原理

深度學習以神經網路為主要的演算法，以多個神經層建構模型，並以優化（Optimization）方式求解，其中求解過程涉及線性代數、偏微分、機率及統計…等數理概念，因此，本章會先討論相關數統理論，並以程式實作，再進而討論求解的核心 -- 梯度下降法。

第 2 章　神經網路（Neural Network）原理

2-1 必備的數學與統計知識

現在每天幾乎都會看到幾則有關人工智慧（AI）的新聞，介紹 AI 的各式研發成果，一般人也許會基於好奇想一窺究竟，了解背後運用的技術與原理，會發現一堆數學符號及統計公式，可能就會產生疑問，要建構 AI 模型，非要搞定數學、統計不可嗎？答案是肯定的，我們都知道機器學習是從資料中學習到知識（Knowledge Discovery from Databases, KDD），如果資料是水果，演算法就是從資料中萃取出知識的果汁機，它必須以數學及統計為理論基礎，才能證明其解法具有公信力與精準度，然而數學/統計理論都有侷限，只有在假設成立的情況下，演算法才是有效的，因此，如果不瞭解演算法各項假設，隨意套用公式，就好像無視交通規則，在馬路上任意飆車一樣的危險。

▲ 圖片來源：Decision makers need more math[1]

因此，以深度學習而言，我們至少需要熟悉以下學科：

1. 線性代數（Linear Algebra）
2. 微積分（Calculus）
3. 統計與機率（Statistics and Probability）
4. 線性規劃（linear programming）

2-1 必備的數學與統計知識

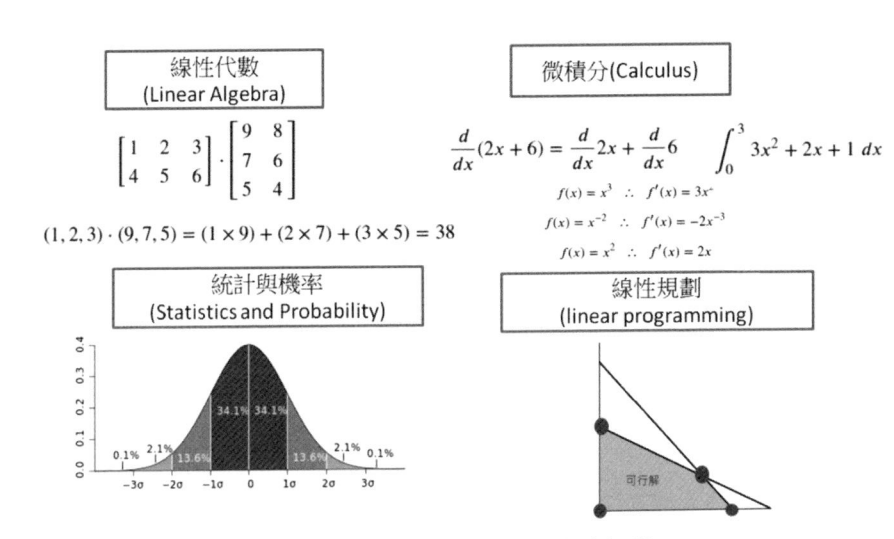

▲ 圖 2.1 必備的數學與統計知識

以神經網路優化求解的過程為例，4 門學科全部用上了，如下圖：

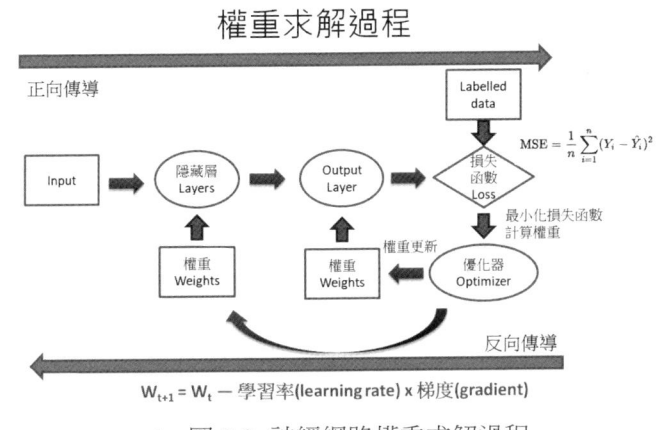

▲ 圖 2.2 神經網路權重求解過程

1. 正向傳導：藉由『線性代數』計算誤差及損失函數。
2. 反向傳導：透過『偏微分』計算梯度，同時，利用『線性規劃』優化技巧尋找最佳解。
3. 『統計』則串聯整個環節，例如資料的探索與分析、損失函數定義、效能衡量指標，通通都基於統計的理論架構而成。
4. 深度學習的推論以『機率』為基礎，預測目標值（Y）。

第 2 章　神經網路（Neural Network）原理

四項學科相互為用，貫穿整個求解過程，因此，要通曉深度學習的運作原理，並且正確選用各種演算法，甚至進而能夠修改或創新演算法，都必須對其背後的數學和統計，有一定基礎的認識，以免誤用/濫用。

四門學科在大學至少都是兩學期的課程，加起來要總共要 24 個學分，對已經離開大學殿堂很久的工程師而言，在上班之餘，還要重修上述課程，相信大部份的人都會萌生退意了吧！想想看我們是不是有速成的捷徑呢？

筆者在這裡借用一個概念 Statistical Programming，原意是『以程式解決統計問題』，換個角度想，我們是不是也能『**以程式設計的方式學統計**』，以縮短學習歷程。

通常要按部就班學習數學及統計，都是從瞭解『假設』➔『定義/定理』➔『證明』➔『應用』，一步一步學習：

- 『假設』是『定義/定理』成立的前提。
- 『證明』是『定理』的驗證。
- 『應用』是『定義/定理』的實踐。

由於『證明』都會有一堆的數學符號及公式推導，常會讓人頭暈腦脹，降低學習動力，因此，筆者大膽建議，工程師將心力著重在假設、定義、定理的理解與應用，並利用程式進行大量個案的驗證，對已進入職場的工程師會是一條較可行的捷徑，雖然忽略證明的作法，會讓學習無法徹底的融會貫通。

接下來我們就以上述的作法，對四項學科作一重點式的介紹，除了說明深度學習需要理解的知識外，更強調如何『以撰寫程式實現相關理論及解題方法』。

▋2-2 線性代數（Linear Algebra）

張量（Tensor）是描述向量空間（vector space）中物體的特徵，包括零維的純量（Scalar）、一維的向量（Vector）、二維的矩陣（Matrix）或更多維度的張量，線性代數則是說明張量如何進行各種運算，它被廣泛應用於各種數值分析的領域。以下就以實例說明張量的概念與運算。

2-2-1 向量（Vector）

向量（Vector）是一維的張量，它與線段的差別是除了長度（Magnitude）以外，還有方向（direction），數學表示法為：

$$\vec{v} = \begin{bmatrix} 2 \\ 1 \end{bmatrix}$$

以圖形表示如下：

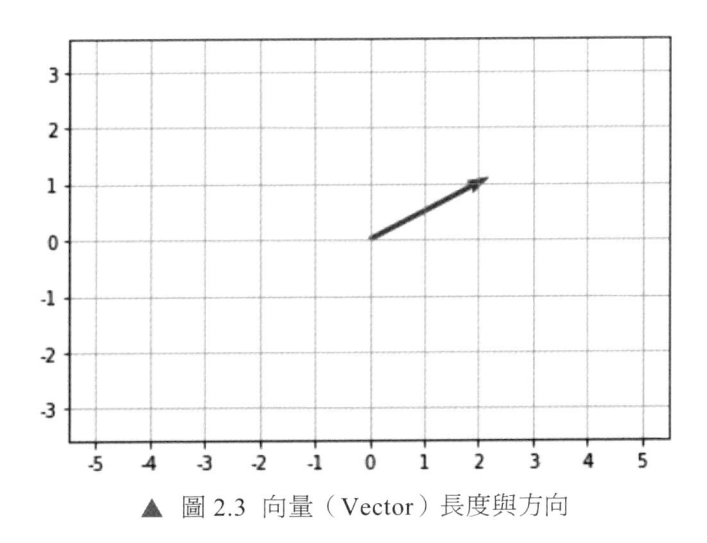

▲ 圖 2.3 向量（Vector）長度與方向

範例．使用程式計算向量的長度與方向。

程式：02_2_1_線性代數_向量.ipynb。

1. 計算長度（magnitude）：為與原點的距離，通常採用歐幾里得距離（Euclidean distance）程式撰寫如下：

$$\| \vec{v} \| = \sqrt{v_1^2 + v_2^2} = \sqrt{5}$$

```
1  # 向量(Vector)
2  v = np.array([2,1])
3
4  # 向量長度(magnitude)計算
5  (v[0]**2 + v[1]**2) ** (1/2)
```

第 2 章　神經網路（Neural Network）原理

2. 也可以使用 np.linalg.norm() 計算向量長度。

```
1  # 使用 np.linalg.norm() 計算向量長度(magnitude)
2  import numpy as np
3
4  magnitude = np.linalg.norm(v)
5  print(magnitude)
```

3. 方向（direction）：使用與 X 軸的夾角（θ）表示，可運用 $\tan^{-1}$ 函數計算。

$$\tan(\theta) = \frac{1}{2}$$

移項如下：

$$\theta = \tan^{-1}\left(\frac{1}{2}\right) \approx 26.57°$$

```
1  import math
2  import numpy as np
3
4  # 向量(Vector)
5  v = np.array([2,1])
6
7  vTan = v[1] / v[0]
8  print ('tan(θ) = 1/2')
9
10 theta = math.atan(vTan)
11 print('弧度(radian) =', round(theta,4))
12 print('角度(degree) =', round(theta*180/math.pi, 2))
13
14 # 也可以使用 math.degrees() 轉換角度
15 print('角度(degree) =', round(math.degrees(theta), 2))
```

4. 向量四則運算規則：

- 加減乘除一個常數：常數直接對每個元素作加減乘除。

- 加減乘除另一個向量：兩個向量的相同位置的元素作加減乘除，所以兩個向量的元素個數須相等。

- 向量加減法：加減一個常數，長度、方向均改變。

```
1  # 載入套件
2  import numpy as np
3  import matplotlib.pyplot as plt
4
5  # 向量(Vector) + 2
6  v = np.array([2,1])
7  v1 = np.array([2,1]) + 2
8  v2 = np.array([2,1]) - 2
```

2-2 線性代數（Linear Algebra）

```
 9
10  # 原點
11  origin = [0], [0]
12
13  # 畫有箭頭的線
14  plt.quiver(*origin, *v1, scale=10, color='r')
15  plt.quiver(*origin, *v, scale=10, color='b')
16  plt.quiver(*origin, *v2, scale=10, color='g')
17
18  plt.annotate('orginal vector',(0.025, 0.01), xycoords='data'
19               , fontsize=16)
20
21  # 作圖
22  plt.axis('equal')
23  plt.grid()
24
25  plt.xticks(np.arange(-0.05, 0.06, 0.01), labels=np.arange(-5, 6, 1))
26  plt.yticks(np.arange(-3, 5, 1) / 100, labels=np.arange(-3, 5, 1))
27  plt.show()
```

- 執行結果：向量加減一個常數，長度、方向均改變。

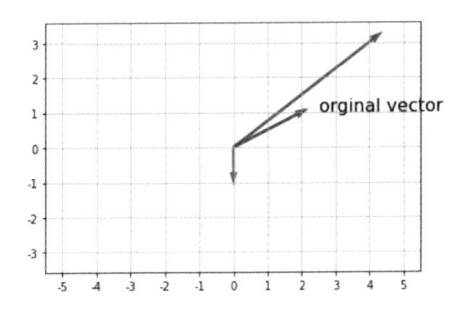

5. 向量乘除法：乘除一個常數，長度改變、方向不改變。

```
 1  # 載入套件
 2  import numpy as np
 3  import matplotlib.pyplot as plt
 4
 5  # 向量(Vector) * 2
 6  v = np.array([2,1])
 7  v1 = np.array([2,1]) * 2
 8  v2 = np.array([2,1]) / 2
 9
10  # 原點
11  origin = [0], [0]
12
13  # 畫有箭頭的線
14  plt.quiver(*origin, *v1, scale=10, color='r')
15  plt.quiver(*origin, *v, scale=10, color='b')
16  plt.quiver(*origin, *v2, scale=10, color='g')
17
```

```
18  plt.annotate('orginal vector',(0.025, 0.008), xycoords='data'
19              , color='b', fontsize=16)
20
21  # 作圖
22  plt.axis('equal')
23  plt.grid()
24
25  plt.xticks(np.arange(-0.05, 0.06, 0.01), labels=np.arange(-5, 6, 1))
26  plt.yticks(np.arange(-3, 5, 1) / 100, labels=np.arange(-3, 5, 1))
27  plt.show()
```

- 執行結果：向量乘除一個常數，長度改變、方向不改變。

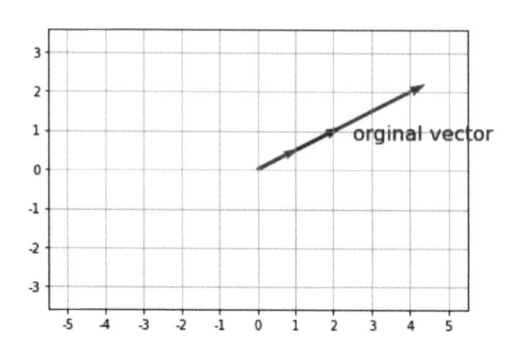

6. 向量加減乘除另一個向量：兩個向量的相同位置的元素作加減乘除。

```
1   # 載入套件
2   import numpy as np
3   import matplotlib.pyplot as plt
4
5   # 向量(Vector) * 2
6   v = np.array([2,1])
7   s = np.array([-3,2])
8   v2 = v+s
9
10  # 原點
11  origin = [0], [0]
12
13  # 畫有箭頭的線
14  plt.quiver(*origin, *v, scale=10, color='b')
15  plt.quiver(*origin, *s, scale=10, color='b')
16  plt.quiver(*origin, *v2, scale=10, color='g')
17
18  plt.annotate('orginal vector',(0.025, 0.008), xycoords='data'
19              , color='b', fontsize=16)
20
21  # 作圖
22  plt.axis('equal')
23  plt.grid()
24
25  plt.xticks(np.arange(-0.05, 0.06, 0.01), labels=np.arange(-5, 6, 1))
```

2-2 線性代數（Linear Algebra）

```
26  plt.yticks(np.arange(-3, 5, 1) / 100, labels=np.arange(-3, 5, 1))
27  plt.show()
```

- 執行結果：

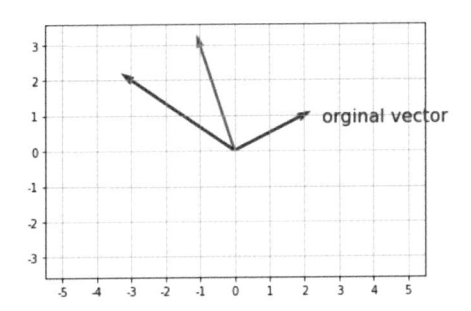

7. 『內積』或稱『點積』（Dot Product）：NumPy 以 @ 作為內積的運算符號，而非 *。

$$\vec{v} \cdot \vec{s} = (v_1 \cdot s_1) + (v_2 \cdot s_2) \ldots + (v_n \cdot s_n)$$

```
1   # 載入套件
2   import numpy as np
3
4   # 向量(Vector)
5   v = np.array([2,1])
6   s = np.array([-3,2])
7
8   # 『內積』或稱『點積乘法』(Dot Product)
9   d = v @ s
10
11  print (d)
```

8. 計算兩個向量的夾角，公式如下：

$$\vec{v} \cdot \vec{s} = \|\vec{v}\| \|\vec{s}\| \cos(\theta)$$

移項：

$$\cos(\theta) = \frac{\vec{v} \cdot \vec{s}}{\|\vec{v}\| \|\vec{s}\|}$$

再利用 $\cos^{-1}()$ 計算夾角 θ。

```
1   # 載入套件
2   import math
3   import numpy as np
```

2-9

```
 4
 5  # 向量(Vector)
 6  v = np.array([2,1])
 7  s = np.array([-3,2])
 8
 9  # 計算長度(magnitudes)
10  vMag = np.linalg.norm(v)
11  sMag = np.linalg.norm(s)
12
13  # 計算 cosine(ϑ)
14  cos = (v @ s) / (vMag * sMag)
15
16  # 計算 ϑ
17  theta = math.degrees(math.acos(cos))
18
19  print(theta)
```

2-2-2 矩陣（Matrix）

矩陣是二維的張量，擁有列（Row）與行（Column），可用以表達一個平面 N 個點（Nx2）、或一個 3D 空間 N 個點（Nx3），例如：

$$A = \begin{bmatrix} 1 & 2 & 3 \\ 4 & 5 & 6 \end{bmatrix}$$

範例. 矩陣運算測試。

程式：02_2_2_ 線性代數 _ 矩陣 .ipynb。

1. 矩陣加法 / 減法與向量相似，相同位置的元素作運算即可，但乘法運算通常是指內積（dot product），使用 @。
2. 兩個矩陣相加。

$$\begin{bmatrix} 1 & 2 & 3 \\ 4 & 5 & 6 \end{bmatrix} + \begin{bmatrix} 6 & 5 & 4 \\ 3 & 2 & 1 \end{bmatrix} = \begin{bmatrix} 7 & 7 & 7 \\ 7 & 7 & 7 \end{bmatrix}$$

```
1  # 載入套件
2  import numpy as np
3
4  # 矩陣
5  A = np.array([[1,2,3],
6                [4,5,6]])
7  B = np.array([[6,5,4],
8                [3,2,1]])
9
```

```
10  # 加法
11  print(A + B)
```

3. 兩個矩陣相乘。

$$\begin{bmatrix} 1 & 2 & 3 \\ 4 & 5 & 6 \end{bmatrix} \cdot \begin{bmatrix} 9 & 8 \\ 7 & 6 \\ 5 & 4 \end{bmatrix} = ?$$

解題：左邊矩陣的第 2 維須等於右邊矩陣的第 1 維，即（m,k）x（k,n）=（m,n）

$$\begin{bmatrix} 1 & 2 & 3 \\ 4 & 5 & 6 \end{bmatrix} \cdot \begin{bmatrix} 9 & 8 \\ 7 & 6 \\ 5 & 4 \end{bmatrix} = \begin{bmatrix} 38 & 32 \\ 101 & 86 \end{bmatrix}$$

其中左上角的計算過程為（1,2,3）·（9,7,5）=（1×9）+（2×7）+（3×5）=38，右上角的計算過程為（1,2,3）·（8,6,4）=（1×8）+（2×6）+（3×4）=32，以此類推，如下圖。

$$\begin{bmatrix} 1 & 2 & 3 \\ 4 & 5 & 6 \end{bmatrix} \cdot \begin{bmatrix} 9 & 8 \\ 7 & 6 \\ 5 & 4 \end{bmatrix} = \begin{bmatrix} \boxed{38} & 32 \\ 101 & 86 \end{bmatrix}$$

$$\begin{bmatrix} 1 & 2 & 3 \\ 4 & 5 & 6 \end{bmatrix} \cdot \begin{bmatrix} 9 & 8 \\ 7 & 6 \\ 5 & 4 \end{bmatrix} = \begin{bmatrix} 38 & \boxed{32} \\ 101 & 86 \end{bmatrix}$$

$$\begin{bmatrix} 1 & 2 & 3 \\ 4 & 5 & 6 \end{bmatrix} \cdot \begin{bmatrix} 9 & 8 \\ 7 & 6 \\ 5 & 4 \end{bmatrix} = \begin{bmatrix} 38 & 32 \\ \boxed{101} & 86 \end{bmatrix}$$

$$\begin{bmatrix} 1 & 2 & 3 \\ 4 & 5 & 6 \end{bmatrix} \cdot \begin{bmatrix} 9 & 8 \\ 7 & 6 \\ 5 & 4 \end{bmatrix} = \begin{bmatrix} 38 & 32 \\ 101 & \boxed{86} \end{bmatrix}$$

```
 1  # 載入套件
 2  import numpy as np
 3
 4  # 矩陣
 5  A = np.array([[1,2,3],
 6                [4,5,6]])
 7  B = np.array([[9,8],
 8                [7,6],
 9                [5,4],
10               ])
11
```

```
12  # 乘法
13  print(A @ B)
```

4. 矩陣（A、B）相乘，A x B 是否等於 B x A？

```
1   # 乘法：A x B != B x A
2
3   A = np.array([[1,2],
4                 [4,5]])
5   B = np.array([[9,8],
6                 [7,6],
7                 ])
8
9   print(A @ B)
10  print()
11  print(B @ A)
12  print()
13  print('A x B != B x A')
```

- 執行結果：A x B 不一定等於 B x A。

$$[[23\ 20]\\ [71\ 62]]$$

$$[[41\ 58]\\ [31\ 44]]$$

A x B != B x A

5. 矩陣在運算時，除了一般的加減乘除外，還有一些特殊的矩陣，包括轉置矩陣（Transpose）、反矩陣（Inverse）、對角矩陣（Diagonal matrix）、單位矩陣（Identity matrix）⋯等。

6. 轉置矩陣：列與行互換。

$$\begin{bmatrix} 1 & 2 & 3 \\ 4 & 5 & 6 \end{bmatrix}^T = \begin{bmatrix} 1 & 4 \\ 2 & 5 \\ 3 & 6 \end{bmatrix}$$

（A^T）^{T} = A：進行兩次轉置，會回復成原來的矩陣。

```
1   import numpy as np
2
3   A = np.array([[1,2,3],
4                 [4,5,6]])
5
6   # 轉置矩陣
7   print(A.T)
```

2-2 線性代數（Linear Algebra）

- 也可以使用 np.transpose（A）。

7. 反矩陣（A^{-1}）：A 必須為方陣，即列數與行數須相等，且必須是非奇異方陣（non-singular），即每一列或行之間不可以相依於其他列或行。

```
1  import numpy as np
2
3  A = np.array([[1,2,3],
4                [4,5,6],
5                [7,8,9],
6               ])
7  print(np.linalg.inv(A))
```

- 執行結果：

```
[[ 3.15251974e+15 -6.30503948e+15  3.15251974e+15]
 [-6.30503948e+15  1.26100790e+16 -6.30503948e+15]
 [ 3.15251974e+15 -6.30503948e+15  3.15251974e+15]]
```

8. 若 A 為非奇異矩陣，則 A@A^{-1} = 單位矩陣（I）。所謂的非奇異矩陣是任一列不能其他列的倍數或多列的組合，包括各種四則運算。矩陣的行也須符合相同的規則。

9. 驗證下列矩陣 A@A^{-1} 是否等於單位矩陣（I）。

$$A = \begin{bmatrix} 9 & 8 \\ 7 & 6 \end{bmatrix}$$

```
1  # A @ A反矩陣 = 單位矩陣(I)
2  A = np.array([[9,8],
3                [7,6],
4               ])
5
6  print(np.around(A @ np.linalg.inv(A)))
```

- 執行結果：結果為單位矩陣，表示 A 為非奇異矩陣。

```
[[1. 0.]
 [0. 1.]]
```

10. 再驗證下列矩陣驗算 A@A^{-1} 是否等於單位矩陣（I）。

$$A = \begin{bmatrix} 1 & 2 & 3 \\ 4 & 5 & 6 \\ 7 & 8 & 9 \end{bmatrix}$$

第 2 章　神經網路（Neural Network）原理

```
1  # A @ A反矩陣 != 單位矩陣(I)
2  # A 為 singular 矩陣
3  # 第二行 = 第一行 + 1
4  # 第三行 = 第一行 + 2
5  A = np.array([[1,2,3],
6                [4,5,6],
7                [7,8,9],
8                ])
9
10 print(np.around(A @ np.linalg.inv(A)))
```

- 執行結果：

$$[[\ 0.\ \ 1.\ -0.]$$
$$[\ 0.\ \ 2.\ -1.]$$
$$[\ 0.\ \ 3.\ \ 2.]]$$

- A 為奇異（Singular）矩陣，因為

  第二行 = 第一行 + 1

  第三行 = 第一行 + 2

  故 A@A^{-1} 不等於單位矩陣。

2-2-3 聯立方程式求解

在國中階段，我們通常會以高斯消去法（Gaussian Elimination）解聯立方程式。以下列方程式為例：

$$x + y = 16$$
$$10x + 25y = 250$$

第一個方程式乘以 -10，加上第二個方程式，即可消去 x，變成：

$$-10(x + y) = -10(16)$$
$$10x + 25y = 250$$

簡化為：

$$15y = 90$$

2-2 線性代數（Linear Algebra）

得到 y=6，再代入任一方程式，得到 x=10。

以上過程，如果以線性代數求解就簡單多了：

- 以矩陣表示如下，A：方程式中未知數（x,y）的係數，B 為等號右邊的常數：
  AX = B，移項後 X = A^{-1} B。

- 其中 A = $\begin{bmatrix} 1 & 1 \\ 10 & 25 \end{bmatrix}$ X = $\begin{bmatrix} x \\ y \end{bmatrix}$ B = $\begin{bmatrix} 16 \\ 250 \end{bmatrix}$

- 證明如下：

    * 兩邊各乘 A^{-1}：

      A^{-1}AX = A^{-1} B

    * A^{-1}A 等於單位矩陣，且任一矩陣乘以單位矩陣，還是等於原矩陣，故

      X = A^{-1} B

- 注意，以上式求得（x,y）的前提是 A 須為非奇異矩陣。

範例. 求解聯立方程式。

程式：02_2_3_ 聯立方程式求解 .ipynb。

1. 以 NumPy 套件求解上述聯立方程式。

```
1  # x+y=16
2  # 10x+25y=250
3
4  # 載入套件
5  import numpy as np
6
7  # 定義方程式的 A、B
8  A = np.array([[1 , 1], [10, 25]])
9  B = np.array([16, 250])
10 print('A=')
11 print(A)
12 print('')
13 print('B=')
14 print(B.reshape(2, 1))
15
16 # np.linalg.solve：線性代數求解
17 print('\n線性代數求解：')
18 print(np.linalg.inv(A) @ B)
```

- 第 18 行的 inv：為反矩陣函數。

- 執行結果：x=10，y=6。

第 2 章　神經網路（Neural Network）原理

2. 也可以直接呼叫 np.linalg.solve() 函數求解。

```
1  print(np.linalg.solve(A, B))
```

3. 畫圖：交叉點即聯立方程式的解，因為它同時符合 2 個方程式。

```
1   # x+y=16
2   # 10x+25y=250
3
4   # 載入套件
5   import numpy as np
6   from matplotlib import pyplot as plt
7
8   # 取第一個方程式的兩端點
9   A1 = [16, 0]
10  A2 = [0, 16]
11
12  # 取第二個方程式的兩端點
13  B1 = [25,0]
14  B2 = [0,10]
15
16  # 畫線
17  plt.plot(A1,A2, color='blue')
18  plt.plot(B1, B2, color="orange")
19  plt.xlabel('x')
20  plt.ylabel('y')
21  plt.grid()
22
23  # 交叉點 (10, 6)
24  plt.scatter([10], [6], color="red", s=100)
25  plt.show()
```

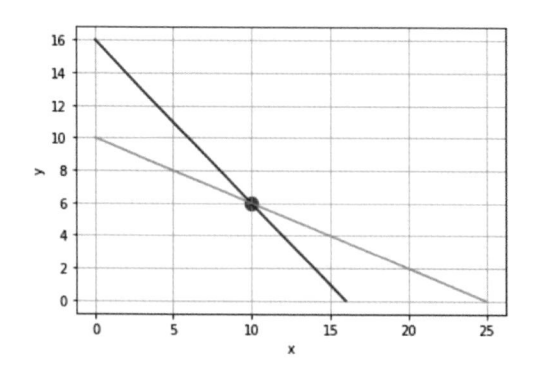

- 執行結果：

4. 求解 3 個變數的聯立方程式。

$$-1x + 3y = -72$$
$$3x + 4y - 4z = -4$$
$$-20x - 12y + 5z = -50$$

```python
1  # 載入套件
2  import numpy as np
3
4  # 定義方程式的 A、B
5  A = np.array([[-1, 3, 0], [3, 4, -4], [-20, -12, 5]])
6  B = np.array([-72, -4, -50])
7  print('A=')
8  print(A)
9  print('')
10 print('B=')
11 print(B.reshape(3, 1))
12
13 # np.linalg.solve：線性代數求解
14 print('\n線性代數求解：')
15 print(np.linalg.solve(A, B))
```

- 執行結果：(x,y,z) = (12,-20,-10)。

5. 也可以使用 SymPy 套件求解，直接將聯立方程式整理至等號左方，呼叫 solve() 函數，參數內的多項式均假設等號（=）右方為 0。

```python
1  from sympy.solvers import solve
2  from sympy import symbols
3
4  # 設定變數
5  x, y, z = symbols('x y z')
6
7  # 解題：設定聯立方程式，等號右邊預設為 0
8  solve([-1*x + 3*y + 72, 3*x + 4*y - 4*z + 4,-20*x + -12*y + 5*z + 50])
```

2-3 微積分（Calculus）

微積分包括微分（Differentiation）與積分（Integration），微分是描述函數某一點的變化率，藉由微分可以得到特定點的斜率（Slope）或是梯度（Gradient），即函數變化的速度，二次微分可以得到加速度。積分則是微分的相反，互為逆運算，積分可以計算長度、面積、體積，也可以用來計算累積分配函數（Cumulative Distribution Function, cdf）。

第 2 章　神經網路（Neural Network）原理

在神經網路求解的過程中，微分佔有舉足輕重的地位，因此深度學習的套件都提供自動微分（Automatic Differentiation）的功能，用以計算各特徵變數的梯度（gradient），進而求得模型的權重（Weight）參數。

2-3-1 微分（Differentiation）

微分是描述函數的變化率（Rate of Change），例如 y=2x+5，表示 x 每增加一單位，y 會增加 2，因此，變化率就等於 2，也稱為『斜率』（Slope），5 為『截距』（Intercept），在深度學習中通常稱為『偏差』（Bias），是函數在 Y 軸與原點的距離，也意謂無法使用特徵 (x) 捕捉到的影響力。

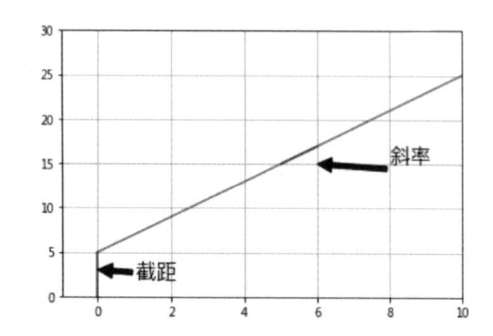

▲ 圖 2.1　斜率（Slope）與截距（Intercept）

斜率定義如下：

$$(y_2-y_1)/(x_2-x_1) = (f(x_2)-f(x_1))/(x_2-x_1)$$

要取得某一點 x 的斜率，就取非常相近的兩個點（x, x+h），h 趨近於 0，以數學式表示如下：

$$f'(x) = \frac{\Delta y}{\Delta x} = \lim_{h \to 0} \left(\frac{f(x+h) - f(x)}{h} \right)$$

這就是微分的定義，但上述極限值（lim）不一定存在，會存在的前提如下：

1. h 為正值時的極限值等於（=）h 為負值時的極限值，亦即函數在該點時是連續（Continuity）的。
2. 上述極限值不等於無窮大（∞）或負無窮大（-∞）。

2-18

2-3 微積分（Calculus）

以下函數在 x=5 的地方是連續的，由上方（5.25）逼近，或由下方（4.75）逼近是相等的，相關彩色圖形可參考 3_1_ 微分 .ipynb。

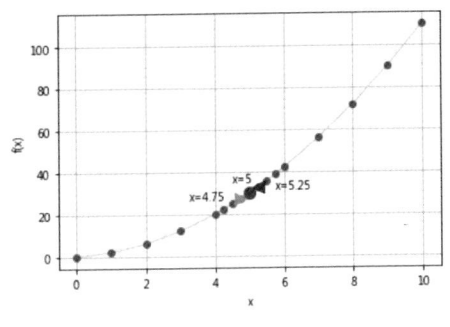

▲ 圖 2.2　連續函數

相反的，下圖在 x=0 時，是不連續的，逼近 x=0 時有兩個解。

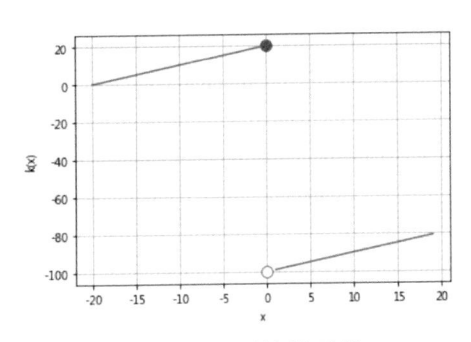

▲ 圖 2.3　不連續函數

接著來看看幾個應用實例。

範例 1. 微分實作。

程式：02_3_1_ 微分 .ipynb。

1. 繪製函數 f(x)=2x+5。

```
1  # 載入套件
2  import numpy as np
3  import matplotlib.pyplot as plt
4
5  # 樣本點
6  x = np.array(range(0, 11))
7  y = 2 * x + 5
8
```

2-19

第 2 章 神經網路（Neural Network）原理

```
 9  # 作圖
10  plt.grid()
11  plt.plot(x, y, color='g')
12  plt.xlim(-1, 10)
13  plt.ylim(0, 30)
14
15  # 截距 (Intercept)
16  x = [0, 0]
17  y = [0, 5]
18  plt.plot(x, y, color="r")
19  plt.annotate("截距", xy=(0,3), xytext=(1, 2), xycoords='data',
20               arrowprops=dict(facecolor='black'), fontsize=18)
21
22  # 斜率(Slope)
23  x = np.array([5, 6])
24  y = 2 * x + 5
25  plt.plot(x, y, color="r")
26  plt.annotate("斜率", xy=(6,15), xytext=(8, 15), xycoords='data',
27               arrowprops=dict(facecolor='black'), fontsize=18)
28
29  plt.show()
```

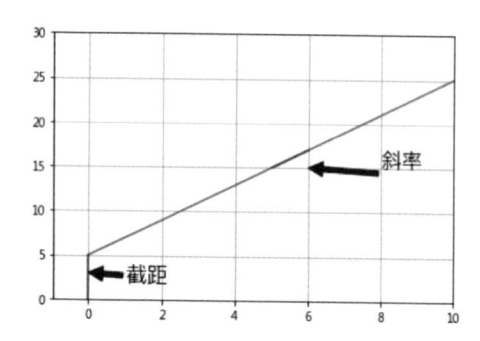

- 執行結果：

- 一次方的函數每一點的斜率均相同

2. 繪製二次方函數 $f(x)=-10x^2+100x+5$，求最大值。

```
 1  from matplotlib import pyplot as plt
 2
 3  # 二次曲線
 4  def f(x):
 5      return -10*(x**2) + (100*x) + 5
 6
 7  # 一階導數
 8  def fd(x):
 9      return -20*x + 100
10
11  # 設定樣本點
12  x = list(range(0, 11))
13  y = [f(i) for i in x]
14
```

2-20

2-3 微積分（Calculus）

```
15  # 一階導數的樣本點
16  yd = [fd(i) for i in x]
17
18  # 畫二次曲線
19  plt.plot(x,y, color='green', linewidth=1)
20
21  # 畫一階導數
22  plt.plot(x,yd, color='purple', linewidth=3)
23
24  # 畫三個點的斜率 x = (2, 5, 8)
25  x1 = 2
26  x2 = 5
27  x3 = 8
28  plt.plot([x1-1,x1+1],[k(x1)-(fd(x1)),f(x1)+(fd(x1))], color='red', linewidth=3)
29  plt.plot([x2-1,x2+1],[k(x2)-(fd(x2)),f(x2)+(fd(x2))], color='red', linewidth=3)
30  plt.plot([x3-1,x3+1],[k(x3)-(fd(x3)),f(x3)+(fd(x3))], color='red', linewidth=3)
31
32  # 最大值
33  plt.axvline(5)
34
35  plt.grid()
36  plt.show()
```

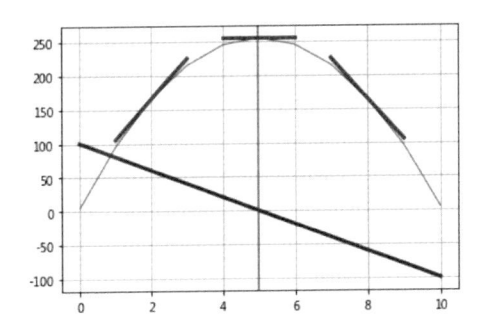

- 執行結果：

- 一次方函數整條線上的每一個點的斜率都相同，但是二次方函數上的每一個點的斜率就都不一樣了，如上圖，相關彩色圖形可參考範例程式。

  ＊ 綠線（細拋物線）：二次曲線，是一條對稱的拋物線。

  ＊ 紫線（斜線）：拋物線的一階導數。

  ＊ 紅線（拋物線的切線）：三個點（2,5,8）的斜率。

- 每一個點的斜率即該點與二次函數的切線（紅線），均不相同，斜率值可透過微分求得一階導數（圖中的斜線），隨著 x 變大，斜率越來越小，二次曲線的最大值就發生在斜率 =0 的地方，當 x=5 時，f(x)=255。

3. 繪製二次方函數 $f(x)=x^2+2x+7$，求最小值。

```
1  from matplotlib import pyplot as plt
2
```

```
 3  # 二次曲線
 4  def f(x):
 5      return (x**2) + (2*x) + 7
 6
 7  # 一階導數
 8  def fd(x):
 9      return 2*x + 2
10
11  # 設定樣本點
12  x = list(range(-10, 11))
13  y = [f(i) for i in x]
14
15  # 一階導數的樣本點
16  yd = [fd(i) for i in x]
17
18  # 畫二次曲線
19  plt.plot(x,y, color='green')
20
21  # 畫一階導數
22  plt.plot(x,yd, color='purple', linewidth=3)
23
24  # 畫三個點的斜率 x = (-7, -1, 5)
25  x1 = 5
26  x2 = -1
27  x3 = -7
28  plt.plot([x1-1,x1+1],[f(x1)-(fd(x1)),f(x1)+(fd(x1))], color='red', linewidth=3)
29  plt.plot([x2-1,x2+1],[f(x2)-(fd(x2)),f(x2)+(fd(x2))], color='red', linewidth=3)
30  plt.plot([x3-1,x3+1],[f(x3)-(fd(x3)),f(x3)+(fd(x3))], color='red', linewidth=3)
31
32  # 最大值
33  plt.axvline(-1)
34
35  plt.grid()
36  plt.show()
```

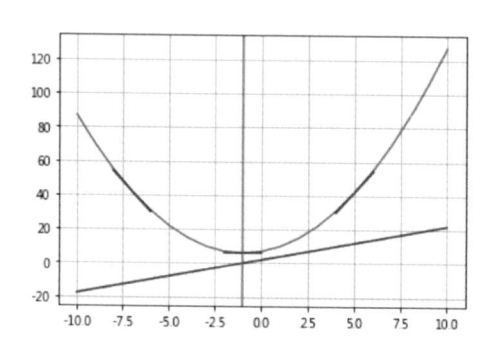

- 執行結果：

- 斜率值可透過微分求得一階導數（圖中的斜線），隨著 x 變大，斜率越來越大，拋物線的最小值就發生在斜率 =0 的地方，當 x=-1 時，f(x)=6。

由上得知微分兩次的**二階導數（f''(x)）為常數**，且為正值時，函數有最小值，反之，為負值時，函數有最大值。但若是 **f(x) 為三次方（以上）的函數**，一階導數 =0 的點，可能只是區域的最佳解（Local Minimum/Maximum），而不是全局最佳解（Global Minimum/Maximum）。

2-3 微積分（Calculus）

4. 繪製三次方曲線 f(x)= $x^3-2x+100$，求最小值。

```python
1  from matplotlib import pyplot as plt
2  import numpy as np
3
4  # f(x)= x^3-2x+100
5  def f(x):
6      return (x**3) - (2*x) + 100
7
8  # 一階導數
9  def fd(x):
10     return 3*(x**2) - 2
11
12 # 設定樣本點
13 x = list(range(-10, 11))
14 y = [f(i) for i in x]
15
16 # 一階導數的樣本點
17 yd = [fd(i) for i in x]
18
19 # 畫二次曲線
20 plt.plot(x,y, color='green')
21
22 # 畫一階導數
23 plt.plot(x,yd, color='purple', linewidth=3)
24
25 # 最小值


26 x1=np.array([sqrt(6)/3, -sqrt(6)/3])
27 plt.scatter([x1], [f(x1)])
28 plt.scatter([-sqrt(6)/3], [f(-1)])
29
30 plt.grid()
31 plt.show()
```

- 執行結果：

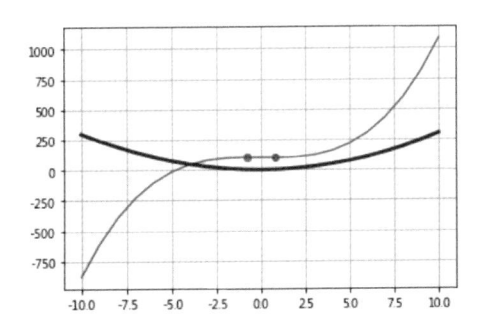

- 三次方函數 f(x)= $x^3-2x+100$ 在斜率 =0 的點只是區域最小值（Local Minimum），而非全局最小值（Global Minimum）。

- 凸函數（Convex）才能順利求得全局最佳解。

第 2 章　神經網路（Neural Network）原理

2-3-2　微分定理

微分相關定理整理如下：

1. f(x) 一階導數的表示法為 f'(x) 或 $\frac{dy}{dx}$。
2. f(x) 為常數（C）➔ f'(x)= 0
3. f(x)= Cg(x)➔ f'(x)= Cg'(x)
4. f(x)=g(x)+h(x)➔ f'(x)=g'(x)+h'(x)
5. 次方的規則：f(x)=x^n ➔ f'(x)=nx^{n-1}
6. 乘積的規則：$\frac{d[f(x)g(x)]}{dx}$ = f'(x)g(x)+f(x)g'(x)
7. 商的規則：若 r(x)=s(x)/t(x)

$$r'(x) = \frac{s'(x)t(x) - s(x)t'(x)}{[t(x)]^2}$$

8. 連鎖律（Chain Rule）

$$\frac{d}{dx}[o(i(x))] = o'(i(x)) \cdot i'(x)$$

以 f(x)=-10x^2+100x+5 為例，針對多項式的每一項個別微分再相加，就得到 f(x) 的一階導數：

$$f'(x)= -20\ x + 100$$

SymPy 套件直接支援微積分函數的計算，可以驗證定理，接下來我們就寫一些程式來練習一下。

範例. 使用 SymPy 套件計算微分。

程式：02_3_1_ 微分 .ipynb 下半部。

1. f(x) 為常數（C）➔ f'(x)= 0

```
1  # 常數微分 f(x) = C ==> f'(x) = 0
2  from sympy import *
3
4  x = Symbol('x')
5  # f(x) 為常數
```

2-24

2-3 微積分（Calculus）

```
6  y = 0 * x + 5
7  yprime = y.diff(x)
8  yprime
```

- 執行結果：0。

2. f(x)= Cg(x) ➔ f(x)= Cg'(x)

```
1  # f(x) = Cg(x) ==> f'(x) = Cg'(x)
2
3  from sympy import *
4
5  x = Symbol('x')
6
7  # Cg(x)
8  y1 = 5 * x ** 2
9  yprime1 = y1.diff(x)
10 print(yprime1)
11
12 # g(x)
13 y2 = x ** 2
14 # Cg'(x)
15 yprime2 = 5 * y2.diff(x)
16 print(yprime2)
17
18 # 比較
19 yprime1 == yprime1
```

- 執行結果均為 10x。

3. 乘積的規則：$\dfrac{d[f(x)g(x)]}{dx}$ = f'(x)g(x)+f(x)g'(x)

```
1  # (d[f(x)g(x)])/dx = f'(x)g(x)+f(x)g'(x)
2
3  from sympy import *
4
5  x = Symbol('x')
6
7  # d[f(x)g(x)])/dx
8  f = x ** 2
9  g = x ** 3
10 y1 = f*g
11 yprime1 = y1.diff(x)
12 print(yprime1)
13
14 # f'(x)g(x)+f(x)g'(x)
15 yprime2 = f.diff(x) * g + f * g.diff(x)
16 print(yprime2)
17
18 # 比較
19 yprime1 == yprime1
```

- 執行結果：$5x^4$。

第 2 章　神經網路（Neural Network）原理

4. 連鎖律（Chain Rule）：

$$\frac{d}{dx}[o(i(x))] = o'(i(x)) \cdot i'(x)$$

```python
from sympy import *

x = Symbol('x')

# d[f(g(x))])/dx
g = x ** 3
f = g ** 2
yprime1 = f.diff(x)
print(yprime1)

# f'(g(x))g'(x)
g = Symbol('g')
f = g ** 2
g1 = x ** 3
yprime2 = f.diff(g) * g1.diff(x)
# 將 f'(g(x)) 的 g 以 x ** 3 取代
print(yprime2.subs({g:x ** 3}))

# 比較
yprime1 == yprime1
```

- 執行結果：$6x^5$。

5. 微分 $f(x)=-10x^2+100x+5$。

```python
from sympy import *

x = Symbol('x')
# f(x)=-10x2 +100x+5
y = -10 * x**2 + 100 * x + 5
yprime = y.diff(x)
yprime
```

- 執行結果：$100-20x$。

6. 利用一階導數 =0，求最大值。

```python
from sympy.solvers import solve

# 一階導數=0
dict1 = solve([yprime])
print(dict1)
x1 = dict1[x]
print(f'x={x1}, 最大值={-10 * x1**2 + 100 * x1 + 5}')
```

- 執行結果:x=5, 最大值 =255。

7. 微分 $f(x)=x^2+2x+7$。

```
1  from sympy import *
2
3  x = Symbol('x')
4  # f(x)=x2+2x+7
5  y = (x**2) + (2*x) + 7
6  yprime = y.diff(x)
7  yprime
```

- 執行結果:2x + 2。

8. 再利用一階導數 =0,求最小值。

```
1  from sympy.solvers import solve
2
3  # 一階導數=0
4  dict1 = solve([yprime])
5  print(dict1)
6  x1 = dict1[x]
7  print(f'x={x1}, 最小值={(x1**2) + (2*x1) + 7}')
```

- 執行結果:x=-1, 最小值 =6。

2-3-3 偏微分(Partial Differentiation)

在神經網路求解的過程中,偏微分(Partial Differentiation)佔有舉足輕重的地位,被用來計算各特徵變數的梯度(gradient),進而求得最佳權重(Weight)。梯度與斜率相似,單一變數的變化率稱為斜率,多變數的斜率稱為梯度。

在上一節中,f(x) 只有單一變數,如果是多個變數 x_1、x_2、x_3…,要如何找到最小值或最大值呢?這時,我們就可以使用偏微分求取每個變數的梯度,讓函數沿著特定方向找最佳解,如下圖,即沿著等高線(Contour),逐步向圓心逼近,這就是所謂的『梯度下降法』(Gradient Descent)。

第 2 章　神經網路（Neural Network）原理

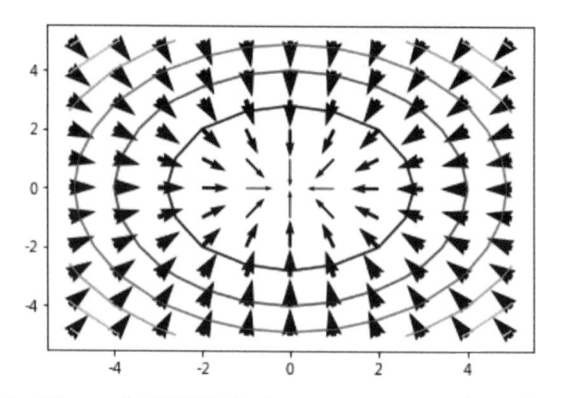

▲ 圖 2.4　梯度下降法（Gradient Descent）示意圖

範例 1. 偏微分計算。

程式：02_3_2_ 偏微分 .ipynb。

1. 分別對 x、y 作偏微分。

    先針對 x 作偏微分：

    $$\frac{\partial f(x, y)}{\partial x} = \frac{\partial (x^2 + y^2)}{\partial x}$$

    → 將其他變數視為常數，對每一項個別微分：

    $$\frac{\partial x^2}{\partial x} = 2x$$

    $$\frac{\partial y^2}{\partial x} = 0$$

    → 加總：

    $$\frac{\partial f(x, y)}{\partial x} = 2x + 0 = 2x$$

    → 再針對 y 作偏微分，同樣將其他變數視為常數：

    $$\frac{\partial f(x, y)}{\partial y} = 0 + 2y = 2y$$

2-3 微積分（Calculus）

→ 總結：

$$\frac{\partial f(x, y)}{\partial x} = 2x$$

$$\frac{\partial f(x, y)}{\partial y} = 2y$$

```
1  # f(x, y) = x^2 + y^2
2
3  from sympy import *
4  x, y = symbols('x y')
5
6  # 對 x 偏微分
7  y1 = x**2 + y**2
8  yprime1 = y1.diff(x)
9  print(yprime1)
10
11 # 對 y 偏微分
12 yprime2 = y1.diff(y)
13 print(yprime2)
14
```

- 第 8、12 行：y1.diff(x)、y1.diff(y) 分別對 x、y 作偏微分。

範例 2. 使用梯度下降法求函數 $f(x)=x^2$ 的最小值。

程式：02_3_2_ 偏微分 .ipynb。

1. 以梯度下降法求函數 $f(x)=x^2$ 的最小值，步驟如下：

    - 任意設定一起始點（x_start）：第 24 行。

    - 以一階導數計算該點的梯度 fd(x)：第 8 行。

    - 沿著梯度更新 x，逐步逼近最佳解，逼近幅度大小以學習率（第 26 行的 lr）控制，公式如下：

      新的 x = 原來的 x - 學習率（learning rate）* 梯度（gradient）

    - 重複前 2 步驟多次（第 25 行的 epochs），若函數值以不再變化或差異很小，即算找到最佳解。

```
1  import numpy as np
2  import matplotlib.pyplot as plt
3
4  # 函數 f(x)=x^2
5  def f(x): return x ** 2
6
```

第 2 章　神經網路（Neural Network）原理

```python
 7  # 一階導數:dy/dx=2*x
 8  def fd(x): return 2 * x
 9
10  def GD(x_start, df, epochs, lr):
11      xs = np.zeros(epochs+1)
12      w = x_start
13      xs[0] = w
14      for i in range(epochs):
15          dx = df(w)
16          # 權重的更新
17          # W_NEW = W - 學習率(learning rate) x 梯度(gradient)
18          w += - lr * dx
19          xs[i+1] = w
20      return xs
21
22
23  # 超參數(Hyperparameters)
24  x_start = 5      # 起始權重
25  epochs = 25      # 執行週期數
26  lr = 0.1         # 學習率
27
28  # 梯度下降法, 函數 fd 直接當參數傳遞
29  w = GD(x_start, fd, epochs, lr=lr)
30  # 顯示每一執行週期得到的權重
31  print (np.around(w, 2))
32
33  # 畫圖
34  color = 'r'
35  from numpy import arange
36  t = arange(-6.0, 6.0, 0.01)
37  plt.plot(t, f(t), c='b')
38  plt.plot(w, f(w), c=color, label='lr={}'.format(lr))
39  plt.scatter(w, f(w), c=color, )
40
41  # 設定中文字型
42  from matplotlib.font_manager import FontProperties
43  font = FontProperties(fname=r"c:\windows\fonts\msjhbd.ttc", size=20)
44  plt.title('梯度下降法', fontproperties=font)
45  plt.xlabel('w', fontsize=20)
46  plt.ylabel('Loss', fontsize=20)
47
48  # 矯正負號
49  plt.rcParams['axes.unicode_minus'] = False
50
51  plt.show()
```

2-3 微積分（Calculus）

- 執行結果：

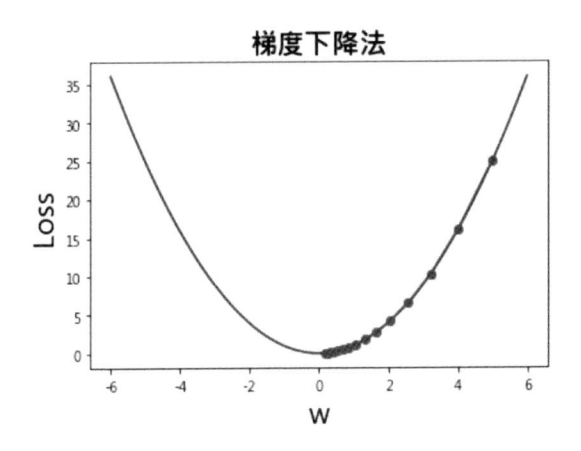

梯度下降法

- 得到 x 每一點的座標如下：

[5.   4.   3.2 2.56 2.05 1.64 1.31 1.05 0.84 0.67 0.54 0.43 0.34 0.27
 0.22 0.18 0.14 0.11 0.09 0.07 0.06 0.05 0.04 0.03 0.02 0.02]

- 我們可以改變第 24~26 行的參數，觀察執行結果。

2. 改變起始點 x_start = -5（第 24 行），重新執行，依然可以找到最小值。

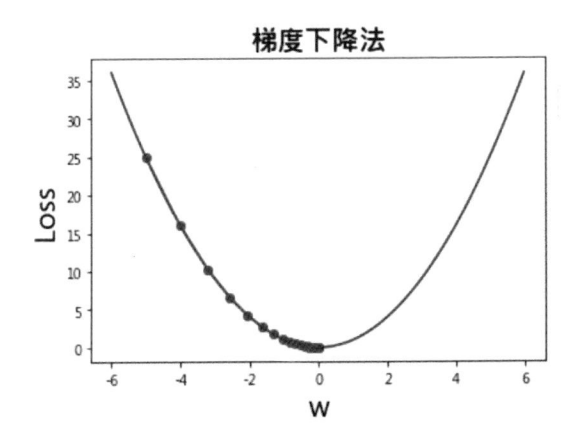

梯度下降法

2-31

3. 設定學習率 lr = 0.9（第 26 行）：更新幅度過大，可能會跳過最小值。

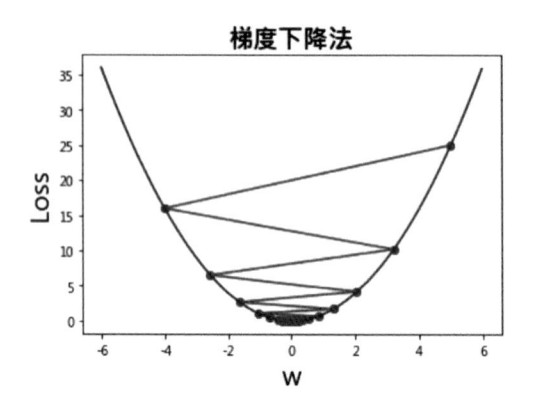

梯度下降法

4. 設定學習率 lr = 0.01：更新幅度過小，還未逼近到最小值，就提前停止了，增加『執行週期數』（第 25 行）可解決問題。

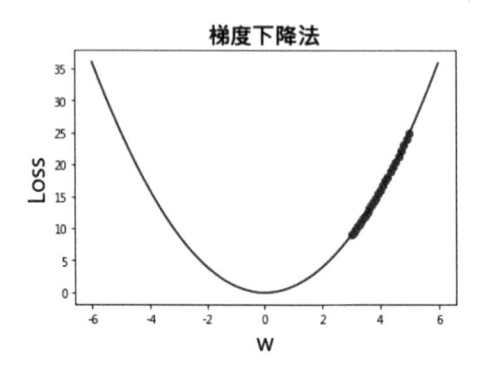

梯度下降法

上述程式是神經網路優化求解的簡化版，在後續的章節會詳細的探討，目前只聚焦在說明偏微分在深度學習的重要性。

2-3-4 線性迴歸求解

再看另一個應用，在優化求解中，也常利用一階導數等於 0 的特性，求取最佳解，例如，以『最小平方法』（Ordinary Least Square, OLS）對簡單線性迴歸 y = wx + b 求解。首先定義『目標函數』（Object Function）或稱『損失函數』（Loss Function）為『均方誤差』（MSE），即預測值與實際值差距的平方和，MSE 當然愈小愈好，所以它是一個最小化的問題，所以，我們可以利用偏微分推導出公式，過程如下：

2-3 微積分（Calculus）

1. MSE= $\Sigma\varepsilon^2 /n = \Sigma(y-\hat{y})^2/n$

   其中 ε：誤差，即實際值（y）與預測值（$\hat{y}$）之差

   　　n：為樣本個數

2. MSE = SSE/n，n 為常數可忽略，不影響最小值求解，只考慮 SSE：

$$SSE = \Sigma\varepsilon^2 = \Sigma(y-\hat{y})^2 = \Sigma(y-wx-b)^2$$

3. 分別對 w 及 b 作偏微分，並令一階導數 =0，可以得到兩個聯立方程式，進而求得 w 及 b 的解。

4. 對 b 偏微分，又因

   f(x)= g(x)g(x)= g'(x)g(x)+ g(x)g'(x)= 2 g(x)g'(x)：

$$\frac{dSSE}{db} = -2\sum_{i=1}^{n}(y-wx-b) = 0$$

➔ 兩邊同除 -2

$$\sum_{i=1}^{n}(y-wx-b) = 0$$

➔ 分解

$$\sum_{i=1}^{n}y - \sum_{i=1}^{n}wx - \sum_{i=1}^{n}b = 0$$

➔ 除以 n，$\bar{x}$、$\bar{y}$ 為 x、y 的平均數

$$\bar{y} - w\bar{x} - b = 0$$

➔ 移項

$$b = \bar{y} - w\bar{x}$$

5. 對 w 偏微分：

$$\frac{dSSE}{dw} = -2\sum_{i=1}^{n}(y-wx-b)x = 0$$

第 2 章　神經網路（Neural Network）原理

→ 兩邊同除 -2

$$\sum_{i=1}^{n}(y-wx-b)x=0$$

→ 分解

$$\sum_{i=1}^{n}yx-\sum_{i=1}^{n}wx-\sum_{i=1}^{n}bx=0$$

→ 代入 4 的計算結果 $b=\overline{y}-w\overline{x}$

$$\sum_{i=1}^{n}yx-\sum_{i=1}^{n}wx-\sum_{i=1}^{n}(\overline{y}-w\overline{x})x=0$$

→ 化簡

$$\sum_{i=1}^{n}(y-\overline{y})x-w\sum_{i=1}^{n}(x^2-\overline{x}x)=0$$

$$w=\sum_{i=1}^{n}(y-\overline{y})x\,/\,\sum_{i=1}^{n}(x^2-\overline{x}x)$$

$$w=\sum_{i=1}^{n}(y-\overline{y})x\,/\,\sum_{i=1}^{n}(x-\overline{x})^2$$

結論：

$$w=\sum_{i=1}^{n}(y-\overline{y})x\,/\,\sum_{i=1}^{n}(x-\overline{x})^2$$

$$b=\overline{y}-w\overline{x}$$

範例 2. 使用最小平方方法（OLS）求解線性迴歸。

程式：02_3_3_ 線性迴歸 .ipynb。

1. 現有一個世界人口統計資料集，以年度（year）為 x，人口數為 y，依上述公式計算迴歸係數 w、b，撰寫程式如下：

```
1  # 使用 OLS 公式計算 w、b
2  # 載入套件
3  import matplotlib.pyplot as plt
4  import numpy as np
5  import math
6  import pandas as pd
```

2-3 微積分（Calculus）

```
 7
 8  # 載入資料集
 9  df = pd.read_csv('./data/population.csv')
10
11  w = ((df['pop'] - df['pop'].mean()) * df['year']).sum() \
12      / ((df['year'] - df['year'].mean())**2).sum()
13  b = df['pop'].mean() - w * df['year'].mean()
14
15  print(f'w={w}, b={b}')
```

- 執行結果：

  w=0.061159358661557375, b=-116.35631056117687

2. 使用 NumPy 的現成函數 polyfit() 驗算：

```
1  # 使用 NumPy 的現成函數 polyfit()
2  coef = np.polyfit(df['year'], df['pop'], deg=1)
3  print(f'w={coef[0]}, b={coef[1]}')
```

- 執行結果：

  w=0.061159358661554586, b=-116.35631056117121

答案相去不遠，Yelp!

假如特徵 (x) 不只一個，亦即多元迴歸，可以使用矩陣求解，將 b 視為 w 的一環：

$$y=wx+b \rightarrow y=wx+b*1 \rightarrow y=[x\ 1]\begin{bmatrix}w\\b\end{bmatrix} \rightarrow y=x_{new}w_{new}$$

x_{new} 如下：

$$\begin{bmatrix} x_1 & x_2 & x_n & 1 \\ .. & .. & .. & 1 \\ .. & .. & .. & 1 \\ .. & .. & .. & 1 \end{bmatrix}$$

一樣對 SSE 偏微分，一階導數 =0 有最小值，公式推導如下：

$$SSE = \sum \varepsilon^2 = (y-\hat{y})^2 = (y-wx)^2 = yy'-2wxy + w'x'xw$$

$$\frac{dSSE}{dw} = -2xy + 2wx'x = 0$$

第 2 章　神經網路（Neural Network）原理

→ 移項、整理

$$(xx')w = xy$$

→ 移項

$$w = (xx')^{-1}xy$$

範例 3. 使用最小平方方法（OLS）求解多元線性迴歸。

程式：02_3_4_ 多元線性迴歸 .ipynb。

1. 一樣使用世界人口統計資料集，以年度（year）為 x，人口數為 y，依上述公式計算迴歸係數 w、b。

```
1  import pandas as pd
2
3  df = pd.read_csv('./data/population.csv')
```

2. 使用矩陣計算：第 14 行依照上述公式計算。

```
1   import numpy as np
2
3   X = df[['year']].values
4
5   # b = b * 1
6   one=np.ones((len(df), 1))
7
8   # 將 x 與 one 合併
9   X = np.concatenate((X, one), axis=1)
10
11  y = df[['pop']].values
12
13  # 求解
14  w = np.linalg.inv(X.T @ X) @ X.T @ y
15  print(f'w={w[0, 0]}, b={w[1, 0]}')
```

- 執行結果：

    w=0.06115935866154644,b=-116.35631056115507

2-3 微積分（Calculus）

3. 使用 polyfit 驗算。

```
1  coef = np.polyfit(df['year'], df['pop'], deg=1)
2  print(f'w={coef[0]}, b={coef[1]}')
```

- 執行結果：幾乎相同。

    w=0.061159358661554586,b=-116.35631056117121

4. 使用 Scikit-Learn LinearRegression 類別驗算。

```
1  from sklearn.linear_model import LinearRegression
2
3  X, y = df[['year']].values, df['pop'].values
4
5  lr = LinearRegression()
6  lr.fit(X, y)
7  lr.coef_, lr.intercept_
```

- 執行結果：幾乎相同。

    w=0.06115936,b=-116.35631056117116）

5. 以計程車小費資料集為例，求解多元線性迴歸。

```
1  # 載入資料集
2  import seaborn as sns
3
4  df= sns.load_dataset('tips')
5  df.head()
```

- 執行結果：tip（小費）為預測目標，其他欄位為特徵(x)。

	total_bill	tip	sex	smoker	day	time	size
0	16.99	1.01	Female	No	Sun	Dinner	2
1	10.34	1.66	Male	No	Sun	Dinner	3
2	21.01	3.50	Male	No	Sun	Dinner	3
3	23.68	3.31	Male	No	Sun	Dinner	2
4	24.59	3.61	Female	No	Sun	Dinner	4

第 2 章　神經網路（Neural Network）原理

6. 文字欄位轉換為數值欄位：使用字典對照文字與數值轉換。

```
1  df.sex = df.sex.map({'Female':0, 'Male':1})
2  df.smoker = df.smoker.map({'No':0, 'Yes':1})
3  df.day = df.day.map({'Thur':0, 'Fri':1, 'Sat':2, 'Sun':3})
4  df.time = df.time.map({'Lunch':0, 'Dinner':1})
5  df.head()
```

- 執行結果：

	total_bill	tip	sex	smoker	day	time	size
0	16.99	1.01	0	0	3	1	2
1	10.34	1.66	1	0	3	1	3
2	21.01	3.50	1	0	3	1	3
3	23.68	3.31	1	0	3	1	2
4	24.59	3.61	0	0	3	1	4

7. 使用矩陣計算：與前面程式碼相同。

```
1  X, y = df.drop('tip', axis=1).values, df.tip.values
2
3  # b = b * 1
4  one=np.ones((X.shape[0], 1))
5
6  # 將 x 與 one 合併
7  X2 = np.concatenate((X, one), axis=1)
8
9  # 求解
10 w = np.linalg.inv(X2.T @ X2) @ X2.T @ y
11 w
```

- 執行結果：最後一個數字為偏差。

    0.09432509,-0.03464496,-0.07566309,0.05273982,0.11247777,

    0.17481962,0.72400336

8. 以 Scikit-Learn 的線性迴歸驗證：與前面程式碼相同。

```
1  from sklearn.linear_model import LinearRegression
2
```

2-38

2-3 微積分（Calculus）

```
3  lr = LinearRegression()
4  lr.fit(X, y)
5
6  lr.coef_, lr.intercept_
```

- 執行結果：最後一個數字為偏差，答案與矩陣計算結果幾乎相同。

    0.09432509,-0.03464496,-0.07566309,0.05273982,-0.11247777,

    0.17481962,0.7240033611886232

由上面應用可以瞭解線性代數與微分的結合，可以幫我們解決許多數學問題。

2-3-5 積分（Integration）

積分則是微分的相反，微分用於斜率或梯度的計算，積分則可以計算長度、面積、體積，也可以用來計算累積的累積分配函數（Cumulative Distribution Function,cdf）。

1. 積分一般數學表示式如下：

$$\int_0^n f(x)dx$$

2. 若 f(x)= x，積分結果為 $\frac{1}{2}x^2$。

3. 限定範圍的積分：先求積分，再將上、下限代入多項式，相減可得結果。

$$\int_0^3 xdx = \frac{1}{2}x^2 \Big|_0^3$$

➔ （(1/2)*3^2）–（(1/2)*0^2）= 4.5

範例 1. 積分計算並作圖。

程式：02_3_5_積分.ipynb。

$$\int_0^3 xdx$$

1. 積分運算：SciPy 套件支援積分運算

```
1  # 載入套件
2  import numpy as np
```

```
3  import scipy.integrate as integrate
4  import numpy as np
5  import math
6
7
8  f = lambda x: x
9  i, e = integrate.quad(f, 0, 3)
10
11 print('積分值: ' + str(i))
12 print('誤差: ' + str(e))
```

- 執行結果如下：

  積分值 :4.5

  誤差 :4.9960036108132044e-14

- 第 3 行載入 SciPy 套件。

- 第 9 行呼叫 integrate.quad()，作限定範圍的積分，參數如下：

  * 函數 f(x)

  * 範圍下限：設為負無窮大（$-\infty$）

  * 範圍上限：設為正無窮大（∞）

  * 輸出：含積分結果及誤差值

2. 作圖

```
1  import matplotlib.pyplot as plt
2
3  # 樣本點
4  x = range(0, 11)
5
6  # 作圖
7  plt.plot(x, f(x), color='purple')
8
9  # 積分面積
10 area = np.arange(0, 3, 1/20)
11 plt.fill_between(area, f(area), color='green')
12
13 # 設定圖形屬性
14 plt.xlabel('x')
15 plt.ylabel('f(x)')
16 plt.grid()
17
18 plt.show()
```

2-3 微積分（Calculus）

- 執行結果：

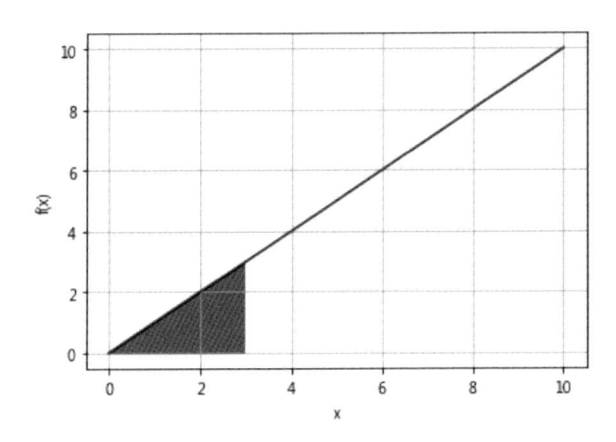

- 第 11 行 fill_between() 會填滿整個積分區域。

範例 2. 以積分計算常態分配（Normal Distribution）的機率並作圖。

程式：02_3_6_計算常態分配機率.ipynb

1. 常態分配的機率密度函數（Probability Density Function, pdf）如下：

$$f(x;\mu,\sigma) = \frac{1}{\sqrt{2\pi*\sigma^2}} * e^{-\frac{1}{2}*(\frac{x-\mu}{\sigma})^2}$$

2. 計算常態分配（-∞,∞）的機率。

```
1  # 載入套件
2  import scipy.integrate as integrate
3  import numpy as np
4  import math
5
6
7  # 平均數(mean)、標準差(std)
8  mean = 0
9  std = 1
10 # 常態分配的機率密度函數(Probability Density Function, pdf)
11 f = lambda x: (1/((2*np.pi*std**2) ** .5)) * np.exp(-0.5*((x-mean)/std)**2)
12
13 # 積分，從負無窮大至無窮大
14 i, e = integrate.quad(f, -np.inf, np.inf)
15
16 print('累積機率:', round(i, 2))
17 print('誤差:', str(e))
```

第 2 章　神經網路（Neural Network）原理

- 執行結果：機率總和為 1.0，任何機率分配的總和必然為 1，即所有事件發生的機率總和必然為 100%。

- 誤差 :10^{-8}。

- 第 7、8 行定義平均數（mean）為 0、標準差（std）為 1，也就是『標準』常態分配，又稱 Z 分配。

- 第 11 行定義常態分配的機率密度函數（pdf）。

3. 常態分配常見的信賴區間包括正 / 負 1、2、3 倍標準差，我們可以計算其機率。有關信賴區間的定義會在機率與統計的章節說明。

```
1  # 1倍標準差區間之機率
2  i, e = integrate.quad(f, -1, 1)
3
4  print('累積機率:', round(i, 3))
5  print('誤差:', str(e))
```

- 執行結果如下：± 1 倍標準差的信賴區間機率 =68.3%，故 integrate.quad() 參數設為 -1 及 1。

- 誤差 :7.579375928402476e-15

4. ± 2 倍標準差的信賴區間機率 =95.4%。

```
1  # 2倍標準差區間之機率
2  i, e = integrate.quad(f, -2, 2)
3
4  print('累積機率:', round(i, 3))
5  print('誤差:', str(e))
```

- 執行結果：± 2 倍標準差的信賴區間機率 =95.4%。

- 誤差 :1.8403560456416134e-11

- integrate.quad() 參數設為 -2 及 2。

5. ± 3 倍標準差的信賴區間機率 =99.7%。

```
1  # 3倍標準差區間之機率
2  i, e = integrate.quad(f, -3, 3)
3
4  print('累積機率:', round(i, 3))
5  print('誤差:', str(e))
```

2-3 微積分（Calculus）

- 執行結果：± 3 倍標準差的信賴區間機率 =99.7%。

- 誤差 :1.1072256503105314e-14

6. 我們常用 ± 1.96 倍標準差的信賴區間，機率 =95%，剛好是個整數。

```python
1  # 95%信賴區間之機率
2  i, e = integrate.quad(f, -1.96, 1.96)
3
4  print('累積機率:', round(i, 3))
5  print('誤差:', str(e))
```

- 執行結果：± 1.96 倍標準差的信賴區間機率 =95%。

7. 可以使用隨機亂數及直方圖，繪製標準常態分配圖。

```python
1  import matplotlib.pyplot as plt
2  import numpy as np
3  import seaborn as sns
4
5  # 使用 randn 產生標準常態分配的亂數 10000 個樣本點
6  x = np.random.randn(10000)
7
8  # 直方圖，參數 hist=False：不畫階梯直方圖，只畫平滑曲線
9  sns.distplot(x, hist=False)
10
11 # 設定圖形屬性
12 plt.xlabel('x')
13 plt.ylabel('f(x)')
14 plt.grid()
15
16 # 對正負1、2、3倍標準差畫虛線
17 plt.axvline(-3, c='r', linestyle=':')
18 plt.axvline(3, c='r', linestyle=':')
19 plt.axvline(-2, c='g', linestyle=':')
20 plt.axvline(2, c='g', linestyle=':')
21 plt.axvline(-1, c='b', linestyle=':')
22 plt.axvline(1, c='b', linestyle=':')
23
24 # 1 倍標準差機率
25 plt.annotate(text='', xy=(-1,0.25), xytext=(1,0.25), arrowprops=dict(arrowstyle='<->'))
26 plt.annotate(text='68.3%', xy=(0,0.26), xytext=(0,0.26))
27 # 2 倍標準差機率
28 plt.annotate(text='', xy=(-2,0.05), xytext=(2,0.05), arrowprops=dict(arrowstyle='<->'))
29 plt.annotate(text='95.4%', xy=(0,0.06), xytext=(0,0.06))
30 # 3 倍標準差機率
31 plt.annotate(text='', xy=(-3,0.01), xytext=(3,0.01), arrowprops=dict(arrowstyle='<->'))
32 plt.annotate(text='99.7%', xy=(0,0.02), xytext=(0,0.02))
33
34 plt.show()
```

- 執行結果如下：

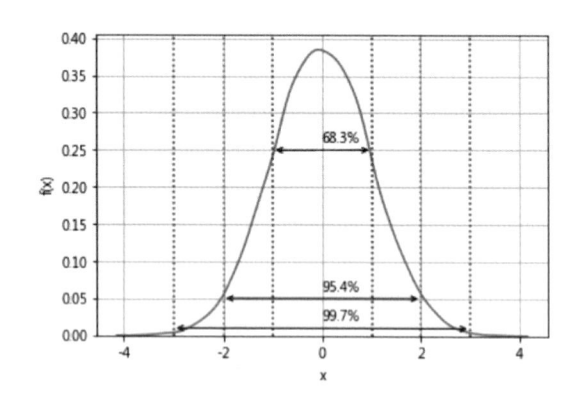

- 程式說明：

  * 第 9 行 sns.distplot（x,hist=False）：使用 Seaborn 畫直方圖，參數 hist=False 表示不畫階梯直方圖，只畫平滑曲線，新版指令為 sns.kdeplot(x)。

  * 第 17 行 axvline()：畫垂直線，標示正負 1、2、3 倍標準差。

2-4 機率（Probability）與統計（Statistics）

統計是從資料所推衍出來的資訊，包括一些描述統計量、機率分佈（or 分配）等，例如平均數、標準差等。而我們蒐集到的資料統稱為『資料集』（dataset），它包括一些觀察值（Observations）或案例（Cases），即資料集的列，以及一些特徵（features）或稱屬性（attributes），即資料集的欄位。譬如，要預測下一任台北市長的選舉，我們作一次問卷調查，相關定義如下：

1. 資料集：全部問卷調查資料。
2. 觀察值：每張問卷。
3. 特徵或屬性：每一個問題，通常以 X 表示，X 通常使用大寫，表可能有多個特徵。
4. 特徵值或屬性值：每一個問題的回答。
5. 目標（Target）欄位：市長候選人，通常以 y 表示。

以下我們會依序介紹以下內容：

1. 抽樣
2. 描述統計量

2-4 機率（Probability）與統計（Statistics）

3. 機率
4. 機率分配函數
5. 假設檢定

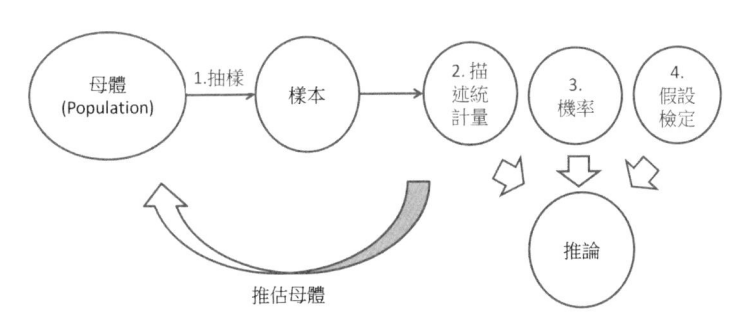

▲ 圖 2.5 基礎統計介紹的範圍及其關聯

2-4-1 資料型態

依照特徵的資料型態，分為定性（qualitative）及定量（quantitative）的欄位。定性（qualitative）欄位為非數值型的欄位，通常包含有限的類別。又可以分為名目資料（Nominal Data）及有序資料（Ordinal Data）。定量（quantitative）欄位為數值型欄位，又可以分為離散型資料（Discrete Data）及連續型資料（Continuous Data）。預測的目標（Target）欄位如為離散型變數，適用分類（Classification）的演算法，反之，目標欄位為連續型變數，適用迴歸（Regression）的演算法。

名目資料（Nominal Data）的欄位值並沒有順序或大小的隱含意義，例如顏色、紅、藍、綠並沒有誰大誰小的隱含意義，轉換為代碼時要特別注意，可使用 One-hot encoding 去除順序或大小的隱含意義，作法是將每一類別轉為獨立的啞變數（Dummy variable），每個啞變數只有 True/False 或 1/0 兩種值，例如下圖，Color 特徵有三種類別，經過 One-hot encoding 處理，會轉換為三個啞變數，轉換後紅、藍、綠就沒有大小之分。

	color	is_blue	is_green	is_red
0	green	0	1	0
1	red	0	0	1
2	green	0	1	0
3	blue	1	0	0

第 2 章　神經網路（Neural Network）原理

範例. 特徵轉換，包含 One-hot encoding。

程式：02_4_1_ 特徵轉換 .ipynb。

1. One-hot encoding：使用 get_dummies 函數，參數依序為資料框（Data Frame）、要轉換的欄位、前置符號（prefix）、分隔符號（prefix_sep）。

```
1  df2 = pd.get_dummies(df["color"], columns=["color"],
2                      prefix='is', prefix_sep='_')
3
4  # 連結轉換前後的欄位，相互比較
5  pd.concat((df["color"], df2), axis=1)
```

- 執行結果：

	color	is_blue	is_green	is_red
0	green	0	1	0
1	red	0	0	1
2	green	0	1	0
3	blue	1	0	0

2. 有序資料（Ordinal Data）：欄位值有順序或大小的隱含意義，例如衣服尺寸，XL > L > M > S，依尺寸大小，分別編碼為 XL（4）、L（3）、M（2）、S（1）。

```
1  # 以字典定義轉換規則
2  size_mapping = {'XL': 4,
3                  'L': 3,
4                  'M': 2,
5                  'S': 1}
6
7  # 使用 map() 轉換
8  df['size'] = df['size'].map(size_mapping)
9  df
```

- 執行結果：

	color	size	price	classlabel
0	green	4	10.1	class1
1	red	2	10.1	class1
2	green	3	13.5	class2
3	blue	1	15.3	class1

2-4 機率（Probability）與統計（Statistics）

- 第 2 行：以字典定義轉換規則，key 為原值，value 為轉換後的值。
- 第 4 行 map()：以字典轉換 size 欄位的每一個值。

2-4-2 抽樣（Sampling）

再以預測下一任台北市長的選舉為例，囿於人力、時間及經費的限制，不太可能調查所有市民的投票傾向，通常我們只會隨機抽樣 1000 份或更多一點的樣本進行調查，以推測全體市民的投票傾向，這種方式稱之為抽樣（Sampling），相關名詞定義如下：

1. 母體（Population）：全體有投票權的市民。
2. 樣本（Sample）：被抽中調查的市民。
3. 分層抽樣（Stratified Sampling）：依母體某些屬性的比例，進行相同比例的抽樣，希望能充分代表母體的特性，例如政黨、年齡、性別比例。

範例. 抽樣測試。

程式：02_4_2_抽樣.ipynb。

1. 簡單抽樣：從一個集合中隨機抽出 5 個。

```
1  import random
2  import numpy as np
3
4  # 1~10 的集合
5  list1 = list(np.arange(1, 10 + 1))


6
7  # 隨機抽出 5 個
8  print(random.sample(list1, 5))
```

- 執行結果：[1,6,2,8,5]
- 第 8 行 random.sample()：自集合中隨機抽樣。
- 每次執行結果均不相同，且每個項目不重複，此種抽樣方法稱為『不放回式抽樣』（Sampling Without Replacement）。

2. 放回式抽樣（Sampling With Replacement）：抽出後再放回集合中，下一次抽樣仍可能抽到相同的樣本。

```
1  import random
2  import numpy as np
```

第 2 章　神經網路（Neural Network）原理

```
3
4  # 1~10 的集合
5  list1 = list(np.arange(1, 10 + 1))
6
7  # 隨機抽出 5 個
8  print(random.choices(list1, k=5))
```

- 執行結果：[8,4,7,4,6]

- 第 8 行 random.choices()：自集合中隨機抽樣。

- 每次執行結果均不相同，但項目會重複，如上，4 被重複抽出兩次，此種抽樣方法稱為『放回式抽樣』。

3. 以 Pandas 套件進行抽樣。

```
1  from sklearn import datasets
2  import pandas as pd
3
4  # 載入鳶尾花(iris)資料集
5  ds = datasets.load_iris()
6
7  # x, y 合成一個資料集
8  df = pd.DataFrame(data=ds.data, columns=ds.feature_names)
9  df['y'] = ds.target
10
11 # 隨機抽出 5 個
12 print(df.sample(5))
```

- 執行結果如下：

	sepal length (cm)	sepal width (cm)	petal length (cm)	petal width (cm)	y
102	7.1	3.0	5.9	2.1	2
40	5.0	3.5	1.3	0.3	0
32	5.2	4.1	1.5	0.1	0
48	5.3	3.7	1.5	0.2	0
77	6.7	3.0	5.0	1.7	1

- 第 12 行 df.sample（5）：自集合中隨機抽樣 5 筆。

- 為『不放回式抽樣』（Sampling Without Replacement）。

4. 以 Pandas 套件進行分層抽樣（Stratified Sampling）。

```
1  from sklearn import datasets
2  from sklearn.model_selection import StratifiedShuffleSplit
3  import pandas as pd
4
5  # 載入鳶尾花(iris)資料集
```

2-48

2-4 機率（Probability）與統計（Statistics）

```
 6  ds = datasets.load_iris()
 7
 8  # x, y 合成一個資料集
 9  df = pd.DataFrame(data=ds.data, columns=ds.feature_names)
10  df['y'] = ds.target
11
12  # 隨機抽出 6 個
13  stratified = StratifiedShuffleSplit(n_splits=1, test_size=6)
14  x = list(stratified.split(df, df['y']))
15
16  print('重新洗牌的全部資料:')
17  print(x[0][0])
18
19  print('\n抽出的索引值:')
20  print(x[0][1])
```

- 執行結果：

```
重新洗牌的全部資料:
[ 38 115 136  80 111  94   1   0  48 100 108 104   8  51 131  78   9 142
 112  11 126  79  95   2  46 128 125  65  55  10  72 145 130  56 138  96
  88  19   7  43   4  82  32  91 127  87 133  73  85  62 129  42  57  84
  40 105  49  75 113 147  99  27   6 135  58  35  26 124  92  70  69 139
  66 101  74  60 110  15  39  59   3  53  89 107  61 143 118  86  71  98
  50  41  34  12 149  77  23  21 117 121  97  54 119  64 120  45  81 141
 122 114  20 144 134 132  17  24  13  22  44 123  31 116  76  18  47 137
  63  83  29  25  36 102  28  37  33  93 148  14 146  16 103  68  90 140]

抽出的索引值:
[ 52   5  30 109 106  67]
```

- 第 13 行 StratifiedShuffleSplit（test_size=6）：將集合重新洗牌，並從中隨機抽樣 6 筆。

- 第 14 行 stratified.split（df,df['y']）：以 y 欄位為 key，分層抽樣，得到的結果是 generator 資料型態，須轉為 list，才能一次取出。

- 第 1 個輸出是重新洗牌的全部資料，第 2 個是抽出的索引值。

- 利用 df.iloc[x[0][1]] 指令，取得資料如下，觀察 y 欄，每個類別各有兩筆與母體比例相同。

	sepal length (cm)	sepal width (cm)	petal length (cm)	petal width (cm)	y
52	6.9	3.1	4.9	1.5	1
5	5.4	3.9	1.7	0.4	0
30	4.8	3.1	1.6	0.2	0
109	7.2	3.6	6.1	2.5	2
106	4.9	2.5	4.5	1.7	2
67	5.8	2.7	4.1	1.0	1

第 2 章　神經網路（Neural Network）原理

- 母體比例可以 df['y'].value_counts() 指令取得資料如下，0/1/2 類別各 50 筆。

```
2    50
1    50
0    50
```

5. 以 Pandas 套件進行不分層抽樣。

```
1  from sklearn import datasets
2  from sklearn.model_selection import train_test_split
3  import pandas as pd
4
5  # 載入鳶尾花(iris)資料集
6  ds = datasets.load_iris()
7
8  # x, y 合成一個資料集
9  df = pd.DataFrame(data=ds.data, columns=ds.feature_names)
10 df['y'] = ds.target
11
12 # 隨機抽出 6 個
13 train, test = train_test_split(df, test_size=6)
14 x = list(stratified.split(df, df['y']))
15
16 print('\n抽出的資料:')
17 print(test)
```

- 執行結果：

	sepal length (cm)	sepal width (cm)	petal length (cm)	petal width (cm)	y
141	6.9	3.1	5.1	2.3	2
147	6.5	3.0	5.2	2.0	2
97	6.2	2.9	4.3	1.3	1
80	5.5	2.4	3.8	1.1	1
22	4.6	3.6	1.0	0.2	0
61	5.9	3.0	4.2	1.5	1

- 第 13 行 train_test_split（test_size=6）：將資料集重新洗牌，並從中隨機抽樣 6 筆為測試資料，其餘為訓練資料。

- 取得資料如上，觀察 y 欄，類別 0 有 1 筆，類別 1 有 3 筆，類別 2 有 2 筆，與母體比例不相同。

2-4-3 基礎統計（Statistics Fundamentals）

經過各種方式蒐集到一組樣本後，我們會先對樣本進行探索，通常會先衡量其集中趨勢（Central Tendency）及資料離散的程度（Measures of Variance），這些指標稱之為描述統計量（Descriptive Statistics），較常見的描述統計量如下：

1. 集中趨勢（Central Tendency）

    1.1 平均數（Mean）：母體以 μ 表示，樣本以 $\bar{x}$ 表示。

    $$\mu = \frac{\sum_{i=1}^{n} x_i}{n}$$

    1.2 中位數（Median）：將所有樣本由小排到大，以中間的樣本值為準。若樣本數為偶數，則取中間樣本值的平均數。中位數可以避免離群值（Outlier）的影響，例如，統計平均年收入，如果加入一位超級富豪，則平均數將大幅提高，但中位數毫不受影響。假設有一組樣本：

    [100,200,300,400,500]

    平均數 = 中位數 = 300

    現在把 500 改為 50000：

    [100,200,300,400,50000]

    平均數 = 100200，被 50000 影響，大幅提升，無法反應大多數資料的的樣態。

    中位數 = 300，仍然不變。

    1.3 眾數（Mode）：頻率發生最高的數值，以大多數的資料為主。

2. 資料離散的程度（Measures of Variance）

    2.1 級距（Range）：最大值 - 最小值。

    2.2 百分位數（Percentiles）與四分位數（Quartiles）：例如 100 為 75 百分位數表示有 75% 的樣本小於 100。

    2.3 變異數（Variance）：母體以 δ 表示，樣本以 s 表示。

    $$\delta = \sqrt{\frac{\sum(x-\mu)^2}{n}}$$

    $$s = \sqrt{\frac{\sum(x-\mu)^2}{n-1}}$$

第 2 章 神經網路（Neural Network）原理

3. 可以使用箱形圖（box plot）或稱盒鬚圖，直接顯示上述相關的統計量。

範例. 計算描述統計量。

程式：02_4_3_ 基礎統計 .ipynb。

1. 載入檔案。

```
1  import random
2  import pandas as pd
3  import numpy as np
4
5  # 讀取檔案
6  df = pd.read_csv('./data/president_heights.csv')
7  df.rename(columns={df.columns[-1]:'height'}, inplace=True)
8
9  # 觀看前 5 筆
10 df.head()
```

- 執行結果：

	order	name	height
0	1	George Washington	189
1	2	John Adams	170
2	3	Thomas Jefferson	189
3	4	James Madison	163
4	5	James Monroe	183

2. 以美國歷屆總統的身高資料，計算描述統計量。

```
1  # 集中趨勢(Central Tendency)
2  print(f"平均數={df['height'].mean()}")
3  print(f"中位數={df['height'].median()}")
4  print(f"眾  數={df['height'].mode()[0]}")
5  print()
6
7  # 資料離散的程度(Measures of Variance)
8  from scipy import stats
9
10 print(f"級距(Range)={df['height'].max() - df['height'].min()}")
11 print(f"182cm 百分位數={stats.percentileofscore(df['height'], 182, 'strict')}")
12 print(f"變異數={df['height'].std():.2f}")
```

2-4 機率（Probability）與統計（Statistics）

- 執行結果：
  * 平均數 =179.73809523809524
  * 中位數 =182.0
  * 眾數 =183
  * 級距（Range）=30
  * 182cm 的百分位數 =47.61904761904761
  * 變異數 =7.02

3. 以美國歷屆總統的身高資料，計算四分位數（Quartiles）。

```
1  print(f"四分位數\n{df['height'].quantile([0.25,0.5,0.75,1])}")
```

- 執行結果：
  * 四分位數
  * 0.25   174.25
  * 0.50   182.00
  * 0.75   183.00
  * 1.00   193.00

4. 顯示描述統計量。

```
1  print(f"{df['height'].describe()}")
```

- 執行結果：
  * count   42.000000
  * mean   179.738095
  * std   7.015869
  * min   163.000000
  * 25%   174.250000
  * 50%   182.000000

第 2 章　神經網路（Neural Network）原理

* 75%　183.000000

* max　193.000000

5. 繪製箱形圖（box plot）或稱盒鬚圖，定義如下。

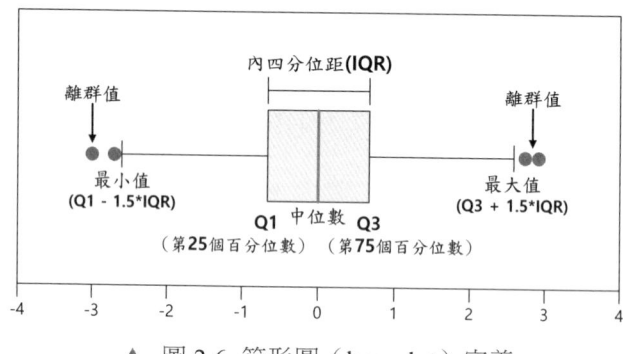

▲ 圖 2.6　箱形圖（box plot）定義

可觀察下列特性：

1. 中間的線：中位數。

2. 中間的箱子：Q1~Q3 共 50% 的資料集中在箱子內。

3. 是否存在離群值（Outlier），注意，離群值發生的可能原因包括記錄或登打錯誤、感測器失常造成量測錯誤或是重要的警訊（Influential Sample），應仔細探究為何發生，不要直接剔除掉。一般而言，會更進一步收集更多樣本，來觀察離群值是否再次出現，另一方面，更多的樣本也可以稀釋離群值的影響力。

```
1  import matplotlib.pyplot as plt
2
3  # 修正中文亂碼
4  plt.rcParams['font.sans-serif'] = ['Microsoft JhengHei']  # 微軟正黑體
5  plt.rcParams['axes.unicode_minus'] = False
6
7  df['height'].plot(kind='box', title='美國歷屆總統的身高資料', figsize=(8,6))
8  plt.show()
```

• 執行結果：

2-4 機率（Probability）與統計（Statistics）

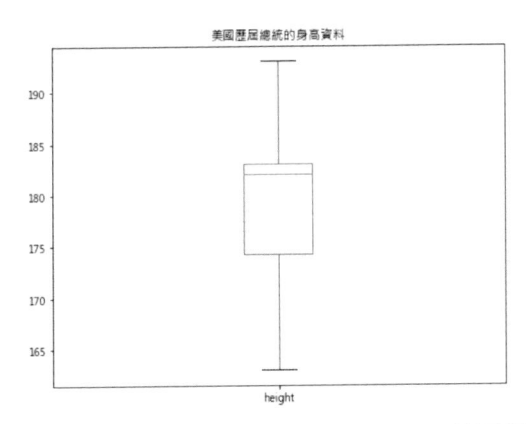

除了單變數的統計量，也會進行多變數分析，了解變數之間的關聯度（Correla-tion），在資料探索與分析（Exploratory Data Analysis，簡稱 EDA）階段，我們還會依據資料欄位的屬性進行不同面向的觀察，如下圖：

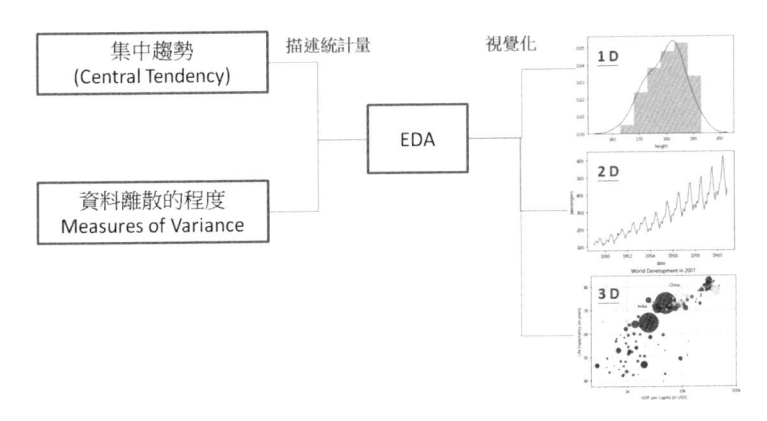

▲ 圖 2.7 資料探索與分析（Exploratory Data Analysis）不同面向的觀察

常用的統計圖功能簡要說明如下，同時也使用程式實作相關圖表：

1. 直方圖（Histogram）：觀察資料集中趨勢、離散的程度、機率分配及偏態（Skewness）/峰態（Kurtosis）。

```
1  # 直方圖
2  import seaborn as sns
3
4  # 繪圖
5  sns.kdeplot(df['height'])
6  plt.show()
```

- 執行結果：

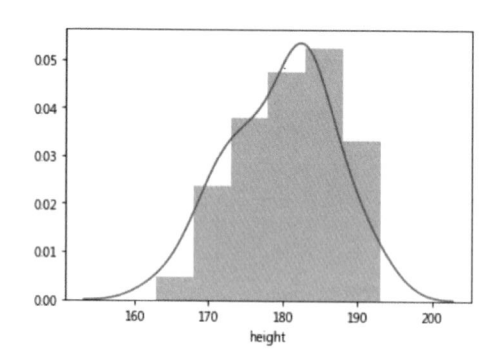

2. 圓餅圖（Pie Chart）：顯示單一變數的各類別所佔比例。

```
1  # 餅圖(Pie Chart)
2  import matplotlib.pyplot as plt
3  import pandas as pd
4  import numpy as np
5
6  # 讀取資料檔
7  df = pd.read_csv('./data/gdp.csv')
8
9  # 轉為整數欄位
10 df.pop = df['pop'].astype(int)
11
12 # 取最大 5 筆
13 df2 = df.nlargest(5, 'pop')
14
15 # 散佈圖(Scatter Chart)
16 plt.pie(df2['pop'], explode=[0, 0,0,0,0.2], labels=['china', 'india', 'USA', 'Rus', 'Other'])
17
18 plt.show()
```

- 執行結果：

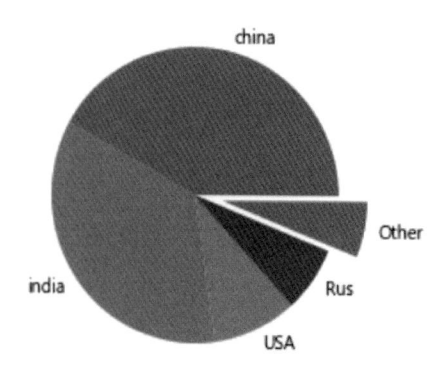

2-4 機率（Probability）與統計（Statistics）

3. 折線圖（Line Chart）：觀察趨勢，尤其是時間的趨勢，譬如股價、氣象溫度、營收、運量分析等。

```
1  # 線圖(Line Chart)
2  import matplotlib.pyplot as plt
3  import pandas as pd
4  import seaborn as sns
5
6  # 讀取資料檔
7  df = pd.read_csv('./data/airline.csv')
8
9  # 轉為日期欄位
10 df['date'] = pd.to_datetime(df['date'])
11
12 # 繪圖
13 sns.lineplot(df['date'], df['passengers'])
14
15 plt.show()
```

- 執行結果：

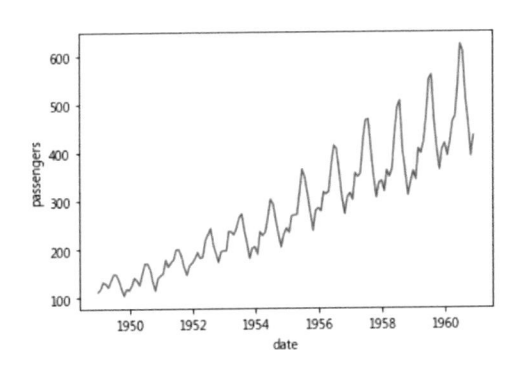

4. 散佈圖（Scatter Chart）：顯示兩個變數的關聯，亦用於觀察是否有離群值。

```
1  # 散佈圖(Scatter Chart)
2  import matplotlib.pyplot as plt
3  import pandas as pd
4  import numpy as np
5
6  # 讀取資料檔
7  df = pd.read_csv('./data/gdp.csv')
8
9  # 轉為整數欄位
10 df.pop = df['pop'].astype(int)
11
12 # 散佈圖(Scatter Chart)
13 plt.scatter(x = df.gdp, y = df.life_exp)
14
15 # 設定繪圖屬性
```

2-57

第 2 章　神經網路（Neural Network）原理

```
16  plt.xscale('log')
17  plt.xlabel('國民生產毛額')
18  plt.ylabel('壽命')
19  plt.title('World Development in 2007')
20  plt.xticks([1000,10000,100000], ['1k','10k','100k'])
21  plt.grid()
22
23  plt.show()
```

- 執行結果：

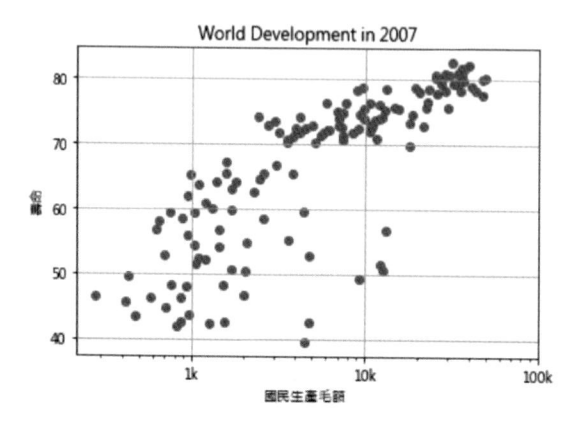

5. 氣泡圖（Bubble Chart）：散佈圖再加一個變數，以該變數值作為點的大小，並且做 3 個維度的分析。

```
1   # 氣泡圖(Bubble Chart)
2   import matplotlib.pyplot as plt
3   import pandas as pd
4   import numpy as np
5
6   # 讀取資料檔
7   df = pd.read_csv('./data/gdp.csv')
8
9   # 轉為整數欄位
10  df.pop = df['pop'].astype(int)
11
12  # 散佈圖(Scatter Chart) + 加一個變數 pop，作為點的大小
13  col=np.resize(['red', 'green', 'blue', 'yellow', 'lightblue'], df.shape[0])
14  plt.scatter(x = df.gdp, y = df.life_exp, s = df.pop * 2, c = col, alpha = 0.8)
15
16  # 設定繪圖屬性
17  plt.xscale('log')
18  plt.xlabel('國民生產毛額')
19  plt.ylabel('壽命')
20  plt.title('World Development in 2007')
21  plt.xticks([1000,10000,100000], ['1k','10k','100k'])
```

2-4 機率（Probability）與統計（Statistics）

```
22  plt.grid()
23
24  # 加註
25  plt.text(1550, 71, 'India')
26  plt.text(5700, 80, 'China')
27
28  plt.show()
```

- 執行結果：

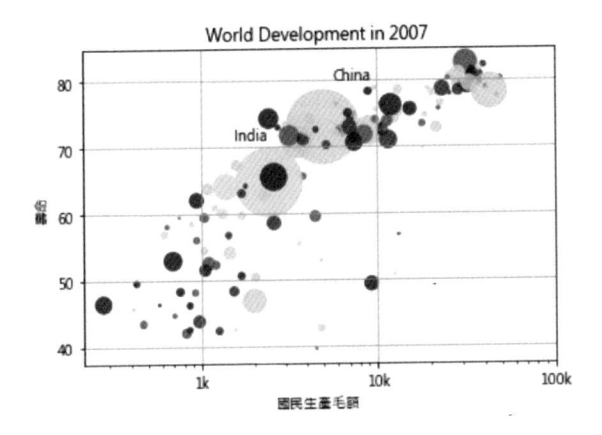

6. 熱力圖（Heatmap）：顯示各變數之間的關聯度（Correlation），得以輕易辨識出較高的關聯度。輸入資料須為各變數之間的關聯係數（Correlation coefficient），公式如下，可使用 Pandas 的 corr() 函數計算。

$$r_{x,y} = \frac{\sum_{i=1}^{n}(x_i - \bar{x})(y_i - \bar{y})}{\sqrt{\sum_{i=1}^{n}(x_i - \bar{x})^2 (y_i - \bar{y})^2}}$$

```
1   # 熱力圖(Heatmap)
2   import matplotlib.pyplot as plt
3   import pandas as pd
4   import seaborn as sns
5   from sklearn import datasets
6
7   # 讀取 sklearn 內建資料檔
8   ds = datasets.load_iris()
9
10  df = pd.DataFrame(ds.data, columns=ds.feature_names)
11  df['y'] = ds.target
```

第 2 章　神經網路（Neural Network）原理

```
12
13  # 繪製熱力圖(Heatmap)
14  # df.corr() : 關聯係數(Correlation coefficient)
15  sns.heatmap(df.corr())
16  plt.show()
```

- 執行結果：

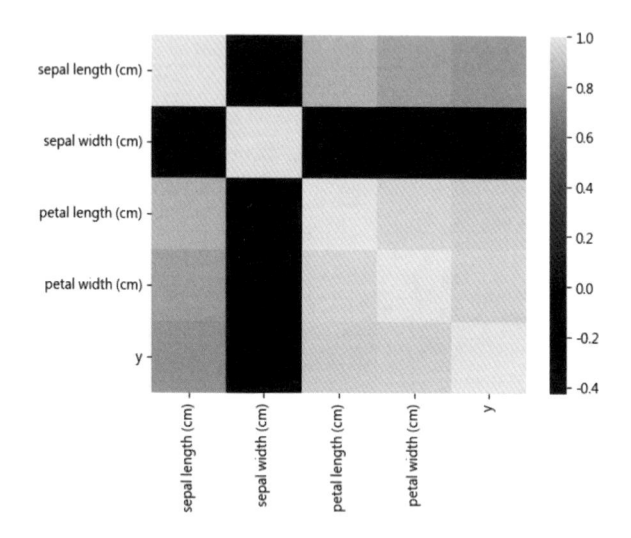

7. 另外，Seaborn 套件還提供許多指令，使用一個指令就可以產生多張的圖表，例如 pairplot（下圖）、facegrid…，可參閱 Seaborn 官網 [2]。

```
1   # pairplot
2   import matplotlib.pyplot as plt
3   import pandas as pd
4   import seaborn as sns
5   from sklearn import datasets
6
7   # 讀取 sklearn 內建資料檔
8   ds = datasets.load_iris()
9
10  df = pd.DataFrame(ds.data, columns=ds.feature_names)
11  df['y'] = ds.target
12
13  # 繪圖
14  sns.pairplot(df)
15  plt.show()
```

- 執行結果：

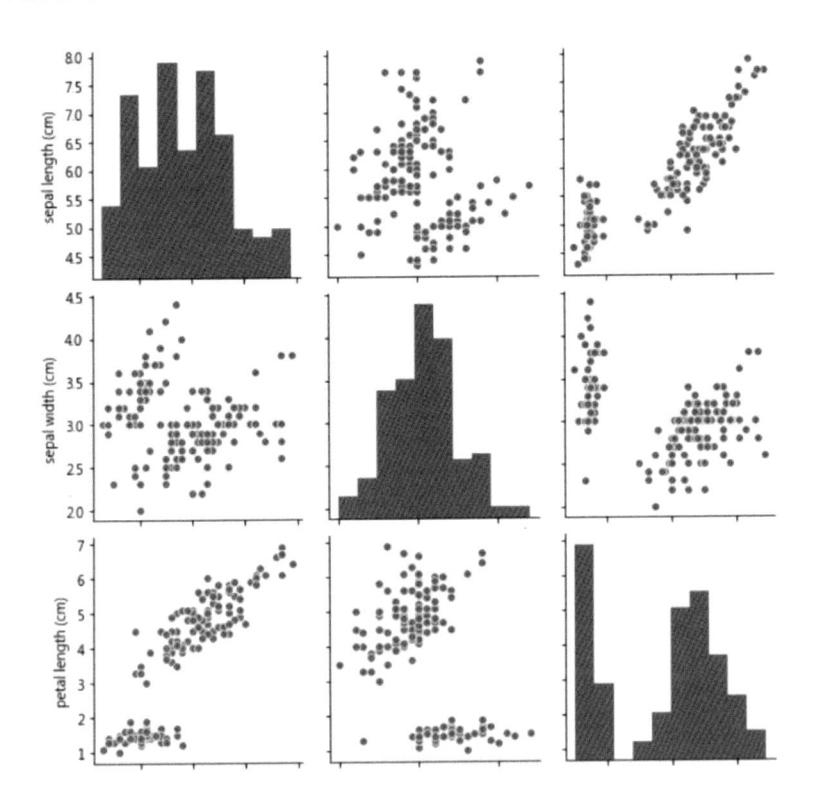

2-4-4 機率（Probability）

以台北市長選舉為例，當我們收集到問卷調查後，接著就可以建立模型來預測誰會當選，預測結果通常是以機率表示各候選人當選的可能性。

2-4-4-1 機率的定義與定理

接下來我們先瞭解機率的相關術語（Terminology）及定義。

1. 實驗（Experiment）或稱試驗（Trial）：針對計畫目標進行一連串的行動，例如擲硬幣、擲骰子、抽撲克牌、買樂透…等。
2. 樣本空間（Sample Space）：一個實驗會出現的所有可能結果，例如擲硬幣一次的樣本空間為 { 正面、反面 }，擲硬幣兩次的樣本空間為 { 正正、正反、反正、反反 }。

第 2 章　神經網路（Neural Network）原理

3. 樣本點（Sample Point）：樣本空間內任一可能的結果，例如擲硬幣兩次會有四種樣本點，抽一張樸克牌，有 52 種樣本點。

4. 事件（Event）：某次實驗發生的結果，一個事件可能含多個樣本點，例如，抽一張樸克牌，出現紅色的事件，該事件就包含 26 種樣本點。

5. **機率**：某次事件發生的可能性，因此

$$機率 = \frac{發生此事件的樣本點}{樣本空間的所有樣本點}$$

舉例來說，擲硬幣兩次出現兩個正面為 { 正正 }1 個樣本點，樣本空間所有樣本點為 { 正正、正反、反正、反反 }4 個樣本點，故機率等於 1/4。同樣的，擲硬幣兩次出現一正一反的機率 = { 正反、反正 }/{ 正正、正反、反正、反反 }= 1/2。

另外，還有兩個重要的觀念『獨立 vs. 相依』與『互斥』，先談『獨立 vs. 相依』。

1. 獨立（Independence）：當兩個事件（A、B）先後發生，A 事件發生，不會影響 B 事件出現的機率，即稱為獨立，例如擲硬幣兩次，不管第一次出現正面或反面，都不會影響第二次擲硬幣時出現正面 / 反面的機率。

2. 相依（Dependence）：與獨立相反，A 事件發生，會影響 B 事件出現的機率，稱為相依，例如抽 2 張樸克牌，第一次抽出不放回，就會影響第二張抽牌的機率。

    * 假設第一張抽出梅花，不放回，第二張再抽出梅花的機率 =12/51

    * 假設第一張抽出黑桃，不放回，第二張再抽出梅花的機率 =13/51。

3. 『條件機率』（Conditional Probability）：考慮兩個事件（A、B）的相依性，我們可以使用條件機率來表示，P（B|A）表已知 A 事件發生，B 事件出現的機率，若 A、B 獨立，則 P（B|A）= P（B）。

再談『互斥』（Mutually Exclusive）。如果 A 事件發生，就不會出現 B 事件，兩事件不可能同時發生，就稱為互斥，例如擲骰子一次，出現 6 點與出現 1 點的機率均為 1/6，若要計算『出現 6 點或 1 點』的機率，就是 1/6+1/6=2/6，因為出現 6 點與出現 1 點互斥，但要計算『出現梅花或 A』的機率，兩者機率就不能直接相加，必須扣掉重疊的樣本（梅花 A），即（1/4）+（4/52）-（1/52）。

依照以上說明，相關機率可使用集合論（Set Theory）定義如下：

1. 交集（Intersection）：A、B 事件獨立，則同時發生 A 及 B 事件的機率

$$P(A \cap B) = P(A) \times P(B) 。$$

2. A、B 事件不獨立,則已知 A 事件發生,B 事件出現的機率

$$P(B|A)= P(A \cap B)/P(A)$$。

3. 聯集(Union):A、B 事件互斥,則發生 A 或 B 事件的機率

$$P(A \cup B)= P(A)+ P(B)$$。

4. A、B 事件不互斥,則發生 A 或 B 事件的機率

$$P(A \cup B)= P(A)+ P(B)-P(\cup A \cap B)$$。

5. 樣本空間內所有互斥事件的機率總和 = 1。

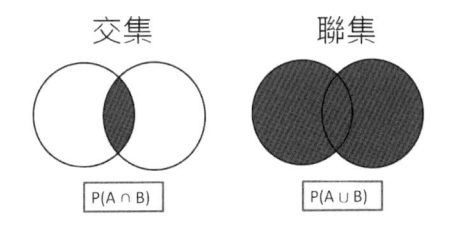

▲ 圖 2.8 交集(Intersection)及聯集(Union)

2-4-4-2 排列與組合

有些事件我們會關心發生的順序,相反的,有些時候則不關心事件發生的順序,計算的公式會有所不同,結果也不一樣,關心事件發生的順序稱為『排列』(Permutation),反之,不關心事件發生順序稱為『組合』(Combination)。例如,擲硬幣兩次的『排列』樣本空間為 { 正正、正反、反正、反反 },但『組合』

樣本空間為 { 正正、一正一反、反反 },因為『正反』、『反正』都是『一正一反』。

範例. 列舉各種案例說明排列與組合的相關計算公式及程式撰寫。

程式:02_4_4_ 機率 .ipynb。

1. 有三顆球,標號各為 1、2、3,它們排列的事件共有幾種?

```
1  from itertools import permutations
2
3  # 測試資料
4  list1 = [1, 2, 3]
5
6  # 宣告排列的類別
```

第 2 章 神經網路（Neural Network）原理

```
7  perm = permutations(list1)
8
9  # 印出所有事件
10 print('所有事件:')
11 list_output = list(perm)
12 for i in list_output:
13     print(i)
14
15 print()
16 print(f'排列數={len(list_output)}')
```

- 執行結果：所有事件如下。

  （1,2,3）

  （1,3,2）

  （2,1,3）

  （2,3,1）

  （3,1,2）

  （3,2,1）

  排列數 =6

1. 有三顆球，抽出兩顆球排列的事件共有幾種？

```
1  from itertools import permutations
2
3  # 測試資料
4  list1 = [1, 2, 3]
5
6  # 宣告排列的類別，三個球抽出兩個球
7  perm = permutations(list1, 2)
8
9  # 印出所有事件
10 print('所有事件:')
11 list_output = list(perm)
12 for i in list_output:
13     print(i)
14
15 print()
16 print(f'排列數={len(list_output)}')
```

- 執行結果：所有事件如下。

  （1,2）

  （1,3）

2-4 機率（Probability）與統計（Statistics）

   （2,1）

   （2,3）

   （3,1）

   （3,2）

   排列數 =6

- 出現（1,2）的機率 =[（1,2），（2,1）]/6 = 2/6 = 1/3

3. 有三顆球，抽出兩顆球組合的事件共有幾種？

```
1  from itertools import combinations
2
3  # 測試資料
4  list1 = [1, 2, 3]
5
6  # 宣告排列的類別，三個球抽出兩個球
7  comb = combinations(list1, 2)
8
9  # 印出所有事件
10 print('所有事件:')
11 list_output = list(comb)
12 for i in list_output:
13     print(i)
14
15 print()
16 print(f'組合數={len(list_output)}')
```

- 執行結果：所有事件如下。

   （1,2）

   （1,3）

   （2,3）

   組合數 =3

- 出現（1,2）的機率 =[（1,2）]/3 = 1/3

4. 再進一步衍生，袋子中有 10 顆不同顏色的球，抽出 3 顆共有幾種排列方式？排列公式如下：

   排列 = $\dfrac{n!}{(n-k)!}$

第 2 章　神經網路（Neural Network）原理

n: 母體數（10）

k: 抽樣數（3）

```
1  import math
2
3  n=10
4  k=3
5
6  # math.factorial: 階乘
7  perm = math.factorial(n) / math.factorial(n-k)
8
9  print(f'排列方式={perm:.0f}')
```

- 執行結果：排列方式共 720 種。

- 說明：抽第一次有 10 種選擇，抽第二次有 9 種選擇，抽第三次有 8 種選擇，故

    $10 \times 9 \times 8 = \dfrac{10!}{(10-3)!}$

5. 擲硬幣 3 次，出現 0、1、2、3 次正面的組合共有幾種？

解題：因為只考慮出現正面的次數，不管出現的順序，故屬於組合的問題，公式如下，因為出現的順序不同，仍視為同一事件，故要多除於 k!：

組合 = $\dfrac{n!}{k!(n-k)!}$

n: 實驗次數，4 次

k: 出現正面次數，0、1、2、3 次

```
1  import math
2
3  n=3
4
5  # math.factorial: 階乘
6  for k in range(4):
7      # 組合公式
8      comb = math.factorial(n) / (math.factorial(k) * math.factorial(n-k))
9      print(f'出現 {k} 次正面的組合共 {comb:.0f} 種')
```

- 執行結果：

    ＊ 出現 0 次正面的組合共 1 種

    ＊ 出現 1 次正面的組合共 3 種

* 出現 2 次正面的組合共 3 種

* 出現 3 次正面的組合共 1 種

2-4-4-3 二項分配（Binomial Distribution）

擲硬幣、擲骰子、抽撲克牌一次，均假設每個單一樣本點發生的機率都均等，如果不是均等，上述的排列／組合的公式需要再考慮機率 p，例如擲硬幣出現正面的機率 =0.4，p^k 代表出現 k 次正面的機率，$(1-p)^{(n-k)}$ 代表出現 n-k 次反面的機率。

$$P(x=k) = \frac{n!}{k!(n-k)!} p^k (1-p)^{(n-k)}$$

範例 1. 擲硬幣 3 次的機率計算。

程式：02_4_4_ 機率 .ipynb。

1. 假設擲硬幣出現正面的機率 =0.4，現擲硬幣 3 次，出現 0、1、2、3 次正面的機率各是多少。

```
1  import math
2
3  n=3
4  p=0.4
5
6  # math.factorial: 階乘
7  for k in range(4):
8      # 組合公式
9      comb = math.factorial(n) / (math.factorial(k) * math.factorial(n-k))
10     prob = comb * (p**k) * ((1-p)**(n-k))
11     print(f'出現 {k} 次正面的機率 = {prob:.4f}')
```

- 執行結果：

    * 出現 0 次正面的機率 = 0.2160

    * 出現 1 次正面的機率 = 0.4320

    * 出現 2 次正面的機率 = 0.2880

    * 出現 3 次正面的機率 = 0.0640

2. 使用 SciPy comb 組合函數，執行結果同上。

```
1  # 使用 scipy comb 組合函數
2  from scipy import special as sps
3
```

第 2 章　神經網路（Neural Network）原理

```
4  n=3
5  p=0.4
6
7  for k in range(4):
8      prob = sps.comb(n, k) * (p**k) * ((1-p)**(n-k))
9      print(f'出現 {k} 次正面的機率 = {prob:.4f}')
10
```

3. 直接使用 SciPy 的統計模組 binom.pmf（k,n,p），執行結果同上。

```
1  from scipy.stats import binom
2
3  n=3
4  p=0.4
5
6  # binom
7  for k in range(4):
8      print(f'出現 {k} 次正面的機率 = {binom.pmf(k,n,p):.4f}')
```

範例 2. 試算今彩 539 平均報酬率 [3]。

程式：02_4_4_機率 .ipynb。

1. 基本玩法：從 39 個號碼中選擇 5 個號碼。各獎項的中獎方式如下表：

獎項	中獎方式	中獎方式圖示
頭獎	與當期五個中獎號碼完全相同者	●●●●●
貳獎	對中當期獎號之其中任四碼	●●●●
參獎	對中當期獎號之其中任三碼	●●●
肆獎	對中當期獎號之其中任二碼	●●

各獎項金額如下表：

獎項	頭獎	貳獎	參獎	肆獎
單注獎金	$8,000,000	$20,000	$300	$50

2. 解題：

- 假定每個號碼出現的機率是相等的（1/39），固定選 5 個號碼，故計算機率時不需考慮 p。

- 計算從 39 個號碼選 5 個號碼的組合數量，作為機率的樣本空間。

2-4 機率（Probability）與統計（Statistics）

```python
1  # 從39個號碼選5個號碼，總共有幾種
2  from scipy import special as sps
3
4  print(f'從39個號碼選5個號碼，總共有{int(sps.comb(39,5))}種')
5
6  # 頭獎號碼的個數，5個中獎號碼全中，不含非中獎號碼
7  first_price_count = int(sps.comb(5,5) * sps.comb(34,0))
8  print(f'頭獎號碼的個數: {first_price_count}')
9
10 # 頭獎號碼的個數，5個中獎號碼中4個，有1個非中獎號碼
11 second_price_count = int(sps.comb(5,4) * sps.comb(34,1))
12 print(f'二獎號碼的個數: {second_price_count}')
13
14 # 頭獎號碼的個數，5個中獎號碼中3個，有3個非中獎號碼
15 third_price_count = int(sps.comb(5,3) * sps.comb(34,2))
16 print(f'三獎號碼的個數: {third_price_count}')
17
18 # 頭獎號碼的個數，5個中獎號碼中2個，有3個非中獎號碼
19 fourth_price_count = int(sps.comb(5,2) * sps.comb(34,3))
20 print(f'四獎號碼的個數: {fourth_price_count}')
21
```

- 執行結果：

    * 從 39 個號碼選 5 個號碼，總共有 575757 種

    * 頭獎號碼的個數 :1

    * 二獎號碼的個數 :170

    * 三獎號碼的個數 :5610

    * 四獎號碼的個數 :59840

3. 計算平均中獎金額

```python
1  # 平均中獎金額=(頭獎金額 * 頭獎組數 + 二獎金額 * 二獎組數 +
2  #             三獎金額 * 三獎組數 + 四獎金額 * 四獎組數) / 全部組合數
3  average_return = (8000000 * 1 + 20000 * 170 + 300 * 5610 + 50 * 59840) / 575757
4  print(f'平均中獎金額: {average_return:.2f}')
5
6  print(f'今彩539平均報酬率={((average_return / 50) - 1) * 100:.2f}%')
```

- 執行結果：

    * 平均中獎金額 :27.92

    * 今彩 539 平均報酬率 =-44.16%

排列與組合的理論看似簡單，然而實務上的應用千變萬化，建議讀者多找一些案例實作，才能運用自如。

第 2 章　神經網路（Neural Network）原理

2-4-5　機率分配（Distribution）

機率分配（Distribution）結合了描述統計量及機率的概念，對資料作進一步的分析，希望推測母體是呈現何種形狀的分佈，例如常態分配、均勻分配、卜瓦松（Poisson）分配或二項分配等，並且依據樣本推估機率分配相關的母數，譬如平均數、變異數等。有了完整機率分配資訊後，就可以進行預測、區間估計、假設檢定等。

機率分配的種類非常多，如下圖，這裡我們僅介紹幾個常見的機率分配。

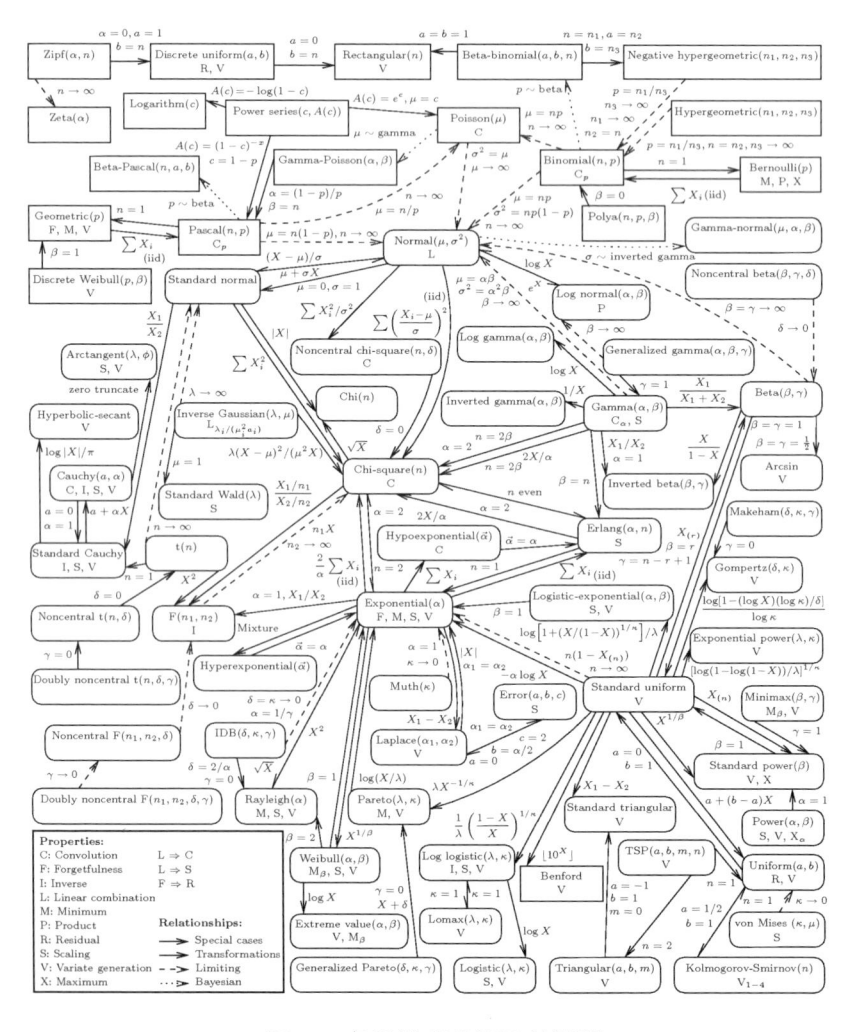

▲ 圖 2.9　各種機率分配及其關聯，

圖片來源：Univariate Distribution Relationships[4]

2-4 機率（Probability）與統計（Statistics）

首先介紹幾個專有名詞：

1. 機率密度函數（Probability Density Function, 簡稱 pdf）：發生各種事件的機率。
2. 累積分配函數（Cumulative Distribution Function, 簡稱 cdf）：等於或低於某一觀察值的機率。
3. 機率質量函數（Probability Mass Function, 簡稱 pmf）：如果是離散型的機率分配，pdf 改稱為 pmf。

接下來看幾個常見的機率分配。

2-4-5-1 常態分配（Normal Distribution）

常態分配因為是 Carl Friedrich Gauss 提出的，又稱為高斯分配（Gauss Distribution），世上大部份的事件都屬於常態分配，例如考試的成績，考低分及高分的學生人數會比較少，中等分數的人數會占大部份。其他例子如全年的溫度、業務員的業績、一堆紅豆的重量…等等。常態分配的機率密度函數定義如下：

$$f(x;\mu,\sigma) = \frac{1}{\sqrt{2\pi*\sigma^2}} * e^{-\frac{1}{2}*(\frac{x-\mu}{\sigma})^2}$$

有兩個參數：

1. μ：平均數（Mean），是全體樣本的平均數。
2. δ：標準差（Standard Deviation），描述全體樣本的離散的程度。

機率密度函數可簡寫成 N（μ,δ）。

我們在『2-3-4 積分』節已經介紹過了，並撰寫一些程式，這裡直接使用 SciPy 的統計模組作一些實驗。

範例 1. 使用 SciPy 繪製機率密度函數（pdf）。

程式：02_4_5_1_ 常態分配 .ipynb。

1. 繪製常態分配的機率密度函數。

```
1  import matplotlib.pyplot as plt
2  import numpy as np
3  from scipy.stats import norm
```

第 2 章　神經網路（Neural Network）原理

```
 4
 5  # 觀察值範圍
 6  z1, z2 = -4, 4
 7
 8  # 樣本點
 9  x = np.arange(z1, z2, 0.001)
10  y = norm.pdf(x,0,1)
11
12  # 繪圖
13  plt.plot(x,y,'black')
14
15  # 填色
16  plt.fill_between(x,y,0, alpha=0.3, color='b')
```

- 執行結果：

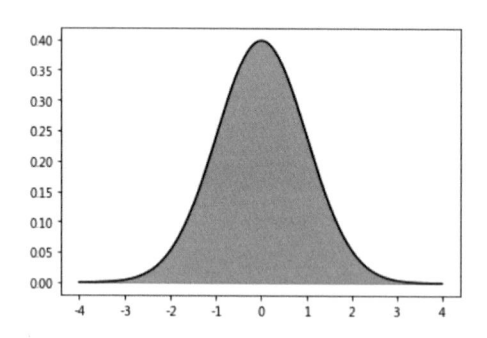

2. 繪製常態分配的正 / 負 1、2、3 倍標準差區間及其機率。

```
 1  import matplotlib.pyplot as plt
 2  import numpy as np
 3  from scipy.stats import norm
 4
 5  z1, z2 = -1, 1
 6  x1 = np.arange(z1, z2, 0.001)
 7  y1 = norm.pdf(x1,0,1)
 8
 9  # 1倍標準差區域
10  plt.fill_between(x1,y1,0, alpha=0.3, color='b')
11  plt.ylim(0,0.5)
12
13  # 觀察值範圍
14  z1, z2 = -4, 4
15
16  # 樣本點
17  x2 = np.arange(z1, z2, 0.001)
18  y2 = norm.pdf(x2,0,1)
19
20  # 繪圖
21  plt.plot(x2,y2,'black')
22
23  # 1倍標準差機率
24  plt.annotate(text='', xy=(-1,0.25), xytext=(1,0.25), arrowprops=dict(arrowstyle='<->'))
```

2-4 機率（Probability）與統計（Statistics）

```
25  plt.annotate(text='68.3%', xy=(0,0.26), xytext=(0,0.26))
26
27  # 填色
28  plt.fill_between(x2,y2,0, alpha=0.1, color='b')
29  plt.show()
```

- 執行結果：

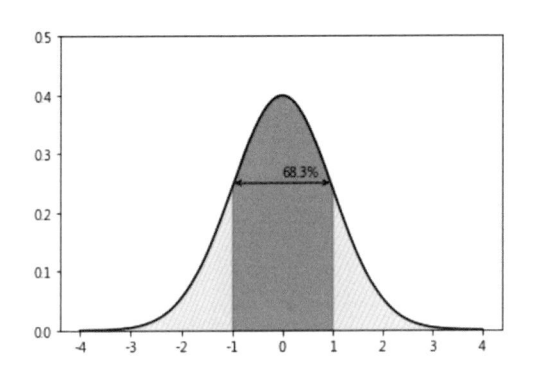

- 第 5 行改為（-2,2）、（-3,3），結果如下：

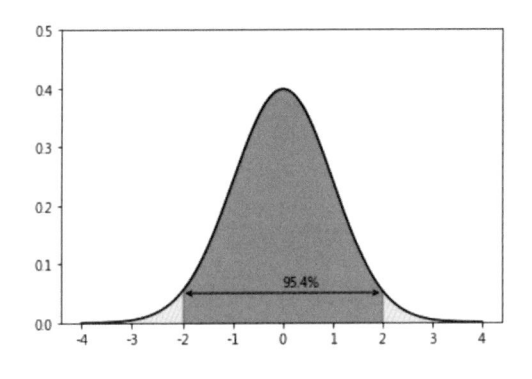

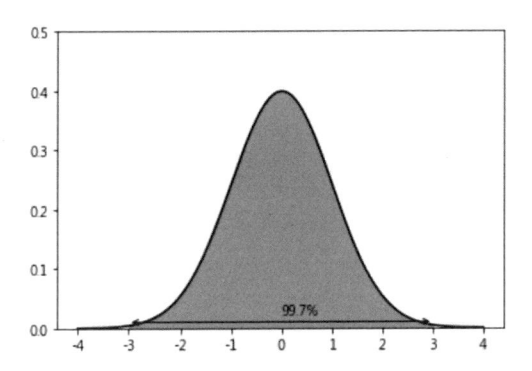

第 2 章 神經網路（Neural Network）原理

3. 繪製累積分配函數（cdf）。

```
1  import matplotlib.pyplot as plt
2  import numpy as np
3  from scipy.stats import norm
4
5  # 觀察值範圍
6  z1, z2 = -4, 4
7
8  # 樣本點
9  x = np.arange(z1, z2, 0.001)
10
11 # norm.cdf：累積機率密度函數
12 y = norm.cdf(x,0,1)
13
14 # 繪圖
15 plt.plot(x,y,'black')
16
17 plt.show()
```

- 第 11 行從 pdf 改為 cdf。
- 執行結果：最後會達到 1.0。

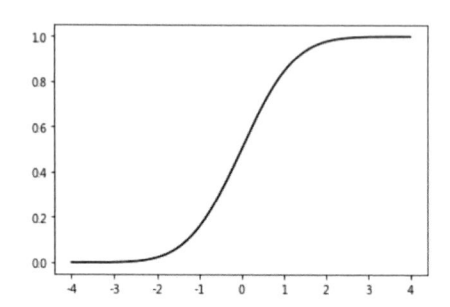

4. 有一個 t 分配，很像常態分配，常被用來作平均數的假設檢定（Hypothesis Test），可使用 SciPy 繪製 t 分配的機率密度函數（pdf），並與常態分配比較。

```
1  import matplotlib.pyplot as plt
2  import numpy as np
3  from scipy.stats import t, norm
4
5  # 觀察值範圍
6  z1, z2 = -4, 4
7
8  # 樣本點
9  x = np.arange(z1, z2, 0.1)
10 dof = 10 #len(x) - 1
```

2-4 機率（Probability）與統計（Statistics）

```
11  y = t.pdf(x, dof)
12
13  # 繪圖
14  plt.plot(x,y,'red', label='t分配', linewidth=5)
15
16  # 繪製常態分配
17  y2 = norm.pdf(x)
18  plt.plot(x,y2, 'green', label='常態分配')
19
20  plt.legend()
21  plt.show()
```

- 第 11 行為 t 分配。

- 執行結果：

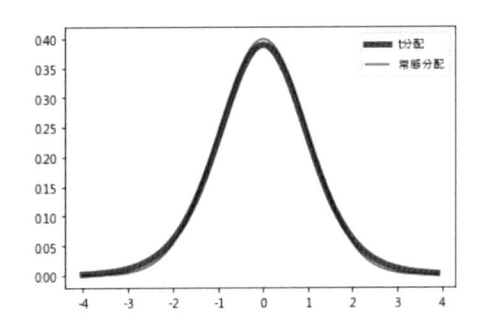

- 較粗的線（紅色）：t 分配的尾巴比較厚。

2-4-5-2 均勻分配（Uniform Distribution）

這也是一種常見的分配，通常是屬於離散型的資料，所有樣本點發生的機率都相同，例如擲硬幣、擲骰子，出現每一面的機會都均等。均勻分配的機率密度函數如下：

$$f(x) = \frac{1}{(b-a)}, a \leq x \leq b$$

累積分配函數如下：

範例 2. 繪製擲骰子的機率質量函數（Probability Mass Function, pmf）。

程式：02_4_5_2_ 其他分配 .ipynb。

第 2 章　神經網路（Neural Network）原理

1. 繪製機率質量函數。

```python
import numpy as np
import matplotlib.pyplot as plt

# 修正中文亂碼
plt.rcParams['font.sans-serif'] = ['Microsoft JhengHei'] # 微軟正黑體
plt.rcParams['axes.unicode_minus'] = False

# 點數與機率
face = [1,2,3,4,5,6]
probs = np.full((6), 1/6)

# 繪製長條圖
plt.figure(figsize=(6,4))
plt.bar(face, probs)
plt.ylabel('機率', fontsize=12)
plt.xlabel('點數', fontsize=12)
plt.title('擲骰子', fontsize=12)
plt.ylim([0,1])
plt.show()
```

- 執行結果：

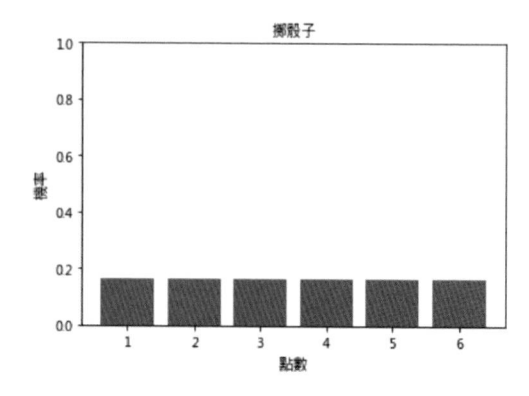

2. 可以使用隨機亂數模擬。

```python
import numpy as np

```

```
3  # 使用 randint 產生亂數
4  x = np.random.randint(1, 6)
5  print(x)
```

2-4-5-3 二項分配（Binomial Distribution）

顧名思義二項分配就是只有兩種觀察值，成功 / 失敗、正面 / 反面、有 / 沒有…等，類似的分配有三種：

1. 伯努利（Bernoulli）分配：做**一次二分類**的實驗，機率密度函數如下：

   $f(x) = p^x (1-p)^{1-x}, x = 0 或 1$ ，p: 成功的機率

2. 二項（Binomial）分配：做**多次二分類**的實驗。

   $f(k, n, p) = \binom{n}{k} p^k (1-p)^{n-k}$, n：試驗次數，k：成功次數 ，p: 成功的機率

3. 多項（Multinomial）分配：做**多次多分類**的實驗。

   $f(x_1, x_2, \ldots x_k, n, p_1, p_2, \ldots, p_k) = \frac{n!}{x_1! x_2! \ldots x_k!} p_1^{x_1} p_2^{x_2} \ldots p_k^{x_k}, \sum x_i = n$

範例 2. 繪製擲硬幣的機率質量函數（Probability Mass Function, pmf）。

程式：02_4_5_2_ 其他分配 .ipynb。

1. 繪製擲硬幣 1 次的機率質量函數。

```
1   import matplotlib.pyplot as plt
2   import numpy as np
3
4   # 正面/反面機率
5   probs = np.array([0.6, 0.4])
6   face = [0, 1]
7   plt.bar(face, probs)
8   plt.title('伯努利(Bernoulli)分配', fontsize=16)
9   plt.ylabel('機率', fontsize=14)
10  plt.xlabel('正面/反面', fontsize=14)
11  plt.ylim([0,1])
12
13  plt.show()
```

• 執行結果：

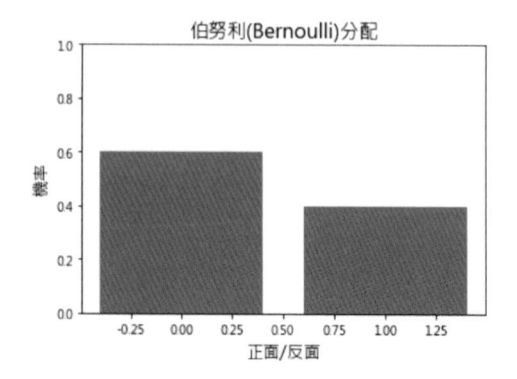

第 2 章　神經網路（Neural Network）原理

伯努利(Bernoulli)分配

2. 繪製擲硬幣 10 次的機率質量函數。

```python
1   import matplotlib.pyplot as plt
2   import numpy as np
3   from scipy.stats import binom
4
5   # 正面/反面機率
6   n = 10
7   p = 0.6
8
9   probs=[]
10  for k in range(11):
11      p_binom = binom.pmf(k,n,p)
12      probs.append(p_binom)
13
14  face=np.arange(11)
15  plt.bar(face, probs)
16  plt.title('二項(Binomial)分配', fontsize=16)
17  plt.ylabel('機率', fontsize=14)
18  plt.xlabel('正面次數', fontsize=14)
19
20  plt.show()
```

• 執行結果：

二項(Binomial)分配

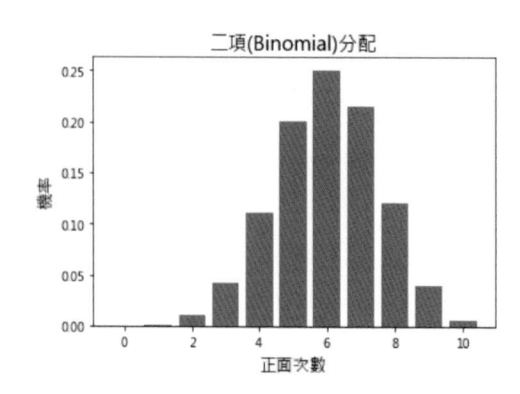

2-4 機率（Probability）與統計（Statistics）

3. 計算擲骰子 10 次的機率。

```
1  import matplotlib.pyplot as plt
2  import numpy as np
3  from scipy.stats import multinomial
4  
5  # 擲骰子次數
6  n = 6
7  
8  # 假設擲出各種點數的機率
9  p = [0.1, 0.2, 0.2, 0.2, 0.2, 0.1]
10 
11 probs=[]
12 
13 # 擲骰子1~6點各出現1次的機率
14 p_multinom = multinomial(n,p)
15 p_multinom.pmf([1,1,1,1,1,1])
```

- 執行結果：0.011520000000000013。

4. 若擲骰子 60 次，1~6 點各出現 10 次的機率。

```
1  # 擲骰子60次
2  n = 60
3  
4  # 擲骰子1~6點各出現10次的機率
5  p_multinom = multinomial(n,p)
6  p_multinom.pmf([10,10,10,10,10,10])
```

- 執行結果：4.006789054168284e-06，機率小很多，由於，擲骰子 60 次出現的樣本空間有更多的組合。

5. 擲骰子 1000 次，1~6 點各出現的次數。

```
1  # 隨機抽樣1000次
2  y = np.random.choice(np.arange(1,7), size=1000, p=p)
3  
4  # np.bincount：0~6點各出現的次數，故要扣掉第一個元素
5  times = np.bincount(y)[1:]
6  
7  # 繪圖
8  plt.bar(np.arange(1,7), times)
9  plt.show()
```

- 執行結果：

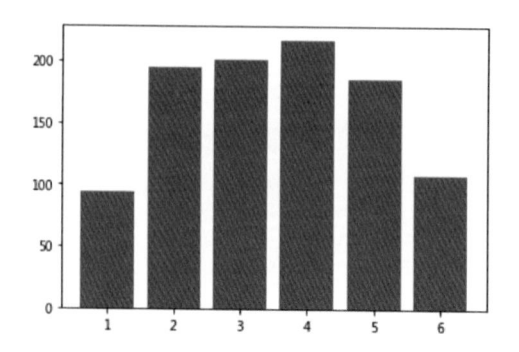

2-4-5-4 卜瓦松分配（Poisson Distribution）

卜瓦松分配經常運用在『給定的時間內發生 k 次事件』的機率分配函數，我們可以建立一套顧客服務的模型，用來估計一個服務櫃台的平均等候人數，進而計算出需要安排幾個服務櫃台，來達成特定的服務水準（Service Level Agreement,SLA），應用層面非常廣，例如售票櫃台、便利商店結帳櫃台、客服中心的電話線數、依個人發生車禍的次數決定車險定價…等。

卜瓦松分配的機率密度函數如下：

$$f(k,\lambda) = \frac{\lambda^k e^{-\lambda}}{k!}, \lambda : 平均發生次數，e: 自然對數，k: 發生次數$$

範例. 繪製各種 λ 值的卜瓦松分配機率密度函數。

程式：02_4_5_2_ 其他分配 .ipynb。

```
1  import matplotlib.pyplot as plt
2  import numpy as np
3  from scipy.stats import poisson
4
5  # λ = 2、4、6、8
6  for lambd in range(2, 10, 2):
7      k = np.arange(0, 10)
8      poisson1 = poisson.pmf(k, lambd)
9
10     # 繪圖
11     plt.plot(n, poisson1, '-o', label=f"λ = {lambd}")
12
13 # 設定圖形屬性
14 plt.xlabel('發生次數', fontsize=12)
15 plt.ylabel('機率', fontsize=12)
16 plt.title("Poisson Distribution (各種λ值)", fontsize=16)
17 plt.legend()
```

• 執行結果：

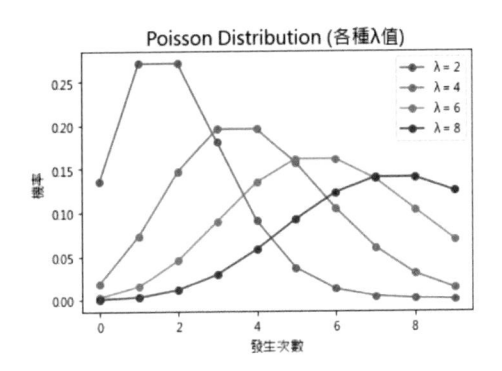

2-4-6 假設檢定（Hypothesis Testing）

根據維基百科的定義[5]，假設檢定（Hypothesis Testing）是用於檢定現有數據是否足以支持特定假設的方法，通常要判定一種新藥是否顯著有效，我們不希望因為隨機抽樣的誤差，造成判定的錯誤，因此治癒人數的比例必須超過某一顯著水準，才能夠認定為具有醫療效果。

2-4-6-1『信賴區間』（Confidence Interval）

在討論假設檢定之前，我們先來了解什麼是『信賴區間』（Confidence Interval），之前談到平均數、中位數都是以單一值表達樣本的集中趨勢，這稱為『點估計』（Point Estimation），這種表達方式並不一定恰當，以常態分配而言，如下圖機率密度函數，樣本剛好等於平均數的機率也只有 0.4，改用區間估計會是一個比較穩健的做法，例如，有 95% 的信心，樣本會落在特定區間內，這個區間我們稱為『信賴區間』（Confidence Interval）。

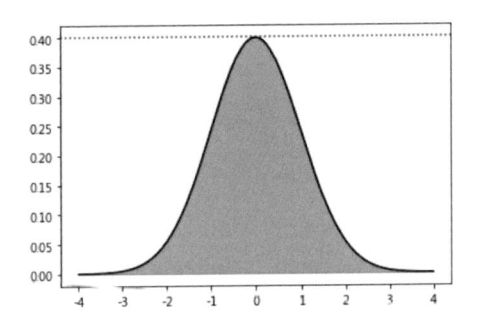

第 2 章　神經網路（Neural Network）原理

以常態分配而言，1 倍的標準差信賴水準（Confidence Level）約為 68.3%，2 倍的標準差信賴水準（Confidence Level）約為 95.4%，3 倍的標準差信賴水準（Confidence Level）約為 99.7%。

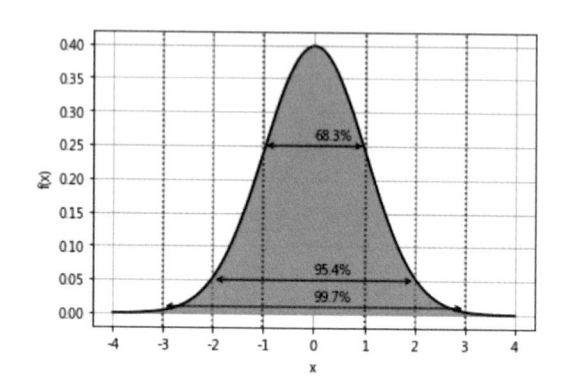

以業界常用的 95% 信賴水準為例，就是『我們有 95% 的信心，樣本會落在（-1.96 δ, 1.96 δ）區間內』，一般醫藥有效性的檢定也是以此概念表達，例如疫苗。

範例. 以美國歷屆總統的身高資料，計算各式描述統計量及 95% 的信賴區間。

程式：02_4_6_1_信賴區間.ipynb。

1. 計算各式描述統計量及 95% 的信賴區間。

```
1  import random
2  import pandas as pd
3  import numpy as np
4
5  # 讀取檔案
6  df = pd.read_csv('./data/president_heights.csv')
7  df.rename(columns={df.columns[-1]:'height'}, inplace=True)
8
9  # 計算信賴區間
10 m = df['height'].mean()
11 sd = df['height'].std()
12 print(f'平均數={m}, 標準差={sd}, 信賴區間=({m-2*sd}, {m+2*sd})')
```

- 執行結果：平均數 =179.74，標準差 =7.02, 信賴區間 =（165.71,193.77）。

- 如上所述，如果單純以平均數 179.74 說明美國總統的身高並不完整，因為其中有多位總統的身高低於 170，若再加上『95% 美國總統的身高介於（165.71,193.77）』，會讓資訊更加完整。

2-82

2-4 機率（Probability）與統計（Statistics）

2. 利用隨機亂數產生常態分配的樣本，再使用 SciPy 的統計模組（stats）計算信賴區間。先利用隨機亂數產生 10000 筆樣本，樣本來自常態分配 N（5,2），即平均數為 5，標準差 =2，使用 norm.interval() 計算信賴區間，參數分別為信賴水準、平均數、標準差。

```
1  import matplotlib.pyplot as plt
2  import numpy as np
3  from scipy.stats import norm
4
5  # 產生隨機亂數的樣本
6  mu = 5        # 平均數
7  sigma = 2     # 標準差
8  n = 10000     # 樣本數
9  data = np.random.normal(mu, sigma, n)
10
11 cl = .95      #信賴水準(Confidence Level)
12
13 # 計算信賴區間
14 m = data.mean()
15 sd = data.std()
16 y1 = norm.interval(cl, m, sd)
17 print(f'平均數={m}, 標準差={sd}, 信賴區間={y1}')
```

- 執行結果：信賴區間 =（1.0487,8.9326），約為（$\mu-1.96\delta$，$\mu+1.96\delta$）。

3. 繪圖。

```
1  import seaborn as sns
2
3  # 直方圖，參數 hist=False：不畫階梯直方圖，只畫平滑曲線
4  sns.distplot(data, hist=False)
5
6  # 畫信賴區間
7  plt.axvline(y1[0], c='r', linestyle=':')
8  plt.axvline(y1[1], c='r', linestyle=':')
9
10 # 標示 95%
11 plt.annotate(text='', xy=(y1[0],0.025), xytext=(y1[1],0.025),
12              arrowprops=dict(arrowstyle='<->'))
13 plt.annotate(text='95%', xy=(mu,0.03), xytext=(mu,0.03))
14
15 plt.show()
```

- 執行結果：

第 2 章　神經網路（Neural Network）原理

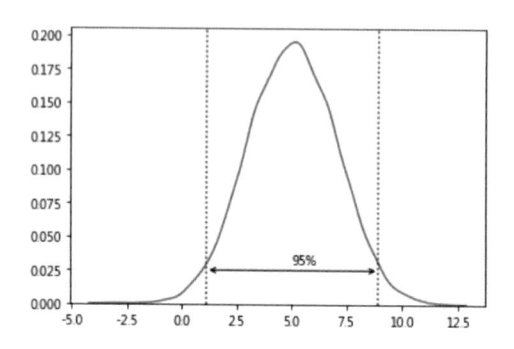

4. 利用隨機亂數產生二項分配的樣本，再使用 SciPy 的統計模組（stats）計算信賴區間。

```python
1  import matplotlib.pyplot as plt
2  import numpy as np
3  from scipy.stats import binom
4
5  # 產生隨機亂數的樣本
6  trials = 1      # 實驗次數
7  p = 0.5         # 出現正面的機率
8  n = 10000       # 樣本數
9  data = np.random.binomial(trials, p, n)
10
11 cl = .95        #信賴水準(Confidence Level)
12
13 # 計算信賴區間
14 m = data.mean()
15 y1 = binom.interval(cl, n, m)
16 print(f'平均數={m}, 信賴區間={y1}')
```

- 執行結果：平均數 =0.5039，標準差 =0.4999847897686489，信賴區間 =（4941.0, 5137.0），約為（$\mu-1.96\delta$，$\mu+1.96\delta$）。

- 二項分配標準差公式 = $\sqrt{p(1-p)}$，可以驗算（p*（1-p））**.5 = 0.5，隨機亂數的樣本與理論值相去不遠 (0.49998 ≅ 0.5)。

2-4-6-2 中央極限定理（Central Limit Theorem）

『中央極限定理』（Central Limit Theorem,CLT）是指當樣本數很大時，不管任何機率分配，每批樣本的平均數之機率分配會近似於常態分配。因此，我們就可以使用常態分配估計任何樣本平均數的信賴區間或進行假設檢定。

例如下圖，母體為 Poisson distribution，如果我們取 10,000 批樣本，每批含 100 個樣本，每批平均值會呈現常態分配。

2-4 機率（Probability）與統計（Statistics）

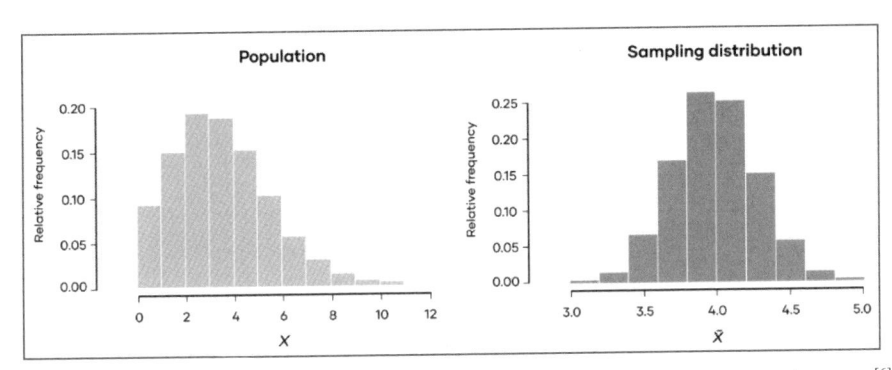

▲ 圖 2.10 中央極限定理示意圖，圖片來源：Scribbr, Central Limit Theorem[6]

然而，因為每一個樣本是一批的資料，假設個數為 n，則每批樣本的平均數所形成的標準差不是原樣本標準差，而是如下：

$$\delta_{\bar{x}} = \frac{\delta}{\sqrt{n}}$$

例如，我們有 1000 批的樣本，每批的樣本有 10 筆資料，則每批樣本的平均數所形成的標準差：

$$\delta_{\bar{x}} = \frac{\delta}{\sqrt{10}}$$

又例如，樣本來自二項分配，標準差 = $\sqrt{p(1-p)}$，則每批樣本的平均數所形成的標準差：

$$\delta_{\bar{x}} = \frac{\sqrt{p(1-p)}}{\sqrt{n}} = \sqrt{\frac{p(1-p)}{n}}$$

平均數的標準差稱為標準誤差（standard error），有別於原樣本的標準差。

範例. 驗證來自二項分配的樣本平均數近似常態分配，並繪圖驗證。

程式：02_4_6_2_ 中央極限定理 .ipynb。

1. 利用隨機亂數產生二項分配的 10000 批樣本，每批含 100 筆資料，計算平均數、標準誤差。

```
1  import matplotlib.pyplot as plt
2  import numpy as np
3  from scipy.stats import binom
4
```

第 2 章　神經網路（Neural Network）原理

```
5  # 產生隨機亂數的樣本
6  trials = 1    # 實驗次數
7  p = 0.5       # 出現正面的機率
8  n = 10000     # 樣本數
9  data = np.random.binomial(trials, p, n)
10
11 cl = .95      #信賴水準(Confidence Level)
12
13 # 計算信賴區間
14 m = data.mean()
15 sd = data.std()
16 y1 = binom.interval(cl, n, m)
17 print(f'平均數={m}, 標準差={sd}, 信賴區間={y1}')
```

- 執行結果：平均數 =0.5003，標準差 =0.04982，信賴區間 =（4906.0,5102.0）。
- 平均數與原樣本的平均數幾乎相同 $(0.5003 \cong 0.5)$。
- 標準差與理論值也很接近 $(0.04982 \cong 0.05)$，理論值 =（p*（1-p）/trials）**0.5=0.04982。

2. 繪圖驗證。

```
1  # 直方圖
2  plt.figure(figsize=(6,4))
3  sns.histplot(data, bins=20)
4  plt.show()
```

- 執行結果：直方圖近似常態分配。

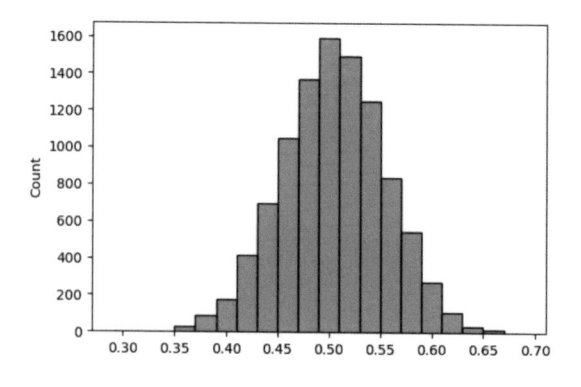

2-4 機率（Probability）與統計（Statistics）

2-4-6-3 假設檢定（Hypothesis Testing）

假設藥廠發明一種新藥，要證明新藥的效果是有效的，就會進行臨床實驗，讓一組病人服用新藥，另一組病人服用安慰劑，前者稱為實驗組（Treatment Group），後者稱為控制組（Control Group），若兩組經實驗後，有明顯差異，就可證明新藥是有效的，這種檢定的方法就稱為假設檢定（Hypothesis Testing）。

假設檢定步驟如下：

1. 建立兩種假設：

    1.1 虛無假設（Null Hypothesis）：以 H_0 表示，實驗組的檢驗平均值未優於控制組，即 $\mu_1 <= \mu_2$。

    1.2 對立假設（Alternative Hypothesis）：以 H_1 或 H_A 表示，實驗組的檢驗平均值優於控制組，即 $\mu_1 > \mu_2$。

2. 決定統計量：例如檢定平均數會採用 t 檢定，還有其他各種統計量及應用時機可參閱『An Interactive Guide to Hypothesis Testing in Python』[7] 及『17 Statistical Hypothesis Tests in Python』[8]。

3. 設定顯著水準（Significance Level）：以 α 表示，一般設定 $\alpha=5\%$，即實驗組的檢驗平均值須落在 95% 信賴區間之外，才算有顯著差異，如果要嚴格一點，可設定更小的顯著水準。

4. 進行抽樣、檢定與下結論。

依檢定範圍會分為單邊（One-side）、雙邊（Two-side），依樣本的設計會分為單樣本（Single-sample）、雙樣本（Two-sample），針對檢定的統計量，可能是平均數、標準差或整個機率分配的檢定。

單邊（One-side）檢定是只關心機率分配的一邊，例如，A 班的成績是否優於 B 班（$\mu_A > \mu_B$），為右尾檢定（Right-tail Test），病毒核酸檢測（PCR），ct 值 30 以下即視為確診（$\mu<30$），為左尾檢定（Left-tail Test），雙邊（Two-side）檢定則關心兩邊，例如，A 班的成績是否與 B 班有顯著差異（$\mu_A \neq \mu_B$）。

單樣本（Single-sample）檢定，只有一組樣本，例如，抽樣一群顧客，調查是否喜歡公司特定的產品，雙樣本（Two-sample）則有兩組樣本互相作比較，例如新藥有效性測試、疫苗測試、產品比較…等，雙樣本設計又稱為 A/B Test。

第 2 章　神經網路（Neural Network）原理

對立假設是虛無假設反面的條件，通常對立假設是我們希望的結果，因此，檢定後，如果虛無假為真時，我們會使用『不能拒絕』（fail to reject）虛無假設，而不會說虛無假設成立。

以下就以實例說明各式的檢定。

範例 1. 川普的身高 190 公分，是否比歷屆的美國總統平均身高有顯著的不同？

程式：02_4_6_3_假設檢定.ipynb

1. 假設歷屆的美國總統身高為常態分配，以顯著水準 5% 檢定，信賴區間為 $[\mu-1.96\delta, \mu+1.96\delta]$。

```python
import random
import pandas as pd
import numpy as np
import seaborn as sns

# 讀取檔案
df = pd.read_csv('./data/president_heights.csv')
df.rename(columns={df.columns[-1]:'height'}, inplace=True)

# 計算信賴區間

m = df['height'].mean()
sd = df['height'].std()
print(f'平均數={m:.2f}, 標準差={sd:.2f}, '
      f'信賴區間=({m-1.96*sd:.2f}, {m+1.96*sd:.2f})')

sns.distplot(df['height'])
plt.axvline(m, color='yellow', linestyle='dashed', linewidth=2)
plt.axvline(m-1.96*sd, color='magenta', linestyle='dashed', linewidth=2)
plt.axvline(m+1.96*sd, color='magenta', linestyle='dashed', linewidth=2)

# 川普的身高190公分
plt.axvline(190, color='red', linewidth=2)

plt.show()
```

- 執行結果：平均數 =179.74, 標準差 =7.02, 信賴區間 =（165.99,193.49），川普的身高為 190 公分，在信賴區間內，表示川普與歷屆的美國總統身高比較，並未顯著的比較高。
- 以圖形看，川普的身高 190 公分在左右兩條虛線（信賴區間）之間。

2-4 機率（Probability）與統計（Statistics）

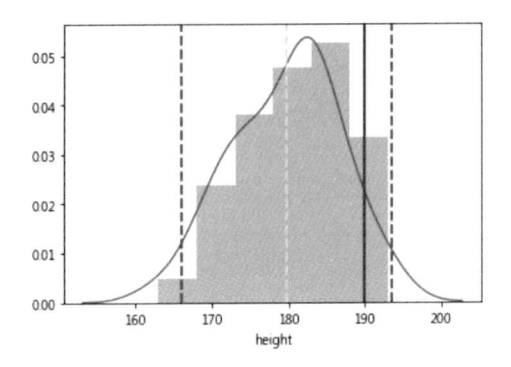

範例 2. 調查顧客是否喜歡公司新上市的食品，公司進行問卷調查，取得客戶的評價，範圍介於 0~10 分，已知母體平均分數為 5 分，標準差 2 分，請使用假設檢定確認顧客是否喜歡公司新上市的食品。

解題：

虛無假設 H_0：$\mu \leq 5$

對立假設 H_1：$\mu > 5$

檢驗平均數，我們會使用 t 檢定，當樣本大於 30 時，會近似於常態分配。t 統計量如下：

$$t = \frac{\bar{x} - \mu}{s/\sqrt{n}}$$

1. 使用隨機亂數模擬問卷調查結果。

```
1  import numpy as np
2  import matplotlib.pyplot as plt
3
4  # 問卷調查，得到客戶的評價
5  np.random.seed(123)
6  lo = np.random.randint(0, 5, 6)      # 0~4分 6筆
7  mid = np.random.randint(5, 7, 38)    # 5~6分 38筆
8  hi = np.random.randint(7, 11, 6)     # 7~10分 6筆
9  sample = np.append(lo,np.append(mid, hi))
10
11 print(f"最小值:{sample.min()}, 最大值:{sample.max()}, 平均數:{sample.mean()}")
12
13 plt.hist(sample)      # 畫直方圖
14 plt.show()
```

- 執行結果：最小值 :1, 最大值 :9, 平均數 :5.46。

第 2 章 神經網路（Neural Network）原理

2. 使用 stats.ttest_1samp 作假設檢定，參數 alternative='greater' 為右尾檢定，SciPy 需 v1.6.0 以上才支援 alternative 參數。

```
1  from scipy import stats
2  import seaborn as sns
3
4  # t檢定，scipy 需 v1.6.0 以上才支援 alternative
5  # https://docs.scipy.org/doc/scipy/reference/generated/scipy.stats.ttest_1samp.html
6  t,p = stats.ttest_1samp(sample, 5, alternative='greater')
7  print(f"t統計量:{t:.4f}, p值:{p}")
8
9  # 單尾檢定，右尾顯著水準 5%，故取信賴區間 90%，兩邊各 5%
10 ci = stats.norm.interval(0.90, pop_mean, pop_std)
11 sns.distplot(pop, bins=100)
12
13 # 畫平均數
14 plt.axvline(pop.mean(), color='yellow', linestyle='dashed', linewidth=2)
15
16 # 畫右尾顯著水準
17 plt.axvline(ci[1], color='magenta', linestyle='dashed', linewidth=2)
18
19 # 畫t統計量
20 plt.axvline(pop.mean() + t*pop.std(), color='red', linewidth=2)
21
22 plt.show()
```

- 執行結果：t 統計量 2.0250, p 值 0.024。90% 的信賴區間約在（μ−1.645δ, μ+1.645δ）之間，因 t 統計量（2.0250）>1.645，故對立假設為真，確認顧客喜歡公司新上市的食品。使用 t 統計量比較，需考慮單 / 雙尾檢定，須比較不同的值，有點麻煩，專家會改用 p 值與顯著水準比較，若小於顯著水準，則對立假設為真，反之，不能拒絕虛無假設，因為 p 值的計算公式 =1-cdf（累積分配函數）。

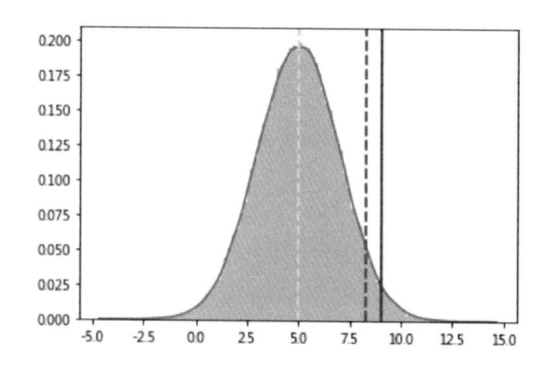

- 上圖實線為『母體平均數 + t 統計量 x 母體標準差』，旁邊虛線為信賴區間的右界，前者大於後者，表示效果顯著，有明顯差異。

2-4 機率（Probability）與統計（Statistics）

- 若使用左尾檢定，也是呼叫 stats.ttest_1samp()，參數 alternative='less'。
- 若使用雙尾檢定，也是呼叫 stats.ttest_1samp()，刪除參數 alternative 即可。

範例 3. 要檢定新藥有效性，將實驗對象分為兩組，一組為實驗組（Treatment Group），讓他們服用新藥，另一組為對照組，又稱控制組（Control Group），讓他們服用安慰劑，檢驗兩組疾病復原狀況是否有明顯差異。

虛無假設 H_0：$\mu_1 <= \mu_2$

對立假設 H_1：$\mu_1 > \mu_2$

1. 使用隨機亂數模擬復原狀況如下：

```
1  import numpy as np
2  import matplotlib.pyplot as plt
3  from scipy import stats
4
5  np.random.seed(123)
6  Control_Group = np.random.normal(66.0, 1.5, 100)
7  Treatment_Group = np.random.normal(66.55, 1.5, 200)
8  print(f"控制組平均數:{Control_Group.mean():.2f}")
9  print(f"實驗組平均數:{Treatment_Group.mean():.2f}")
```

- 執行結果：控制組平均數:66.04, 實驗組平均數:66.46。

2. 使用 stats.ttest_ind 作假設檢定，參數放入兩組數據及 alternative='greater' 表示右尾檢定。

```
1  # t檢定，scipy 需 v1.6.0 以上才支援 alternative
2  # https://docs.scipy.org/doc/scipy/reference/generated/scipy.stats.ttest_1samp.html
3  t,p = stats.ttest_ind(Treatment_Group, Control_Group, alternative='greater')
4  print(f"t統計量:{t:.4f}, p值:{p:.4f}")
5
6  # 單尾檢定，右尾顯著水準 5%，故取信賴區間 90%，兩邊各 5%
7  pop = np.random.normal(Control_Group.mean(), Control_Group.std(), 100000)
8  ci = stats.norm.interval(0.90, Control_Group.mean(), Control_Group.std())
9  sns.distplot(pop, bins=100)
10
11 # 畫平均數
12 plt.axvline(pop.mean(), color='yellow', linestyle='dashed', linewidth=2)
13
14 # 畫右尾顯著水準
15 plt.axvline(ci[1], color='magenta', linestyle='dashed', linewidth=2)
16
17 # 畫t統計量
18 plt.axvline(pop.mean() + t*pop.std(), color='red', linewidth=2)
19
20 plt.show()
```

第 2 章　神經網路（Neural Network）原理

- 執行結果：t 統計量 2.2390,p 值 0.0129，p 值若小於顯著水準（0.05），對立假設為真，表示新藥有顯著療效。

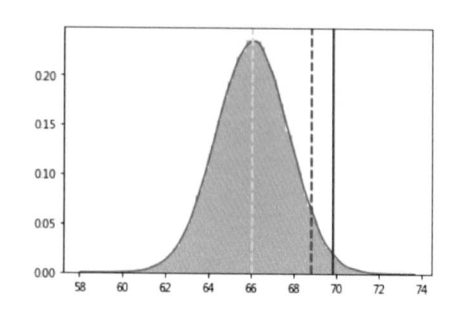

- 上圖實線為『母體平均數 + t 統計量 x 母體標準差』，旁邊虛線為信賴區間的右界，前者大於後者，表示效果顯著，有明顯差異。

範例 4. 另外有一種配對檢定（Paired Tests），例如學生同時參加期中及期末考試，我們希望檢驗期末時學生是否有顯著進步，因兩次考試都是以同一組學生作實驗，故稱為配對檢定。

虛無假設 H_0：$\mu_1 = \mu_2$

對立假設 H_1：$\mu_1 \neq \mu_2$

1. 使用隨機亂數模擬學生兩次考試成績如下：

```python
import numpy as np
import matplotlib.pyplot as plt
from scipy import stats

np.random.seed(123)
midTerm = np.random.normal(60, 5, 100)
endTerm = np.random.normal(61, 5, 100)
print(f"期中考:{midTerm.mean():.2f}")
print(f"期末考:{endTerm.mean():.2f}")
```

- 執行結果：期中考 :60.14, 期末考 :60.90。

2. 使用 stats.ttest_rel 作假設檢定，參數放入兩組數據及 alternative='greater' 表右尾檢定。

```python
# t檢定，scipy 需 v1.6.0 以上才支援 alternative
# https://docs.scipy.org/doc/scipy/reference/generated/scipy.stats.ttest_1samp.html
t,p = stats.ttest_rel(endTerm, midTerm, alternative='greater')
print(f"t統計量:{t:.4f}, p值:{p:.4f}")
```

2-92

2-4 機率（Probability）與統計（Statistics）

```python
 5
 6  # 單尾檢定，右尾顯著水準 5%，故取信賴區間 90%，兩邊各 5%
 7  pop = np.random.normal(Control_Group.mean(), Control_Group.std(), 100000)
 8  ci = stats.norm.interval(0.90, Control_Group.mean(), Control_Group.std())
 9  sns.distplot(pop, bins=100)
10
11  # 畫平均數
12  plt.axvline(pop.mean(), color='yellow', linestyle='dashed', linewidth=2)
13
14  # 畫右尾顯著水準
15  plt.axvline(ci[1], color='magenta', linestyle='dashed', linewidth=2)
16
17  # 畫t統計量
18  plt.axvline(pop.mean() + t*pop.std(), color='red', linewidth=2)
19
20  plt.show()
```

- 執行結果：t 統計量 1.0159,p 值 0.1561，p 值大於顯著水準（0.05），對立假設為假，表示學生成績沒有顯著進步。

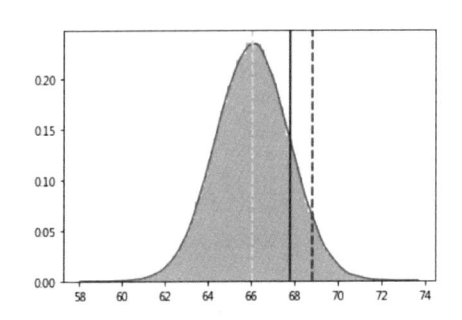

- 上圖實線為『母體平均數 + t 統計量 x 母體標準差』，右側虛線為信賴區間的右界，前者小於後者，表示效果不顯著。

範例 5. 使用假設檢定進行特徵選取。

程式：02_4_6_4_ 特徵選取 .ipynb

1. 載入套件。

```python
1  import pandas as pd
2  import numpy as np
3  import statsmodels.api as sm
4  from sklearn import datasets
5  import matplotlib.pyplot as plt
```

第 2 章　神經網路（Neural Network）原理

2. 載入 Scikit_learn 內建的糖尿病診斷資料集。

```
1  X, y = datasets.load_diabetes(return_X_y=True, as_frame=True)
2  X.head()
```

- 執行結果：含有 10 個特徵。

	age	sex	bmi	bp	s1	s2	s3	s4	s5	s6
0	0.038076	0.050680	0.061696	0.021872	-0.044223	-0.034821	-0.043401	-0.002592	0.019907	-0.017646
1	-0.001882	-0.044642	-0.051474	-0.026328	-0.008449	-0.019163	0.074412	-0.039493	-0.068332	-0.092204
2	0.085299	0.050680	0.044451	-0.005670	-0.045599	-0.034194	-0.032356	-0.002592	0.002861	-0.025930
3	-0.089063	-0.044642	-0.011595	-0.036656	0.012191	0.024991	-0.036038	0.034309	0.022688	-0.009362
4	0.005383	-0.044642	-0.036385	0.021872	0.003935	0.015596	0.008142	-0.002592	-0.031988	-0.046641

3. 使用 statsmodels 套件進行線性迴歸及分析報告。

```
1  model = sm.OLS(y, X)
2  res = model.fit()
3  print(res.summary())
```

- 執行結果：觀察【P>|t|】欄位，若 P<0.05 表示特徵效果顯著，只有 bmi 效果較顯著。

```
==============================================================================
                 coef    std err          t      P>|t|      [0.025      0.975]
------------------------------------------------------------------------------
age          -10.0099    179.967     -0.056      0.956    -363.729     343.709
sex         -239.8156    184.404     -1.300      0.194    -602.255     122.624
bmi          519.8459    200.401      2.594      0.010     125.964     913.728
bp           324.3846    197.053      1.646      0.100     -62.917     711.687
s1          -792.1756   1255.052     -0.631      0.528   -3258.944    1674.593
s2           476.7390   1021.170      0.467      0.641   -1530.341    2483.819
s3           101.0433    640.151      0.158      0.875   -1157.155    1359.242
s4           177.0632    486.370      0.364      0.716    -778.883    1133.009
s5           751.2737    517.768      1.451      0.148    -266.384    1768.931
s6            67.6267    198.747      0.340      0.734    -323.004     458.257
==============================================================================
Omnibus:                        1.506   Durbin-Watson:                   0.223
Prob(Omnibus):                  0.471   Jarque-Bera (JB):                1.404
Skew:                           0.017   Prob(JB):                        0.496
Kurtosis:                       2.726   Cond. No.                         21.7
==============================================================================
```

2-4 機率（Probability）與統計（Statistics）

4. 使用 Scikit_learn 隨機森林演算法驗證，先訓練模型。

```
1  from sklearn.ensemble import RandomForestClassifier
2
3  forest = RandomForestClassifier()
4  forest.fit(X, y)
```

5. 依特徵重要性排序。

```
1  importances = forest.feature_importances_
2  indices = np.argsort(importances)[::-1]
3  column_list = X.columns.to_list().copy()
4  for f in indices:
5      print(f"{f+1:02d} {column_list[f]:20s} {importances[f]:.4f}")
```

- 執行結果：bmi 貢獻度排名第一。

```
03 bmi                  0.1273
09 s5                   0.1228
06 s2                   0.1160
04 bp                   0.1110
05 s1                   0.1104
01 age                  0.1095
07 s3                   0.1080
10 s6                   0.1077
08 s4                   0.0601
02 sex                  0.0271
```

6. 繪圖。

```
1   plt.figure(figsize=(6,3))
2   plt.title('Feature Importance')
3   plt.bar(range(len(importances)),
4           importances[indices],
5           align='center')
6
7   plt.xticks(range(len(importances)),
8              np.array(column_list)[indices], rotation=45)
9   plt.tight_layout()
10  plt.show()
```

- 執行結果：bmi 貢獻度排名第一。

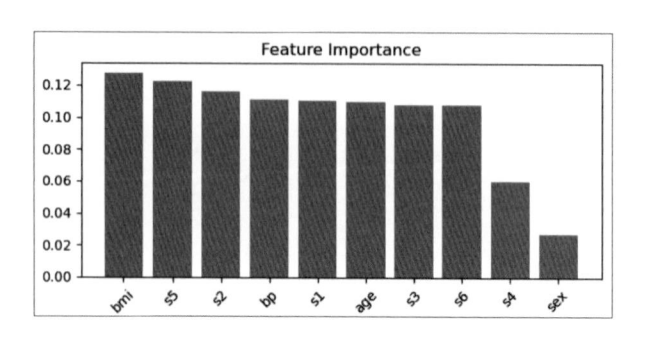

Feature Importance 圖表（由高至低）：bmi、s5、s2、bp、s1、age、s3、s6、s4、sex

7. 結論：當特徵很多時，可用假設檢定篩選重要的特徵，簡化模型。

另外，假設檢定有一個活生生的案例，2024 年總統大選藍白合之爭，候選人柯文哲提議若民調未贏過侯友宜 3% 以上，即禮讓侯友宜參選，經由馬英九基金會進行藍白合民意調查，雙方對統計結果發生認知的差異，之後還引發學者論辯，藍營認為採用單尾檢定（柯文哲 > 侯友宜），統計誤差應為 6%，白營主張就是 3%，讀者認為是 3% 或 6% 呢？可參閱筆者『藍白合民調之統計誤差計算，3% 或 6%？』[9] 一文。

2-4-7 小結

以上基礎統計的介紹，環環相扣，以台北市長選舉為例：

1. 抽樣：使用抽樣方法，從有投票權的市民中隨機抽出一批市民意見作為樣本。
2. 計算描述統計量：推估母體的平均數、標準差、變異數等描述統計量，描繪出大部分市民的想法與差異。
3. 機率：推測候選人當選的機率，再進而推估機率分配函數。
4. 假設檢定：依據機率分配函數，進行假設檢定，推論候選人當選的可信度或確定性，這就是一般古典統計的基礎流程，與機器學習以資料為主的預測相輔相成，可彼此驗證對方。

2-5 線性規劃（Linear Programming）

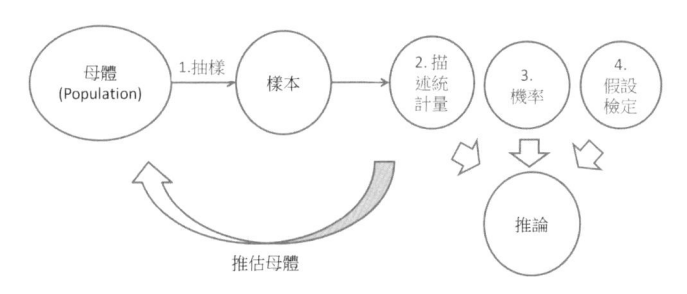

▲ 圖 2.11 上述介紹的基礎統計範圍及其關聯

2-5 線性規劃（Linear Programming）

線性規劃（Linear Programming）是給定一些線性的限制條件，求取目標函數的最大值或最小值。

範例 1. 最大化目標函數 $z = 3x+2y$。

程式：02_5_1_ 線性規劃 .ipynb。

限制條件： $2x+y \leq 100$

$x+y \leq 80$

$x \leq 40$

$x \geq 0, y \geq 0$

1. 先畫個圖，塗色區域（黃色）為可行解（Feasible Solutions），即符合限制條件的區域：

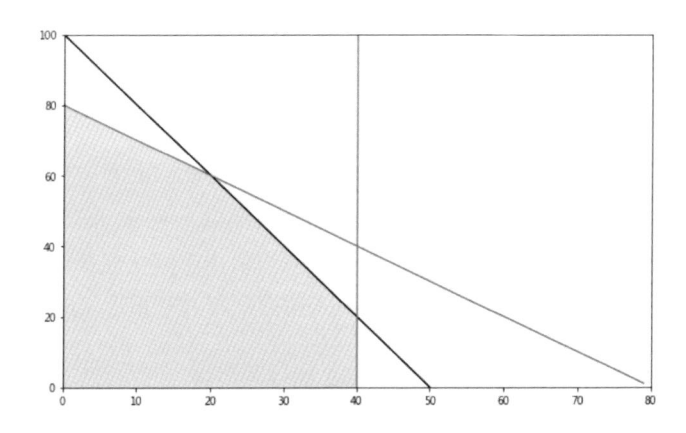

程式碼如下：

```python
1  import numpy as np
2  import matplotlib.pyplot as plt
3
4  plt.figure(figsize=(10,6))
5
6  # 限制式 2x+y = 100
7  x = np.arange(0,80)
8  y = 100 - 2 * x
9
10 # 限制式 x+y = 80
11 y_2 = 80 - x
12 plt.plot(x, y, 'black', x, y_2, 'g')
13
14 # 限制式 x = 40
15 plt.axvline(40)
16
17 # 座標軸範圍
18 plt.xlim(0,80)
19 plt.ylim(0,100)
20
21 # 限制式 x+y = 80 取邊界線樣本點
22 x1 = np.arange(0,21)
23 y1 = 80 - x1
24
25 # 限制式 2x+y = 100 取邊界線樣本點
26 x2 = np.arange(20,41)
27 y2 = 100 - 2 * x2
28
29 # 限制式 x = 40 取邊界線樣本點
30 x3 = np.array([40]*20)
31 y3 = np.arange(0,20)
32
33 # 整合邊界線樣本點
34 x1 = np.concatenate((x1, x2, x3))
35 y1 = np.concatenate((y1, y2, y3))
36
37 # 可行解(Feasible Solutions)
38 plt.fill_between(x1, y1, color='yellow')
39
40 plt.show()
```

2. 以上的線性規劃求解可使用『單形法』（Simplex Method），上述問題比較簡單，凸集合的最佳解發生在可行解的頂點（Vertex），所以依上圖，我們只要求每一個頂點對應的目標函數值，比較並找到最大的數值即可。不過，深度學習的變數動則幾百個，且神經層及神經元又很多，無法使用單形法求解，而是採取優化的方式，逐漸逼近找到近似解。以下直接運用程式來解題，不介紹單形法的原理。

2-5 線性規劃（Linear Programming）

首先安裝 pulp 套件：

- pip install pulp

3. 以套件 pulp 求解，定義目標函數及限制條件。

```python
 1  from pulp import LpMaximize, LpProblem, LpStatus, lpSum, LpVariable
 2
 3  # 設定題目名稱及最大化(LpMaximize)或最小化(LpMinimize)
 4  model = LpProblem("範例1. 最大化目標函數", LpMaximize)
 5
 6  # 變數初始化，x >= 0, y >= 0
 7  x = LpVariable(name="x", lowBound=0)
 8  y = LpVariable(name="y", lowBound=0)
 9
10  # 目標函數
11  objective_function = 3 * x + 2 * y
12
13  # 限制條件
14  constraint = 2 * x + 4 * y >= 8
15  model += (2 * x + y <= 100, "限制式1")
16  model += (x + y <= 80, "限制式2")
17  model += (x <= 40, "限制式3")
18
19  model += objective_function
20  model
```

- 執行結果：顯示問題定義。

```
範例1._最大化目標函數:
MAXIMIZE
3*x + 2*y + 0
SUBJECT TO
限制式1: 2 x + y <= 100

限制式2: x + y <= 80

限制式3: x <= 40

VARIABLES
x Continuous
y Continuous
```

4. 呼叫 model.solve 求解。

```python
1  status = model.solve()
2  status = 'yes' if status == 1 else 'no'
3  print(f'有解嗎? {status}')
4
5  print(f"目標函數: {model.objective.value()}")
6  for var in model.variables():
```

第 2 章 神經網路（Neural Network）原理

```
 7      print(f"{var.name}: {var.value()}")
 8
 9  print(f'\n限制式的值(不太重要)')
10  for name, constraint in model.constraints.items():
11      print(f"{name}: {constraint.value()}")
```

- 執行結果：當 x=20、y=60 時，目標函數最大值 =180。

```
有解嗎? yes
目標函數: 180.0
x: 20.0
y: 60.0

限制式的值(不太重要)
限制式1: 0.0
限制式2: 0.0
限制式3: -20.0
```

範例 2. 運用線性規劃來安排客服中心各時段的人力配置，請參考拙著的部落文 [10]。

程式：02_5_2_ 客服人力配置 .ipynb。

1. 定義問題。

```python
 1  import pulp as p
 2
 3  # 建立線性規劃 求取目標函數的最小值
 4  Lp_prob = p.LpProblem('Problem', p.LpMinimize)
 5
 6  # 宣告變數(Variables)
 7  x0 = p.LpVariable("x0", 0,None,p.LpInteger) # Create a variable x >= 0
 8  x4 = p.LpVariable("x4", 0,None,p.LpInteger) # Create a variable x >= 0
 9  x8 = p.LpVariable("x8",0,None,p.LpInteger) # Create a variable x >= 0
10  x12 = p.LpVariable("x12",0,None,p.LpInteger) # Create a variable x >= 0
11  x16 = p.LpVariable("x16",0,None,p.LpInteger) # Create a variable x >= 0
12  x20 = p.LpVariable("x20",0,None,p.LpInteger) # Create a variable x >= 0
13
14  # 定義目標函數(Objective Function)
15  Lp_prob += x0 + x4 + x8 + x12 + x16 + x20
16
17  # 定義限制條件(Constraints)
18  Lp_prob += x20+x0>=400
19  Lp_prob += x0+x4>=800
20  Lp_prob += x4+x8>=1000
21  Lp_prob += x8+x12>=800
22  Lp_prob += x12+x16>=1200
23  Lp_prob += x16+x20>=2000
24
```

2-5 線性規劃（Linear Programming）

```
25  # 顯示問題的定義
26  print(Lp_prob)
```

- 執行結果：顯示問題的數學式。

```
MINIMIZE
1*x0 + 1*x12 + 1*x16 + 1*x20 + 1*x4 + 1*x8 + 0
SUBJECT TO
_C1: x0 + x20 >= 400

_C2: x0 + x4 >= 800

_C3: x4 + x8 >= 1000

_C4: x12 + x8 >= 800

_C5: x12 + x16 >= 1200

_C6: x16 + x20 >= 2000

VARIABLES
0 <= x0 Integer
0 <= x12 Integer
0 <= x16 Integer
0 <= x20 Integer
0 <= x4 Integer
0 <= x8 Integer
```

1. 求解。

```
1  status = Lp_prob.solve() # Solver
2  print(p.LpStatus[status]) # The solution status
3
4  # 顯示答案
5  print('x0={}'.format(p.value(x0)))
6  print('x4={}'.format(p.value(x4)))
7  print('x8={}'.format(p.value(x8)))
8  print('x12={}'.format(p.value(x12)))
9  print('x16={}'.format(p.value(x16)))
10 print('x20={}'.format(p.value(x20)))
11
12 print('需求總人數={}'.format(p.value(Lp_prob.objective)))
```

- 執行結果：顯示各時段人力及總需求人力。

```
x0=0.0
x4=800.0
x8=200.0
x12=600.0
x16=600.0
x20=1400.0
需求總人數=3600.0
```

2-101

第 2 章　神經網路（Neural Network）原理

以上是簡單的線性規劃，還有其他的演算法，例如整數規劃（Integer Program-ming）、二次規劃（Quadratic Programming）、非線性規劃（Non-linear Pro-gramming），除了應用在機器學習求解外，也廣泛被應用於財務工程、工業工程⋯等領域。

▎2-6 最大概似法（MLE）

最小平方法（Ordinary Least Squares,OLS）與最大概似法（Maximum Likelihood Estimation MLE）是優化常見的方法，用以估算參數值，例如迴歸係數、常態分配的平均數/變異數，筆者戲稱兩者是優化求解的倚天劍與屠龍刀，許多演算法憑藉這兩把寶刀迎刃而解。

最小平方法在『2-3-3 簡單線性迴歸求解』已介紹過，主要目標是最小化均方誤差（MSE），再利用一階導數等於 0 的特性，求取最佳解。最大概似法（Maximum Likelihood Estimation,MLE）也是估算參數值的方法，目標是找到一個參數值，使出現目前事件的機率最大。如下圖，圓點是樣本點，曲線是四個可能的常態機率分配（平均數/變異數不同），我們希望利用最大概似法找到最適配（fit）的一個常態機率分配。

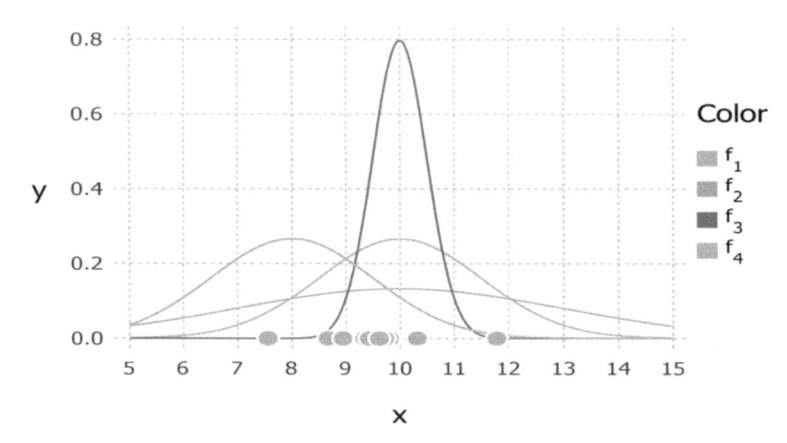

▲ 圖 2.12　最大概似法（Maximum Likelihood Estimation,MLE）示意圖

以下使用最大概似法求算常態分配的參數 μ 及 δ。

- 常態分配

$$P(x; \mu, \sigma) = \frac{1}{\sigma\sqrt{2\pi}} \exp\left(-\frac{(x-\mu)^2}{2\sigma^2}\right)$$

2-6 最大概似法（MLE）

- 假設來自常態分配的多個樣本互相獨立，聯合機率就等於個別的機率相乘。

$$\mathcal{L}(\mathcal{D}|\mu, \sigma^2) = \prod_{i=1}^{N} \frac{1}{\sqrt{2\pi\sigma^2}} e^{-\frac{(x_i-\mu)^2}{2\sigma^2}}$$

- 目標就是求取參數值 μ 及 δ，最大化機率值，即最大化目標函數。

$$\underset{\mu,\sigma^2}{\mathrm{argmax}}\ \mathcal{L}(\mathcal{D}|\mu, \sigma^2)$$

- N 個樣本機率全部相乘，不易計算，通常我們會取 log，變成一次方，所有的機率值取 log 後，大小順序並不會改變。

$$\underset{\mu,\sigma^2}{\mathrm{argmax}}\ \log \mathcal{L}(\mathcal{D}|\mu, \sigma^2)$$

- 取 Log 後，化簡如下：

$$\log \mathcal{L}(\mathcal{D}|\mu, \sigma^2) = -\frac{N}{2}\log(2\pi\sigma^2) - \frac{1}{2\sigma^2}\sum_{i=1}^{N}(x_i-\mu)^2$$

- 對 μ 及 δ 分別偏微分。

$$\frac{\partial \log \mathcal{L}}{\partial \mu} = -\frac{1}{\sigma^2}\sum_{i=1}^{N}(x_i-\mu)$$

$$\frac{\partial \log \mathcal{L}}{\partial \sigma} = -\frac{N}{\sigma} + \frac{1}{\sigma^3}\sum_{i=1}^{N}(x_i-\mu)^2$$

- 一階導數為 0 時，有最大值，得到目標函數最大值下的 μ 及 δ。所以，使用最大概似法算出的 μ 及 δ 就如同我們常見的公式。

$$\mu = \frac{1}{N}\sum_{i=1}^{N} x_i$$

$$\sigma^2 = \frac{1}{N}\sum(x_i-\mu)^2$$

以下範例請參閱 6_ 最大概似法 .ipynb 程式。

第 2 章　神經網路（Neural Network）原理

範例 1. 如果樣本點 x=1，計算來自常態分配 N（0,1）的機率。

1. 使用 NumPy 計算機率密度函數（pdf）。

```
1  # 載入套件
2  import numpy as np
3  import math
4
5  # 常態分配的機率密度函數(Probability Density Function, pdf)
6  def f(x, mean, std):
7      return (1/((2*np.pi*std**2) ** .5)) * np.exp(-0.5*((x-mean)/std)**2)
8
9  f(1, 0, 1)
```

- 執行結果：0.24。

2. 也可以使用 scipy.stats 模組計算。

```
1  from scipy.stats import norm
2
3  # 平均數(mean)、標準差(std)
4  mean = 0
5  std = 1
6
7  # 計算來自常態分配N(0,1)的機率
8  norm.pdf(1, mean, std)
```

- 執行結果：0.24。

3. 繪製機率密度函數（pdf）。

```
1   import matplotlib.pyplot as plt
2   import numpy as np
3   from scipy.stats import norm
4
5   # 觀察值範圍
6   z1, z2 = -4, 4
7
8   # 平均數(mean)、標準差(std)
9   mean = 0
10  std = 1
11
12  # 樣本點
13  x = np.arange(z1, z2, 0.001)
14  y = norm.pdf(x, mean, std)
15
16  # 繪圖
17  plt.plot(x,y,'black')
18
19  # 填色
20  plt.fill_between(x,y,0, alpha=0.3, color='b')
```

```
21
22  plt.plot([1, 1], [0, norm.pdf(1, mean, std)], color='r')
23  plt.show()
```

- 執行結果：

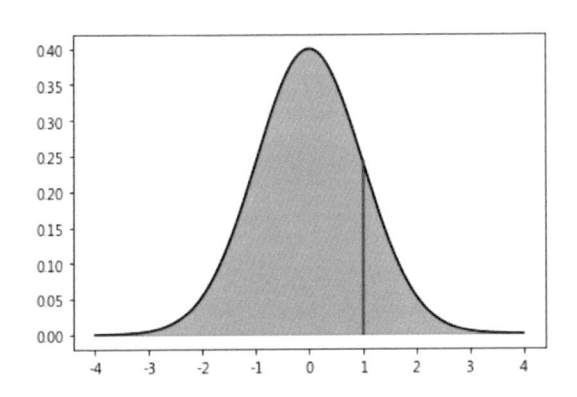

範例 2. 如果有兩個樣本點 x=1、3,來自常態分配 N（1,1）及 N（2,3）的可能性,哪一個比較大？

1. 假設兩個樣本是獨立的,故聯合機率為兩個樣本機率相乘,使用 scipy.stats 模組計算機率。

```
1   # 載入套件
2   import numpy as np
3   import math
4   from scipy.stats import norm
5
6   # 計算來自常態分配 N(1,1)的機率
7   mean = 1    # 平均數(mean)
8   std = 1     # 標準差(std)
9   print(f'來自 N(1,1)的機率：{norm.pdf(1, mean, std) * norm.pdf(3, mean, std)}')
10
11
12  # 計算來自常態分配 N(2,3)的機率
13  mean = 2    # 平均數(mean)
14  std = 3     # 標準差(std)
15  print(f'來自 N(2,3)的機率：{norm.pdf(1, mean, std) * norm.pdf(3, mean, std)}')
```

- 執行結果如下,來自 N（1,1）可能性比較大。

來自 N(1,1)的機率：0.021539279301848634
來自 N(2,3)的機率：0.01582423339377573

第 2 章　神經網路（Neural Network）原理

範例 3. 如果有一組樣本，計算來自哪一個常態分配 N（μ,δ）機率最大，請依上面證明計算 μ、δ？

1. 對常態分配的 pdf 取對數（log）：

```python
1   # 載入套件
2   from scipy.stats import norm
3   from sympy import symbols, pi, exp, log
4   from sympy.stats import Probability, Normal
5
6   # 樣本
7   data = [1,3,5,3,4,2,5,6]
8
9   # x變數、平均數(m)、標準差(s)
10  x, m, s = symbols('x m s')
11
12  # 常態分配的機率密度函數(Probability Density Function, pdf)
13  pdf = (1/((2*pi*s**2) ** .5)) * exp(-0.5*((x-m)/s)**2)
14  # 顯示 log(pdf) 函數
15  log_p = log(pdf)
16  log_p
```

- 執行結果如下，Jupyter Notebook 可顯示 Latex 數學式。

$$\log\left(\frac{0.707106781186547 e^{-\frac{0.5(-m+x)^2}{s^2}}}{\pi^{0.5}\left(s^2\right)^{0.5}}\right)$$

2. 帶入樣本資料。

```python
1   # 帶入資料
2   logP = 0
3   for xi in data:
4       logP += log_p.subs({x: xi})
5
6   logP
```

3. 上述函數使用 diff() 各對平均數、變異數偏微分

```python
1   from sympy import diff
2
3   logp_diff_m = diff(logP, m)  # 對平均數(m)偏微分
4
5   logp_diff_s = diff(logP, s)  # 對變異數(s)偏微分
6
7   print('m 偏導數:', logp_diff_m)
8   print('s 偏導數:', logp_diff_s)
```

2-7 神經網路（Neural Network）求解

4. 使用 simplify() 簡化偏導數

```
1  from sympy import simplify
2
3  # 簡化 m 偏導數
4  logp_diff_m = simplify(logp_diff_m)
5  print(logp_diff_m)
6  print()
7
8  # 簡化 s 偏導數
9  logp_diff_s = simplify(logp_diff_s)
10 logp_diff_s
```

5. 令一階導數 =0，有最大值，可得到聯立方程式

```
1  from sympy import solve
2
3  funcs = [logp_diff_s, logp_diff_m]
4  solve(funcs, [m, s])
```

- 執行結果：[（3.62,-1.57）,（3.62,1.57）]。

6. 使用 NumPy 驗證。

```
1  np.mean(data), np.std(data)
```

- 執行結果：（3.62,1.57），與 MLE 求解的結果相符。

2-7 神經網路（Neural Network）求解

有了以上的基礎後，我們將憑藉微積分、矩陣、機率等數學知識，針對神經網路求解，這是進入深度學習領域非常重要的概念。

2-7-1 神經網路（Neural Network）

神經網路是深度學習最重要的演算法，它主要是模擬生物神經傳導系統，透過層層解析，進行推論（Inference）。

第 2 章　神經網路（Neural Network）原理

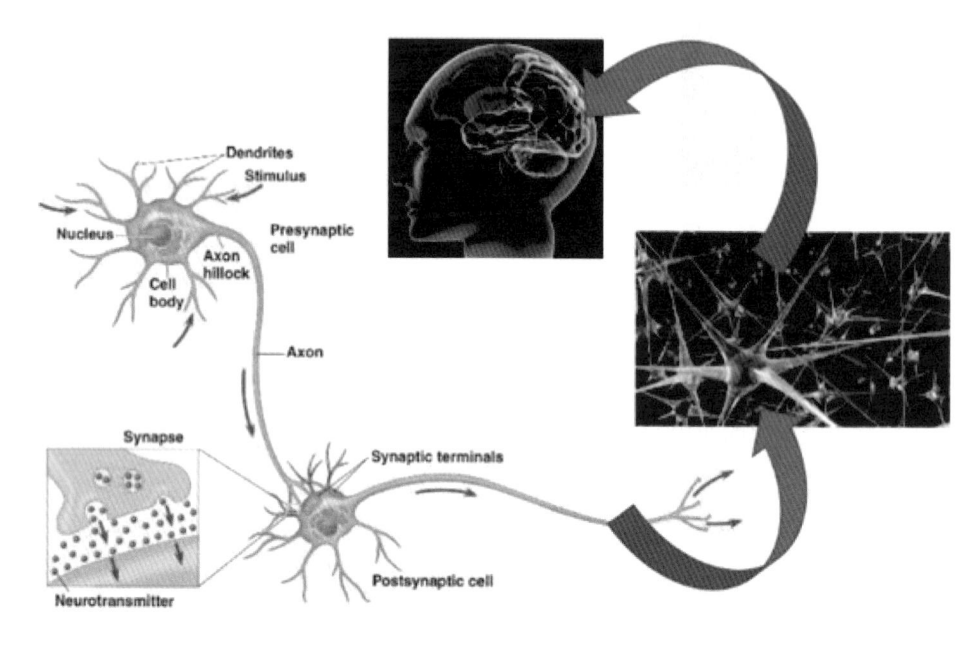

▲ 圖 2.13　生物神經網路的傳導系統

神經網路的運行原理如下，首先是生物的表層神經元接收到外界訊號，經細胞體歸納分析後，再透過突觸將分析結果傳給下一層的每個神經元，下一層神經元進行相同的動作，再往後傳導，最後傳至大腦，大腦作出最後的判斷與反應。

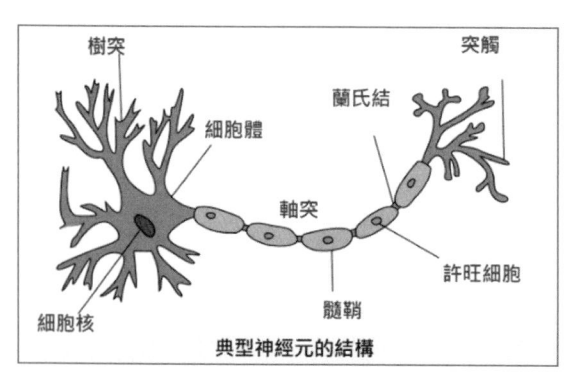

▲ 圖 2.14　單一神經元結構，圖片來源：維基百科, 神經元 [11]

AI 科學家將上述生物神經網路簡化成下列的網路結構：

2-7 神經網路（Neural Network）求解

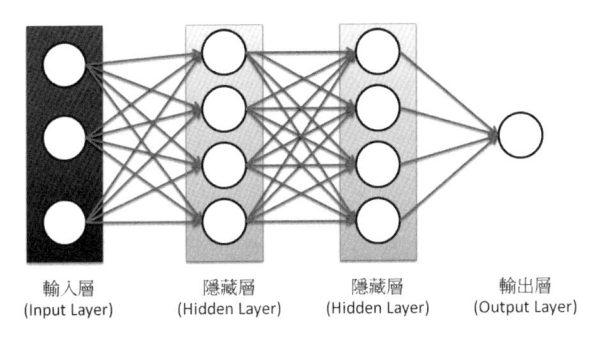

▲ 圖 2.15 AI 神經網路

AI 神經網路最簡單的連接方式稱為完全連接（Full connected, FC），上一層的每個神經元均連接至下一層的每個神經元，因此，我們可以把第二層以後的神經元均視為一條迴歸線的 y，它的特徵變數 (x) 就是前一層的每一個神經元，如下圖，例如 y_1、z_1 兩條迴歸線。

$$y_1 = w_1 x_1 + w_2 x_2 + w_3 x_3 + b$$

$$z_1 = w_1 y_1 + w_2 y_2 + w_3 y_3 + b$$

所以，簡單的講，一個神經網路可視為多條迴歸線組合而成的模型。

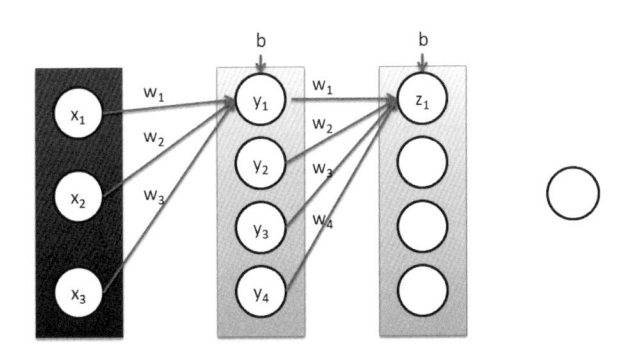

▲ 圖 2.16 AI 神經網路

以上的迴歸線是線性的，為了支援更通用性的解決方案（Generic Solution），模型還會乘上一個非線性函數，稱為『激勵函數』（Activation Function），期望模型也能解決非線性的問題，如下圖。因激勵函數並不能明確表達原意，以下直接以 Activation Function 表示。

第 2 章 神經網路（Neural Network）原理

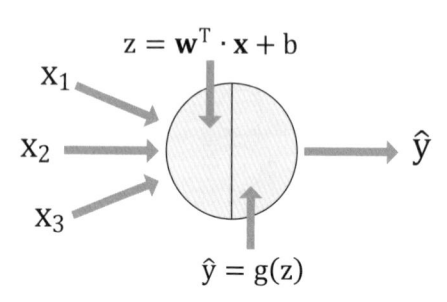

$$z = \mathbf{w}^T \cdot \mathbf{x} + b$$

$$\hat{y} = g(z)$$

g：Activation Function

▲ 圖 2.17 激勵函數（Activation Function）

如果不考慮 Activation Function，每一條線性迴歸線的權重（Weight）及偏差（Bias）可以透過最小平方法（OLS）求解，請參閱 2-3-3 說明，但乘以非線性的 Activation Function，就很難用純粹的數學求解了，因此，學者就利用優化（Optimization）方法，分別對權重、偏差各參數作偏微分，沿著切線（即梯度）逐步逼近，找到最佳解，這種演算法就稱為『梯度下降法』（Gradient Descent）。

有人用了一個很好的比喻，『當我們在山頂時，不知道下山的路，於是，就沿路往下走，遇到叉路時，就選擇坡度最大的叉路走，直到已無叉路往下走，我們就認為已抵達平地』，梯度下降法就是利用『偏微分』（Partial Differential），求算梯度，依梯度的方向，逐步往下走，直到損失函數沒有顯著改善為止，這時我們就認為已經找到最佳解了。筆者開個玩笑，所謂『最佳解』就是找不到真正的解，只是找到近似解而已。

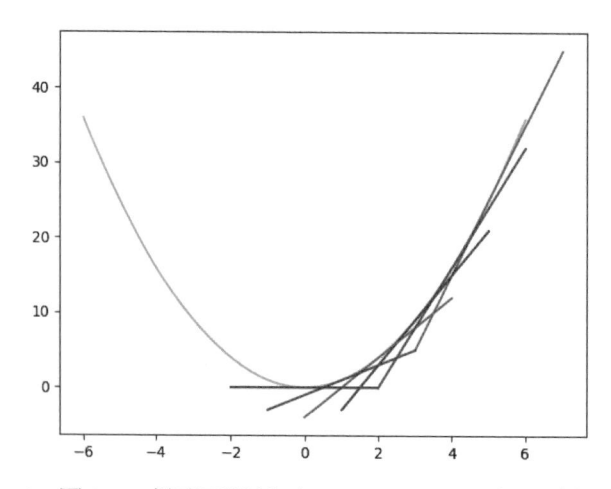

▲ 圖 2.18 梯度下降法（Gradient Descent）示意圖

2-7-2 梯度下降法（Gradient Descent）

梯度其實就是斜率，單變數迴歸線的權重稱為斜率，多變數迴歸線時，須個別作偏微分求取權重值，就稱為梯度。以下，先針對單變數求解，示範如何使用梯度下降法（Gradient Descent）求取最小值。

範例 1. 假定損失函數 f(x)= x^2，而非 MSE，請使用梯度下降法求取最小值。

注意，損失函數又稱為目標函數（Objective function）或成本函數（Cost function），在神經網路相關文獻中大多稱為損失函數，本書將統一使用『損失函數』。

程式：02_7_梯度下降法.ipynb。

1. 定義函數（func）及其一階導數（dfunc）：

```
1  # 載入套件
2  import numpy as np
3  import matplotlib.pyplot as plt
4
5  # 目標函數(損失函數):y=x^2
6  def func(x): return x ** 2 #np.square(x)
7
8  # 目標函數的一階導數:dy/dx=2*x
9  def dfunc(x): return 2 * x
```

2. 定義梯度下降法函數，反覆更新 x，更新的公式如下，後面章節我們會說明公式的由來。

    - 新的 x = 目前的 x - 學習率（learning_rate）* 梯度（gradient）

```
1  # 梯度下降
2  # x_start: x的起始點
3  # df: 目標函數的一階導數
4  # epochs: 執行週期
5  # lr: 學習率
6  def GD(x_start, df, epochs, lr):
7      xs = np.zeros(epochs+1)
8      x = x_start
9      xs[0] = x
10     for i in range(epochs):
11         dx = df(x)
12         # x更新 x_new = x – learning_rate * gradient
13         x += - dx * lr
14         xs[i+1] = x
15     return xs
```

3. 設定起始點、學習率（lr）、執行週期數（epochs）等參數後，呼叫梯度下降法求解。

```
1   # 超參數(Hyperparameters)
2   x_start = 5        # 起始權重
3   epochs = 15        # 執行週期數
4   lr = 0.3           # 學習率
5
6   # 梯度下降法
7   # *** Function 可以直接當參數傳遞 ***
8   w = GD(x_start, dfunc, epochs, lr=lr)
9   print (np.around(w, 2))
10
11  color = 'r'
12  from numpy import arange
13  t = arange(-6.0, 6.0, 0.01)
14  plt.plot(t, func(t), c='b')
15  plt.plot(w, func(w), c=color, label='lr={}'.format(lr))
16  plt.scatter(w, func(w), c=color, )
17
18  # 設定中文字型
19  from matplotlib.font_manager import FontProperties
20  font = FontProperties(fname=r"c:\windows\fonts\msjhbd.ttc", size=20)
21  # 矯正負號
22  plt.rcParams['axes.unicode_minus'] = False
23
24  plt.title('梯度下降法', fontproperties=font)
25  plt.xlabel('X', fontsize=20)
26  plt.ylabel('損失函數', fontsize=20)
27  plt.show()
```

- 執行結果：

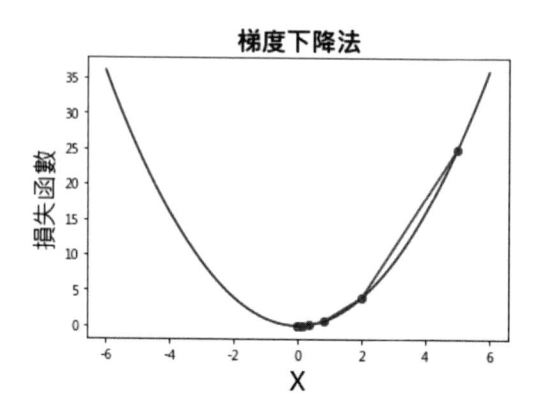

- 每一執行週期的損失函數如下，隨著 x 變化，損失函數逐漸收斂，即前後週期的損失函數差異逐漸縮小，最後當 x=0 時，損失函數 f(x) 等於 0，為函數的最小值，與最小平方法（OLS）的計算結果相同。

2-7 神經網路（Neural Network）求解

[5.2,0.8,0.32,0.13,0.05,0.02,0.01,0,0,0,0,0,0,0,0]

- 可以改變起始點（x_start）為任意值，例如 -5，依然可以找到相同的最小值。

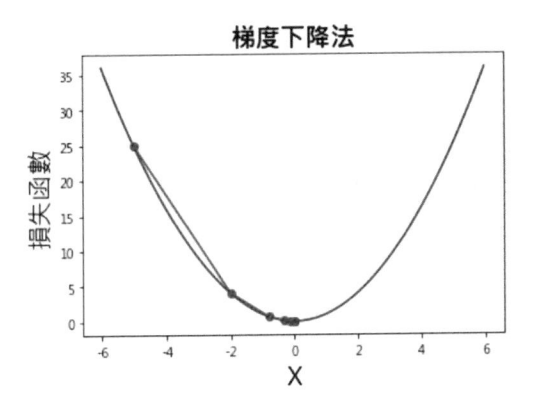

梯度下降法

範例 2. 假定損失函數 f(x)= $2x^4 - 3x - 20$，請使用梯度下降法求取最小值。

1. 定義函數及其微分。

```
1  # 損失函數
2  def func(x): return 2*x**4-3*x+2*x-20
3
4  # 損失函數一階導數
5  def dfunc(x): return 8*x**3-6*x+2
```

2. 繪製損失函數。

```
1   from numpy import arange
2   t = arange(-6.0, 6.0, 0.01)
3   plt.plot(t, func(t), c='b')
4
5   # 設定中文字型
6   from matplotlib.font_manager import FontProperties
7   font = FontProperties(fname=r"c:\windows\fonts\msjhbd.ttc", size=20)
8   # 矯正負號
9   plt.rcParams['axes.unicode_minus'] = False
10
11  plt.title('梯度下降法', fontproperties=font)
12  plt.xlabel('X', fontsize=20)
13  plt.ylabel('損失函數', fontsize=20)
14  plt.show()
```

2-113

第 2 章　神經網路（Neural Network）原理

- 執行結果：

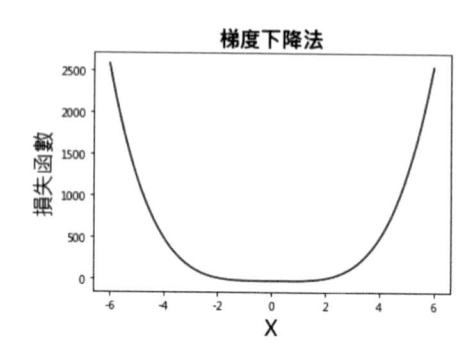

2. 梯度下降法函數（GD）不變，執行程式，如果學習率不變（lr = 0.3），會出現錯誤訊息：『Result too large』，原因是學習率過大，梯度下降過程錯過最小值，往函數左方逼近，造成損失函數值越來越大，最後導致溢位。

```
OverflowError                             Traceback (most recent call last)
<ipython-input-20-e4de47c74444> in <module>
      6 # 梯度下降法
      7 # *** Function 可以直接當參數傳遞 ***
----> 8 w = GD(x_start, dfunc, epochs, lr=lr)
      9 print (np.around(w, 2))
     10

<ipython-input-3-32de2a0b3ea7> in GD(x_start, df, epochs, lr)
      9     xs[0] = w
     10     for i in range(epochs):
---> 11         dx = df(w)
     12         # 權重的更新 W_new = W — learning_rate * gradient
     13         w += - dx * lr

<ipython-input-17-2c5b4a8cd0cb> in dfunc(x)
      3
      4 # 損失函數一階導數
----> 5 def dfunc(x): return 8*x**3-6*x+2
      6

OverflowError: (34, 'Result too large')
```

3. 修改學習率（lr = 0.001），同時增加執行週期數（epochs = 150,000），避免還未逼近到最小值，就先提早結束。

```
# 超參數(Hyperparameters)
x_start = 5        # 起始權重
epochs = 15000     # 執行週期數
lr = 0.001         # 學習率
```

2-7 神經網路（Neural Network）求解

- 執行結果：當 x=0.5 時，函數有最小值。

  [5.   4.03 3.53...0.5 0.5 0.5]

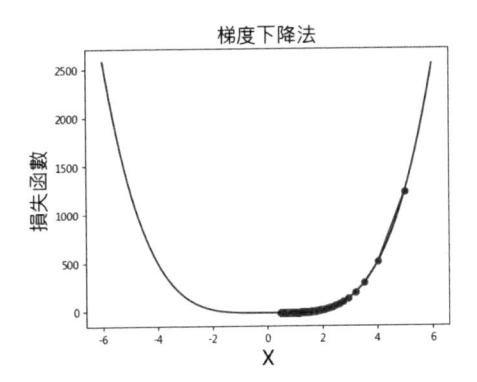

梯度下降法

4. 計算函數最小值。

```
1  func(0.5)
```

- 執行結果：函數最小值 =-20.375。

5. SciPy 也有提供最小平方法的函數 leastsq，測試看看。

```
1  from scipy.optimize import leastsq
2
3  # 最小平方法
4  def func(x): return 2*x**4-3*x+2*x-20
5  leastsq(func, 5, full_output=True)
```

- 執行結果：x= 1.81736891 時有函數最小值（fvec）=10^{-14}，結果遠比上述自行開發差，應該是因為執行週期（nfev）只有 17 次，學習率設定太大的關係。

```
(array([1.81736891]),
 array([[0.00045231]]),
 {'fvec': array([1.42108547e-14]),
  'nfev': 17,
  'fjac': array([[-47.01968293]]),
  'ipvt': array([0], dtype=int32),
  'qtf': array([-4.80690783e-07])},
 'The relative error between two consecutive iterates is at most 0.000000',
 2)
```

2-115

觀察上述範例，不管函數為何，我們以相同的梯度下降法（GD 函數）都能夠找到函數最小值，最重要的關鍵是『x 的更新公式』：

$$\text{新的 x = 目前的 x - 學習率（learning\_rate）} * \text{梯度（gradient）}$$

接著來探討公式的由來，也就是神經網路求解的精華所在。

2-7-3 神經網路求解

神經網路求解是一個正向傳導與反向傳導反覆執行的過程，如下圖所示。

1. 由於神經網路是多條迴歸線的組合，建立模型的主要任務就是計算出每條迴歸線的權重（w）與偏差（b）。

2. 依上述範例的邏輯，一開始我們指定 w、b 為任意值，將訓練資料的特徵值 (x) 帶入迴歸方程式 y=wx+b，可以求得預測值（$\hat{y}$），進而計算出損失函數，例如 MSE，這個過程稱為『正向傳導』（Forward Propagation）。

3. 透過最小化 MSE 的目標和偏微分，可以找到更好的 w、b，並依學習率來更新每一層神經網路的 w、b，此過程稱之為『反向傳導』（Backpropagation）。這部份可以藉由微分的連鎖率（Chain Rule），一次計算出每層神經元對應的 w、b，公式為：

$$W_{t+1} = W_t - \text{學習率（learning rate）} \times \text{梯度（gradient）}$$

其中：

$$\text{梯度} = -2*x*(y - \hat{y}) \quad [\text{後續證明}]$$

學習率：優化器事先設定的固定值或可動態調整的函數。

4. 重複正向傳導與反向傳導，直到損失函數不再有明顯改善為止，最後得到的 w、b 就是我們認定的最佳解。

2-7 神經網路（Neural Network）求解

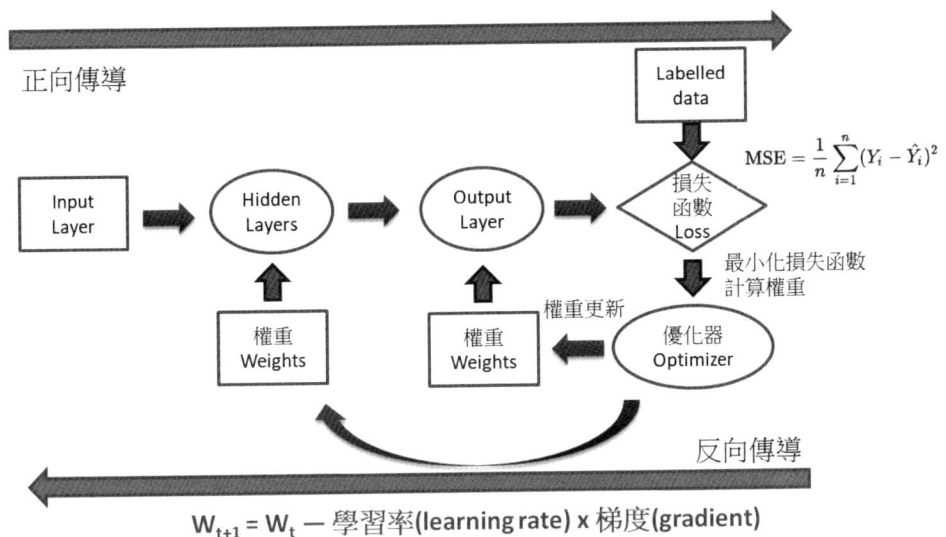

▲ 圖 2.19 神經網路權重求解過程

梯度（gradient）公式證明如下：

1. 損失函數 MSE = $\frac{\sum(y-\hat{y})^2}{n}$ ，因 n 為常數，故僅考慮分子，即 SSE。
2. SSE = $\sum(y-\hat{y})^2 = \sum(y-wx)^2 = \sum(y^2 - 2ywx + w^2x^2)$
3. 以矩陣表示，SSE = $y^2 - 2ywx + w^2x^2$
4. $\frac{\partial SSE}{\partial w}$ = $-2yx + 2wx^2$ = -2 x（y−wx）= -2 x（y- $\hat{y}$）
5. 同理，$\frac{\partial SSE}{\partial b}$ = $-2x^0$（y- $\hat{y}$）= -2（y- $\hat{y}$）
6. 為了簡化公式，常把係數 2 拿掉。
7. 最後公式為：調整後權重 = 原權重 **+**（學習率 * 梯度）
8. 有些文章將梯度的負號拿掉，公式就修正為：

   調整後權重 = 原權重 -（學習率 * 梯度）。

以上是以 MSE 為損失函數時的梯度公式，若是採用其他損失函數，梯度公式也會隨之不同，如果再加上 Activation Function，梯度公式計算就更加複雜了，幸好，所有的深度學習套件均提供自動微分（Automatic Differentiation）功能，不管多複雜的函數，都有辦法算出梯度，我們就不用煩惱了。後續有些演算法會自訂損失函數，因而產生意想不到的功能，例如『風格轉換』（Style Transfer）可以合成兩張圖像，生成對抗

第 2 章　神經網路（Neural Network）原理

網路（Generative Adversarial Network, GAN），可以產生幾可亂真的圖像，也因為如此關鍵，我們才花費了這麼多的篇幅鋪陳『梯度下降法』。

除了梯度下降法，還有其他神經網路求解的演算法，2024 年 4 月劉子鳴（Ziming Liu）等學者發表『KAN:Kolmogorov-Arnold Networks』[12]，不固定 Activation Function，直接求解非線性函數，曾引發非常多的討論，有興趣的讀者可參閱筆者的部落文『Hello KAN, 建構深度學習模型的另一種思維』[13]。

基礎的數學與統計就介紹到此，告一段落，下一章起，我們將開始以 Tensor-Flow 實現各種演算法，並介紹相關的應用。

參考資料（References）

[1]　Keith McNulty,《Decision makers need more math》, 2018
　　　(https://blog.keithmcnulty.org/decision-makers-need-more-math-78ba07306193)

[2]　Seaborn 官網
　　　(https://seaborn.pydata.org/examples/index.html)

[3]　台灣彩券官網
　　　(https://www.taiwanlottery.com.tw/DailyCash/index.asp)

[4]　 Univariate Distribution Relationships
　　　(http://www.math.wm.edu/~leemis/chart/UDR/UDR.html)

[5]　維基百科中關於假設檢定的定義
　　　(https://zh.wikipedia.org/wiki/%E5%81%87%E8%AA%AA%E6%AA%A2%E5%AE%9A)

[6]　Scribbr, Central Limit Theorem
　　　(https://www.scribbr.com/statistics/central-limit-theorem/)

[7]　An Interactive Guide to Hypothesis Testing in Python
　　　(https://towardsdatascience.com/an-interactive-guide-to-hypothesis-testing-in-python-979f4d62d85/)

[8]　17 Statistical Hypothesis Tests in Python
　　　(https://machinelearningmastery.com/statistical-hypothesis-tests-in-python-cheat-sheet/)

參考資料（References）

[9] 藍白合民調之統計誤差計算，3% 或 6%?
(https://ithelp.ithome.com.tw/articles/10340531)

[10] 陳昭明，《Day 14：客服人力規劃 (Workforce Planning) -- 線性規劃求解》，2019
(https://ithelp.ithome.com.tw/articles/10222877)

[11] 維基百科, 神經元
(https://zh.wikipedia.org/wiki/ 神經元)

[12] 劉子鳴 (Ziming Liu) 等學者 , KAN: Kolmogorov-Arnold Networks
(https://arxiv.org/abs/2404.19756)

[13] Hello KAN, 建構深度學習模型的另一種思維
(https://ithelp.ithome.com.tw/articles/10342873)

第 2 章　神經網路（Neural Network）原理

MEMO

第二篇

TensorFlow 基礎篇

TensorFlow 是 Google Brain 於 2015 年發布的深度學習框架，它與 PyTorch 是目前深度學習框架佔有率最高的套件，TensorFlow 在 1.4 版納入 Keras，兼顧簡易性與效能，是深度學習最佳的入門套件。

本篇將介紹 TensorFlow 的整體架構，包含下列內容：

1. 從張量（Tensor）運算、自動微分（Automatic Differentiation）、神經層，最後構建完整的神經網路。

2. 說明神經網路的各項函數，例如 Activation Function、損失函數（Loss Function）、優化器（Optimizer）、效能衡量指標（Metrics），並介紹運用梯度下降法找到最佳解的原理與過程。

3. 示範 TensorFlow 各項工具的使用，包含回呼函數（Callback）、TensorBoard 視覺化工具、TensorFlow Dataset、TensorFlow Serving 佈署等。

4. 神經網路完整流程的實踐。

5. 卷積神經網路（Convolutional Neural Network, CNN）。

6. 預先訓練的模型（Pre-trained Model）。

7. 轉移學習（Transfer Learning）。

第 3 章

TensorFlow 架構與主要功能

▌ 3-1 常用的深度學習套件

維基百科[1]列舉近 20 種的深度學習套件，有許多已經逐漸被淘汰了，以 Python /C++ 語言的套件而言，目前比較流行的只剩以下這四種：

第 3 章　TensorFlow 架構與主要功能

套件名稱	程式語言	優點	缺點
TensorFlow/Keras	Python, C++, js, Java, Go	1.Keras 簡單、易入門。支援多種程式語言。 2. 效能佳。	與 Python 未完全整合，例如模型結構不能直接使用 if。
PyTorch	Python, C++, Java	1. 與 Python 整合較好，有彈性。 2. 學界採用佔比有增加的趨勢。	執行速度較慢。
Caffe	C++	1. 執行速度非常快。 2. 專長於影像應用，NVidia 有一改良版。	1. 複雜。 2. 無 NLP 模組。
Jax	Python	加速陣列運算。 高效數值分析。 適合大規模專案。	須從較低階的運算建構模型。

TensorFlow、PyTorch 分居佔有率的前兩名，其他套件均望塵莫及，如下圖所示，因此推薦讀者若是使用 Python 語言，熟悉 TensorFlow、PyTorch 就綽綽有餘了，假如偏好 C/C++ 的話，則可以使用 Caffe 框架。

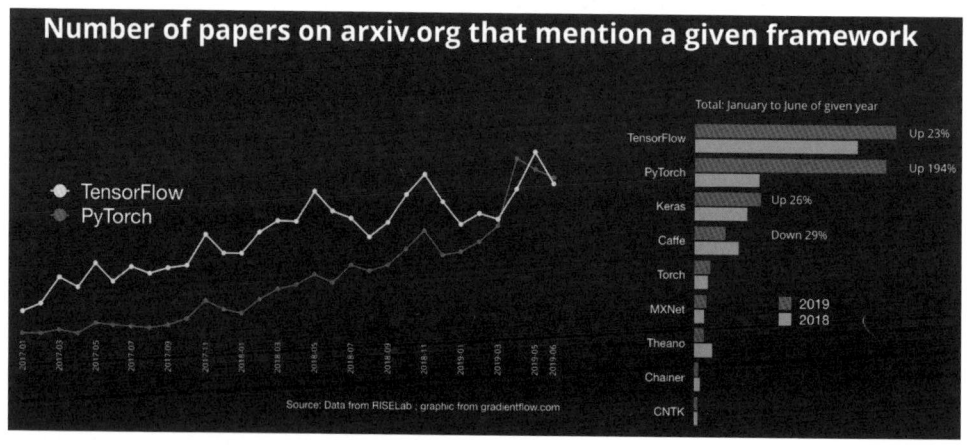

▲ 圖 3.1 深度學習套件在 Arxiv 網站的論文採用比率，
　　圖片來源：Which deep learning framework is the best？[2]

以目前（2025 年）發展的狀況觀察，TensorFlow 與 PyTorch 比較如下：

比較項目	佔優勢者	說明
學習門檻	TensorFlow	TensorFlow/Keras 提供簡易指令，例如模型訓練，只需 fit 指令。 文件較易閱讀。
效能	TensorFlow	TensorFlow 提供靜態運算圖，效能較佳。 TensorFlow 自動判斷是否使用 GPU 運算。 TensorFlow 預設的超參數設定較佳。
生態環境	TensorFlow	TensorFlow Lite 支援手機模型。 TensorFlow.js 可使用 JavaScript 開發。 TensorFlow Extended 提供模型佈署。
彈性	PyTorch	PyTorch 只在 Python 基礎加上類別 / 函數，較靈活。
版本相容性	PyTorch	PyTorch 語法較有一致性，TensorFlow 函數用法變動較大。
市占率	TensorFlow/PyTorch 各擅勝場	比較進階的模型及研發採用 PyTorch 較多。

本書的範例以 TensorFlow/Keras 為主，主要是它較易入門，如果希望同時學習 PyTorch，可參閱筆者撰寫的另一本著作『開發者傳授 PyTorch 秘笈』[3]，TensorFlow 與 PyTorch 基本概念是相通的，只要按照本書的學習路徑研讀，一次學會兩種套件應該不是難事，相關差異對照可參閱筆者部落文『同時搞定 TensorFlow、PyTorch』[4]。另外有一些較進階的課題會使用 PyTorch，例如 YOLO、Transformers，因為這些套件以 PyTorch 模型為主。

3-2 TensorFlow 架構

梯度下降法是神經網路求解的主要方法，求解過程需要使用張量（Tensor）運算，另外，在反向傳導的過程中，要透過偏微分計算梯度，並利用連鎖律一次求算多層神經網路的參數（即權重），因此，多數深度學習套件至少都會具備下列功能：

第 3 章　TensorFlow 架構與主要功能

1. 張量運算：包括各種向量、矩陣運算。
2. 自動微分（Auto Differentiation）。
3. 各種神經層（Layers）組合而成的神經網路（Neural Network）。

學習路徑可以從簡單的張量運算開始，再逐漸熟悉高階的神經層函數，以奠定扎實的基礎。

接著再來瞭解 TensorFlow 執行環境（Runtime），如下圖，支援 GPU、分散式運算、低階 C API 等，也支援多種網路協定。

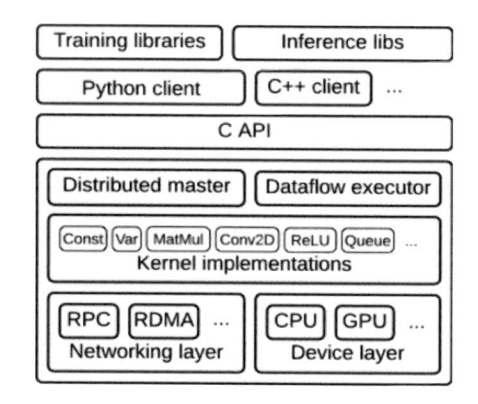

▲ 圖 3.2　TensorFlow 執行環境（Runtime），圖片來源：TensorFlow GitHub [5]

TensorFlow 執行環境包括三種版本：

1. TensorFlow：一般電腦版本。
2. TensorFlowjs：使用 JavaScript 開發的網頁版本。
3. TensorFlow Lite：輕量版 TensorFlow，適合運算速度及記憶體有限的手機和物聯網裝置，現已改名為 LiteRT。

程式設計堆疊（Programming Stack）如下圖，在 2.x 版後以 Keras 為主軸，TensorFlow 團隊依照 Keras 的規格重新開發，整合既有模組，比獨立開發的 Keras 套件功能更加強大，因此現在要撰寫 Keras 程式，應以 TensorFlow 為主，避免再使用 Keras 獨立套件了。但 2024 年 Keras 大神 François Chollet 又重新升級 Keras 獨立套件至 v3.0，TensorFlow 安裝時也會同時安裝對應的 Keras 版本，並在某些時機呼叫 Keras 獨立套件，發展方向有些不明確。目前 TensorFlow v2.18 對應的 Keras v3.0 版本，是一個較穩定的版本。

3-3 張量（Tensor）運算

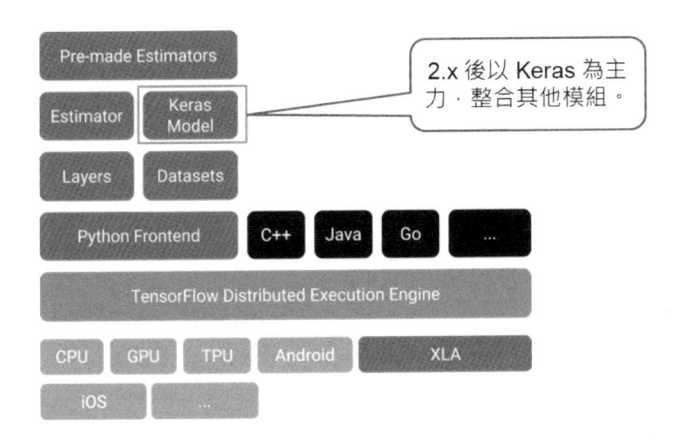

▲ 圖 3.3 TensorFlow 程式設計堆疊（Programming Stack），
圖片來源：Explained: Deep Learning in Tensorflow [6]

除了 Keras 的引進，特別要注意的是，**TensorFlow 2.x 版之後，預設模式改為 Eager Execution Mode**，捨棄 Session 語法，若未調回原先的 Graph Execution Mode，1.x 版的程式執行時將會產生錯誤。由於目前網路上許多範例的程式新舊雜陳，還是有很多 1.x 版的程式，讀者應特別注意。

3-3 張量（Tensor）運算

張量（Tensor）是一維向量、二維矩陣及更高維度空間的統稱，主要是提供一個簡明的數學框架，簡化運算代數的運算，例如聯立方程式的計算，可以很容易的使用矩陣表示，並利用反矩陣相乘求解。

TensorFlow 顧名思義就是提供張量（Tensor）的定義、計算與變數值的傳遞，因此，它最底層的功能即支援各項張量的資料型態與運算函數，基本上都與 NumPy 套件相容，遵循 NumPy 的設計理念與語法，包括傳播（Broadcasting）機制的支援。

範例. TensorFlow 基本測試與張量運算。

程式：**03_1_ 張量運算 .ipynb**。

1. 顯示 TensorFlow 版本：

```
1  # 載入套件
2  import tensorflow as tf
3
```

```
4  # 顯示版本
5  print(tf.__version__)
```

2. 檢查 GPU 是否存在：若在 Windows 作業系統下安裝最新版的 TensorFlow，無法使用 GPU 顯卡，以下指令可在雲端 Colaboratory 或 WSL 環境下測試。

```
1  # check cuda available
2  tf.config.list_physical_devices('GPU')
```

- 執行結果如下，可顯示 GPU 簡略資訊：

  [PhysicalDevice（name='/physical_device:GPU:0',device_type='GPU'）]

3. 如果要知道 GPU 卡的詳細規格，可以安裝 PyCuda 套件：

   pip install pycuda

   - 接著就可以執行本書所附的範例：python GpuQuery.py。
   - 執行結果如下，即可顯示 GPU 的詳細規格，重要資訊排列在前面：

```
偵測 1 個CUDA裝置
裝置 0: GeForce GTX 1050 Ti
        計算能力: 6.1
        GPU記憶體: 4096 MB
        6 個處理器，各有 128 個CUDA核心數，共 768 個CUDA核心數

        ASYNC_ENGINE_COUNT: 5
        CAN_MAP_HOST_MEMORY: 1
        CLOCK_RATE: 1392000
        COMPUTE_CAPABILITY_MAJOR: 6
        COMPUTE_CAPABILITY_MINOR: 1
        COMPUTE_MODE: DEFAULT
        CONCURRENT_KERNELS: 1
        ECC_ENABLED: 0
        GLOBAL_L1_CACHE_SUPPORTED: 1
        GLOBAL_MEMORY_BUS_WIDTH: 128
        GPU_OVERLAP: 1
        INTEGRATED: 0
        KERNEL_EXEC_TIMEOUT: 1
        L2_CACHE_SIZE: 1048576
        LOCAL_L1_CACHE_SUPPORTED: 1
        MANAGED_MEMORY: 1
        MAXIMUM_SURFACE1D_LAYERED_LAYERS: 2048
        MAXIMUM_SURFACE1D_LAYERED_WIDTH: 32768
```

4. 宣告常數（constant）：參數可以是常數、list、NumPy array。

```
1  # 宣告常數(constant)，參數可以是常數、list、numpy array
2  x = tf.constant([[1, 2]])
```

5. 支援四則運算。

```
1  print(x+10)
2  print(x-10)
```

3-3 張量（Tensor）運算

```
3  print(x*2)
4  print(x/2)
```

- 執行結果如下：

    tf.Tensor（[[11 12]], shape=（1, 2）, dtype=int32）

    tf.Tensor（[[-9 -8]], shape=（1, 2）, dtype=int32）

    tf.Tensor（[[2 4]], shape=（1, 2）, dtype=int32）

    tf.Tensor（[[0.5 1.]], shape=（1, 2）, dtype=float64）

- 如果要顯示數值，須轉回 NumPy 資料型態，因為 TensorFlow 資料型態雖與 NumPy 相容，但還是各屬不同的資料型態，另外，TensorFlow 資料有可能存在 GPU 記憶體內，需轉回一般記憶體內才能交由 CPU 顯示。

    （x+10）.numpy()

6. 四則運算也可以使用 TensorFlow 函數。

```
1  # 轉為負數
2  print(tf.negative(x))
3
4  # 常數、List、Numpy array 均可運算
5  print(tf.add(1, 2))
6  print(tf.add([1, 2], [3, 4]))
7
8  print(tf.square(5))
9  print(tf.reduce_sum([1, 2, 3]))
10
11 # 混用四則運算符號及TensorFlow 函數
12 print(tf.square(2) + tf.square(3))
```

- reduce_sum 是沿著特定軸加總，輸出會少一維，故 [1, 2, 3] 套用函數運算後等於 6。

7. TensorFlow 會自動決定在 CPU 或 GPU 運算，可由下列指令偵測。一般而言，常數（tf.constant）放在 CPU，變數（Variable）則放在 GPU 上，兩者加總時，會將常數搬到 GPU 上再加總，過程不需人為操作。但請注意，**PyTorch 套件稍嫌麻煩，必須手動將變數搬移至 CPU 或 GPU 運算，不允許一個變數在 CPU，另一個變數在 GPU**。

```
1  x1 = tf.constant([[1, 2, 3]], dtype=float)
2  print("x1 是否在 GPU #0 上: ", x1.device.endswith('GPU:0'))
```

第 3 章　TensorFlow 架構與主要功能

```
3
4   # 設定 x 為均勻分配亂數 3x3
5   x2 = tf.random.uniform([3, 3])
6   print("x2 是否在 GPU #0 上: ", x2.device.endswith('GPU:0'))
7
8   x3=x1+x2
9   print("x3 是否在 GPU #0 上: ", x3.device.endswith('GPU:0'))
```

- 執行結果：

    x1 是否在 GPU #0 上： False

    x2 是否在 GPU #0 上： True

    x3 是否在 GPU #0 上： True

8. 也能夠使用 with tf.device（"CPU:0"）或 with tf.device（"GPU:0"），強制指定在 CPU 或 GPU 運算。

```
1   import time
2
3   # 計算 10 次的時間
4   def time_matmul(x):
5       start = time.time()
6       for loop in range(10):
7           tf.matmul(x, x)
8
9       result = time.time()-start
10      print("{:0.2f}ms".format(1000*result))
11
12  # 強制指定在cpu運算
13  print("On CPU:", end='')
14  with tf.device("CPU:0"):
15      x = tf.random.uniform([1000, 1000])
16      assert x.device.endswith("CPU:0")
17      time_matmul(x)
18
19  # 強制指定在gpu運算
20  if tf.config.list_physical_devices("GPU"):
21      print("On GPU:", end='')
22      with tf.device("GPU:0"):
23          x = tf.random.uniform([1000, 1000])
24          assert x.device.endswith("GPU:0")
25          time_matmul(x)
```

- 第一次執行結果如下，CPU 運算比 GPU 快速：

    On CPU: 64.00ms

    On GPU: 311.49ms

3-3 張量（Tensor）運算

- 多次執行後，GPU 反而比 CPU 運算快很多：

    On CPU:58.00ms

    On GPU:1.00ms

9. 稀疏矩陣（Sparse Matrix）運算：稀疏矩陣是指矩陣內只有很少數的非零元素，如果依一般的矩陣儲存會非常浪費記憶體，運算也是如此，因為大部份項目為零，不需浪費時間計算，所以，科學家針對稀疏矩陣設計特殊的資料儲存結構及運算演算法，TensorFlow 也支援此類資料型態。

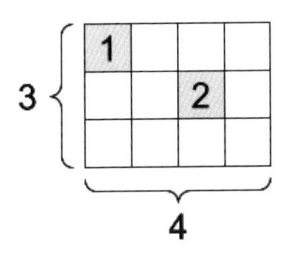

▲ 圖 3.4 稀疏矩陣（Sparse Matrix）

10. TensorFlow 稀疏矩陣只需設定有值的位置（indices）和數值（values），並設定維度（dense_shape）如下：

```
1  # 稀疏矩陣只需設定有值的位置及數值
2  sparse_tensor = tf.sparse.SparseTensor(indices=[[0, 0], [1, 2]],
3                                          values=[1, 2],
4                                          dense_shape=[3, 4])
5  print(sparse_tensor)
```

- 執行結果：

```
SparseTensor(indices=tf.Tensor(
[[0 0]
 [1 2]], shape=(2, 2), dtype=int64), values=tf.Tensor([1 2], shape=(2,), dtype=int32), dense_shape=tf.Tensor([3 4], shape=(2,), dtype=int64))
```

11. 轉為正常的矩陣格式。

```
1  # 轉為正常的矩陣格式
2  x = tf.sparse.to_dense(sparse_tensor)
3  print(type(x))
4
5  # 2.31 以前版本會出錯
6  x.numpy()
```

3-9

- 執行結果：

```
<class 'tensorflow.python.framework.ops.EagerTensor'>
array([[1, 0, 0, 0],
       [0, 0, 2, 0],
       [0, 0, 0, 0]])
```

12. 如果要執行 TensorFlow 1.x 版 Graph Execution Mode 的程式，須禁用（Dis-able）2.x 版的功能，並改變載入套件的命名空間（Namespace）。

```
1  if tf.__version__[0] != '1':           # 是否為 TensorFlow 1.x版
2      import tensorflow.compat.v1 as tf  # 改變載入套件的命名空間(Namespace)
3      tf.disable_v2_behavior()           # 使 2.x 版功能失效(Disable)
```

TensorFlow 大刀闊斧的改良反而造成與舊版相容性差的困擾，1.x 版的程式在 2.x 版的預設模式下均無法執行，雖然可以把 2.x 版的預設模式切換回 1.x 版的模式，但是要自行修改的地方還是蠻多的，而且也缺乏未來性，因此，在這裡建議大家：

- Eager Execution Mode 已是 TensorFlow 的主流，不要再使用 1.x 版的 Session 或 TFLearn 等舊的架構，套句電視劇對白，**已經回不去了**。

- 手上有許多 1.x 版的程式，如果很重要，非用不可，可利用官網移轉（Migration）指南[7] 進行修改，單一檔案比較可行，但若是複雜的套件，可能就要花上大半天的時間了。

- 官網也有提供指令，能夠一次升級整個目錄的所有程式，如下：**tf_up-grade_v2 --intree <1.x 版程式目錄> --outtree <輸出目錄>**

  詳細使用方法可參閱 TensorFlow 官網升級指南[8]。

13. 禁用 2.x 版的功能後，測試 1.x 版程式，Graph Execution Mode 程式須使用 tf.Session：

```
1  # 測試1.x版程式
2  x = tf.constant([[1, 2]])
3  neg_x = tf.negative(x)
4
5  with tf.Session() as sess:     # 使用 session
6      result = sess.run(neg_x)
7      print(result)
```

14. GPU 記憶體管理：由於 TensorFlow 對於 GPU 記憶體的垃圾回收（Garbage Collection）機制並不完美，因此，常會出現下列 GEMM 錯誤：

```
InternalError:  Blas GEMM launch failed : a.shape=(32, 784), b.shape=(784, 256), m=32, n=256, k=784
        [[node sequential/dense/MatMul (defined at <ipython-input-3-9d42ad511782>:5) ]] [Op:__inference_train_function_581]
```

此訊息表示 GPU 記憶體不足，尤其是使用 Jupyter Notebook 時，因為 Jupyter Notebook 是一個網頁程式，關掉某一個 Notebook 檔案，網站仍然在執行中，所以不代表該檔案的資源會被回收，通常要選擇『**Kernel > Restart**』選單才會回收資源。另一個方法，就是限制 GPU 的使用配額，以下是 TensorFlow 2.x 版的方式，1.x 版並不適用：

```
1   # 限制 TensorFlow 只能使用 GPU 2GB 記憶體
2   gpus = tf.config.experimental.list_physical_devices('GPU')
3   if gpus:
4       try:
5           # 限制 第一顆 GPU 只能使用 2GB 記憶體
6           tf.config.experimental.set_virtual_device_configuration(gpus[0],
7               [tf.config.experimental.VirtualDeviceConfiguration(memory_limit=1024*2)])
8
9           # 顯示 GPU 個數
10          logical_gpus = tf.config.experimental.list_logical_devices('GPU')
11          print(len(gpus), "Physical GPUs,", len(logical_gpus), "Logical GPUs")
12      except RuntimeError as e:
13          # 顯示錯誤訊息
14          print(e)
```

15. 也可以不使用 GPU：

```
1   import os
2
3   os.environ["CUDA_VISIBLE_DEVICES"] = "-1"
```

3-4 自動微分（Automatic Differentiation）

反向傳導時，會更新每一層的權重，這時就輪到偏微分運算派上用場，所以，深度學習套件的第二項主要功能就是自動微分（Automatic Differentiation）。

第 3 章　TensorFlow 架構與主要功能

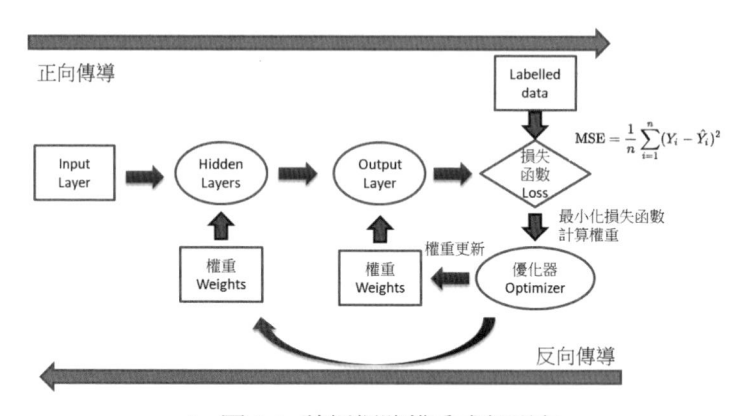

▲ 圖 3.5　神經網路權重求解過程

範例 1. 自動微分測試。

程式：**03_2_ 自動微分 .ipynb**。

1. 呼叫 tf.GradientTape() 函數可自動微分，再呼叫 g.gradient（y, x）可取得 y 對 x 作偏微分的梯度。

```python
import numpy as np
import tensorflow as tf

x = tf.Variable(3.0)            # 宣告TensorFlow變數(Variable)

with tf.GradientTape() as g:    # 自動微分
    y = x * x                   # y = x^2

dy_dx = g.gradient(y, x)        # 取得梯度，f'(x) = 2x, x=3 ==> 6

print(dy_dx.numpy())            # 轉換為 NumPy array 格式
```

- 執行結果：

    $f(x) = x^2$

    $f(x) = 2x$

    f'（3）= 2 * 3 = 6。

2. 宣告為變數（tf.Variable）時，該變數會自動參與自動微分，但宣告為常數（tf.constant）時，如欲參與自動微分，須額外設定 g.watch()。

3-4 自動微分（Automatic Differentiation）

```
1  import numpy as np
2  import tensorflow as tf
3
4  x = tf.constant(3.0)              # 宣告 TensorFlow 常數
5
6  with tf.GradientTape() as g:      # 自動微分
7      g.watch(x)                    # 設定常數參與自動微分
8      y = x * x                     # y = x^2
9
10 dy_dx = g.gradient(y, x)          # 取得梯度，f'(x) = 2x, x=3 ==> 6
11
12 print(dy_dx.numpy())              # 轉換為 NumPy array 格式
```

- 執行結果：與上面相同。

3. 計算二階導數：呼叫 tf.GradientTape()、g.gradient（y, x）函數兩次，即能取得二階導數。

```
1  x = tf.constant(3.0)                      # 宣告 TensorFlow 常數
2  with tf.GradientTape() as g:
3      g.watch(x)
4      with tf.GradientTape() as gg:         # 自動微分
5          gg.watch(x)                       # 設定常數參與自動微分
6          y = x * x                         # y = x^2
7
8      dy_dx = gg.gradient(y, x)             # 一階導數
9  d2y_dx2 = g.gradient(dy_dx, x)            # 二階導數
10
11 print(f'一階導數={dy_dx.numpy()}, 二階導數={d2y_dx2.numpy()}')
```

- 執行結果：一階導數 =6.0, 二階導數 =2.0

    $f(x) = x^2$

    $f(x) = 2x$

    $f''(x) = 2$。

    $f''(3) = 2$。

4. 多變數計算導數：各自呼叫 g.gradient（y, x）函數，可取得每一個變數的梯度。若呼叫 g.gradient() 兩次或以上，tf.GradientTape() 須加參數 persistent=True，使 tf.GradientTape() 不會被自動回收，用完之後，可使用『del g』刪除 GradientTape 物件。

第 3 章　TensorFlow 架構與主要功能

```
1  x = tf.Variable(3.0)              # 宣告 TensorFlow 常數
2  with tf.GradientTape(persistent=True) as g:  # 自動微分
3      y = x * x                     # y = x^2
4      z = y * y                     # z = y^2
5
6  dz_dx = g.gradient(z, x)          # 4*x^3
7  dy_dx = g.gradient(y, x)          # 2*x
8
9  del g                             # 不用時可刪除 GradientTape 物件
10
11 print(f'dy/dx={dy_dx.numpy()}, dz/dx={dz_dx.numpy()}')
```

- 執行結果：dy/dx=6, dz/dx=108。

    $z = f(x) = y^2 = x^4$

    ➔ $f(x) = 4x^3$

    ➔ f'（3）= 108。

5. 順便認識一下 PyTorch 自動微分的語法，與 TensorFlow 稍有差異。

```
1  import torch          # 載入套件
2
3  x = torch.tensor(3.0, requires_grad=True)  # 設定 x 參與自動微分
4  y=x*x                 # y = x^2
5
6  y.backward()          # 反向傳導
7
8  print(x.grad)         # 取得梯度
```

- requires_grad=True 參數宣告 x 參與自動微分。
- 呼叫 y.backward()，要求作反向傳導。
- 呼叫 x.grad 取得梯度。

範例 2. 利用 TensorFlow 自動微分求解簡單線性迴歸的參數（w、b）。

程式：**03_3_ 簡單線性迴歸 .ipynb**。

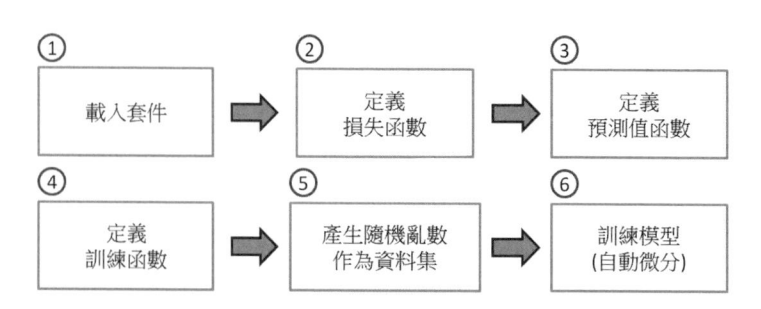

① 載入套件 → ② 定義 損失函數 → ③ 定義 預測值函數

④ 定義 訓練函數 → ⑤ 產生隨機亂數 作為資料集 → ⑥ 訓練模型 (自動微分)

3-4 自動微分（Automatic Differentiation）

1. 載入套件。

```
1  # 載入套件
2  import numpy as np
3  import tensorflow as tf
```

2. 定義損失函數 $MSE = \dfrac{\sum(y - \hat{y})^2}{n}$。

```
1  # 定義損失函數
2  def loss(y, y_pred):
3      return tf.reduce_mean(tf.square(y - y_pred))
```

3. 定義預測值函數 y = w x + b。

```
1  # 定義預測值函數
2  def predict(X):
3      return w * X + b
```

4. 定義訓練函數：在自動微分中需重新計算損失函數值，assign_sub 函數相當於『 -= 』。

```
1   # 定義訓練函數
2   def train(X, y, epochs=40, lr=0.0001):
3       current_loss=0                              # 損失函數值
4       for epoch in range(epochs):                 # 執行訓練週期
5           with tf.GradientTape() as t:            # 自動微分
6               t.watch(tf.constant(X))             # 宣告 TensorFlow 常數參與自動微分
7               current_loss = loss(y, predict(X))  # 計算損失函數值
8
9           dw, db = t.gradient(current_loss, [w, b])  # 取得 w, b 個別的梯度
10
11          # 更新權重：新權重 = 原權重 – 學習率(learning_rate) * 梯度(gradient)
12          w.assign_sub(lr * dw) # w -= lr * dw
13          b.assign_sub(lr * db) # b -= lr * db
14
15          # 顯示每一訓練週期的損失函數
16          print(f'Epoch {epoch}: Loss: {current_loss.numpy()}')
```

5. 產生隨機亂數作為資料集，進行測試。

```
1  # 產生線性隨機資料100筆，介於 0-50
2  n = 100
3  X = np.linspace(0, 50, n)
4  y = np.linspace(0, 50, n)
5
6  # 資料加一點雜訊(noise)
7  X += np.random.uniform(-10, 10, n)
8  y += np.random.uniform(-10, 10, n)
```

6. 執行訓練。

```
1  # w、b 初始值均設為 0
2  w = tf.Variable(0.0)
3  b = tf.Variable(0.0)
4
5  # 執行訓練
6  train(X, y)
7
8  # w、b 的最佳解
9  print(f'w={w.numpy()}, b={b.numpy()}')
```

- 執行結果：w=0.9464, b=0.0326。

- 損失函數值隨著訓練週期越來越小，如下圖。

```
Epoch 0: Loss: 890.1063232421875
Epoch 1: Loss: 607.071533203125
Epoch 2: Loss: 419.7763671875
Epoch 3: Loss: 295.8358459472656
Epoch 4: Loss: 213.81948852539062
Epoch 5: Loss: 159.5459747314453
Epoch 6: Loss: 123.63106536865234
Epoch 7: Loss: 99.86465454101562
Epoch 8: Loss: 84.1374282836914
Epoch 9: Loss: 73.7300033569336
Epoch 10: Loss: 66.84292602539062
Epoch 11: Loss: 62.28538513183594
Epoch 12: Loss: 59.269386291503906
Epoch 13: Loss: 57.27349090576172
Epoch 14: Loss: 55.95263671875
Epoch 15: Loss: 55.078487396240234
Epoch 16: Loss: 54.49993133544922
Epoch 17: Loss: 54.116981506347656
Epoch 18: Loss: 53.86347579956055
Epoch 19: Loss: 53.69562911987305
Epoch 20: Loss: 53.584468841552734
```

4. 顯示結果：迴歸線確實居於樣本點中線。

```
1  import matplotlib.pyplot as plt
2
3  plt.scatter(X, y, label='data')
4  plt.plot(X, predict(X), 'r-', label='predicted')
5  plt.legend()
```

- 執行結果：

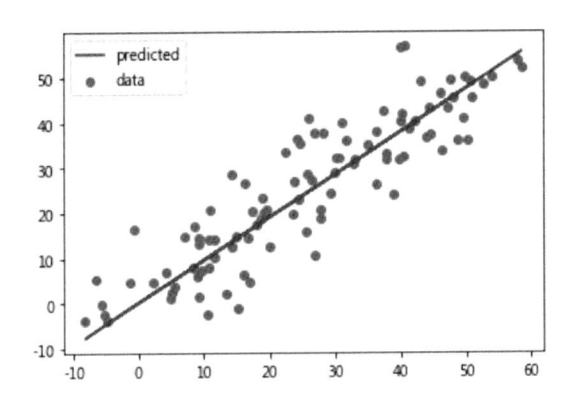

有了 TensorFlow 自動微分的功能，正向與反向傳導變的非常簡單，只要熟悉運作的架構，後續複雜的模型也可以運用自如。

注意，若出現以下錯誤：

OMP: Error #15: Initializing libiomp5md.dll, but found libiomp5md.dll already initialized.

請在第一格執行：

import os

os.environ["KMP_DUPLICATE_LIB_OK"]="TRUE"

3-5 神經層（Neural Network Layer）

上一節運用自動微分實現一條簡單線性迴歸線的求解，然而神經網路是多條迴歸線的組合，並且每一條迴歸線可能再乘上非線性的 Activation Function，假如使用自動微分函數逐一定義每條公式，層層串連，程式可能要很多個迴圈才能完成。所以為了簡化程式開發的複雜度，TensorFlow/Keras 直接建構各式各樣的神經層（Layer）函數，可以使用神經層組合神經網路的結構，我們只需要專注在演算法的設計即可，輕鬆不少。

神經網路是多個神經層組合而成的，如下圖，包括輸入層（Input Layer）、隱藏層（Hidden Layer）及輸出層（Output Layer），其中隱藏層可以有任意多層，一般而言，隱藏層大於或等於兩層，即稱為『深度』（Deep）學習。

神經網路(Neural Network)

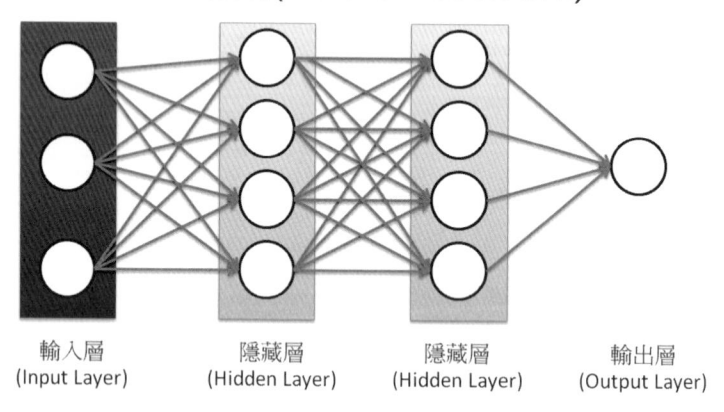

▲ 圖 3.6 神經網路示意圖

TensorFlow/Keras 提供數十種神經層，分成下列類別，可參閱 Keras 官網說明（https://keras.io/api/layers/）：

1. 核心類別（Core Layer）：包括完全連接層（Full Connected Layer）、Activation layer、嵌入層（Embedding layer）⋯等。
2. 卷積層（Convolutional Layer）。
3. 池化層（Pooling Layer）。
4. 循環層（Recurrent Layer）。
5. 前置處理層（Preprocessing layer）：提供 one-hot encoding、影像前置處理、資料增補（Data augmentation）⋯等。

我們先來測試兩個最簡單的『完全連接層』範例。

範例 1. 使用完全連接層估算簡單線性迴歸的參數（w、b）。

程式：03_4_ 簡單的完全連階層 .ipynb。

1. 產生隨機資料，與上一節範例相同。

```
1  # 載入套件
2  import numpy as np
3  import tensorflow as tf
4
5  # 產生線性隨機資料100筆，介於 0-50
6  n = 100
```

3-5 神經層（Neural Network Layer）

```
 7  X = np.linspace(0, 50, n)
 8  y = np.linspace(0, 50, n)
 9
10  # 資料加一點雜訊(noise)
11  X += np.random.uniform(-10, 10, n)
12  y += np.random.uniform(-10, 10, n)
```

2. 建立模型：神經網路僅使用 1 層完全連接層，而且輸入只有 1 個神經元，即 X，輸出也只有 1 個神經元，即 y。Dense 本身有一個參數 use_bias，即是否有偏差項，預設值為 True，除了一個神經元輸出外，還會有一個偏差項。以上設定其實就等於 y=wx+b。為聚焦概念的說明，暫時不解釋其他參數，後面章節會有詳盡說明。

```
1  # 定義完全連接層(Dense)
2  # units：輸出神經元個數，input_shape：輸入神經元個數
3  layer1 = tf.keras.layers.Input([1])
4  layer2 = tf.keras.layers.Dense(units=1)
5
6  # 神經網路包含一層完全連接層
7  model = tf.keras.Sequential([layer1, layer2])
```

3. 定義模型的損失函數（loss）及優化器（optimizer）。

```
1  # 定義模型的損失函數(loss)為 MSE，優化器(optimizer)為 adam
2  model.compile(loss='mean_squared_error',
3                optimizer=tf.keras.optimizers.Adam())
```

4. 模型訓練：只需一個指令 model.fit（X, y）即可，訓練過程的損失函數變化都會存在 history 變數中。

```
1  history = model.fit(X, y, epochs=500, verbose=False)
```

5. 訓練過程繪圖。

```
1  import matplotlib.pyplot as plt
2
3  # 修正中文亂碼
4  plt.rcParams['font.sans-serif'] = ['Arial Unicode MS']
5  plt.rcParams['axes.unicode_minus'] = False
6
7  plt.xlabel('訓練週期', fontsize=20)
8  plt.ylabel("損失函數(loss)", fontsize=20)
9  plt.plot(history.history['loss'])
```

第 3 章　TensorFlow 架構與主要功能

- 執行結果：損失函數值隨著訓練週期越來越小。

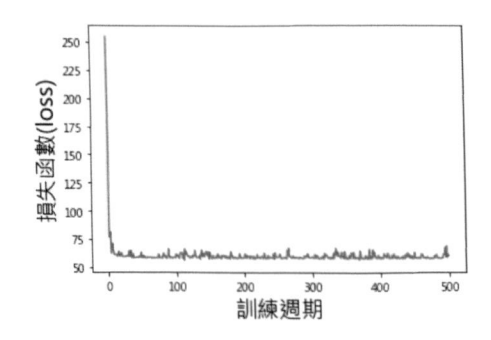

6. 取得模型參數 w= 第 2 層的第一個參數，b 為輸出層的第一個參數。

```
1  w = layer2.get_weights()[0][0][0]
2  b = layer2.get_weights()[1][0]
3
4  print(f"w : {w:.4f} , b : {b:.4f}")
```

- 執行結果：w：0.8798，b：3.5052，因輸入資料為隨機亂數。

7. 繪圖顯示迴歸線。

```
1  import matplotlib.pyplot as plt
2
3  plt.scatter(X, y, label='data')
4  plt.plot(X, X * w + b, 'r-', label='predicted')
5  plt.legend()
```

- 執行結果：

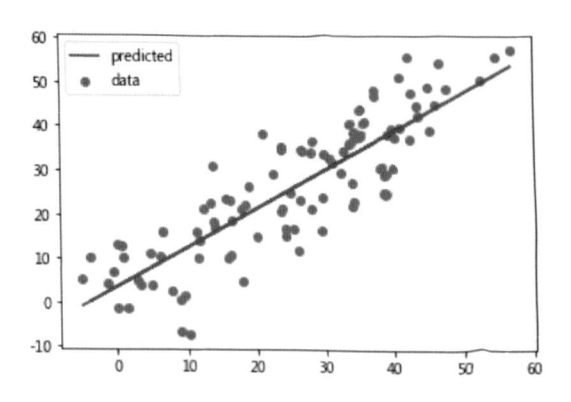

3-20

3-5 神經層（Neural Network Layer）

與自動微分比較，程式更簡單，只要設定模型結構、損失函數、優化器後，呼叫訓練（fit）函數即可。

再看一個有趣的例子，利用神經網路自動求出華氏與攝氏溫度的換算公式。

範例 2. 使用完全連接層推算華氏與攝氏溫度的換算公式。

華氏（F）= 攝氏（C）*（9/5）+ 32

程式：03_5_ 華氏與攝氏溫度換算 .ipynb。

1. 利用換算公式，隨機產生 151 筆資料。

```
1  # 載入套件
2  import numpy as np
3  import tensorflow as tf
4
5  # 隨機產生151筆資料
6  n = 151
7  C = np.linspace(-50, 100, n)
8  F = C * (9/5) + 32
9
10 for i, x in enumerate(C):
11     print(f"華氏(F):{F[i]:.2f} , 攝氏(C):{x:.0f}")
```

2. 建立模型：神經網路只有一層完全連接層，而且輸入只有一個神經元，即攝氏溫度，輸出只有一個神經元，即華氏溫度。

```
1  # 定義完全連接層(Dense)
2  # units：輸出神經元個數，input_shape：輸入神經元個數
3  layer1 = tf.keras.layers.Input([1])
4  layer2 = tf.keras.layers.Dense(units=1)
5
6  # 神經網路包含一層完全連接層
7  model = tf.keras.Sequential([layer1, layer2])
8
9  # 定義模型的損失函數(Loss)為 MSE，優化器(optimizer)為 adam
10 model.compile(loss='mean_squared_error',
11               optimizer=tf.keras.optimizers.Adam(0.1))
```

3. 模型訓練。

```
1  history = model.fit(X, y, epochs=500, verbose=False)
```

4. 訓練過程繪圖。

```
1  import matplotlib.pyplot as plt
2
3  # 修正中文亂碼
4  plt.rcParams['font.sans-serif'] = ['Microsoft JhengHei'] # 微軟正黑體
5  plt.rcParams['axes.unicode_minus'] = False
6
7  plt.figure(figsize=(6,4))
8  plt.xlabel('訓練週期', fontsize=14)
9  plt.ylabel("損失函數(loss)", fontsize=14)
10 plt.plot(history.history['loss'])
11 plt.show()
```

- 執行結果如下圖：損失函數值隨著訓練週期越來越小。

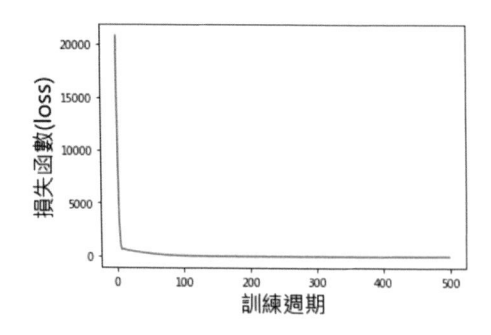

5. 測試：輸入攝氏 100 度及 0 度預測華氏溫度，答案完全正確。

```
1  y_pred = model.predict(np.array([100.0]), verbose=False)[0][0]
2  print(f"華氏(F)：{y_pred:.2f} , 攝氏(C)：100")
3
4  y_pred = model.predict(np.array([0.0]), verbose=False)[0][0]
5  print(f"華氏(F)：{y_pred:.2f} , 攝氏(C)：0")
```

- 執行結果：

  華氏（F）：212.00 , 攝氏（C）：100

  華氏（F）：32.00 , 攝氏（C）：0

6. 取得模型參數 w、b。

```
1  w = layer2.get_weights()[0][0][0]
2  b = layer2.get_weights()[1][0]
3
4  print(f"w：{w:.4f} , b：{b:.4f}")
```

- 執行結果：w：1.8000，b：31.9999，近似下列華氏與攝氏溫度的換算公式，實在太神奇了。

  華氏（F）＝攝氏（C）＊（9/5）＋32

- 其實換算公式也是一條迴歸線，想通了，對於結果就不足為奇了。

讀者讀到這裡，應該會好奇想知道如何使用更多的神經元和神經層，甚至更複雜的神經網路結構，下一章我們將正式邁入深度學習的殿堂，學習如何用 TensorFlow 解決各種實際的案例，並且詳細剖析各個函數的用法及參數說明。

參考資料 (References)

[1]　維基百科中針對常見的深度學習套件之比較圖表
(https://en.wikipedia.org/wiki/Comparison_of_deep-learning_software)

[2]　Amol Mavuduru ,《Which deep learning framework is the best?》, 2020
(https://towardsdatascience.com/which-deep-learning-framework-is-the-best-eb51431c39a)

[4]　開發者傳授 PyTorch 秘笈

[5]　同時搞定 TensorFlow、PyTorch
(https://ithelp.ithome.com.tw/articles/10285361)

[5]　TensorFlow 官方 GitHub
(https://github.com/tensorflow/docs/blob/master/site/en/r1/guide/extend/architecture.md)

[6]　Sonu Sharma ,《Explained: Deep Learning in Tensorflow》, 2019
(https://towardsdatascience.com/explained-deep-learning-in-tensorflow-chapter-1-9ab389fe90a1)

第 3 章　TensorFlow 架構與主要功能

[7]　TensorFlow 官網移轉指南
　　　(https://www.tensorflow.org/guide/migrate)

[8]　TensorFlow 官網升級指南
　　　(https://www.tensorflow.org/guide/upgrade)

第 4 章

神經網路實作

接下來以神經網路實作各種應用,可以暫時跟數學／統計說再見,會著重於概念的澄清與程式的撰寫,筆者盡可能的運用大量圖解,幫助讀者迅速掌握各種演算法的原理,避免流於長篇大論。

同時也會藉由『手寫阿拉伯數字辨識』的案例,實作機器學習流程的 10 大步驟,並詳細解說構建神經網路的函數用法及各項參數代表的意義,到最後我們會撰寫一個完整的視窗介面程式及網頁程式,讓終端使用者(End User)親身體驗 AI 應用程式,期望激發使用者對『企業導入 AI』有更多的發想。

第 4 章　神經網路實作

4-1 撰寫第一支神經網路程式

▲ 圖 4.1　MNIST 手寫阿拉伯數字資料集

辨識手寫阿拉伯數字辨識的程序如下：

1. 讀取手寫阿拉伯數字的影像，將圖片中的每個像素當成一個特徵。資料來源為 MNIST 機構所收集的 60000 筆訓練資料，另含 10000 筆測試資料，每筆資料是一個阿拉伯數字的圖片，寬高均為 28 個點。

2. 建立神經網路模型，利用梯度下降法，求解模型的參數值，一般稱為權重（Weight）。

3. 依照模型推算每一個圖片是 0~9 的機率，再以最大機率者為預測結果。

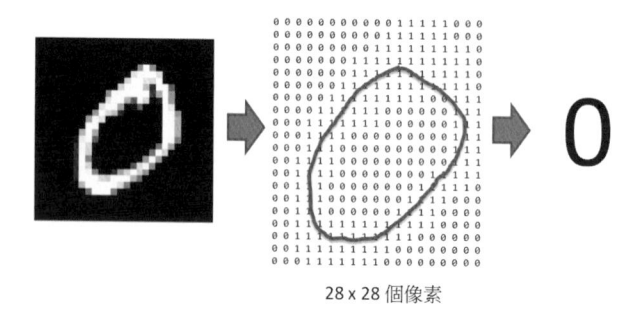

28 x 28 個像素

▲ 圖 4.2　手寫阿拉伯數字辨識，左圖為輸入的圖形，
中間為圖形的像素，右圖為預測結果

4-1-1 最簡短的程式

Tensorflow 官網在首頁展示一支超短程式，示範如何撰寫手寫阿拉伯數字的辨識，要證明 TensorFlow 簡單好用，現在我們就來看看這支程式。

4-1 撰寫第一支神經網路程式

範例 1. TensorFlow 官網的手寫阿拉伯數字辨識。

程式：**04_01_ 手寫阿拉伯數字辨識.ipynb**：

```python
import tensorflow as tf
mnist = tf.keras.datasets.mnist

# 匯入 MNIST 手寫阿拉伯數字 訓練資料
(x_train, y_train),(x_test, y_test) = mnist.load_data()

# 特徵縮放至 (0, 1) 之間
x_train, x_test = x_train / 255.0, x_test / 255.0

# 建立模型
model = tf.keras.models.Sequential([
    tf.keras.layers.Input((28, 28)),
    tf.keras.layers.Flatten(),
    tf.keras.layers.Dense(128, activation='relu'),
    tf.keras.layers.Dropout(0.2),
    tf.keras.layers.Dense(10, activation='softmax')
])

# 設定優化器(optimizer)、損失函數(loss)、效能衡量指標(metrics)
model.compile(optimizer='adam',
              loss='sparse_categorical_crossentropy',
              metrics=['accuracy'])

# 模型訓練，epochs：執行週期，validation_split：驗證資料佔 20%
model.fit(x_train, y_train, epochs=5, validation_split=0.2)

# 模型評估
model.evaluate(x_test, y_test)
```

- 執行結果如下：

```
Epoch 1/5
1500/1500 [==============================] - 5s 3ms/step - loss: 0.5336 - accuracy: 0.8432 - val_loss: 0.1558 - val_accuracy: 0.9555
Epoch 2/5
1500/1500 [==============================] - 5s 3ms/step - loss: 0.1676 - accuracy: 0.9505 - val_loss: 0.1134 - val_accuracy: 0.9657
Epoch 3/5
1500/1500 [==============================] - 5s 3ms/step - loss: 0.1203 - accuracy: 0.9646 - val_loss: 0.0975 - val_accuracy: 0.9704
Epoch 4/5
1500/1500 [==============================] - 5s 3ms/step - loss: 0.0981 - accuracy: 0.9703 - val_loss: 0.0968 - val_accuracy: 0.9717
Epoch 5/5
1500/1500 [==============================] - 5s 3ms/step - loss: 0.0786 - accuracy: 0.9758 - val_loss: 0.0958 - val_accuracy: 0.9713
313/313 [==============================] - 1s 3ms/step - loss: 0.0807 - accuracy: 0.9752

[0.08072374016046524, 0.9751999974250793]
```

第 4 章　神經網路實作

上述程式扣除註解僅 10 多行，辨識的準確率高達 97~98%，真的厲害！每一行程式碼會在後續逐一詳細說明。

4-1-2　程式強化

上一節的範例是官網為了炫技，刻意縮短了程式，本節將會按照機器學習流程的 10 大步驟，撰寫完整程式，針對每個步驟仔細解析，請務必理解每一行程式背後代表的意涵。

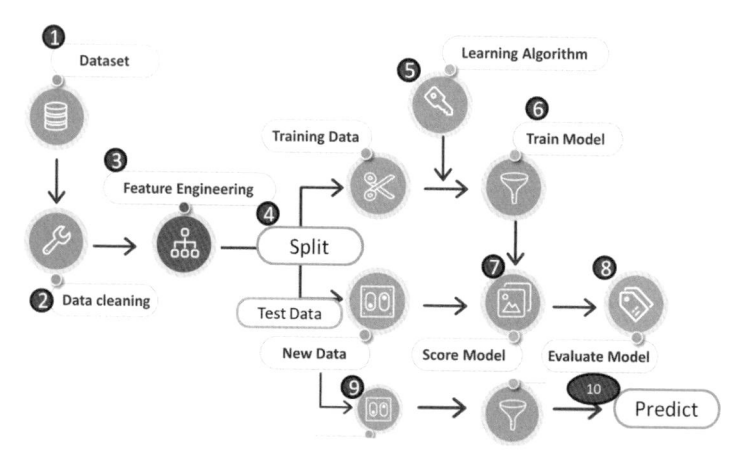

▲ 圖 4.2　機器學習流程 10 大步驟

範例 2. 依據上圖 10 大步驟撰寫手寫阿拉伯數字辨識。

程式：**04_02_ 手寫阿拉伯數字辨識 _ 完整版 .ipynb**。

1. 步驟 1：載入 MNIST 手寫阿拉伯數字資料集

```
1  import tensorflow as tf
2  mnist = tf.keras.datasets.mnist
3
4  # 載入 MNIST 手寫阿拉伯數字資料
5  (x_train, y_train),(x_test, y_test) = mnist.load_data()
6
7  # 訓練/測試資料的 X/y 維度
8  print(x_train.shape, y_train.shape,x_test.shape, y_test.shape)
```

- 執行結果：取得 60000 筆訓練資料，10000 筆測試資料，每筆資料是一個阿拉伯數字，寬高各為 (28, 28) 的點陣圖形，要注意資料的維度及其大小，必須與模型的輸入規格契合。

4-1 撰寫第一支神經網路程式

```
(60000, 28, 28) (60000,) (10000, 28, 28) (10000,)
```

2. 步驟 2：對資料集進行探索與分析（EDA），先觀察訓練資料的目標值（y），即圖片的真實答案（Ground truth）。

```
1  # 訓練資料前10筆圖片的數字
2  y_train[:10]
```

- 執行結果如下，每筆資料答案是一個阿拉伯數字。

```
array([5, 0, 4, 1, 9, 2, 1, 3, 1, 4], dtype=uint8)
```

3. 列印第一筆訓練資料的像素 (x)。

```
1  # 顯示第1張圖片內含值
2  x_train[0]
```

- 執行結果如下，每筆像素的值介於（0, 255）之間，為灰階影像，0 為白色，255 為最深的黑色，**注意，這與 RGB 色碼剛好相反，RGB 黑色為 0，白色為 255**。

```
       [  0,   0,   0,   0,   0,   0,   0,   0,   0,   0,   0,   0,   0,
          0,   0,   0,   0,   0,   0,   0,   0,   0,   0,   0,   0,   0,
          0,   0],
       [  0,   0,   0,   0,   0,   0,   0,   0,   0,   0,   0,   0,   3,
         18,  18,  18, 126, 136, 175,  26, 166, 255, 247, 127,   0,   0,
          0,   0],
       [  0,   0,   0,   0,   0,   0,   0,   0,  30,  36,  94, 154, 170,
        253, 253, 253, 253, 253, 225, 172, 253, 242, 195,  64,   0,   0,
          0,   0],
       [  0,   0,   0,   0,   0,   0,   0,  49, 238, 253, 253, 253, 253,
        253, 253, 253, 253, 251,  93,  82,  82,  56,  39,   0,   0,   0,
          0,   0],
       [  0,   0,   0,   0,   0,   0,   0,  18, 219, 253, 253, 253, 253,
        253, 198, 182, 247, 241,   0,   0,   0,   0,   0,   0,   0,   0,
          0,   0],
       [  0,   0,   0,   0,   0,   0,   0,   0,  80, 156, 107, 253, 253,
        205,  11,   0,  43, 154,   0,   0,   0,   0,   0,   0,   0,   0,
          0,   0],
```

4. 為了看清楚圖片的手寫的數字，將非 0 的數值轉為 1，變為黑白兩色（Binary）的圖片。

```
1  # 將非0的數字轉為1，顯示第1張圖片
2  data = x_train[0].copy()
3  data[data>0]=1
4
```

第 4 章 神經網路實作

```
5  # 將轉換後二維內容顯示出來，隱約可以看出數字為 5
6  text_image=[]
7  for i in range(data.shape[0]):
8      text_image.append(''.join(str(data[i])))
9  text_image
```

- 執行結果如下，筆者以筆描繪 1 的範圍，隱約可以看出是 5。

```
['[0 0 0 0 0 0 0 0 0 0 0 0 0 0 0 0 0 0 0 0 0 0 0 0 0 0 0 0]',
 '[0 0 0 0 0 0 0 0 0 0 0 0 0 0 0 0 0 0 0 0 0 0 0 0 0 0 0 0]',
 '[0 0 0 0 0 0 0 0 0 0 0 0 0 0 0 0 0 0 0 0 0 0 0 0 0 0 0 0]',
 '[0 0 0 0 0 0 0 0 0 0 0 0 0 0 0 0 0 0 0 0 0 0 0 0 0 0 0 0]',
 '[0 0 0 0 0 0 0 0 0 0 0 0 0 0 0 0 0 0 0 0 0 0 0 0 0 0 0 0]',
 '[0 0 0 0 0 0 0 0 0 0 0 1 1 1 1 1 1 1 1 1 1 1 0 0 0 0 0 0]',
 '[0 0 0 0 0 0 0 1 1 1 1 1 1 1 1 1 1 1 1 1 1 1 0 0 0 0 0 0]',
 '[0 0 0 0 0 0 1 1 1 1 1 1 1 1 1 1 1 1 1 1 0 0 0 0 0 0 0 0]',
 '[0 0 0 0 0 0 1 1 1 1 1 1 1 1 0 1 1 0 0 0 0 0 0 0 0 0 0 0]',
 '[0 0 0 0 0 0 0 1 1 1 1 1 1 0 0 0 0 0 0 0 0 0 0 0 0 0 0 0]',
 '[0 0 0 0 0 0 0 0 1 1 1 1 0 0 0 0 0 0 0 0 0 0 0 0 0 0 0 0]',
 '[0 0 0 0 0 0 0 0 0 1 1 1 0 0 0 0 0 0 0 0 0 0 0 0 0 0 0 0]',
 '[0 0 0 0 0 0 0 0 0 0 1 1 1 1 0 0 0 0 0 0 0 0 0 0 0 0 0 0]',
 '[0 0 0 0 0 0 0 0 0 0 0 1 1 1 1 0 0 0 0 0 0 0 0 0 0 0 0 0]',
 '[0 0 0 0 0 0 0 0 0 0 0 0 1 1 1 1 0 0 0 0 0 0 0 0 0 0 0 0]',
 '[0 0 0 0 0 0 0 0 0 0 0 0 0 1 1 1 1 0 0 0 0 0 0 0 0 0 0 0]',
 '[0 0 0 0 0 0 0 0 0 0 0 0 0 0 1 1 1 0 0 0 0 0 0 0 0 0 0 0]',
 '[0 0 0 0 0 0 0 0 0 0 0 0 1 1 1 1 1 0 0 0 0 0 0 0 0 0 0 0]',
 '[0 0 0 0 0 0 0 0 0 1 1 1 1 1 1 1 0 0 0 0 0 0 0 0 0 0 0 0]',
 '[0 0 0 0 0 0 1 1 1 1 1 1 1 1 0 0 0 0 0 0 0 0 0 0 0 0 0 0]',
 '[0 0 0 0 1 1 1 1 1 1 1 1 0 0 0 0 0 0 0 0 0 0 0 0 0 0 0 0]',
 '[0 0 0 0 1 1 1 1 1 1 0 0 0 0 0 0 0 0 0 0 0 0 0 0 0 0 0 0]',
 '[0 0 0 0 0 0 0 0 0 0 0 0 0 0 0 0 0 0 0 0 0 0 0 0 0 0 0 0]',
 '[0 0 0 0 0 0 0 0 0 0 0 0 0 0 0 0 0 0 0 0 0 0 0 0 0 0 0 0]',
 '[0 0 0 0 0 0 0 0 0 0 0 0 0 0 0 0 0 0 0 0 0 0 0 0 0 0 0 0]']
```

5. 以 Matplotlib 套件顯示第一筆訓練資料圖片，確認是 5。

```
1  # 顯示第1張圖片圖像
2  import matplotlib.pyplot as plt
3
4  # 第一筆資料
5  X2 = x_train[0,:,:]
6
7  # 繪製點陣圖，cmap='gray'：灰階
8  plt.imshow(X2.reshape(28,28), cmap='gray')
9
10 # 隱藏刻度
11 plt.axis('off')
12
13 # 顯示圖形
14 plt.show()
```

- 執行結果：

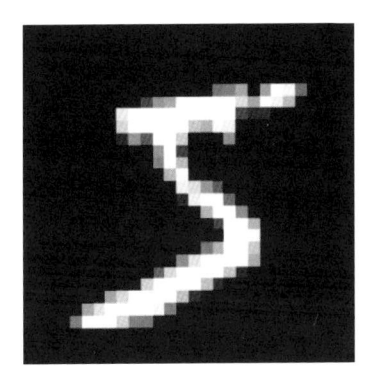

6. 再多觀察一下，0~9 各顯示一張圖片。

```
1  # 0~9各顯示一張圖片
2  fig, ax = plt.subplots(nrows=2, ncols=5, sharex=True, sharey=True,)
3  ax = ax.flatten()
4  for i in range(10):
5      img = x_train[y_train == i][0].reshape(28, 28)
6      ax[i].imshow(img, cmap='Greys')
7
8  ax[0].set_xticks([])
9  ax[0].set_yticks([])
10 plt.tight_layout()
11 plt.show()
```

- 執行結果：

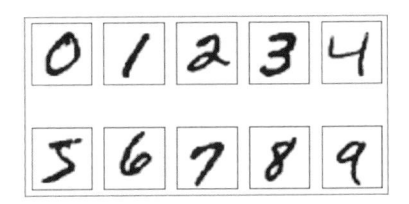

7. 同一個數字顯示 10 張圖片。

```
1  # 同一個數字顯示10張圖片
2  no=9
3  import matplotlib.pyplot as plt
4
5  fig, ax = plt.subplots(nrows=2, ncols=5, sharex=True, sharey=True,)
6  ax = ax.flatten()
```

第 4 章 神經網路實作

```
 7  for i in range(10):
 8      img = x_train[y_train == no][i].reshape(28, 28)
 9      ax[i].imshow(img, cmap='Greys')
10
11  ax[0].set_xticks([])
12  ax[0].set_yticks([])
13  plt.tight_layout()
14  # plt.savefig('images/12_5.png', dpi=300)
15  plt.show()
```

- 執行結果：差異蠻大的，這是由美國高中生及郵局員工撰寫的數字，與台灣人寫法明顯不同。如用傳統像素比對方法，辨識難度應該不小。

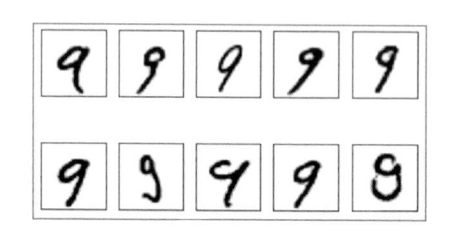

8. 步驟 3：進行特徵工程，將特徵縮放成 [0, 1] 之間，特徵縮放可提高模型準確度，並加快收斂速度，讀者可以省略此一步驟，比較模型準確度差異。特徵縮放採 MinMaxScaler，公式如下：

（x - 樣本最小值）/（樣本最大值 - 樣本最小值）。

```
1  # 特徵縮放，使用常態化(Normalization)，公式 = (x - min) / (max - min)
2  # 顏色範圍：0~255，所以，公式簡化為 x / 255
3  # 注意，顏色0為白色，與RGB顏色不同,(0,0,0) 為黑色。
4  x_train_norm, x_test_norm = x_train / 255.0, x_test / 255.0
5  x_train_norm[0]
```

- 執行結果：

```
        [[0.        , 0.        , 0.        , 0.        , 0.        ,
          0.        , 0.        , 0.        , 0.        , 0.        ,
          0.        , 0.        , 0.00392157, 0.00392157, 0.00392157,
          0.00392157, 0.00392157, 0.00392157, 0.00392157, 0.00392157,
          0.00392157, 0.00392157, 0.00392157, 0.00392157, 0.        ,
          0.        , 0.        , 0.        ],
         [0.        , 0.        , 0.        , 0.        , 0.        ,
          0.        , 0.        , 0.        , 0.00392157, 0.00392157,
          0.00392157, 0.00392157, 0.00392157, 0.00392157, 0.00392157,
          0.00392157, 0.00392157, 0.00392157, 0.00392157, 0.00392157,
          0.00392157, 0.00392157, 0.00392157, 0.00392157, 0.        ,
          0.        , 0.        , 0.        ],
```

9. 步驟 4：資料分割為訓練及測試資料，此步驟無需進行，因為載入 MNIST 資料時，已經切割好了。

10. 步驟 5：建立模型結構如下：

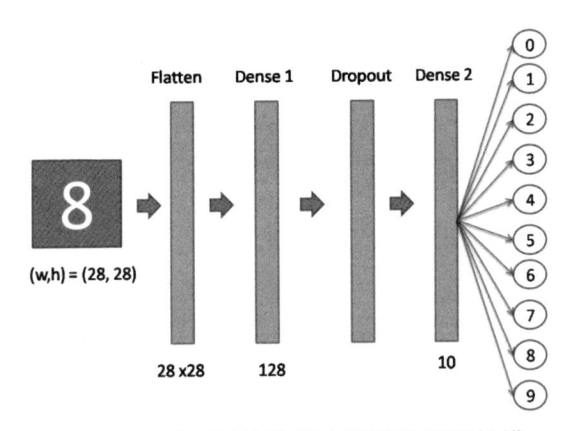

▲ 圖 4.3　手寫阿拉伯數字辨識的模型結構

TensorFlow/Keras 提供兩類模型結構，包括順序型模型（Sequential Model）及 Functional API 模型，順序型模型函數為 tf.keras.models.Sequential，適用於簡單的結構，神經層一層接一層的順序執行，使用 Functional API 可以設計成較複雜的模型結構，包括多個輸入層或多個輸出層，也允許分叉（Branch）、合併（Concatenate），後續使用到再詳細說明。這裡採用簡單的順序型模型，內含各種神經層如下：

```
# 建立模型
model = tf.keras.models.Sequential([
  tf.keras.layers.Input((28, 28)),
  tf.keras.layers.Flatten(),
  tf.keras.layers.Dense(128, activation='relu'),
  tf.keras.layers.Dropout(0.2),
  tf.keras.layers.Dense(10, activation='softmax')
])
```

- 輸入層（Input Layer）：制定輸入維度及大小，維度不含筆數，寬高各 28 個像素。
- 扁平層（Flatten Layer）：將寬高各 28 個像素的圖壓扁成一維陣列（28 x 28 = 784 個特徵）。
- 完全連接層（Dense Layer）：輸入為上一層的輸出，輸出為 128 個神經元，即構成 128 條迴歸線，每一條迴歸線有 784 個特徵。輸出通常訂為 4 的倍數，並無建議值，可經由實驗調校取得較佳的參數值。

- Dropout層：與正則化（Regularization）一樣，希望避免過度擬合（Overfitting），在訓練週期隨機拋棄一定比例的神經元，0.2 表隨機丟棄 20% 的神經元，可以破除特定群組的神經元造成的影響力，另一方面也能使神經網路接收更多神經元的影響力，避免受到局部特徵的影響，藉以矯正過度擬合的現象。通常會在每一層 Dense 後面加一個 Dropout，比例建議在 20%~50%，Dropout 層只對訓練階段有影響，測試及預測階段會自動省略 Dropout 層。

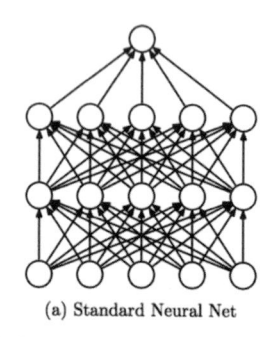

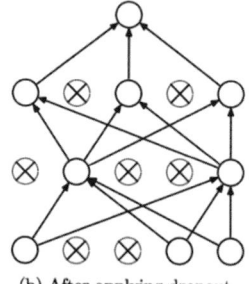

▲ 圖 4.2 左邊為標準的神經網路，右邊是 Dropout Layer 形成的的神經網路。

- 第二個完全連接層（Dense）：為輸出層，因為要辨識 0~9 十個數字，故輸出要設成 10，透過 Softmax Activation Function，可以將輸出轉為機率形式，即預測 0~9 的個別機率，再從中選擇最大機率者為預測值。

11. 編譯指令（model.compile）需設定參數，優化器（optimizer）為 adam，損失函數（loss）為 sparse_categorical_crossentropy（交叉熵），而非 MSE，因為 MSE 適用於迴歸，交叉熵適用於分類，另外優化器最好以物件指定，例如 adam = tf.keras.optimizers.Adam()，再將 adam 作為 compile 的參數，使用字串 'adam'，會造成模型載入時出現警告訊息。

```
1  # 設定優化器(optimizer)、損失函數(loss)、效能衡量指標(metrics)的類別
2  model.compile(optimizer='adam',
3                loss='sparse_categorical_crossentropy',
4                metrics=['accuracy'])
```

12. 步驟 6：結合訓練資料及模型結構，進行模型訓練。

```
1  # 模型訓練
2  history = model.fit(x_train_norm, y_train, epochs=5, validation_split=0.2)
```

4-1 撰寫第一支神經網路程式

- validation_split：將訓練資料切割一部份為驗證資料，0.2 表示驗證資料佔 20%，在訓練過程中，會用驗證資料計算準確度及損失函數值，以利比較訓練資料與驗證資料是否有顯著差異，確認訓練過程有無異常。
- 訓練週期（epochs）：設定訓練要執行的週期數，『所有』訓練資料經過一次正向和反向傳導，稱為一個訓練週期。
- 執行結果如下所示，每一個訓練週期（epoch），都包含訓練的損失（loss）、準確率（accuracy）及驗證資料的損失（val_loss）、準確率（val_accuracy），這些資訊都會儲存在 history 變數內，它是字典（dict）資料型態。

```
Train on 48000 samples, validate on 12000 samples
Epoch 1/5
48000/48000 [==============================] - 4s 77us/sample - loss: 0.3264 - accuracy: 0.9055 - val_loss: 0.1576 - val_accura
cy: 0.9572
Epoch 2/5
48000/48000 [==============================] - 3s 71us/sample - loss: 0.1593 - accuracy: 0.9534 - val_loss: 0.1187 - val_accura
cy: 0.9654
Epoch 3/5
48000/48000 [==============================] - 3s 71us/sample - loss: 0.1188 - accuracy: 0.9649 - val_loss: 0.1039 - val_accura
cy: 0.9682
Epoch 4/5
48000/48000 [==============================] - 3s 72us/sample - loss: 0.0969 - accuracy: 0.9704 - val_loss: 0.1071 - val_accura
cy: 0.9668
Epoch 5/5
48000/48000 [==============================] - 3s 71us/sample - loss: 0.0829 - accuracy: 0.9740 - val_loss: 0.0876 - val_accura
cy: 0.9739
```

13. 對訓練過程的準確率繪圖：

```
1  # 對訓練過程的準確率繪圖
2  plt.figure(figsize=(8, 6))
3  plt.plot(history.history['accuracy'], 'r', label='訓練準確率')
4  plt.plot(history.history['val_accuracy'], 'g', label='驗證準確率')
5  plt.legend()
```

- 執行結果：觀察三個面向。

    * 訓練後的模型準確率是否可接受。

    * 準確率是否隨著訓練週期趨於收斂。

    * 驗證資料與訓練資料的準確率最後是否趨於相近。

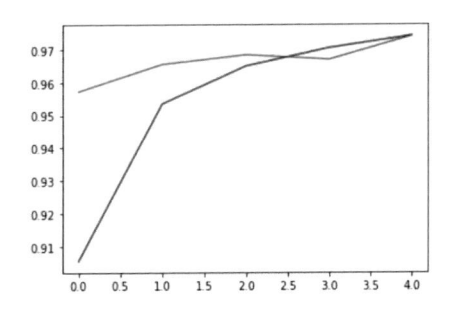

第 4 章　神經網路實作

14. 對訓練過程的損失繪圖：

```
1  # 對訓練過程的損失繪圖
2  import matplotlib.pyplot as plt
3
4  plt.figure(figsize=(8, 6))
5  plt.plot(history.history['loss'], 'r', label='訓練損失')
6  plt.plot(history.history['val_loss'], 'g', label='驗證損失')
7  plt.legend()
```

- 執行結果：隨著訓練週期（epoch）次數的增加，損失越來越低，觀察重點與準確率類似。

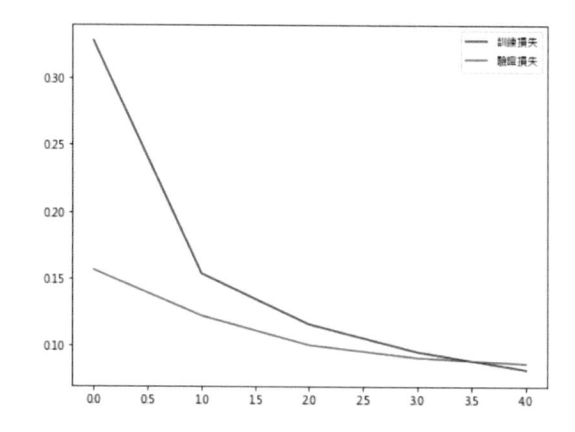

15. 步驟 7：評分（Score Model），呼叫 evaluate()，輸入測試資料，會計算出損失及準確率，由於測試資料未參與訓練，相關數值較客觀。

```
1  # 評分(Score Model)
2  score=model.evaluate(x_test_norm, y_test, verbose=0)
3
4  for i, x in enumerate(score):
5      print(f'{model.metrics_names[i]}: {score[i]:.4f}')
```

- 執行結果：loss: 0.0833，accuracy: 0.9743。

16. 實際比對測試資料的前 20 筆，呼叫 predict 可以得到預測 0~9 每一類別的機率，再使用 np.argmax 可取得每一筆資料最大機率的索引值，即類別。

```
1  # 實際預測 20 筆資料
2  import numpy as np
3
4  # model.predict_classes 已在新版廢除
```

```
5  #predictions = model.predict_classes(x_test_norm)
6  predictions = np.argmax(model.predict(x_test_norm), axis=-1)
7
8  # 比對
9  print('actual    :', y_test[0:20])
10 print('prediction:', predictions[0:20])
```

- 執行結果如下,全部正確。

```
actual    : [7 2 1 0 4 1 4 9 5 9 0 6 9 0 1 5 9 7 3 4]
prediction: [7 2 1 0 4 1 4 9 5 9 0 6 9 0 1 5 9 7 3 4]
```

17. 如果要顯示某一筆的機率,例如第 9 筆,程式如下。

```
1  # 顯示第 9 筆的機率
2  import numpy as np
3
4  predictions = model.predict(x_test_norm[8:9])
5  print(f'0~9預測機率: {np.around(predictions[0], 2)}')
```

- 第 9 筆圖片如下,像 5 又像 6。

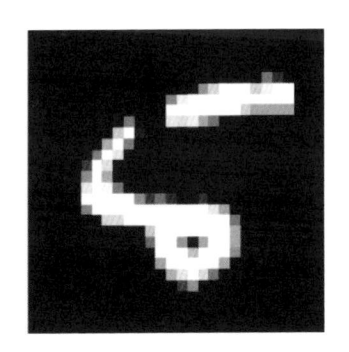

18. 步驟 8:效能評估,通常會以不同的模型結構或超參數(Hyperparameter)組合建立多個模型,再選取其中最佳的模型,目前暫不進行,後續再詳細討論。超參數是指在模型訓練前可以調整的參數,例如學習率、訓練週期、權重初始值、訓練批量等,但不含模型求算的參數如權重(Weight)或偏差項(Bias)。

19. 步驟 9:模型佈署,將最佳模型存檔,再開發使用者介面或提供 API,連同模型檔一併佈署到上線環境(Production Environment)。

```
1  # 模型存檔
2  model.save('model.keras')
```

第 4 章　神經網路實作

```
3
4  # 模型載入
5  model = tf.keras.models.load_model('model.keras')
```

20. 步驟 10：使用新資料預測，之前都是使用 MNIST 內建資料測試，嚴格說並不可靠，因為這些都是出自同一機構所收集的資料，因此，建議讀者自己利用繪圖軟體（例如小畫家）製作圖片測試。我們準備一些圖檔，放在 myDigits 目錄內，讀者可自行修改，再利用下列程式碼測試，注意，**從圖檔讀入影像後要反轉顏色**，顏色 0 為白色，與 RGB 色碼不同，它的 0 為黑色。

```
1  # 使用小畫家，繪製 0~9，實際測試看看
2  from skimage import io
3  from skimage.transform import resize
4  import numpy as np
5
6  # 讀取影像並轉為單色
7  uploaded_file = '../myDigits/8.png'
8  image1 = io.imread(uploaded_file, as_gray=True)
9
10 # 縮為 (28, 28) 大小的影像
11 image_resized = resize(image1, (28, 28))
12 X1 = image_resized.reshape(1,28, 28) #/ 255
13
14 # 反轉顏色，顏色0為白色，與 RGB 色碼不同，它的 0 為黑色
15 X1 = np.abs(1-X1)
16
17 # 預測
18 predictions = np.argmax(model.predict(X1, verbose=False), axis=-1)
19 print(predictions[0])
```

- 執行結果：8，辨識無誤。

21. 上一步驟只測試一個數字，以下使用迴圈一次測試 10 個數字。

```
1  # 使用小畫家，繪製 0~9，實際測試看看
2  from skimage import io
3  from skimage.transform import resize
4  import numpy as np
5
6  # 讀取影像並轉為單色
```

```
 7
 8  for i in range(0, 10):
 9      uploaded_file = f'../myDigits/{i}.png'
10      image1 = io.imread(uploaded_file, as_gray=True)
11
12      # 縮為 (28, 28) 大小的影像
13      image_resized = resize(image1, (28, 28))
14      X1 = image_resized.reshape(1,28, 28) #/ 255
15
16      # 反轉顏色，顏色0為白色，與 RGB 色碼不同，它的 0 為黑色
17      X1 = np.abs(1-X1)
18
19      # 預測
20      predictions = np.argmax(model.predict(X1, verbose=False), axis=-1)
21      print(predictions[0])
```

- 執行結果可能沒有 MNIST 測試資料那麼準確，為什麼呢？後續我們會仔細探究原因。

22. 另外，要瞭解模型結構及輸出入神經元個數，可使用下列指令顯示模型彙總資訊（summary）。

```
1  # 顯示模型的彙總資訊
2  model.summary()
```

- 執行結果：包括每一神經層的名稱及輸出入參數的個數。

```
Model: "sequential_2"
_____
 Layer (type)               Output Shape              Param #
=================================================================
 flatten_2 (Flatten)        (None, 784)               0

 dense_4 (Dense)            (None, 128)               100480

 dropout_2 (Dropout)        (None, 128)               0

 dense_5 (Dense)            (None, 10)                1290
=================================================================
Total params: 101,770
Trainable params: 101,770
Non-trainable params: 0
```

- 計算參數個數：舉例來說 dense_5，輸出參數為 1290，意思是共有 10 條迴歸線，每一條迴歸線都有 128 個特徵對應的權重（w）與一個偏差項（b），所以總共有 10 x（128 +1）= 1290 個參數。

第 4 章　神經網路實作

23. 也可以繪製圖形，顯示模型結構：要繪製圖形顯示模型結構，需先完成以下步驟，才能順利繪製圖形。

    - 安裝 graphviz 軟體，網址為 https://www.graphviz.org/download，再把安裝目錄下的 bin 路徑加到環境變數 Path 中。

    - 安裝兩個套件：

      pip install graphviz pydotplus

    - 執行 plot_model 指令，可以同時顯示圖形和存檔。

```
1  tf.keras.utils.plot_model(model, to_file='model.png')
```

    - 執行結果：

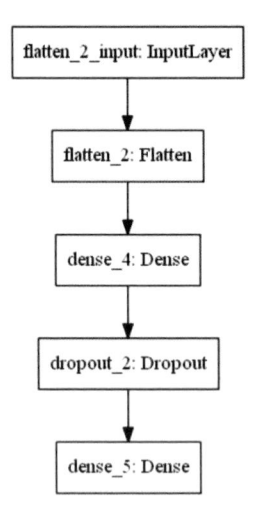

      flatten_2_input: InputLayer
      ↓
      flatten_2: Flatten
      ↓
      dense_4: Dense
      ↓
      dropout_2: Dropout
      ↓
      dense_5: Dense

以上我們依機器學習流程的 10 大步驟撰寫了一支完整的程式，雖然篇幅很長，讀者應該還是有些疑問，針對許多細節的描述，將於下一節登場，我們會做些實驗來說明建構模型的考量，同時解答教學現場同學們常提出的問題。

4-1-3　實驗

上一節我們完成了第一支深度學習的程式，也見識到它的威力，扣除說明，短短 10 幾行的程式就能夠辨識手寫阿拉伯數字，且準確率達到 97%。然而，仔細思考後會產生許多疑問：

4-1 撰寫第一支神經網路程式

1. 模型結構為什麼要設計成兩層 Dense？更多層準確率會提高嗎？
2. 第一層 Dense 輸出為什麼要設為 128？設為其他值會有何影響？
3. 目前第一層 Dense 的 Activation Function 設為 relu，代表什麼意義？設為其他值又會有何不同？
4. 優化器（optimizer）、損失函數（loss）、效能衡量指標（metrics）有哪些選擇？設為其他值會有何影響？
5. Dropout 比例為 0.2，設為其他值會更好嗎？
6. 影像為單色灰階，若是彩色可以辨識嗎？怎麼修改？
7. 訓練週期（epoch）設為 5，設為其他值會更好嗎？
8. 準確率可以達到 100% 嗎？這樣企業才可以安心導入。
9. 如果要辨識其他物件，程式要修改那些地方？
10. 如果要辨識多個數字，例如輸入 4 位數，要如何辨識？
11. 希望了解更詳細的相關資訊，有哪些資源可以參閱？

以上問題是這幾年來授課時學員常提出的疑惑，我們就來逐一實驗，試著尋找答案。

問題 1. 模型結構為什麼要設計成兩層 Dense？更多層準確率會提高嗎？

解答：

1. 前面曾經說過，神經網路是多條迴歸線的組合，而且每一條迴歸線可能還會包在 Activation Function 內，變成非線性的函數，因此，要單純以數學求解幾乎不可能，只能以優化方法求得近似解，但是，只有凸集合（Convex set）的資料集，才保證有全局最佳解（Global Minimization），以 MNIST 為例，總共有 784 個特徵，即 784 度空間，根本無從知道它是否為凸集合，因此嚴格來講，目前神經網路依然是一個黑箱（Black Box）科學，我們只知道它威力強大，但如何達到較佳的準確率，仍舊需要經驗與實驗，因此，模型結構並沒有明確規定要設計成幾層，須依問題及資料不同進行各種實驗，case by case 進行效能調校，找尋較佳的參數值。

2. 理論上，越多層架構，迴歸線就越多，預測應當越準確，像是 ResNet 模型就高達 150 層，但是，經過實驗證實，超過某一界限後，準確率可能會不升反降，這跟訓練資料量有關，如果只有少量的資料，要估算過多的參數（w、b），自然準確率不高。

3. 我們就來小小實驗一下，多一層 Dense，準確率是否會提高？請參閱程式 **04_03_手寫阿拉伯數字辨識 _ 實驗 1.ipynb**。

4. 修改模型結構如下，加一對 Dense/Dropout，其餘程式碼不變。

   - 執行結果如下，準確率不見提升，反而微降。注意，由於，訓練資料隨機抽樣，故每次訓練的準確率都不會相同。

     accuracy: 0.9733

問題 2. 第一層 Dense 輸出為什麼要設為 128？設為其他值會有何影響？

1. 輸出的神經元個數可以任意設定，一般來講，會使用 4 的倍數，以下我們修改為 256，參閱程式 **04_03_ 手寫阿拉伯數字辨識 _ 實驗 2.ipynb**。

```
# 建立模型
model = tf.keras.models.Sequential([
    tf.keras.layers.Input((28, 28)),
    tf.keras.layers.Flatten(),
    tf.keras.layers.Dense(256, activation='relu'),
    tf.keras.layers.Dropout(0.2),
    tf.keras.layers.Dense(10, activation='softmax')
])
```

   - 執行結果如下，準確率略為提高，但不明顯。

     accuracy: 0.9764

2. 同問題 1，照理來說，神經元個數越多，迴歸線就越多，特徵也越多，預測應該會越準確，但經過驗證，準確率並未顯著提高。依『Deep Learning with TensorFlow 2.0 and Keras』一書測試如下圖，也是有一個極限，超過就會不升反降。下圖橫坐標為神經元，縱坐標為準確率。

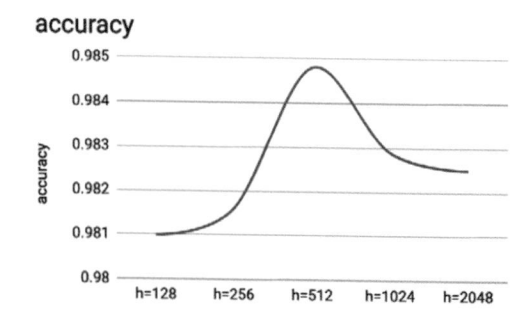

3. 神經元個數越多，訓練時間就越長，如下圖：

4-1 撰寫第一支神經網路程式

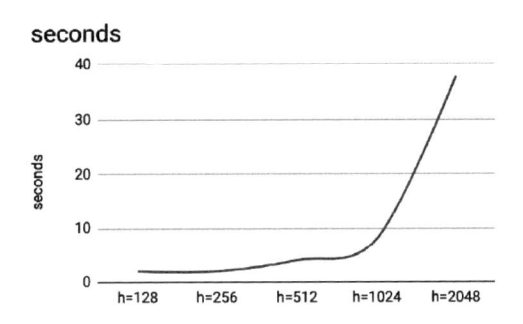

```
seconds
40
30
20
10
0
    h=128   h=256   h=512   h=1024   h=2048
```

問題 3. 目前第一層 Dense 的 Activation Function 設為 relu,代表什麼意義?設為其他值會有何不同?

解答:

Activation Function 有很多種,後面會有詳盡介紹,可先參閱維基百科 [1],表格部份內容如下,欄位包括函數名稱、機率分配圖形、公式及一階導數:

Name	Plot	Function, $f(x)$	Derivative of f, $f'(x)$
Identity		x	1
Binary step		$\begin{cases} 0 & \text{if } x < 0 \\ 1 & \text{if } x \geq 0 \end{cases}$	$\begin{cases} 0 & \text{if } x \neq 0 \\ \text{undefined} & \text{if } x = 0 \end{cases}$
Logistic, sigmoid, or soft step		$\sigma(x) = \dfrac{1}{1+e^{-x}}$ [1]	$f(x)(1-f(x))$
tanh		$\tanh(x) = \dfrac{e^x - e^{-x}}{e^x + e^{-x}}$	$1 - f(x)^2$
Rectified linear unit (ReLU)[11]		$\begin{cases} 0 & \text{if } x \leq 0 \\ x & \text{if } x > 0 \end{cases}$ $= \max\{0, x\} = x\mathbf{1}_{x>0}$	$\begin{cases} 0 & \text{if } x < 0 \\ 1 & \text{if } x > 0 \\ \text{undefined} & \text{if } x = 0 \end{cases}$
Gaussian error linear unit (GELU)[6]		$\dfrac{1}{2}x\left(1 + \operatorname{erf}\left(\dfrac{x}{\sqrt{2}}\right)\right)$ $= x\Phi(x)$	$\Phi(x) + x\phi(x)$
Softplus[12]		$\ln(1 + e^x)$	$\dfrac{1}{1+e^{-x}}$

第 4 章　神經網路實作

早期隱藏層大都使用 Sigmoid 函數，近幾年發現 ReLU 準確率較高，我們先試著比較這兩種，請參閱程式 **04_03_ 手寫阿拉伯數字辨識 _ 實驗 3.ipynb**。

1. 將 relu 改為 sigmoid，如下所示：

```
1  # 建立模型
2  model = tf.keras.models.Sequential([
3      tf.keras.layers.Input((28, 28)),
4      tf.keras.layers.Flatten(),
5      tf.keras.layers.Dense(128, activation='sigmoid'),
6      tf.keras.layers.Dropout(0.2),
7      tf.keras.layers.Dense(10, activation='softmax')
8  ])
```

- 執行結果如下，準確率確實略低於 ReLU。

  accuracy: 0.9652

問題 4. 優化器（Optimizer）、損失函數（Loss）、效能衡量指標（Metrics）有哪些選擇？設為其他值會有何影響？

解答：

1. 優化器有很多種，從最簡單的固定值的學習率，到動態改變的學習率，甚至能夠自訂優化器。優化器的選擇，主要會影響收斂的速度，大多數的狀況下，Adam 優化器都有不錯的表現。詳細的說明請參考『TensorFlow 優化器介紹』[2] 或『Keras 優化器介紹』[3]。

2. 損失函數也有很多選擇，包括常見的均方誤差（MSE）、交叉熵（Cross Entropy），其他更多的請參考『TensorFlow 損失函數介紹』[4] 或『Keras 損失函數介紹』[5]。損失函數的選擇，主要分為兩類，連續型的目標變數選擇 MSE 類型的損失函數，因為預測值與實際值不可能完全吻合，而離散型的目標變數選擇交叉熵類型的損失函數，適用於分類，另外某些損失函數有特殊用途，例如風格轉換（Style Transfer），它能夠產生影像合成的效果，生成對抗網路（GAN）更是發揚光大，可以生成真偽難辨的圖片，後續章節會有詳細的介紹。

3. 效能衡量指標（metrics）：除了準確率（Accuracy），還可以計算精確率（Precision）、召回率（Recall）、F1…，可同時設定多個效能衡量指標，如下程式碼，完整程式可參閱程式 **04_03_ 手寫阿拉伯數字辨識 _ 實驗 4.ipynb**，更詳細的說明請參考『TensorFlow 效能衡量指標介紹』[6]。

```
10  model.compile(optimizer='adam',
11               loss='categorical_crossentropy',
12               metrics=[tf.keras.metrics.CategoricalAccuracy(),
13                        tf.keras.metrics.Precision(),
14                        tf.keras.metrics.Recall()])
```

- 注意,設定多個效能衡量指標時,準確率請不要使用 Accuracy,數值會非常低,需使用 CategoricalAccuracy,表示分類的準確率,而非迴歸的準確率。

- 執行結果如下:

    loss: 0.0757

    categorical_accuracy: 0.9781

    precision_3: 0.9810

    recall_3: 0.9751

問題 5. Dropout 比例為 0.2,設為其他值會更好嗎?

解答:

設定 Dropout 比例為 0.1,測試看看,參閱程式 **04_03_ 手寫阿拉伯數字辨識 _ 實驗 5.ipynb**。

```
2  model = tf.keras.models.Sequential([
3      tf.keras.layers.Input((28, 28)),
4      tf.keras.layers.Flatten(),
5      tf.keras.layers.Dense(128, activation='relu'),
6      tf.keras.layers.Dropout(0.1),
7      tf.keras.layers.Dense(10, activation='softmax')
8  ])
```

- 執行結果如下,準確率略為提高。

    loss: 0.0816

    accuracy: 0.9755

- 拋棄比例過高時,準確率會陡降。

第 4 章　神經網路實作

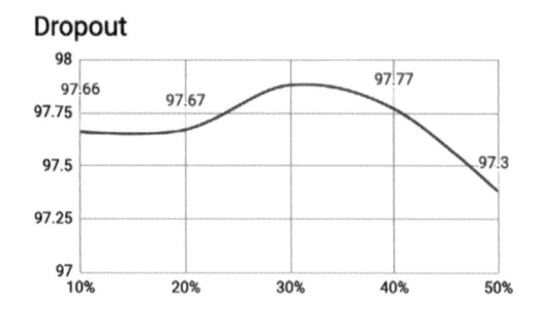

<!-- Dropout 準確率圖: 10%→97.66, 20%→97.67, 30%→97.77, 40%→97.77, 50%→97.3 -->

問題 6. 目前 MNIST 影像為單色灰階，若是彩色可以辨識嗎？怎麼修改？

解答：可以，若顏色有助於辨識，可以將 RGB 三通道分別輸入辨識，後面我們討論卷積神經網路（Convolutional Neural Networks，CNN）時會有範例說明。

問題 7. 訓練週期（epoch）設為 5，設為其他值會更好嗎？

解答：訓練週期（epoch）改為 10，參閱程式 **04_03_ 手寫阿拉伯數字辨識 _ 實驗 6.ipynb**。

```
15  history = model.fit(x_train_norm, y_train, epochs=10, validation_split=0.2)
```

- 執行結果如下，準確率略為提高。

  accuracy: 0.9785

- 理論上，訓練週期越多，準確率越高，但是，過多的訓練週期會造成過度擬合（Overfitting），反而會使測試資料的準確率降低。

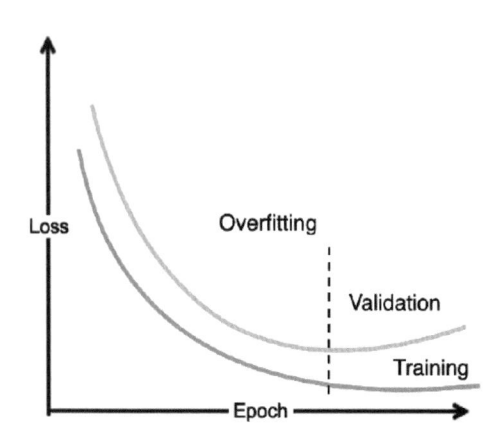

<!-- Loss vs Epoch 圖：Training loss 持續下降，Validation loss 先下降後上升，出現 Overfitting -->

問題 8. 準確率可以達到 100% 嗎？

解答：很少模型準確率能夠達到 100%，因為神經網路使用機率預測，大量資料會有差異，很難有 100% 的機率，另一方面，神經網路是從訓練資料中學習到知識，但是，測試或預測資料並不參與訓練，若與訓練資料分佈有所差異，甚至來自不同的機率分配，很難確保準確率能達到 100%。

問題 9. 如果要辨識其他物件，程式要修改那些地方？

解答：我們只需修改很少的程式碼，就可以辨識其他物件，例如，MNIST 有另一個資料集 FashionMnist，它包含女士身上的 10 種配件，請參閱 **04_03_FashionMnist_ 實驗.ipynb**，除了載入資料的指令不同之外，其他的程式碼幾乎不變。這也說明了一點，神經網路並不是真的認識 **0~9** 或女士身上的 **10** 個配件，它只是從像素資料中推估出的模型，即所謂的 『從資料中學習到知識』（**Knowledge Discovery from Data, KDD**），以 MNIST 而言，模型只是統計 0~9 個個數字，他們的像素大部份分布在那些位置而已。

問題 10. 如果要辨識多個數字，例如輸入 4 位數，要如何辨識？

解答：可以使用影像處理分割數字，再分別依序輸入模型預測即可，更簡單的方法，直接將視覺介面（UI）設計成 4 格，規定使用者只能在每格子內各輸入一個數字。請參閱 split_digits.py 程式，可將 images\multi_digits.png 分割成單一數字。

問題 11. 希望瞭解更詳細的相關資訊，有哪些資源可以參閱？

解答：可以參考 TensorFlow 官網 [7] 或 Keras 官網 [8]，版本快速的更新已經使網路上的資訊新舊雜陳，官網才是最新資訊的正確來源。

以上的實驗大多只針對單一參數作比較，假如要同時比較多個變數，就必須跑遍所有參數組合，這樣程式豈不是很複雜嗎？別擔心，有一些套件可以幫忙，包括 Keras Tuner、hyperopt、Ray Tune、Ax…等，在後續『超參數調校』有較詳細的介紹。

由於這個模型的辨識率很高，要觀察超參數調整對模型的影響，並不容易，建議找一些辨識率較低的模型進行相關實驗，例如 FashionMnist[9]、CiFar 資料集，才能有比較顯著的效果，筆者針對 FashionMnist 作了另一次實驗 **04_03_FashionMnist_ 實驗.ipynb**。

第 4 章　神經網路實作

4-2　Keras 模型種類

TensorFlow/Keras 提供兩類模型結構：

1. Sequential model：順序型模型，按神經層的排列順序，由上往下執行，每一層的輸出都是下一層的輸入，所以，除了第一層要設定輸入的維度外，其他層都不需要設定。
2. Functional API：提供較有彈性的結構，允許非順序型的結構、共享的神經層及多個輸入 / 輸出，亦即模型結構可以有分叉或合併（Split/Merge）。

4-2-1　順序型模型（Sequential Model）

範例. 順序型模型（Sequential Model）測試。

程式：**04_05_ Sequential_model.ipynb**。

1. 模型內可包含各式的神經層，簡潔的寫法如下，以 List 包住神經層。

```
1  model = tf.keras.models.Sequential([
2      tf.keras.layers.Input((28, 28)),
3      tf.keras.layers.Flatten(),
4      tf.keras.layers.Dense(128, activation='relu'),
5      tf.keras.layers.Dropout(0.2),
6      tf.keras.layers.Dense(10, activation='softmax')
7  ])
```

2. 除了第一層要設定輸入的維度（input_shape）外，其他層都不需要設定，只要在第一個參數指定輸出維度。

3. 可以變換另一種寫法，將 input_shape 拿掉，在 model 內設定輸入層及維度參數（shape）。

```
1  model = tf.keras.models.Sequential([
2      tf.keras.layers.Flatten(),
3      tf.keras.layers.Dense(128, activation='relu'),
4      tf.keras.layers.Dropout(0.2),
5      tf.keras.layers.Dense(10, activation='softmax')
6  ])
7  
8  x = tf.keras.layers.Input(shape=(28, 28))
```

4-2 Keras 模型種類

```
 9  # 或 x = tf.Variable(tf.random.truncated_normal([28, 28]))
10  y = model(x)
```

4. 不使用 Sequential 指令，最後一行指令直接串連神經層。

```
1  layer1 = tf.keras.layers.Dense(2, activation="relu", name="layer1")
2  layer2 = tf.keras.layers.Dense(3, activation="relu", name="layer2")
3  layer3 = tf.keras.layers.Dense(4, name="layer3")
4
5  # Call layers on a test input
6  x = tf.ones((3, 3))
7  y = layer3(layer2(layer1(x)))
```

5. 可以在建模後，再加減神經層：pop() 會刪減最上層（Top），注意，神經層是堆疊（Stack），後進先出，即最後一層 Dense。

```
 1  model = tf.keras.models.Sequential([
 2      tf.keras.layers.Input((28, 28)),
 3      tf.keras.layers.Flatten(),
 4      tf.keras.layers.Dense(128, activation='relu'),
 5      tf.keras.layers.Dropout(0.2),
 6      tf.keras.layers.Dense(10, activation='softmax')
 7  ])
 8
 9  # 刪減一層
10  model.pop()
11  print(f'神經層數: {len(model.layers)}')
12  model.layers
```

- 執行結果：

```
神經層數: 3
[<tensorflow.python.keras.layers.core.Flatten at 0x1826cfb8580>,
 <tensorflow.python.keras.layers.core.Dense at 0x1826cf94340>,
 <tensorflow.python.keras.layers.core.Dropout at 0x1826cf94b20>]
```

6. 增加一層神經層。

```
1  # 增加一層
2  model.add(tf.keras.layers.Dense(10))
3  print(f'神經層數: {len(model.layers)}')
4  model.layers
```

4-25

第 4 章　神經網路實作

- 執行結果：

```
神經層數: 4
[<tensorflow.python.keras.layers.core.Flatten at 0x1826cfb8580>,
 <tensorflow.python.keras.layers.core.Dense at 0x1826cf94340>,
 <tensorflow.python.keras.layers.core.Dropout at 0x1826cf94b20>,
 <tensorflow.python.keras.layers.core.Dense at 0x18270b86040>]
```

7. 取得模型各神經層資訊。

```
1  # 建立 3 Layers
2  layer0 = tf.keras.layers.Input((28, 28))
3  layer1 = tf.keras.layers.Dense(2, activation="relu", name="layer1")
4  layer2 = tf.keras.layers.Dense(3, activation="relu", name="layer2")
5  layer3 = tf.keras.layers.Dense(4, name="layer3")
6
7  # 建立模型
8  model = tf.keras.models.Sequential([
9      layer0,
10     layer1,
11     layer2,
12     layer3
13 ])
14
15 # 讀取模型權重
16 print(f'神經層參數類別總數: {len(model.weights)}')
17 model.weights
```

- 執行結果：有 3 層共 6 類資訊，包括 3 層權重（weight）及 3 層偏差（bias），共 6 類。

```
神經層參數類別總數: 6
[<tf.Variable 'layer1/kernel:0' shape=(28, 2) dtype=float32, numpy=
array([[-0.40281397,  0.2205016 ],
       [-0.03376389,  0.15561956],
       [ 0.4151038 ,  0.02849016],
       [-0.15993604,  0.03800485],
       [ 0.06474555,  0.39933097],
       [ 0.41804487, -0.35671836],
       [-0.09304568, -0.12987521],
       [-0.07738718,  0.4214089 ],
       [-0.03624141, -0.15986142],
       [ 0.41953433, -0.14494684],
       [-0.16713199, -0.36726573],
       [ 0.28147936, -0.07306954],
       [ 0.06702495,  0.29314017],
       [-0.4076838 , -0.22369348],
```

4-26

4-2 Keras 模型種類

8. 取得特定神經層資訊。

```
1  print(f'{layer2.name}: {layer2.weights}')
```

9. 取得模型彙總資訊。

```
1  model.summary()
```

10. 可以一邊加神經層,一邊顯示模型彙總資訊,這樣有利於除錯,查看中間處理結果。

```
1   from tensorflow.keras import layers
2
3   model = tf.keras.models.Sequential()
4   model.add(tf.keras.Input(shape=(250, 250, 3)))  # 250x250 RGB images
5   model.add(layers.Conv2D(32, 5, strides=2, activation="relu"))
6   model.add(layers.Conv2D(32, 3, activation="relu"))
7   model.add(layers.MaxPooling2D(3))
8
9   # 顯示目前模型彙總資訊
10  model.summary()
11
12  # The answer was: (40, 40, 32), so we can keep downsampling...
13
14  model.add(layers.Conv2D(32, 3, activation="relu"))
15  model.add(layers.Conv2D(32, 3, activation="relu"))
16  model.add(layers.MaxPooling2D(3))
17  model.add(layers.Conv2D(32, 3, activation="relu"))
18  model.add(layers.Conv2D(32, 3, activation="relu"))
19  model.add(layers.MaxPooling2D(2))
20
21  # 顯示目前模型彙總資訊
22  model.summary()
23
24  # Now that we have 4x4 feature maps, time to apply global max pooling.
25  model.add(layers.GlobalMaxPooling2D())
26
27  # Finally, we add a classification layer.
28  model.add(layers.Dense(10))
```

11. 取得每一層神經層的 output:可設定模型的 input/output。

```
1   # 設定模型
2   initial_model = tf.keras.Sequential(
3       [
4           tf.keras.Input(shape=(250, 250, 3)),
5           layers.Conv2D(32, 5, strides=2, activation="relu"),
6           layers.Conv2D(32, 3, activation="relu"),
7           layers.Conv2D(32, 3, activation="relu"),
8       ]
```

4-27

```
 9  )
10
11  # 設定模型的input/output
12  feature_extractor = tf.keras.Model(
13      inputs=initial_model.inputs,
14      outputs=[layer.output for layer in initial_model.layers],
15  )
16
17  # 呼叫 feature_extractor 取得 output
18  x = tf.ones((1, 250, 250, 3))
19  features = feature_extractor(x)
20  features
```

12. 取得特定神經層的輸出（output）：設定模型的 output 為特定的神經層。

```
 1  # 設定模型
 2  initial_model = tf.keras.Sequential(
 3      [
 4          tf.keras.Input(shape=(250, 250, 3)),
 5          layers.Conv2D(32, 5, strides=2, activation="relu"),
 6          layers.Conv2D(32, 3, activation="relu", name="my_intermediate_layer"),
 7          layers.Conv2D(32, 3, activation="relu"),
 8      ]
 9  )
10
11  # 設定模型的input/output
12  feature_extractor = tf.keras.Model(
13      inputs=initial_model.inputs,
14      outputs=initial_model.get_layer(name="my_intermediate_layer").output,
15  )
16
17  # 呼叫 feature_extractor 取得 output
18  x = tf.ones((1, 250, 250, 3))
19  features = feature_extractor(x)
20  features
```

4-2-2 Functional API

Functional API 提供較有彈性的結構，適用於相對複雜的模型結構，允許非順序型的結構，結構可以分叉或合併（Split/Merge），允許多個輸入/輸出神經層。

範例 1. 先看一個簡單的程式語法，除了第一層之外，每一層均須設定前一層，同時，Model 函數必須指定輸入/輸出（input/output）是那些神經層，都是 List 資料型態，允許多個輸入/輸出。

程式：**04_06_ Functional_API.ipynb**。

4-2 Keras 模型種類

1. Functional API 測試。

```
1  # Functional API
2
3  # 建立第一層 InputTensor
4  InputTensor = layers.Input(shape=(100,))
5
6  # H1 接在 InputTensor 後面
7  H1 = layers.Dense(10, activation='relu')(InputTensor)
8
9  # H2 接在 H1 後面
10 H2 = layers.Dense(20, activation='relu')(H1)
11
12 # Output 接在 H2 後面
13 Output = layers.Dense(1, activation='softmax')(H2)
14
15 # 建立模型，必須指定 inputs / outputs
16 model = tf.keras.Model(inputs=InputTensor, outputs=Output)
17
18 # 顯示模型彙總資訊
19 model.summary()
```

2. 模型包括 3 個輸入、2 個輸出，先不管模型用途，只觀察程式語法，layers.concatenate 可合併神經層。

```
6  # 建立第一層 InputTensor
7  title_input = tf.keras.Input(shape=(None,), name="title")
8  body_input = tf.keras.Input(shape=(None,), name="body")
9  tags_input = tf.keras.Input(shape=(num_tags,), name="tags")
10
11 # 建立第二層
12 title_features = layers.Embedding(num_words, 64)(title_input)
13 body_features = layers.Embedding(num_words, 64)(body_input)
14
15 # 建立第三層
16 title_features = layers.LSTM(128)(title_features)
17 body_features = layers.LSTM(32)(body_features)
18
19 # 合併以上神經層
20 x = layers.concatenate([title_features, body_features, tags_input])
21
22 # 建立第四層，連接合併的 x
23 priority_pred = layers.Dense(1, name="priority")(x)
24 department_pred = layers.Dense(num_departments, name="department")(x)
25
26 # 建立模型，必須指定 inputs / outputs
27 model = tf.keras.Model(
28     inputs=[title_input, body_input, tags_input],
29     outputs=[priority_pred, department_pred],
30 )
31
32 # 繪製模型
33 # show_shapes=True：Layer 含 Input/Output 資訊
34 tf.keras.utils.plot_model(model, "multi_input_and_output_model.png",
35                           show_shapes=True)
```

- 最後一行程式碼繪製的模型圖如下：

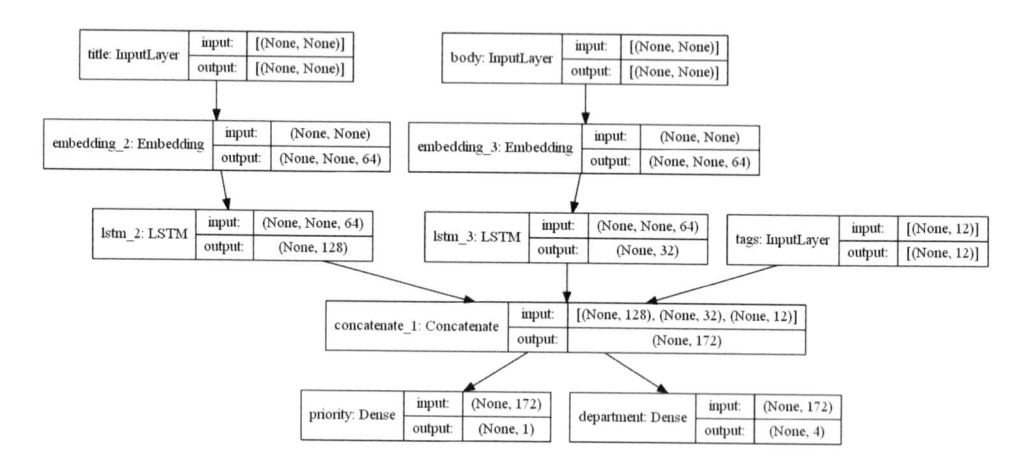

title: InputLayer	input:	[(None, None)]
	output:	[(None, None)]

body: InputLayer	input:	[(None, None)]
	output:	[(None, None)]

embedding_2: Embedding	input:	(None, None)
	output:	(None, None, 64)

embedding_3: Embedding	input:	(None, None)
	output:	(None, None, 64)

lstm_2: LSTM	input:	(None, None, 64)
	output:	(None, 128)

lstm_3: LSTM	input:	(None, None, 64)
	output:	(None, 32)

tags: InputLayer	input:	[(None, 12)]
	output:	[(None, 12)]

concatenate_1: Concatenate	input:	[(None, 128), (None, 32), (None, 12)]
	output:	(None, 172)

priority: Dense	input:	(None, 172)
	output:	(None, 1)

department: Dense	input:	(None, 172)
	output:	(None, 4)

- concatenate 合併了 3 個 layers，它們的輸出維度大小分別為 128/32/12，故合併後，輸出維度大小 =128+32+12=172。

- 最後一行程式碼的參數 show_shapes=True，結構圖會額外添加含有 Input/Output 資訊。

4-3 神經層（Layer）

神經層是神經網路的主要成員，TensorFlow 有各式各樣的神經層，詳情可參閱 Keras 官網神經層介紹 [10]，主要類別如下，隨著演算法的發明，會不斷的增加，還可以自訂神經層（Custom layer）。

- 核心神經層（Core Layers）
- 卷積神經層（Convolution layers）
- 池化神經層（Pooling layers）
- 循環神經層（Recurrent layers）
- 前置神經層（Preprocessing layers）
- 常態化神經層（Normalization layers）
- 正則神經層（Regularization layers）
- 注意力神經層（Attention layers）

4-3 神經層（Layer）

- 維度重置神經層（Reshaping layers）
- 合併神經層（Merging layers）
- 激勵神經層（Activation layers）

由於中文翻譯大部份都不能望文生義，後面的內容均使用英文術語。現階段僅介紹之前用到的核心神經層，其他類型的神經層在後續演算法用到時再說明。

4-3-1 完全連接神經層（Dense Layer）

Dense 是最常見的神經層，上一層神經層每個輸出的神經元都會完全連接到下一層神經層的每個輸入神經元，如下圖。

神經網路(Neural Network)

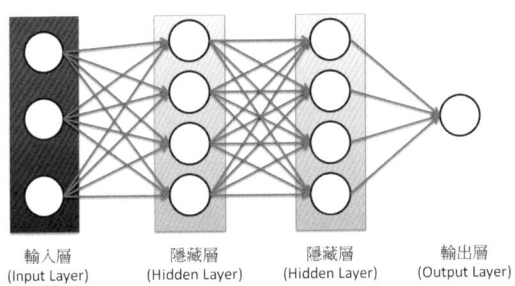

輸入層　　隱藏層　　　隱藏層　　輸出層
(Input Layer)　(Hidden Layer)　(Hidden Layer)　(Output Layer)

範例. Dense 測試。

程式：**04_07_ 神經層 .ipynb**。

1. 模型結構如下：

```
1  import tensorflow as tf
2  from tensorflow.keras import layers
3
4  # 建立模型
5  model = tf.keras.models.Sequential([
6      tf.keras.layers.Input((28, 28)),
7      tf.keras.layers.Flatten(),
8      tf.keras.layers.Dense(128, activation='relu', name="layer1"),
9      tf.keras.layers.Dropout(0.2),
10     tf.keras.layers.Dense(10, activation='softmax', name="layer2")
11 ])
12
```

第 4 章 神經網路實作

```
13  # 設定優化器(optimizer)、損失函數(Loss)、效能衡量指標(metrics)的類別
14  model.compile(optimizer='adam',
15                loss='sparse_categorical_crossentropy',
16                metrics=['accuracy'])
17
18  # 顯示模型彙總資訊
19  model.summary()
```

- 執行結果如下，顯示各層的 Output 及參數個數。

```
Model: "sequential_2"
_____
Layer (type)                 Output Shape              Param #
=================================================================
flatten_2 (Flatten)          (None, 784)               0

layer1 (Dense)               (None, 128)               100480

dropout_3 (Dropout)          (None, 128)               0

layer2 (Dense)               (None, 10)                1290
=================================================================
Total params: 101,770
Trainable params: 101,770
Non-trainable params: 0
_____
```

2. 設定模型的 output 為第一層 Dense：任意設定 X 的數值，只要符合模型的 Input 維度及大小，注意第一個維度是筆數，不需在模型定義，執行結果為 TensorShape（[1, 128]），1 是筆數，128 為第一層 Dense output 個數。

```
1   # 設定模型的 input/output
2   feature_extractor = tf.keras.Model(
3       inputs=model.inputs,
4       outputs=model.get_layer(name="layer1").output,
5   )
6
7   # 呼叫 feature_extractor 取得 output
8   x = tf.ones((1, 28, 28))
9   features = feature_extractor(x)
10  features.shape
```

3. 第一層 Dense 的參數個數計算：28*28*128 個權重（Weight）+ 128 個偏差（Bias）。

4-3 神經層（Layer）

```python
# 第一層 Dense 參數個數計算
parameter_count = (28 * 28) * features.shape[1] + features.shape[1]
print(f'參數(parameter)個數：{parameter_count}')
```

- 執行結果：參數個數共有 100,480 個與模型彙總資訊一致。

4. 第二層 Dense 的參數個數計算：只要修改 outputs 為 layer2。

```python
# 設定模型的 input/output
feature_extractor = tf.keras.Model(
    inputs=model.inputs,
    outputs=model.get_layer(name="layer2").output,
)

# 呼叫 feature_extractor 取得 output
x = tf.ones((1, 28, 28))
features = feature_extractor(x)

parameter_count = 128 * features.shape[1] + features.shape[1]
print(f'參數(parameter)個數：{parameter_count}')
```

- 執行結果：參數個數共有 1290 個與模型彙總資訊一致。

Dense 神經層的參數說明如下：

1. units：輸出神經元個數。
2. activation：指定要使用 activation function，也可以獨立使用 activation layer。
3. use_bias：權重參數估計是否要含偏差項。
4. bias_initializer：偏差初始值。
5. kernel_initializer：權重初始值，預設值是 glorot_uniform，它是均勻分配的隨機亂數。
6. kernel_regularizer：權重是否要使用防止過度擬合的正則函數（regularizer），預設值是無，也可設為 L1 或 L2。
7. bias_regularizer：偏差是否要使用防止過度擬合的正則函數，預設值是無，也可設為 L1 或 L2。
8. activity_regularize：activation function 是否要使用防止過度擬合的正則函數，預設值是無，也可設為 L1 或 L2。
9. kernel_constraint：權重是否有限制範圍。
10. bias_constraint：偏差是否有限制範圍。

第 4 章　神經網路實作

4-3-2 Dropout Layer

Dropout layer 在每一 epoch/step 訓練時，會隨機丟棄設定比例的輸入神經元，避免過度擬合，只會在訓練時運作，預測時會忽視 Dropout，不會有任何作用。參數說明如下：

1. rate：丟棄的比例，介於（0, 1）之間。
2. noise_shape：可以設定時序（time step）間的 Dropout 比例，適用於 RNN 演算法。

根據大部份學者的經驗，在神經網路中使用 Dropout 會比 Regularizer 效果來的好。

4-4 激勵函數（Activation Function）

Activation Function 是將線性方程式轉為非線性，目的是希望能提供更通用的模型。

$$Output = activation\ function(x_1w_1 + x_2w_2 + \cdots + x_nw_n + bias)$$

Activation Function 有非常多種函數，可以參考維基百科表格 [3]，如下圖：

Name	Plot	Function, $f(x)$	Derivative of f, $f'(x)$	Range
Identity		x	1	$(-\infty, \infty)$
Binary step		$\begin{cases} 0 & \text{if } x < 0 \\ 1 & \text{if } x \geq 0 \end{cases}$	$\begin{cases} 0 & \text{if } x \neq 0 \\ \text{undefined} & \text{if } x = 0 \end{cases}$	$\{0, 1\}$
Logistic, sigmoid, or soft step		$\sigma(x) = \dfrac{1}{1+e^{-x}}$ [1]	$f(x)(1-f(x))$	$(0, 1)$
tanh		$\tanh(x) = \dfrac{e^x - e^{-x}}{e^x + e^{-x}}$	$1 - f(x)^2$	$(-1, 1)$
Rectified linear unit (ReLU)[11]		$\begin{cases} 0 & \text{if } x \leq 0 \\ x & \text{if } x > 0 \end{cases}$ $= \max\{0, x\} = x\mathbf{1}_{x>0}$	$\begin{cases} 0 & \text{if } x < 0 \\ 1 & \text{if } x > 0 \\ \text{undefined} & \text{if } x = 0 \end{cases}$	$[0, \infty)$
Gaussian error linear unit (GELU)[6]		$\dfrac{1}{2}x\left(1 + \text{erf}\left(\dfrac{x}{\sqrt{2}}\right)\right)$ $= x\Phi(x)$	$\Phi(x) + x\phi(x)$	$(-0.17\ldots, \infty)$
Softplus[12]		$\ln(1 + e^x)$	$\dfrac{1}{1+e^{-x}}$	$(0, \infty)$

4-4 激勵函數（Activation Function）

Name	Plot	Function, $f(x)$	Derivative of f, $f'(x)$	Range
Exponential linear unit (ELU)[13]		$\begin{cases} \alpha(e^x - 1) & \text{if } x \leq 0 \\ x & \text{if } x > 0 \end{cases}$ with parameter α	$\begin{cases} \alpha e^x & \text{if } x < 0 \\ 1 & \text{if } x > 0 \\ 1 & \text{if } x = 0 \text{ and } \alpha = 1 \end{cases}$	$(-\alpha, \infty)$
Scaled exponential linear unit (SELU)[14]		$\lambda \begin{cases} \alpha(e^x - 1) & \text{if } x < 0 \\ x & \text{if } x \geq 0 \end{cases}$ with parameters $\lambda = 1.0507$ and $\alpha = 1.67326$	$\lambda \begin{cases} \alpha e^x & \text{if } x < 0 \\ 1 & \text{if } x \geq 0 \end{cases}$	$(-\lambda\alpha, \infty)$
Leaky rectified linear unit (Leaky ReLU)[15]		$\begin{cases} 0.01x & \text{if } x < 0 \\ x & \text{if } x \geq 0 \end{cases}$	$\begin{cases} 0.01 & \text{if } x < 0 \\ 1 & \text{if } x \geq 0 \end{cases}$	$(-\infty, \infty)$
Parameteric rectified linear unit (PReLU)[16]		$\begin{cases} \alpha x & \text{if } x < 0 \\ x & \text{if } x \geq 0 \end{cases}$ with parameter α	$\begin{cases} \alpha & \text{if } x < 0 \\ 1 & \text{if } x \geq 0 \end{cases}$	$(-\infty, \infty)$[2]
ElliotSig,[17][18] softsign[19][20]		$\dfrac{x}{1 + |x|}$	$\dfrac{1}{(1 + |x|)^2}$	$(-1, 1)$
Square nonlinearity (SQNL)[21]		$\begin{cases} 1 & \text{if } x > 2.0 \\ x - \frac{x^2}{4} & \text{if } 0 \leq x \leq 2.0 \\ x + \frac{x^2}{4} & \text{if } -2.0 \leq x < 0 \\ -1 & \text{if } x < -2.0 \end{cases}$	$1 \mp \dfrac{x}{2}$	$(-1, 1)$
S-shaped rectified linear activation unit (SReLU)[22]		$\begin{cases} t_l + a_l(x - t_l) & \text{if } x \leq t_l \\ x & \text{if } t_l < x < t_r \\ t_r + a_r(x - t_r) & \text{if } x \geq t_r \end{cases}$ where t_l, a_l, t_r, a_r are parameters.	$\begin{cases} a_l & \text{if } x \leq t_l \\ 1 & \text{if } t_l < x < t_r \\ a_r & \text{if } x \geq t_r \end{cases}$	$(-\infty, \infty)$
Bent identity		$\dfrac{\sqrt{x^2 + 1} - 1}{2} + x$	$\dfrac{x}{2\sqrt{x^2 + 1}} + 1$	$(-\infty, \infty)$
Sigmoid linear unit (SiLU,[6] SiL,[23] or Swish-1[24])		$\dfrac{x}{1 + e^{-x}}$	$\dfrac{1 + e^{-x} + xe^{-x}}{(1 + e^{-x})^2}$	$[-0.278\ldots, \infty)$
Gaussian		e^{-x^2}	$-2xe^{-x^2}$	$(0, 1]$
SQ-RBF		$\begin{cases} 1 - \frac{x^2}{2} & \text{if } |x| \leq 1 \\ \frac{1}{2}(2 - |x|)^2 & \text{if } 1 < |x| < 2 \\ 0 & \text{if } |x| \geq 2 \end{cases}$	$\begin{cases} -x & \text{if } |x| \leq 1 \\ x - 2\,\text{sgn}(x) & \text{if } 1 < |x| < 2 \\ 0 & \text{if } |x| \geq 2 \end{cases}$	$[0, 1]$

▲ 表 4.1 Activation Function 列表，資料來源：維基百科

Tensorflow 支援大部份的函數，如果找不到的話，也能夠自訂函數，它們可以直接設定在神經層的參數，也可以是獨立的神經層。

第 4 章　神經網路實作

範例 1. 常用的 Activation Function 測試。

程式：**04_08_Activation_Function.ipynb**。

1. ReLU（Rectified linear unit）：是目前隱藏層最常用的函數，公式請參考上表，函數名稱為 relu。

```
1  # 設定 x = -10, -9, ..., 10 測試
2  x= np.linspace(-10, 10, 21)
3  x_tf = tf.constant(x, dtype = tf.float32)
4
5  # ReLU
6  y = activations.relu(x_tf).numpy()
7
8  # 繪圖
9  plt.plot(x, y)
10 plt.show()
```

- 執行結果：會忽視過小的外部輸入，比如說，我們輕輕碰一下皮膚，大腦可能不會做出反應。

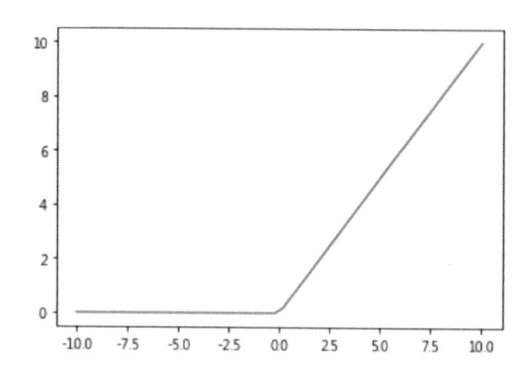

- relu 函數有三個參數：

  * threshold：超過此門檻值，y 才會 >0。例如 threshold=5，如下圖。

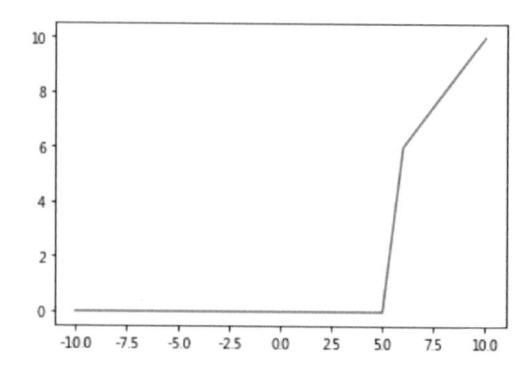

4-4 激勵函數（Activation Function）

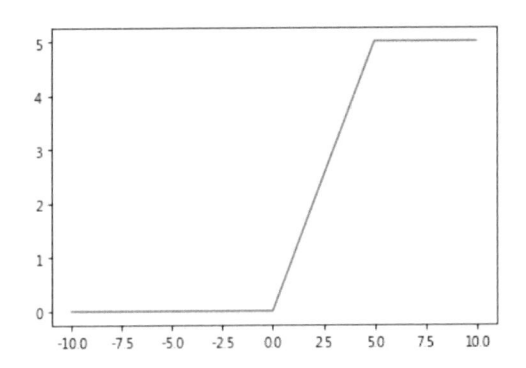

* max_value：y 的上限。例如 max_value=5，如下圖。

* alpha：小於門檻值，y 會等於 x * alpha。例如 alpha=0.5，如下圖，又稱為 Parameteric rectified linear unit（PReLU），若 alpha=0.01，則稱為 Leaky rectified linear unit（Leaky ReLU）。

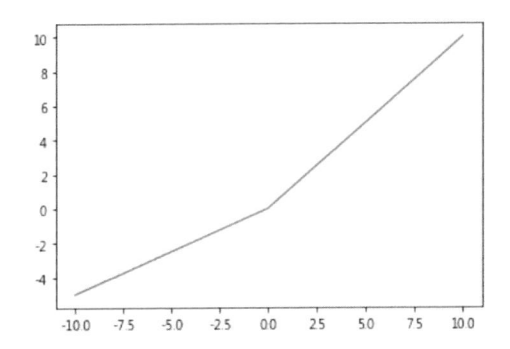

- 相關測試請參閱程式。

2. Sigmoid：即羅吉斯迴歸，因為函數為 S 型而得名，適用於二分類，可加在最後一層 Dense 內，表二分類。

```
1  x= np.linspace(-10, 10, 21)
2  x_tf = tf.constant(x, dtype = tf.float32)
3
4  # sigmoid
5  y = activations.sigmoid(x_tf).numpy()
6
7  # 模糊地帶
8  plt.axvline(-4, color='r')
9  plt.axvline(4, color='r')
```

```
10
11  plt.plot(x, y)
12  plt.show()
```

- 執行結果：函數最小值為 0，最大值為 1，只有兩條直線中間是模糊地帶，但也是一個平滑改變的過程，而非階梯形的函數，可降低預測的變異性（Variance）。

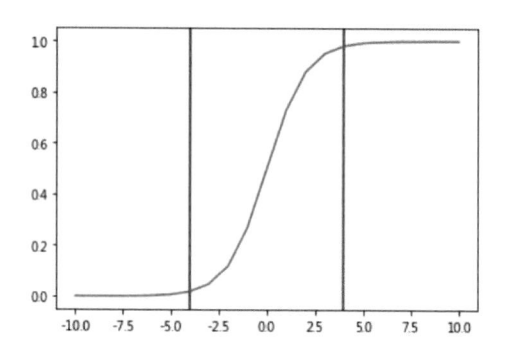

3. tanh：與 Sigmoid 類似，但最小值是 -1。

```
1   x= np.linspace(-10, 10, 21)
2   x_tf = tf.constant(x, dtype = tf.float32)
3
4   # tanh
5   y = activations.tanh(x_tf).numpy()
6
7   # 模糊地帶
8   plt.axvline(-3, color='r')
9   plt.axvline(3, color='r')
10
11  plt.plot(x, y)
12  plt.show()
```

- 執行結果：函數最小值為 -1，最大值為 1，只有兩條直線中間是模糊地帶，平滑改變的過程與 Sigmoid 相比較為陡峭。

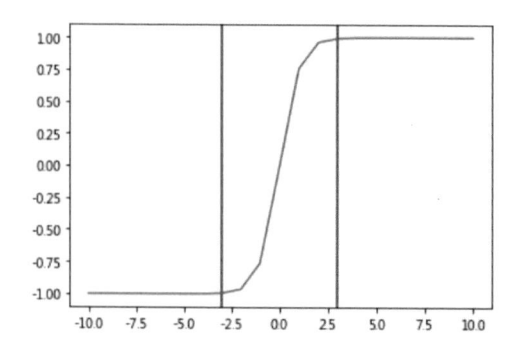

4-4 激勵函數（Activation Function）

4. Softmax：這個函數會將輸入轉為機率，即所有值介於 [0, 1] 之間，總和為 1，適用於多分類，可加在最後一層 Dense 內。

```
1  # activations.softmax 輸入需為 2 維資料，設定 x 為均勻分配，轉換後每一列加總為 1
2  x = np.random.uniform(1, 10, 40).reshape(10, 4)
3  print('輸入：\n', x)
4  x_tf = tf.constant(x, dtype = tf.float32)
5
6  # Softmax
7  y = activations.softmax(x_tf).numpy()
8  print('加總：', np.round(np.sum(y, axis=1)))
```

- 執行結果：softmax 輸入參數必須是 2 維資料，上述程式設定 x 為均勻分配的隨機亂數，轉換後每一列總和為 1。

```
輸入：
[[1.87635979 5.65431617 5.0807942  7.84746665]
 [5.71430835 1.0332283  9.15220976 4.59581463]
 [2.91586061 1.01252361 9.21877442 5.51475778]
 [4.22146643 1.18823067 6.42331145 6.23998232]
 [2.61397256 7.75011603 2.62736731 9.18788281]
 [4.07362709 1.12741729 8.18113136 8.01828864]
 [9.20565064 7.19991452 1.14951624 6.12306693]
 [3.77131905 3.75599114 9.43088289 5.65098351]
 [7.56332186 3.4039989  2.22646493 8.14256405]
 [5.1090939  2.1567948  9.0736152  8.6324396 ]]
加總： [1. 1. 1. 1. 1. 1. 1. 1. 1. 1.]
```

- 使用 NumPy 計算 Softmax。

```
1  # 設定 x = 1, 2, ..., 10 測試
2  x= np.random.uniform(1, 10, 40)
3
4  # Softmax
5  y = np.e ** (x) / tf.reduce_sum(np.e ** (x))
6  print(sum(y))
```

5. 自訂函數，可以使用 TensorFlow 張量函數，只要傳回與輸入 / 輸出相符的維度和資料型態即可，例如：

model.add（layers.Dense（64, activation=tf.nn.tanh））

6. 其他的函數請參見 Keras 官網，包括這兩個網址：

- Keras 官網 Activation Function 說明 [11]。
- Keras 官網 Activation Layer 說明 [12]。

4-39

第 4 章　神經網路實作

一般來說，Activation Function 會接在神經層後面，如下所示：

- x = layers.Dense（10）(x)
- x = layers.LeakyReLU()(x)

TensorFlow/Keras 為簡化語法，允許將 Activation Function 當作參數使用，直接包在神經層的定義中，如下：

- tf.keras.layers.Dense（128, activation='relu'）
- tf.keras.layers.Dense（10, activation='softmax'）

▋4-5　損失函數（Loss Functions）

損失函數（Loss Functions）又稱為目標函數（Objective Function）或成本函數（Cost Function），通常是定義為預測值與實際值的誤差平方和，再加以平均，演算法以損失函數最小化為目標，求算權重的近似解，演算法會因目的不同定義各種損失函數，讓求解的過程隨著損失函數改變，例如迴歸定義損失函數為均方誤差（MSE），使平均誤差最小化，而 MidJourney 採用擴散模型（Diffusion）定義特殊的損失函數，讓生成的圖像不失真，ChatGPT 採用 Transformer 演算法，讓生成的回答與提問最吻合，因此我們可以說損失函數的定義是深度學習演算法中最關鍵的要素。

TensorFlow 損失函數分成三類，請參閱官網[7]：

1. 機率相關的損失函數（Probabilistic Loss）：主要用於分類，例如二分類的交叉熵（BinaryCrossentropy）、多分類的交叉熵（CategoricalCrossentropy）。
2. 迴歸相關的損失函數（Regression Loss）：主要用於預測連續型的目標變數，例如均方誤差（MSE）、平均絕對誤差（MAE）。
3. 鉸鏈損失函數（Hinge Loss）：經常用在『最大間隔分類』（Maximum-margin Classification），適用於支援向量機（SVM）等演算法。

TensorFlow 損失函數一般在 model.compile 中設定，如下：

- model.compile（loss='mean_squared_error', optimizer='sgd'）

上面直接使用字串，如果擔心粗心大意拼錯字的話，也可以使用函數：

- from keras import losses
- model.compile（loss=losses.mean_squared_error, optimizer='sgd'）

4-5 損失函數（Loss Functions）

範例. 實際測試幾個常用的損失函數。

程式：**04_09_Loss_Function.ipynb**。

- BinaryCrossentropy：二分類的交叉熵，熵（Entropy）是指分類的不純度（Impurity），愈低愈好，公式如下：

$$s = -\int p(x) \log p(x)\, dx \quad \rightarrow \text{連續型分配}$$
$$s = -\sum p(x) \log p(x) \quad \rightarrow \text{離散型分配}$$

- 若是二分類 y=0 或 1，則離散型分配的二分類交叉熵等於：

$$s = -y \log(p) - (1-y)\log(1-p) \quad \rightarrow \text{公式 4.1}$$

當

y=0 時 ➔ $s = -\log(1-p)$

y=1 時 ➔ $s = -\log(p)$

- 使用 Sigmoid 計算 p，就可得到 BinaryCrossentropy。

1. 兩筆資料的實際值和預測值如下，計算 BinaryCrossentropy。

```
1  # 兩筆資料實際及預測值
2  y_true = [[0., 1.], [0., 0.]]      # 實際值
3  y_pred = [[0.6, 0.4], [0.4, 0.6]]  # 預測值
4
5  # 二分類交叉熵(BinaryCrossentropy)
6  bce = tf.keras.losses.BinaryCrossentropy()
7  bce(y_true, y_pred).numpy()
```

- 執行結果：0.8149。

- 依照公式 4.1 驗算：

```
1  # 驗算
2  import math
3
4  ((0-math.log(1-0.6) - math.log(0.4)) + (0-math.log(1-0.6) - math.log(0.6)) )/4
```

- 執行結果：0.8149，與 BinaryCrossentropy() 一致。

2. CategoricalCrossentropy：多分類的交叉熵。兩筆資料的實際值和預測值如下，計算 BinaryCrossentropy。

```
1  # 兩筆資料實際及預測值
2  y_true = [[0, 1, 0], [0, 0, 1]]        # 實際值
3  y_pred = [[0.05, 0.95, 0], [0.1, 0.8, 0.1]]  # 預測值
4
5  # 多分類交叉熵(CategoricalCrossentropy)
6  cce = tf.keras.losses.CategoricalCrossentropy()
7  cce(y_true, y_pred).numpy()
```

- 執行結果：1.1769。

3. SparseCategoricalCrossentropy：稀疏矩陣的多分類交叉熵，當目標變數（y）是單一值，而非 one-hot encoding 的資料型態，可使用此損失函數，它會自動執行 one-hot encoding，再與預測值作比較。例如兩筆資料的實際值和預測值如下，計算 SparseCategoricalCrossentropy。

```
1  # 兩筆資料實際及預測值
2  y_true = [1, 2]       # 實際值
3  y_pred = [[0.05, 0.95, 0], [0.1, 0.8, 0.1]]  # 預測值
4
5  # 多分類交叉熵(CategoricalCrossentropy)
6  cce = tf.keras.losses.SparseCategoricalCrossentropy()
7  cce(y_true, y_pred).numpy()
```

- 執行結果：1.1769。

4. MeanSquaredError：計算實際值和預測值的均方誤差。兩筆資料的實際值和預測值如下，計算 MeanSquaredError。

```
1  # 兩筆資料實際及預測值
2  y_true = [[0., 1.], [0., 0.]]      # 實際值
3  y_pred = [[1., 1.], [1., 0.]]      # 預測值
4
5  # 多分類交叉熵(CategoricalCrossentropy)
6  mse = tf.keras.losses.MeanSquaredError()
7  mse(y_true, y_pred).numpy()
```

- 執行結果：（$(1-1)^2 + (0-1)^2$）/ 2 = 0.5。

5. 鉸鏈損失函數（Hinge Loss）：常用於支援向量機（SVM），詳細可參閱維基百科的鉸鏈損失函數介紹[13]：

    - total loss = $\sum$ maximum（1 - y_true * y_pred, 0）
    - 不考慮負值的損失，所以也被稱作單邊損失函數，真實值（y_true）通常是 -1 或 1，如果訓練資料的 y 是 0/1，Hinge Loss 會自動將其轉成 -1/1，再計算損失。

4-6 優化器（Optimizer）

- 兩筆資料的實際值和預測值如下，計算 Hinge Loss。

```
1  # 兩筆資料實際及預測值
2  y_true = [[0., 1.], [0., 0.]]      # 實際值
3  y_pred = [[0.6, 0.4], [0.4, 0.6]]  # 預測值
4
5  # Hinge Loss
6  loss_function = tf.keras.losses.Hinge()
7  loss_function(y_true, y_pred).numpy()
```

- 執行結果：1.3。

- 驗算：

```
1  # 驗算
2  # loss = sum (maximum(1 - y_true * y_pred, 0))
3  (max(1 - (-1) * 0.6, 0) + max(1 - 1 * 0.4, 0) +
4      max(1 - (-1) * 0.4, 0) + max(1 - (-1) * 0.6, 0)) / 4
```

- 執行結果與 Hinge 函數相同。

6. 自訂損失函數（Custom Loss）：撰寫一個函數，輸入為 y 的實際值及預測值，輸出為常數即可，下列程式碼自訂損失為 MSE。

```
1  # 自訂損失函數(Custom Loss)
2  def my_loss_fn(y_true, y_pred):
3      # MSE
4      squared_difference = tf.square(y_true - y_pred)
5      return tf.reduce_mean(squared_difference, axis=-1)  # axis=-1 須設為 -1
6
7  model.compile(optimizer='adam', loss=my_loss_fn)
```

4-6 優化器（Optimizer）

優化器是神經網路中反向傳導的求解方法，著重在兩方面：

1. 動態調整學習率的大小，以加速求解的收斂速度，且不會跳過最佳解。

2. 想辦法跨過區域最小值（Local Minimum）或避開馬鞍點（Saddle Point），找到全局最小值（Global Minimum）。

第 4 章　神經網路實作

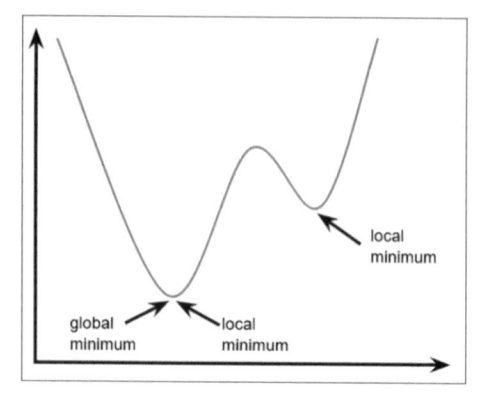

▲ 圖 4.3　區域最小值（Local Minimum）與全局最小值（Global Minimum）

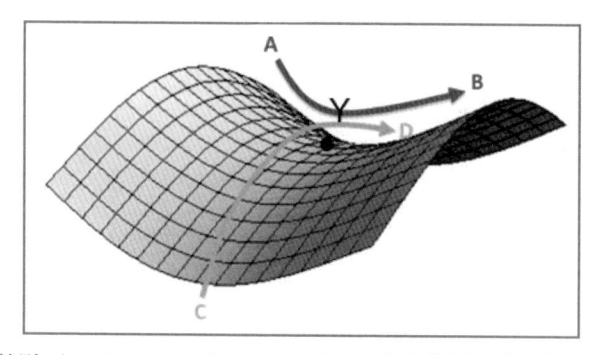

▲ 圖 4.4　馬鞍點（Saddle Point）：由 A 至 B 方向觀察馬鞍點（Y）是最小值，由 C 至 D 方向觀察馬鞍點是最大值

我們在第二章的梯度下降法**範例 7**_ 梯度下降法 .ipynb，採用固定的學習率，雖然可以找到最佳解，但太沒有效率，因為剛開始求解時距離最佳解還很遙遠時，應該可以加大學習率，快速逼近最佳解，之後再逐步縮小學習率，避免錯過最佳解，因此最簡單的方式就是初始化較大的學習率，之後隨著訓練週期或時間，逐步縮小學習率，可加速求解的收斂速度。除了這種方式，還有各種控制學習率的方法，例如考慮動能（Momentum）、衰退（Decay）…等。

優化器的類別也在 model.compile 設定，TensorFlow 支援很多種不同的優化器如下，可參閱 Keras 官網優化器（Optimizers）介紹，瞭解相關學習率公式。

- SGD
- RMSprop

4-6 優化器（Optimizer）

- Adam
- Adadelta
- Adagrad
- Adamax
- Nadam
- Ftrl
- ⋯

範例. 實際測試幾個常用的優化器。

程式：**04_10_Optimizer.ipynb**。

1. 隨機梯度下降法（Stochastic Gradient Decent, SGD）

    依據權重更新的時機差別，梯度下降法分為下面三種：

    - 批量梯度下降法（Batch Gradient Descent, BGD）：以『全部』樣本計算梯度，更新權重。
    - 隨機梯度下降法（Stochastic Gradient Descent, SGD）：一次抽取一小批樣本計算梯度，並立即更新權重。優點是更新速度快，但收斂會較曲折，因為訓練過程中，可能抽到好樣本，也可能抽到壞樣本。

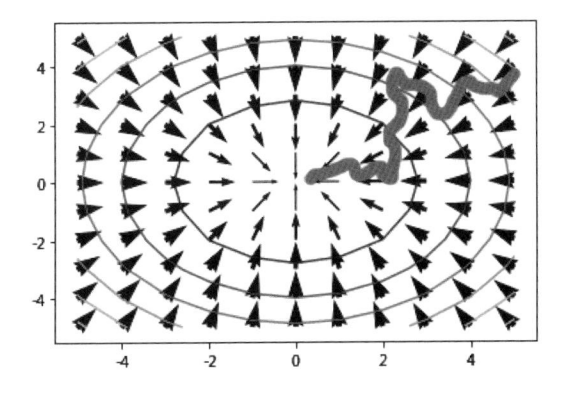

    ▲ 圖 4.5 隨機梯度下降法（Stochastic Gradient Descent, SGD）求解圖示

第 4 章　神經網路實作

SGD 語法：

```
1  # SGD
2  tf.keras.optimizers.SGD(
3      learning_rate=0.01, momentum=0.0, nesterov=False, name="SGD"
4  )
```

- 權重更新公式：w = w - learning_rate * g

  其中 w：權重，g：梯度，learning_rate：學習率

- 動能（momentum）：公式為

  velocity = momentum * velocity - learning_rate * g
  w = w + velocity

  動能通常介於（0, 1），若等於 0，表示學習率為固定值。一般而言，剛開始訓練時，離最小值較遠的時候，學習率可以放膽邁大步，越接近最小值時，學習率變動幅度就要變小，以免錯過最小值，這種動態調整學習率的方式，能夠使求解收斂速度加快，又不會錯過最小值。

- nesterov：是否使用 Nesterov momentum，預設值是 False。要瞭解技術細節可參閱『Understanding Nesterov Momentum（NAG）』[14]。

1.1　隨機梯度下降法的簡單測試。

```
1   # SGD
2   opt = tf.keras.optimizers.SGD(learning_rate=0.1)
3
4   # 任意變數
5   var = tf.Variable(1.0)
6
7   # 損失函數
8   loss = lambda: (var ** 2)/2.0
9
10  # step_count：優化的步驟
11  for i in range(51):
12      step_count = opt.minimize(loss, [var]).numpy()
13      if i % 10 == 0 and i > 0:
14          print(f'優化的步驟:{step_count}, 變數:{var.numpy()}')
```

- 損失函數 $x^2/2$。

- 執行結果：每 10 步列印結果，越來越接近最小值 0。

4-6 優化器（Optimizer）

```
優化的步驟:11, 變數:0.3138105869293213
優化的步驟:21, 變數:0.10941897332668304
優化的步驟:31, 變數:0.03815203905105591
優化的步驟:41, 變數:0.01330279465764761
優化的步驟:51, 變數:0.0046383971348404884
```

1.2 優化三次測試隨機梯度下降法的動能。

```
1  opt = tf.keras.optimizers.SGD(learning_rate=0.1, momentum=0.9)
2  var = tf.Variable(1.0)
3
4  # 損失函數起始值
5  val0 = var.value()
6
7  # 損失函數
8  loss = lambda: (var ** 2)/2.0
9
10 # 優化第一次
11 step_count = opt.minimize(loss, [var]).numpy()
12 val1 = var.value()
13 print(f'優化的步驟:{step_count}, 變化值:{(val0 - val1).numpy()}')
14

15 # 優化第二次
16 step_count = opt.minimize(loss, [var]).numpy()
17 val2 = var.value()
18 print(f'優化的步驟:{step_count}, 變化值:{(val1 - val2).numpy()}')
19
20 # 優化第三次
21 step_count = opt.minimize(loss, [var]).numpy()
22 val3 = var.value()
23 print(f'優化的步驟:{step_count}, 變化值:{(val2 - val3).numpy()}')
```

- 執行結果：

```
val0:1.0
優化的步驟:1, val1:0.8999999761581421, 變化值:0.10000002384185791
優化的步驟:2, val2:0.7199999690055847, 變化值:0.18000000715255737
優化的步驟:3, val3:0.4860000014305115, 變化值:0.23399996757507324
```

2. Adam（Adaptive Moment Estimation）是常用的優化器，這裡就引用 Kingma 等學者於 2014 年發表的『Adam: A Method for Stochastic Optimization』[15] 一文所作的評論『Adam 計算效率高、記憶體耗費少，適合大資料集及參數個數很多的模型』。

Adam 語法：

```
1  # Adam
2  tf.keras.optimizers.Adam(
```

```
3        learning_rate=0.001,
4        beta_1=0.9,
5        beta_2=0.999,
6        epsilon=1e-07,
7        amsgrad=False,
8        name="Adam",
9   )
```

- beta_1：一階動能衰減率（exponential decay rate for the 1st moment estimates）。

- beta_2：二階動能衰減率（exponential decay rate for the 2nd moment estimates）。

- epsilon：誤差值，小於這個值，優化即停止。

- amsgrad：是否使用 AMSGrad，預設值是 False。技術細節可參閱『一文告訴你 Adam、AdamW、Amsgrad 區別和聯繫』[16]。

2.1　Adam 簡單測試。

```
1   # Adam
2   opt = tf.keras.optimizers.Adam(learning_rate=0.1)
3
4   # 任意變數
5   var = tf.Variable(1.0)
6
7   # 損失函數
8   loss = lambda: (var ** 2)/2.0
9
10  # step_count：優化的步驟
11  for i in range(11):
12      step_count = opt.minimize(loss, [var]).numpy()
13      if i % 2 == 0 and i > 0:
14          print(f'優化的步驟:{step_count-1}, 變數:{var.numpy()}')
```

- 執行結果：SGD 執行 50 步，Adam 只需要執行 10 步，就已收斂。

```
優化的步驟:2, 變數:0.7015826106071472
優化的步驟:4, 變數:0.5079597234725952
優化的步驟:6, 變數:0.3234168291091919
優化的步驟:8, 變數:0.15358148515224457
優化的步驟:10, 變數:0.005128741264343262
```

3. 另外還有幾種常用的優化器：

- Adagrad（Adaptive Gradient-based optimization）：設定每個參數的學習率更新頻率不同，較常變動的特徵使用較小的學習率，較少調整，反之，使用較大的學習率，比較頻繁的調整，主要是針對稀疏的資料集。

- RMS-Prop：每次學習率更新是除以均方梯度（average of squared gradients），以指數的速度衰減。
- ADAM：是 Adagrad 改良版，學習率更新會配合過去的平均梯度調整。

官網還有介紹其他的優化器，網路上也有許多優化器的比較和動畫，有興趣的讀者可參閱『Alec Radford's animations for optimization algorithms』[17]。

4-7 效能衡量指標（Performance Metrics）

效能衡量指標是衡量模型優劣的標準，要了解各種效能衡量指標，先要理解混淆矩陣（Confusion Matrix），以二分類而言，如下圖。

	真實 真(True)	真實 假(False)
預測 陽性(Positive)	TP	FP
預測 陰性(Negative)	TN	FN

▲ 圖 4.6 混淆矩陣（Confusion Matrix）

1. 橫軸為預測結果，分為陽性（Positive, 簡稱 P）、陰性（Negative, 簡稱 N）。
2. 縱軸為真實狀況，分為真（True, 簡稱 T）、假（False, 簡稱 F）。
3. 依預測結果及真實狀況的組合，共分為四種狀況：
   - TP（真陽性）：預測為陽性，且預測正確。
   - TN（真陰性）：預測為陰性，且預測正確。
   - FP（偽陽性）：預測為陽性，但預測錯誤，又稱型一誤差（Type I Error），或 α 誤差。
   - FN（偽陰性）：預測為陰性，但預測錯誤，又稱型二誤差（Type II Error），或 β 誤差。
4. 有了 TP/TN/FP/FN 之後，我們就可以定義各種效能衡量指標，常見的有四種：
   - 準確率（Accuracy）=（TP+TN）/（TP+FP+FN+TN），即

     『預測正確數 / 總數』。

- 精確率（Precision）= TP/（TP+FP），即

  『正確預測陽性數 / 總陽性數』。

- 召回率（Recall）= TP/（TP+FN），即

  『正確預測陽性數 / 實際為真的總數』。

- F1 = 精確率與召回率的調和平均數，即

  1 /（（1 / Precision）+（1 / Recall））。

5. FP（偽陽性）與 FN（偽陰性）是相衝突的，以 Covid-19 檢驗為例，如果降低陽性認定值，可以盡最大可能找到所有的確診者，減少偽陰性，避免傳染病擴散，但有些沒病的人會被誤判，偽陽性會因而相對增加，導致資源的浪費，更嚴重可能造成醫療體系崩潰，得不償失，所以，疾病管制署（CDC）會因應疫情的發展，隨時調整陽性認定值。

6. 除了準確率之外，為什麼還需要參考其他指標？

    - 以醫療檢驗設備來舉例，假設某疾病實際染病的比率為 1%，這時我們拿一個故障的檢驗設備，它不管有無染病，都判定為陰性，這時候計算設備準確率，結果竟然是 99%。會有這樣離譜的統計，是因為在此案例中，驗了 100 個樣本，確實只錯一個。所以，碰到真假比例懸殊的不平衡（Imbalanced）樣本，必須使用其他指標來衡量效能。

    - 精確率：再以醫療檢驗設備為例，我們只關心被驗出來的陽性病患，有多少比例是真的染病，而不去關心驗出為陰性者，因為驗出為陰性，通常不會再被複檢，或者不放心又跑到其他醫院複檢，醫院其實很難追蹤他們是否真的沒病。

    - 召回率：比方 Covid-19，我們關心的是所有的染病者有多少比例被驗出陽性，因為一旦有漏網之魚（偽陰性），可能就會造成重大的傷害，如社區傳染。

7. 針對二分類，還有一種較客觀的指標稱為 ROC/AUC 曲線，它是在各種檢驗門檻值下，以假陽率為 X 軸，真陽率為 Y 軸，繪製出來的曲線，稱為 ROC。覆蓋的面積（AUC）越大，表示模型在各種門檻值下的平均效能越好，這個指標有別於一般預測固定以 0.5 當作判斷真假的基準。

8. 有趣的是，損失函數也可視為效能衡量指標，因為當損失函數越小，就表示預測值與實際值越接近，所以，TensorFlow 也可以把損失函數當作效能衡量指標來使用。

4-7 效能衡量指標（Performance Metrics）

TensorFlow 的效能衡量指標可參閱 Keras 官網[18]。

程式：**04_11_Metrics.ipynb**。

範例 1. 假設有 8 筆資料如下，請計算混淆矩陣（Confusion Matrix）。

實際值 = [0, 0, 0, 1, 1, 1, 1, 1]

預測值 = [0, 1, 0, 1, 0, 1, 0, 1]

1. 載入相關套件

```
1  import tensorflow as tf
2  from tensorflow.keras import metrics
3  import numpy as np
4  import matplotlib.pyplot as plt
5  from sklearn.metrics import accuracy_score, classification_report
6  from sklearn.metrics import precision_score, recall_score, confusion_matrix
```

2. Scikit-learn 提供混淆矩陣（Confusion Matrix）函數，程式碼如下。

```
1  from sklearn.metrics import confusion_matrix
2
3  y_true = [0, 0, 0, 1, 1, 1, 1, 1] # 實際值
4  y_pred = [0, 1, 0, 1, 0, 1, 0, 1] # 預測值
5
6  # 混淆矩陣(Confusion Matrix)
7  tn, fp, fn, tp = confusion_matrix(y_true, y_pred).ravel()
8  print(f'TP={tp}, FP={fp}, TN={tn}, FN={fn}')
```

- 注意，Scikit-learn 提供的混淆矩陣，傳回值與圖 4.6 位置不同。
- 實際值與預測值上下比較，TP 為（1, 1）、FP 為（0, 1）、TN 為（0, 0）、FN 為（1, 0）。
- 執行結果：TP=3, FP=1, TN=2, FN=2。

3. 繪圖

```
1  # 顯示矩陣
2  fig, ax = plt.subplots(figsize=(2.5, 2.5))
3
4  # 1:藍色, 0:白色
5  ax.matshow([[1, 0], [0, 1]], cmap=plt.cm.Blues, alpha=0.3)
6
7  # 標示文字
8  ax.text(x=0, y=0, s=tp, va='center', ha='center')
9  ax.text(x=1, y=0, s=fp, va='center', ha='center')
```

4-51

```
10  ax.text(x=0, y=1, s=tn, va='center', ha='center')
11  ax.text(x=1, y=1, s=fn, va='center', ha='center')
12
13  plt.xlabel('實際', fontsize=20)
14  plt.ylabel('預測', fontsize=20)
15
16  # x/y 標籤
17  plt.xticks([0,1], ['T', 'F'])
18  plt.yticks([0,1], ['P', 'N'])
19  plt.show()
```

- 執行結果：

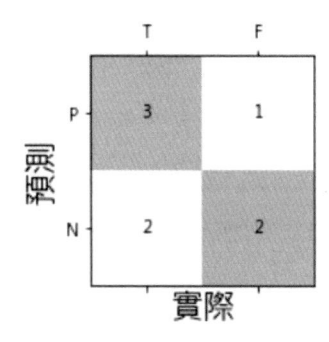

範例 2. 依上述資料計算效能衡量指標。

1. 準確率。

```
1  m = metrics.Accuracy()
2  m.update_state(y_true, y_pred)
3
4  print(f'準確率:{m.result().numpy()}')
5  print(f'驗算={(tp+tn) / (tp+tn+fp+fn)}')
```

- 執行結果：0.625。

2. 計算精確率。

```
1  m = metrics.Precision()
2  m.update_state(y_true, y_pred)
3
4  print(f'精確率:{m.result().numpy()}')
5  print(f'驗算={(tp) / (tp+fp)}')
```

- 執行結果：0.75。

4-7 效能衡量指標（Performance Metrics）

3. 計算召回率。

```
1  m = metrics.Recall()
2  m.update_state(y_true, y_pred)
3
4  print(f'召回率:{m.result().numpy()}')
5  print(f'驗算={(tp) / (tp+fn)}')
```

- 執行結果：0.6。

範例 3. 依資料檔 data/auc_data.csv 計算 AUC。

1. 讀取資料檔

```
1  # 讀取資料
2  import pandas as pd
3  df=pd.read_csv('./data/auc_data.csv')
4  df
```

- 執行結果：

	predict	actual
0	0.11	0
1	0.35	0
2	0.72	1
3	0.10	1
4	0.99	1
5	0.44	1
6	0.32	0
7	0.80	1
8	0.22	1
9	0.08	0
10	0.56	1

2. 以 Scikit-learn 函數計算 AUC

```
1  from sklearn.metrics import roc_curve, roc_auc_score, auc
2
3  # fpr：假陽率, tpr：真陽率, threshold：各種決策門檻
4  fpr, tpr, threshold = roc_curve(df['actual'], df['predict'])
5  print(f'假陽率={fpr}\n\n真陽率={tpr}\n\n決策門檻={threshold}')
```

- 執行結果：

```
假陽率=[0.         0.         0.         0.14285714 0.14285714 0.28571429
 0.28571429 0.57142857 0.57142857 0.71428571 0.71428571 1.        ]
真陽率=[0.         0.09090909 0.27272727 0.27272727 0.63636364 0.63636364
 0.81818182 0.81818182 0.90909091 0.90909091 1.         1.        ]
決策門檻=[1.99 0.99 0.8  0.73 0.56 0.48 0.42 0.32 0.22 0.11 0.1  0.03]
```

3. 繪製 AUC

```
1  # 繪圖
2  auc1 = auc(fpr, tpr)
3  ## Plot the result
4  plt.title('ROC/AUC')
5  plt.plot(fpr, tpr, color = 'orange', label = 'AUC = %0.2f' % auc1)
6  plt.legend(loc = 'lower right')
7  plt.plot([0, 1], [0, 1],'r--')
8  plt.xlim([0, 1])
9  plt.ylim([0, 1])
10 plt.ylabel('True Positive Rate')
11 plt.xlabel('False Positive Rate')
12 plt.show()
```

- 執行結果：

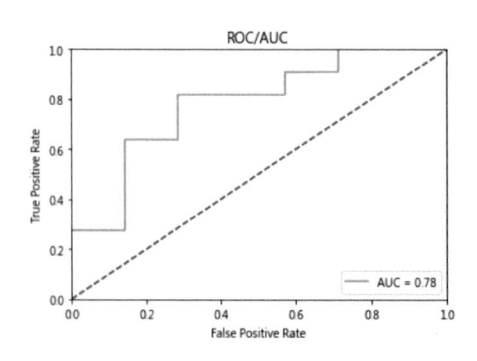

4. 以 TensorFlow 函數計算 AUC

```
1  m = metrics.AUC()
2  m.update_state(df['actual'], df['predict'])
3
4  print(f'AUC:{m.result().numpy()}')
```

- 執行結果：0.7792，與 Scikit-learn 的結果相去不遠。

4-8 超參數調校（Hyperparameter Tuning）

這一節來研究超參數（Hyperparameters）對效能的影響。在 4-1-3 只對單一變數進行調校，假如要同時調校多個超參數，有一些套件可以幫忙，包括 Keras Tuner、hyperopt、Ray Tune、Ax…等。

本節介紹 Keras Tuner 的用法。先安裝套件：

pip install keras-tuner hiplot

範例. 超參數調校。

程式：**04_04_keras_tuner_ 超參數調校 .ipynb**。

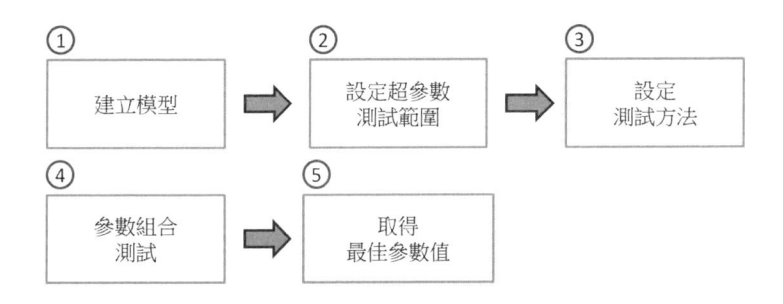

① 建立模型 → ② 設定超參數測試範圍 → ③ 設定測試方法
④ 參數組合測試 → ⑤ 取得最佳參數值

1. 首先要建立模型，並設定超參數測試的範圍：

- 學習率（learning rate）測試選項：0.01, 0.001, 0.0001。
- 第一層 Dense 輸出神經元數：32、64、96、…、512。

```
1  import keras_tuner as kt
2
3  # 建立模型
4  def model_builder(hp):
5      # 學習率(Learning rate)選項：0.01, 0.001, or 0.0001
6      hp_learning_rate = hp.Choice('learning_rate', values = [0.01, 0.001, 0.0001])
7      # 第一層Dense 輸出選項：32、64、…、512
8      hp_units = hp.Int('units', min_value = 32, max_value = 512, step = 32)
9
10     model = tf.keras.models.Sequential([
11         tf.keras.layers.Input((28, 28)),
12         tf.keras.layers.Flatten(),
13         tf.keras.layers.Dense(hp_units, activation='relu'),
14         tf.keras.layers.Dropout(0.1),
```

```
15      tf.keras.layers.Dense(10, activation='softmax')
16    ])
17
18    # 設定優化器(optimizer)、損失函數(Loss)、效能衡量指標(metrics)的類別
19    model.compile(optimizer=tf.keras.optimizers.Adam(learning_rate = hp_learning_rate),
20                  loss='sparse_categorical_crossentropy',
21                  metrics=['accuracy'])
22
23    return model
```

2. 調校設定:呼叫 Hyperband(),設定下列參數。

   - 目標函數(objective):準確率。
   - 最大訓練週期(max_epochs)為 5。
   - 訓練週期數的遞減因子(factor)為 3。
   - 存檔目錄(directory):my_dir,注意,存檔路徑不可有中文,否則程式會出現無法建立目錄的錯誤。
   - 專案名稱(project_name):test1。

```
1  # 調校設定,Hyperband:針對所有參數組合進行測試
2  tuner = kt.Hyperband(model_builder,              # 模型定義
3                       objective = 'val_accuracy', # 目標函數
4                       max_epochs = 5,             # 最大執行週期
5                       factor = 3,                 # 執行週期數的遞減因子
6                       directory = 'my_dir',       # 存檔目錄
7                       project_name = 'test1')     # 專案名稱
```

3. 執行參數調校:

   - 設定 callbacks:每個參數組合測試完成後,清除輸出顯示。
   - get_best_hyperparameters:取得最佳參數值。

```
1  # 參數調校
2  import IPython
3
4  # 每個參數組合測完後,清除顯示
5  class ClearTrainingOutput(tf.keras.callbacks.Callback):
6      def on_train_end(*args, **kwargs):
7          IPython.display.clear_output(wait = True)
8
9
10 # 調校執行
11 tuner.search(x_train_norm, y_train, epochs = 5,
12              validation_data = (x_test_norm, y_test), # 驗證資料
13              callbacks = [ClearTrainingOutput()])     # 執行每個參數組合後回呼
```

4-8 超參數調校（Hyperparameter Tuning）

```
14
15  # 顯示最佳參數值
16  best_hps = tuner.get_best_hyperparameters(num_trials = 1)[0]
17
18  print(f"最佳參數值\n第一層Dense輸出：{best_hps.get('units')}\n",
19         "學習率：{best_hps.get('learning_rate')}")
```

- 最佳參數組合為：
  * 第一層 Dense 輸出：160。
  * 學習率：0.001。

除此之外，Keras Tuner 還有很多的功能，包括：

5. 超參數測試範圍的設定，參閱[19]。

    - Boolean：真 / 假。
    - Choice：多個設定選項。
    - Int/Float：整數 / 浮點數的連續範圍。
    - Fixed：測試所有參數（tune_new_entries=True），除了目前的參數，也可依賴其他參數（parent_name）的設定。只有當其他參數值為特定值時，這個參數才會生效。
    - conditional_scope：條件式，類似 Fixed，依賴其他參數，只有當其他參數值為特定值時，這個條件才會生效。

6. 測試方法（Tuners） 請參閱[20]。

    - Hyperband：測試所有組合。
    - RandomSearch：若測試範圍過大，可隨機抽樣部份組合，加以測試。
    - BayesianOptimization：搭配高斯過程（Gaussian process），依照前次的測試結果，決定下次的測試內容。

7. Oracle：超參數調校的演算法，為測試方法（Tuners）的參數，可決定測試方法的下次測試組合，可參閱[21]。

    另外再加碼推薦，可搭配 Hiplot 視覺化套件，顯示每一種參數組合的設定值與損失 / 準確度。

第 4 章　神經網路實作

8. 解析 Keras Tuner 測試的日誌檔

```python
1  # 解析 Keras Tuner 測試的日誌檔
2  import os
3  import json
4
5  vis_data = []
6  # 掃描目錄內每一個檔案
7  rootdir = 'my_dir/test1'
8  for subdirs, dirs, files in os.walk(rootdir):
9      for file in files:
10         if file.endswith("trial.json"):
11             with open(subdirs + '/' + file, 'r') as json_file:
12                 data = json_file.read()
13                 vis_data.append(json.loads(data))
```

9. 顯示參數組合與測試結果

```python
1  # 顯示參數組合與測試結果
2  import hiplot as hip
3
4  # 建立字典，含參數組合與測試結果
5  data = [{'units': vis_data[idx]['hyperparameters']['values']['units'],
6           'learning_rate': vis_data[idx]['hyperparameters']['values']['learning_rate'],
7           'loss': vis_data[idx]['metrics']['metrics']['loss']['observations'][0]['value'],
8           'val_loss': vis_data[idx]['metrics']['metrics']['val_loss']['observations'][0]['value'],
9           'accuracy': vis_data[idx]['metrics']['metrics']['accuracy']['observations'][0]['value'],
10          'val_accuracy': vis_data[idx]['metrics']['metrics']['val_accuracy']['observations'][0]['value']}
11         for idx in range(len(vis_data))]
12
13 # 顯示
14 hip.Experiment.from_iterable(data).display()
```

- 執行結果：uid 為執行代碼，可以看出第一行（uid=7）有最高的準確率，獲選為最佳參數組合。

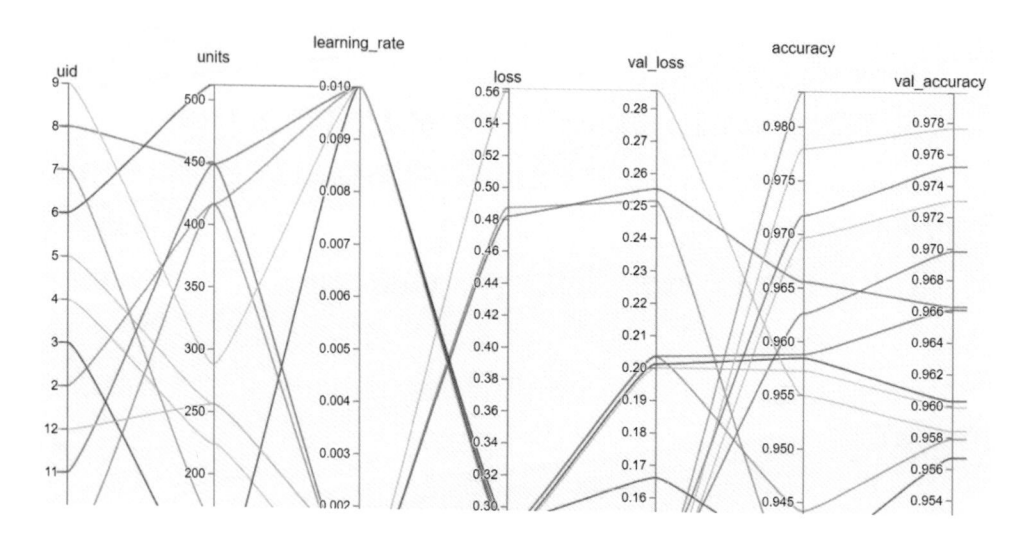

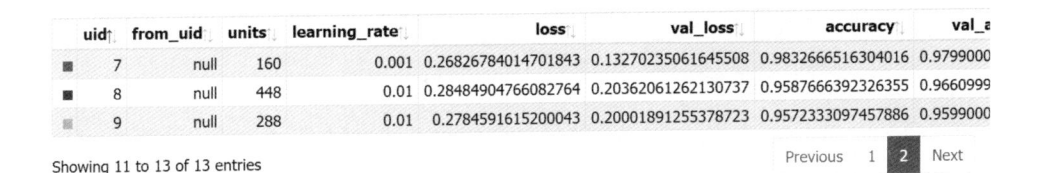

uid	from_uid	units	learning_rate	loss	val_loss	accuracy	val_a
7	null	160	0.001	0.26826784014701843	0.13270235061645508	0.9832666516304016	0.9799000
8	null	448	0.01	0.28484904766082764	0.20362061262130737	0.9587666392326355	0.9660999
9	null	288	0.01	0.2784591615200043	0.20001891255378723	0.9572333097457886	0.9599000

Showing 11 to 13 of 13 entries　　　　　　　　　　　　Previous　1　2　Next

參數調校是深度學習中非常重要的步驟，因為深度學習是一個黑箱科學，加上我們對於高維資料的聯合機率分配也是一無所知，唯有透過大量的實驗，才能獲得較佳的模型。但困難的是，模型訓練非常耗時，如何透過各種方法或套件的協助，縮短調校時間，是建構 AI 模型時須思考的重要課題。

參考資料 (References)

[1]　維基百科 Activation Function 的介紹
　　　(https://en.wikipedia.org/wiki/Activation_function)

[2]　TensorFlow 優化器介紹
　　　(https://www.tensorflow.org/api_docs/python/tf/keras/optimizers)

[3]　Keras 優化器介紹
　　　(https://keras.io/api/optimizers/)

[4]　TensorFlow 損失函數介紹
　　　(https://www.tensorflow.org/api_docs/python/tf/keras/optimizers)

[5]　Keras 損失函數介紹
　　　(https://keras.io/api/losses/)

[6]　TensorFlow 效能衡量指標介紹
　　　(https://www.tensorflow.org/api_docs/python/tf/keras/metrics)

[7]　TensorFlow 官網
　　　(https://www.tensorflow.org)

[8]　Keras 官網
　　　(https://keras.io)

[9]　TensorFlow 官網中 FashionMnist 的介紹
　　　(https://www.tensorflow.org/datasets/catalog/fashion_mnist)

[10] Keras 官網神經層介紹
(https://keras.io/api/layers/)

[11] Keras 官網 Activation Function 說明
(https://keras.io/api/layers/activations/)

[12] Keras 官網 Activation Layers 說明
(https://keras.io/api/layers/activation_layers/)

[13] 維基百科關於鉸鏈損失函數的介紹
(https://zh.wikipedia.org/wiki/Hinge_loss)

[14] 《Understanding Nesterov Momentum (NAG)》, 2018
(https://dominikschmidt.xyz/nesterov-momentum/)

[15] Diederik P. Kingma、Jimmy Ba,《Adam: A Method for Stochastic Optimization》, 2014
(https://arxiv.org/abs/1412.6980)

[16] 深度學習於 NLP,《一文告訴你 Adam、AdamW、Amsgrad 區別和聯繫》, 2019
(https://zhuanlan.zhihu.com/p/39543160)

[17] Alec Radford's animations for optimization algorithms
(https://www.denizyuret.com/2015/03/alec-radfords-animations-for.html)

[18] Keras 官網效能衡量指標的介紹
(https://keras.io/api/metrics/)

[19] Keras 官網超參數測試範圍的設定
(https://keras-team.github.io/keras-tuner/documentation/hyperparameters/)

[20] Keras 官網效能調校 (Tuners) 的介紹
(https://keras-team.github.io/keras-tuner/documentation/tuners/)

[21] Keras 官網 Oracle 的介紹
(https://keras-team.github.io/keras-tuner/documentation/oracles/)

第 5 章

TensorFlow
常用指令與功能

除了建構模型外，TensorFlow 還貼心的提供各種的工具和指令，方便在程式開發流程中使用，包括前置處理、模型存檔 / 載入 / 繪製、除錯等功能。

第 5 章　TensorFlow 常用指令與功能

5-1　特徵轉換

One-hot encoding 可將類別變數轉為多個啞變數（Dummy variable），每個啞變數只含真 / 假值（1/0），避免被演算法誤認該變數類別有順序大小之分，例如顏色，紅 / 藍 / 綠會被轉換成三個變數：『是紅色嗎』、『是藍色嗎』、『是綠色嗎』，而非 1/2/3。

範例. One-hot encoding 測試。

程式 05_01_ 特徵轉換 .ipynb。

1. TensorFlow 的 One-hot encoding 指令如下。

```
1  # One-hot encoding
2  # num_classes：類別個數，可不設定
3  tf.keras.utils.to_categorical([0, 1, 2, 3], num_classes=9)
```

- 執行結果如下，指定 9 種類別，即產生 9 個變數。
- num_classes：類別個數，此參數可以不設定，函數會從資料中找出類別個數。

```
array([[1., 0., 0., 0., 0., 0., 0., 0., 0.],
       [0., 1., 0., 0., 0., 0., 0., 0., 0.],
       [0., 0., 1., 0., 0., 0., 0., 0., 0.],
       [0., 0., 0., 1., 0., 0., 0., 0., 0.]], dtype=float32)
```

2. 修改 MNIST 手寫阿拉伯數字辨識程式。

```
1   mnist = tf.keras.datasets.mnist
2
3   # 載入 MNIST 手寫阿拉伯數字資料
4   (x_train, y_train),(x_test, y_test) = mnist.load_data()
5
6   # 特徵縮放，使用常態化(Normalization)，公式 = (x - min) / (max - min)
7   x_train_norm, x_test_norm = x_train / 255.0, x_test / 255.0
8
9   # One-hot encoding
10  y_train = tf.keras.utils.to_categorical(y_train)
11  y_test = tf.keras.utils.to_categorical(y_test)
12
13  # 建立模型
14  model = tf.keras.models.Sequential([
15      tf.keras.layers.Input((28, 28)),
16      tf.keras.layers.Flatten(),
17      tf.keras.layers.Dense(256, activation='relu'),
```

```
18      tf.keras.layers.Dropout(0.2),
19      tf.keras.layers.Dense(10, activation='softmax')
20  ])
21
22  # 設定優化器(optimizer)、損失函數(loss)、效能衡量指標(metrics)的類別
23  model.compile(optimizer='adam',
24                loss='categorical_crossentropy',
25                metrics=['accuracy'])
26
27  # 模型訓練
28  history = model.fit(x_train_norm, y_train, epochs=5, validation_split=0.2)
29
30  # 評分(Score Model)
31  score=model.evaluate(x_test_norm, y_test, verbose=0)
```

- 第 10~11 行將 y 作 One-hot encoding 轉換。

- 第 23 行損失函數改用 categorical_crossentropy，而非 sparse_categorical_crossentropy，因我們已自行對 y 實施 One-hot encoding 轉換。

3. 特徵縮放：之前使用 MinMaxScaler，將訓練及測試資料轉換為 [0, 1] 之間，是假設特徵 (x) 資料為均勻分配（Uniform distribution），每個可能的值發生機率均等。另一種特徵縮放方法為標準化（Standardization），假定資料為常態分配或其他機率不均等的分配，轉換後資料的平均數為 0，標準差為 1，若資料是常態分配，以 N（0, 1）表示，下列程式碼類似 scikit-learn 的 StandardScaler 類別。

```
1  import numpy as np
2  import tensorflow as tf
3  from tensorflow.keras.layers.experimental import preprocessing
4
5  # 測試資料
6  data = np.array([[0.1, 0.2, 0.3], [0.8, 0.9, 1.0], [1.5, 1.6, 1.7],])
7  layer = preprocessing.Normalization()   # 常態化
8  layer.adapt(data)                        # 訓練
9  normalized_data = layer(data) # 轉換
10
11 # 顯示平均數、標準差
12 print(f"平均數: {normalized_data.numpy().mean():.2f}")
13 print(f"標準差: {normalized_data.numpy().std():.2f}")
```

- 執行結果：平均數：0.00，標準差：1.00。

第 5 章　TensorFlow 常用指令與功能

▋5-2 模型存檔與載入（Model Saving and Loading）

模型存檔包括下列資訊：

1. 模型結構與組態。
2. 權重，含偏差項（bias）。
3. 模型 compile 參數：優化器（Optimizer）、損失函數（Loss）及效能衡量指標（Metrics）的定義。

目前支援 3 種方式：

1. 存檔（Saving）：儲存上述所有資訊，副檔名為 .keras，為一 zip 壓縮檔，內含權重資訊（.h5 檔）、模型結構（Json 檔）及 Metadata 等資訊。注意，TensorFlow v2.18/Keras v3 已經改變預設存檔格式，目前要存成 Keras 2 舊有格式（h5）會出現警告訊息，未來可能不支援。
2. 序列化（Serialization）：只儲存模型結構與組態，不含狀態與權重。
3. 匯出（Export）：只儲存推論所需的正向傳導模型資訊（Forward pass），最小化檔案尺寸，減少佈署時所需的記憶體，取代舊版 State_dict 的儲存方式。

範例 1. 測試模型存檔與載入。

程式：**05_02_ 模型存檔與載入 .ipynb**。

1. 建立及訓練模型。

```
1  import tensorflow as tf
2
3  # 載入 MNIST 手寫阿拉伯數字資料
4  mnist = tf.keras.datasets.mnist
5  (x_train, y_train),(x_test, y_test) = mnist.load_data()
6
7  # 特徵縮放，使用常態化(Normalization)，公式 = (x - min) / (max - min)
8  x_train_norm, x_test_norm = x_train / 255.0, x_test / 255.0
9
10 # 建立模型
11 model = tf.keras.models.Sequential([
12     tf.keras.layers.Input((28, 28)),
13     tf.keras.layers.Flatten(),
14     tf.keras.layers.Dense(256, activation='relu'),
15     tf.keras.layers.Dropout(0.2),
16     tf.keras.layers.Dense(10, activation='softmax')
```

5-2 模型存檔與載入（Model Saving and Loading）

```
17 ])
18
19 # 設定優化器(optimizer)、損失函數(loss)、效能衡量指標(metrics)的類別
20 model.compile(optimizer='adam',
21               loss='sparse_categorical_crossentropy',
22               metrics=['accuracy'])
23
24 # 模型訓練
25 history = model.fit(x_train_norm, y_train, epochs=5, validation_split=0.2)
```

2. 儲存模型，檔名為 model.keras，若使用 model.h5，會出現錯誤。

```
1 model.save('model.keras')
```

3. 載入模型，以另一個變數 model2 承接：

```
1 # 模型載入
2 model2 = tf.keras.models.load_model('model.keras')
3
4 # 評分(Score Model)
5 score=model2.evaluate(x_test_norm, y_test, verbose=0)
6
7 for i, x in enumerate(score):
8     print(f'{model2.metrics_names[i]}: {score[i]:.4f}')
```

4. 可以使用 np.allclose 比較預測結果：執行結果為真。

```
1 # 模型比較
2 import numpy as np
3
4 # 比較，若結果不同，會出現錯誤
5 np.allclose(
6     model.predict(x_test_norm), model2.predict(x_test_norm)
7 )
```

範例 2. 序列化（Serialization）：只取得模型結構，不含權重。

程式：**05_03_ 模型序列化 .ipynb**。

1. 使用 get_config 序列化模型。

```
1 # 取得模型結構
2 config = model.get_config()
```

2. 使用 from_config 反序列化（Deserialization）模型，使用 summary 函數確認是否正確載入模型結構。

第 5 章　TensorFlow 常用指令與功能

```
1  # 載入模型結構
2  new_model = tf.keras.Sequential.from_config(config)
3
4  new_model.summary()
```

3. 可將序列化結果存檔：joblib 是將序列化資料存檔的套件。

```
1  import joblib
2  joblib.dump(config, 'model.joblib')
```

4. 將序列化檔案載入，使用 summary 函數確認是否正確載入模型結構。

```
1  # 從檔案載入模型結構
2  config = joblib.load('model.joblib')
3  new_model2 = tf.keras.Sequential.from_config(config)
4
5  new_model2.summary()
```

5. tf.keras.Sequential.from_config 只適用於順序型模型，若是 Functional API 模型且有多個 Input 或 Output，則會出現錯誤。以下定義一個 Functional API 模型且有多個 Output。

```
1  # Model definded by functional API
2  inputs = tf.keras.layers.Input(shape=(784,))
3  x = tf.keras.layers.Flatten()(inputs)
4  x = tf.keras.layers.Dense(256, activation='relu')(x)
5  x = tf.keras.layers.Dropout(0.2)(x)
6  outputs = tf.keras.layers.Dense(10, activation='softmax')(x)
7  outputs_2 = tf.keras.layers.Dense(10)(x)
8  model_functional_api = tf.keras.Model(inputs=inputs, outputs=[outputs, outputs_2])
9  config = model_functional_api.get_config()
```

6. 使用 tf.keras.Sequential.from_config 載入 Functional API 模型。

```
1  # ** the error happens, we should repplace tf.keras.Sequential.from_config by tf.keras.Model.from_config **
2  new_model_functional_api = tf.keras.Sequential.from_config(config)
3  new_model_functional_api.summary()
```

7. 執行結果：

```
ValueError: Input 0 of layer "dense_34" is incompatible with the layer: expected axis -1 of input shape to have value 256, but received input with shape (None, 10)
```

8. 改用 tf.keras.Model.from_config。

```
1  new_model_functional_api = tf.keras.Model.from_config(config)
2  new_model_functional_api.summary()
```

5-2 模型存檔與載入（Model Saving and Loading）

9. 執行結果：最後兩層均為 output。

Layer (type)	Output Shape	Param #	Connected to
input_layer_13 (InputLayer)	(None, 784)	0	-
flatten_13 (Flatten)	(None, 784)	0	input_layer_13[0][0]
dense_32 (Dense)	(None, 256)	200,960	flatten_13[0][0]
dropout_13 (Dropout)	(None, 256)	0	dense_32[0][0]
dense_33 (Dense)	(None, 10)	2,570	dropout_13[0][0]
dense_34 (Dense)	(None, 10)	2,570	dropout_13[0][0]

10. 繪圖：進一步證實。

```
1  tf.keras.utils.plot_model(new_model_functional_api)
```

- 執行結果：output 有 2 層。

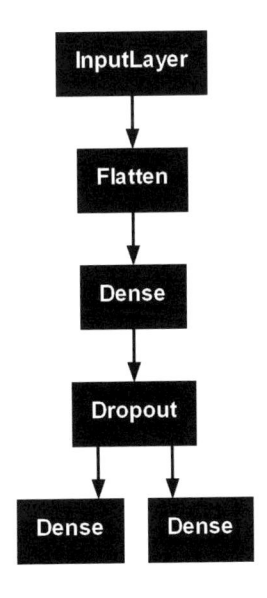

11. 也可以轉換為 JSON 格式。

```
1  # 取得模型結構
2  json_config = model.to_json()
3
4  # 載入模型結構
5  new_model = tf.keras.models.model_from_json(json_config)
```

12. 取得權重：

```
# 取得模型權重
weights = model.get_weights()
weights
```

13. 合併模型結構及權重，並進行預測，驗證模型已含權重。

```
# 設定模型權重
new_model.set_weights(weights)

# predict
score=new_model.evaluate(x_test_norm, y_test, verbose=0)
score
```

- 之前在研討會時，有聽眾問『如何保護模型，不讓他人直接使用？』，這裡提供一個解法，可以利用 model.get_weights 取得模型權重，將其中部份的權重修改後存檔，之後從檔案載入後，先還原遭修改的權重，再使用 model.set_weights（weights）載入權重，他人不知如何還原遭修改的權重，即使從檔案載入，也不能正確預測。

範例 3. 若模型中含有自訂神經層（Custom layer），存檔或序列化均無法成功，必須在自訂神經層加工。

程式：**05_04_ 自訂神經層序列化 .ipynb**。

1. 建立自訂神經層（Custom layer）：單純把 Input 乘以 factor。

```
import tensorflow as tf

class CustomLayer(tf.keras.layers.Layer):
    def __init__(self, factor):
        super().__init__()
        self.factor = factor
        
    def call(self, x):
        return x * self.factor
```

2. 將手寫阿拉伯數字模型加上自訂神經層。

```
model = tf.keras.models.Sequential([
    tf.keras.layers.Input((28, 28)),
    CustomLayer(2),
    tf.keras.layers.Flatten(),
    tf.keras.layers.Dense(256, activation='relu'),
    tf.keras.layers.Dropout(0.2),
```

5-2 模型存檔與載入（Model Saving and Loading）

```
15      tf.keras.layers.Dense(10, activation='softmax')
16 ])
```

3. 存檔：無錯誤訊息。

```
1 model.save('model_custom_layer.keras')
```

4. 但載入發生錯誤。

```
1 model2 = tf.keras.models.load_model('model_custom_layer.keras')
```

- 執行結果：自訂神經層（Custom layer）要加上宣告（Decorator）。

```
Exception encountered: Could not locate class 'CustomLayer'. Make sure custom classes are decorated with `@keras.saving.register_keras_serializable()`.
Full object config: {'module': None, 'class_name': 'CustomLayer', 'config': {'factor': 2, 'trainable': True, 'dtype': {'module': 'keras', 'class_name': 'DTypePolicy', 'config': {'name': 'float32'}, 'registered_name': None}}, 'registered_name': 'CustomLayer', 'build_config': {'input_shape': [None, 28, 28]}}
```

5. 加上宣告（Decorator）：使用 @tf.keras.utils.register_keras_serializable() 或獨立套件 Keras 的 @keras.saving.register_keras_serializable()，並加上 get_config 方法。

```
1  # 清除之前的註冊
2  tf.keras.utils.get_custom_objects().clear()
3
4  @tf.keras.utils.register_keras_serializable()
5  class CustomLayer(tf.keras.layers.Layer):
6      def __init__(self, factor):
7          super().__init__()
8          self.factor = factor
9
10     def call(self, x):
11         return x * self.factor
12
13     def get_config(self):
14         return {"factor": self.factor}
```

6. 重新訓練模型，並存檔，之後載入模型就沒有問題了。

```
1  model.save('model_custom_layer.keras')
2  model2 = tf.keras.models.load_model('model_custom_layer.keras')
3  model2.summary()
```

- 執行結果：第一層為自訂神經層。

Layer (type)	Output Shape	Param #
custom_layer_2 (CustomLayer)	(None, 28, 28)	0
flatten_1 (Flatten)	(None, 784)	0
dense_2 (Dense)	(None, 256)	200,960
dropout_1 (Dropout)	(None, 256)	0
dense_3 (Dense)	(None, 10)	2,570

7. 繪圖驗證。

```
tf.keras.utils.plot_model(model2, "model2.png")
```

- 執行結果：

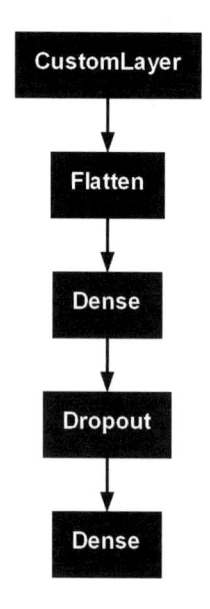

```
CustomLayer
   ↓
Flatten
   ↓
Dense
   ↓
Dropout
   ↓
Dense
```

範例 4. 測試模型匯出與載入（Exporting and Loading）。

程式：**05_05_ 模型匯出 .ipynb**。

1. 建立及訓練模型：與 05_02_ 模型存檔與載入 .ipynb 相同，不再贅述。

2. 匯出模型：會提示模型存在 exported_model 資料夾。

```
model.export("exported_model")
```

3. 載入模型，以另一個變數 model2 承接：

```
1  model2 = tf.saved_model.load("exported_model")
```

4. 模型預測：需使用 serve 方法，若使用 predict 方法，會出現『AttributeError: '_UserObject' object has no attribute 'predict'』錯誤。以下使用前 10 筆資料測試。

```
1  model2.serve(x_test_norm[:10])
```

5. 載入的模型只含推論的資訊，如果使用 fit 方法，重新訓練，也會出現『AttributeError: '_UserObject' object has no attribute 'fit'』錯誤。

要瞭解更詳細的資訊可參考『TensorFlow Save, serialize, and export models』[11] 及『Keras Save, serialize, and export models』[2]，兩篇內容略有差異，應是 TensorFlow 內的 Keras 不完全相容 Keras 獨立套件，在此特別提醒。

5-3 模型彙總與結構圖（Summary and Plotting）

如果設計一個複雜的模型，通常會希望有視覺化的呈現，讓我們一目瞭然，利於檢查結構是否 OK，Tensorflow 提供彙總資訊，以表格顯示，也提供模型結構圖，幫助我們驗證模型是否正確。

範例. 顯示模型彙總與繪製結構圖。

程式：**05_06_ 模型彙總與結構圖 .ipynb**。

1. 顯示模型彙總：下列指令可取得彙總資訊表格，包含每一層的輸入/輸出/參數個數。

```
1  model.summary()
```

- 執行結果：請讀者自行驗算參數個數（Param #），測試自己的理解程度。

第 5 章　TensorFlow 常用指令與功能

```
Model: "sequential"
_____
Layer (type)                 Output Shape              Param #
=================================================================
flatten (Flatten)            (None, 784)               0
_____
dense (Dense)                (None, 256)               200960
_____
dropout (Dropout)            (None, 256)               0
_____
dense_1 (Dense)              (None, 10)                2570
=================================================================
Total params: 203,530
Trainable params: 203,530
Non-trainable params: 0
_____
```

2. 使用索引（index）取得神經層資訊。

```
1  # 以 index 取得神經層資訊
2  model.get_layer(index=0)
```

3. 使用名稱取得神經層資訊：可由 model.summary() 取得名稱，也可以在定義模型時指定神經層名稱。

```
1  # 以名稱取得神經層資訊
2  model.get_layer(name='dense_2')
```

4. 取得神經層權重。

```
1  # 取得神經層權重
2  model.get_layer(name='dense_2').get_weights()
```

5. 繪製結構圖：要繪製模型結構，需先完成以下步驟，才能順利繪製圖形。

- 須安裝 graphviz 軟體，網址為：https://www.graphviz.org/download，再把安裝目錄下的 bin 路徑加到環境變數 path 中。

- 安裝兩個 Python 套件：

  pip install graphviz pydotplus

6. 以下繪圖指令會先產生描述向量圖的文字檔（.dot），再呼叫 graphviz 的 dot.exe，將 .dot 檔案裡的內容轉化為影像檔案（.png）。

```
1  # 繪製結構圖
2  tf.keras.utils.plot_model(model)
```

5-3 模型彙總與結構圖（Summary and Plotting）

7. 加上不同的參數，可顯示各種的額外資訊，例如：

- show_shapes=True：顯示輸入 / 輸出的神經元個數。
- show_dtype=True：顯示輸入 / 輸出的資料型態。
- to_file="model.png"：同時存檔。

```
1  # 繪製結構圖
2  # show_shapes=True：可顯示輸入/輸出的神經元個數
3  # show_dtype=True：可顯示輸入/輸出的資料型態
4  # to_file：可同時存檔
5  tf.keras.utils.plot_model(model, show_shapes=True, show_dtype=True,
6                            to_file="model.png")
```

- 執行結果：

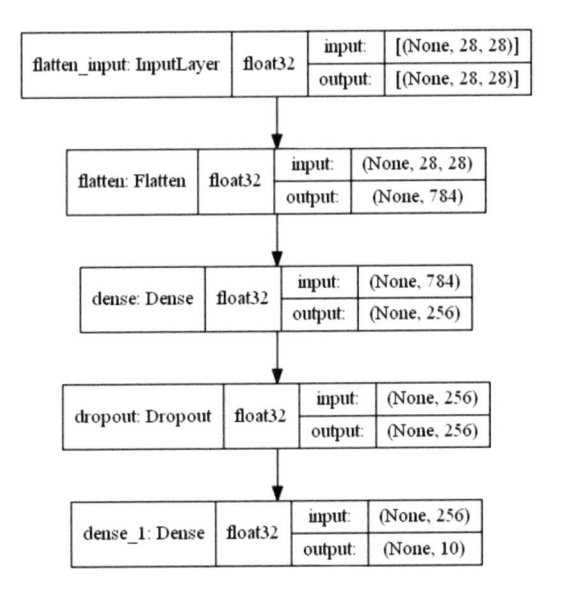

| flatten_input: InputLayer | float32 | input: | [(None, 28, 28)] |
| | | output: | [(None, 28, 28)] |

| flatten: Flatten | float32 | input: | (None, 28, 28) |
| | | output: | (None, 784) |

| dense: Dense | float32 | input: | (None, 784) |
| | | output: | (None, 256) |

| dropout: Dropout | float32 | input: | (None, 256) |
| | | output: | (None, 256) |

| dense_1: Dense | float32 | input: | (None, 256) |
| | | output: | (None, 10) |

也可以產生 dot 格式：

```
1  # 產生 dot 格式及 png 檔
2  import pydotplus as pdp
3  from IPython.display import display, Image
4
5  # 產生 dot 格式
6  dot1 = tf.keras.utils.model_to_dot(model, show_shapes=True, show_dtype=True)
7  # 產生 png 檔
8  display(Image(dot1.create_png()))
```

5-4 回呼函數（Callbacks）

TensorFlow 模型訓練只需要一個指令 fit，非常簡便，但也有壞處，我們無法觀察到訓練過程中發生了甚麼事，為解決這個問題，TensorFlow 提供回呼函數（Callbacks），在訓練指令中可放入要觸發的函數，在每一個週期或批次執行之前與之後，都可以呼叫 Callback 函數，TensorFlow 提供許多類型的 Callbacks，功能如下：

1. 在訓練過程中記錄任何資訊。
2. 在每個查核點（Checkpoint）進行模型存檔。
3. 設定訓練提前結束的條件。
4. 結合 TensorBoard 視覺化工具，即時監看訓練過程。
5. 將訓練過程產生的資訊寫入 CSV 檔案。
6. 使用其他電腦遠端監控訓練過程。
7. 還有更多的其他功能，請參閱 Keras Callbacks API 介紹 [3]。除了內建 Callback，也能夠自訂 Callback。

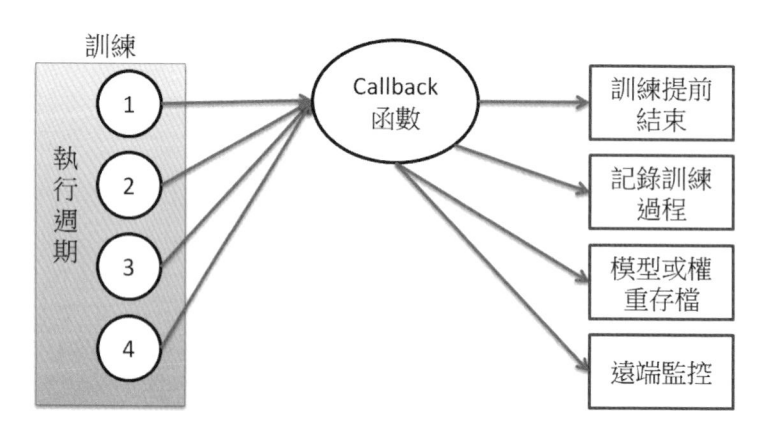

▲ 圖 5.1 回呼函數（Callbacks）

透過 Callback 可以完全解構模型訓練的過程，我們使用一些範例，說明各類型 Callback 的用法。

5-4-1 EarlyStopping Callback

EarlyStopping 是用來設定訓練提前結束的條件，實務上我們可以設定較大的執行週期數，並搭配效能檢視條件，一段時間內效能沒有顯著改善時，就提前結束訓練，這樣就能兼顧效能與訓練時間了。

範例 1. EarlyStopping Callback 測試。

程式：**05_07_EarlyStopping_Callback.ipynb**。

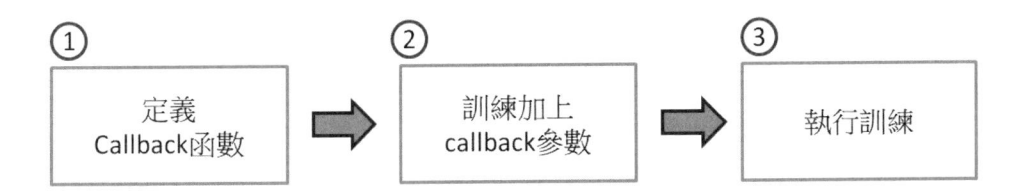

① 定義 Callback函數 → ② 訓練加上 callback參數 → ③ 執行訓練

1. 定義 Callback 函數為『若連續 3 個執行週期 validation accuracy 沒改善就停止訓練』。

```
1  # validation loss 三個執行週期沒改善就停止訓練
2  my_callbacks = [
3      tf.keras.callbacks.EarlyStopping(patience=3, monitor = 'val_accuracy'),
4  ]
```

2. 訓練指令的 callback 參數設定為上述函數。

```
1  # 訓練 20 次，但實際只訓練 12次就停止了
2  history = model.fit(x_train_norm, y_train, epochs=20, validation_split=0.2,
3                     callbacks=my_callbacks)
```

- 執行結果：預計訓練 20 次，但實際只訓練 12 次就停止了，注意，每次執行結果可能不同。

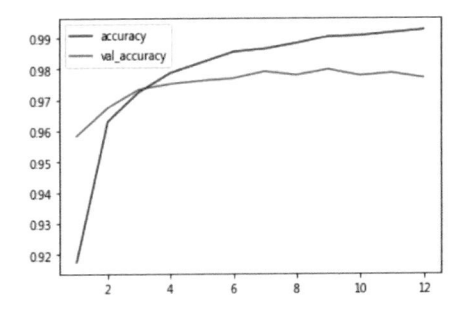

3. 效能指標也可以改為驗證的損失（val_loss），只要連續 3 次沒改善停止訓練。

```
1  # validation loss 三個執行週期沒改善就停止訓練
2  my_callbacks = [
3      tf.keras.callbacks.EarlyStopping(patience=3, monitor = 'val_loss'),
4  ]
```

- 執行結果：

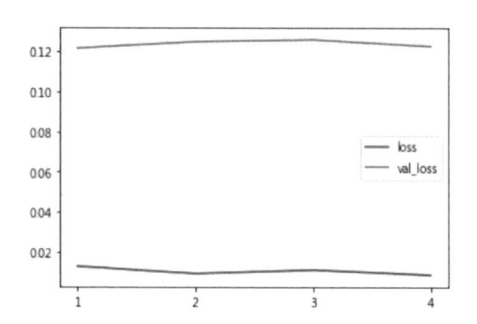

5-4-2 ModelCheckpoint Callback

若訓練過程過長，可能會發生訓練到一半就當掉的狀況，我們可以利用 Model-Checkpoint Callback，在每一個檢查點（Checkpoint）存檔，當掉後再次執行時，可以從中斷點繼續訓練。

範例 2. ModelCheckpoint Callback 測試。

程式：**05_08_ModelCheckpoint_Callback.ipynb**。

1. 定義 ModelCheckpoint callback：副檔名需為『.weights.h5』，並使用 f-string，以執行週期為變數，各個執行週期的儲存才可以有不同的檔名。

```
1  checkpoint_filepath = 'model.{epoch:02d}.weights.h5'  # 使用 f-string 變數
2  model_checkpoint_callback = tf.keras.callbacks.ModelCheckpoint(
3      filepath=checkpoint_filepath,  # 設定存檔名稱
4      save_weights_only=True,        # 只存權重
5      monitor='val_accuracy',        # 監看驗證資料的準確率
6      mode='max',                    # 設定save_best_only=True時，best是指 max or min
7      save_best_only=True)           # 只存最佳模型
8
9  EPOCHS = 3  # 訓練 3 次
10 model.fit(x_train_norm, y_train, epochs=EPOCHS, validation_split=0.2,
11          callbacks=[model_checkpoint_callback])
```

- 執行結果如下，最後的準確率等於 0.9754。

```
Epoch 1/3
1500/1500 ──────── 4s 2ms/step - accuracy: 0.8665 - loss: 0.4583 - val_accuracy: 0.9608 - val_loss: 0.1365
Epoch 2/3
1500/1500 ──────── 2s 2ms/step - accuracy: 0.9611 - loss: 0.1318 - val_accuracy: 0.9703 - val_loss: 0.1001
Epoch 3/3
1500/1500 ──────── 3s 2ms/step - accuracy: 0.9733 - loss: 0.0873 - val_accuracy: 0.9754 - val_loss: 0.0839
```

2. 再執行 3 個週期，準確率會接續上一次的結果，繼續改善。

```
1  # 再訓練 3 次，觀察 accuracy，會接續上一次，繼續改善 accuracy。
2  model.fit(x_train_norm, y_train, epochs=EPOCHS, validation_split=0.2,
3            callbacks=[model_checkpoint_callback])
```

- 執行結果如下，最後的準確率等於 0.9781。

```
Epoch 1/3
1500/1500 ──────── 2s 2ms/step - accuracy: 0.9785 - loss: 0.0685 - val_accuracy: 0.9749 - val_loss: 0.0794
Epoch 2/3
1500/1500 ──────── 3s 2ms/step - accuracy: 0.9828 - loss: 0.0551 - val_accuracy: 0.9770 - val_loss: 0.0740
Epoch 3/3
1500/1500 ──────── 3s 2ms/step - accuracy: 0.9861 - loss: 0.0442 - val_accuracy: 0.9781 - val_loss: 0.0756
```

5-4-3 TensorBoard Callback

TensorBoard 是 Tensorflow 提供的視覺化診斷工具，功能非常強大，除了可以顯示訓練過程之外，也能夠顯示變數、圖片、語音及文字訊息。將 TensorBoard 整合至 Callback 事件裡，就可以在訓練的過程中啟動 TensorBoard 網站，即時觀看訓練資訊。

範例 3. TensorBoard Callback 測試。

程式：**05_09_TensorBoard_Callback.ipynb**。

1. 定義 TensorBoard Callback，並進行模型訓練。

```
1  # 定義 tensorboard callback
2  tensorboard_callback = [tf.keras.callbacks.TensorBoard(log_dir='.\\logs')]
3
4  # 訓練 5 次
5  history = model.fit(x_train_norm, y_train, epochs=5, validation_split=0.2,
6                      callbacks=tensorboard_callback)
```

2. 啟動 TensorBoard 網站：先載入 TensorBoard notebook extension，在 Jupyter notebook 啟動 Tensorboard。

第 5 章　TensorFlow 常用指令與功能

```
1  # 載入 TensorBoard notebook extension
2  %load_ext tensorboard
3
4  # 啟動 Tensorboard
5  %tensorboard --logdir ./logs
```

3. 上述指令可在模型訓練後，觀看訓練過程，若要即時觀看訓練過程，可在訓練開始時使用下列指令啟動 Tensorboard。

    - tensorboard --logdir=.\logs

4. 在瀏覽器輸入以下網址，即可觀看訓練資訊。

    - 瀏覽 http://localhost:6006/

5. 訓練資訊包括：

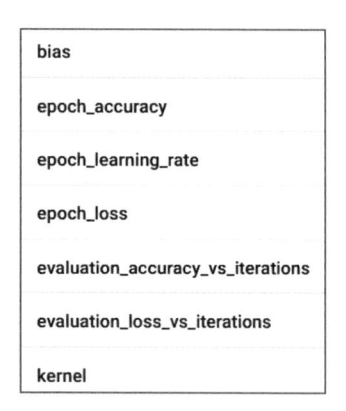

bias
epoch_accuracy
epoch_learning_rate
epoch_loss
evaluation_accuracy_vs_iterations
evaluation_loss_vs_iterations
kernel

6. 例如各週期準確度，畫面如下：

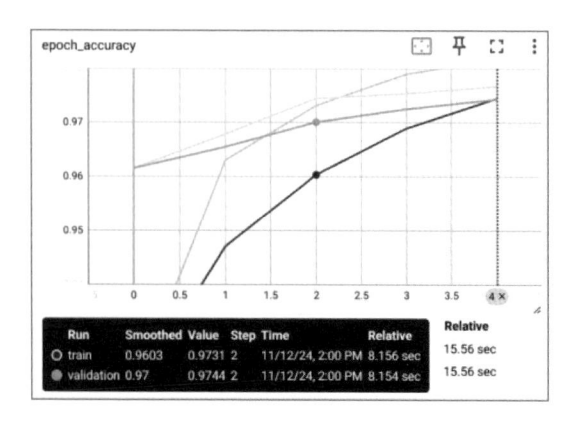

5-4 回呼函數（Callbacks）

- 點選『Graph』頁籤，可以觀察模型的運算圖，顯示模型的運算順序。

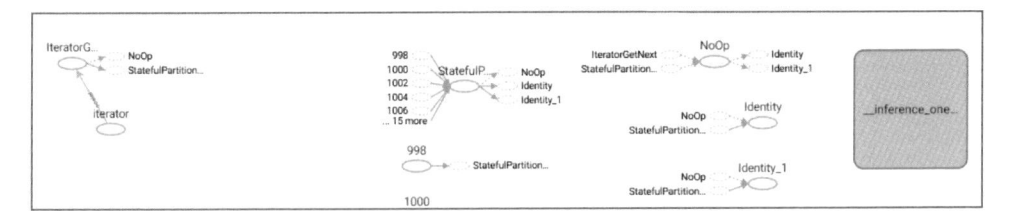

- 請注意，model.fit 的 Callback 參數值是 list，可加入多個 Callback，在訓練時一併觸發，如下，同時設定提前結束訓練、檢查點及 TensorBoard。

```
1  # 可同時定義多個Callback事件
2  my_callbacks = [
3      tf.keras.callbacks.EarlyStopping(patience=3),
4      tf.keras.callbacks.ModelCheckpoint(filepath='model.{epoch:02d}.h5'),
5      tf.keras.callbacks.TensorBoard(log_dir='./logs'),
6  ]
7  model.fit(dataset, epochs=10, callbacks=my_callbacks)
```

5-4-4 自訂 Callback

如果內建的 Callback 不能滿足需求，也可以自訂 Callback，觸發時機可含訓練、測試、預測階段的之前（Before）與之後（After）：

1. on_（train|test|predict）_begin：訓練、測試及預測**開始**前，可觸發事件。
2. on_（train|test|predict）_end：訓練、測試及預測**結束**後，可觸發事件。
3. on_（train|test|predict）_batch_begin：**每批**訓練、測試及預測**開始**前，可觸發事件。
4. on_（train|test|predict）_batch_end：**每批**訓練、測試及預測**結束**後，可觸發事件。
5. on_epoch_begin：**每個執行週期開始**前，可觸發事件。
6. on_epoch_end：**每個執行週期結束**後，可觸發事件。

範例 4. 自訂 Callback 實作。

程式：**05_10_Custom_Callback.ipynb**。

1. 定義觸發的時機及動作（Action），以下動作只單純顯示文字訊息，實際運用時可取得當時的狀態或統計量寫入工作記錄檔。

```python
class CustomCallback(tf.keras.callbacks.Callback):
    def __init__(self):
        self.task_type=''
        self.epoch=0
        self.batch=0

    def on_train_begin(self, logs=None):
        self.task_type='訓練'
        print("訓練開始...")

    def on_train_end(self, logs=None):
        print("訓練結束.")

    def on_epoch_begin(self, epoch, logs=None):
        self.epoch=epoch
        print(f"{self.task_type}第 {epoch} 執行週期開始...")

    def on_epoch_end(self, epoch, logs=None):
        print(f"{self.task_type}第 {epoch} 執行週期結束.")

    def on_test_begin(self, logs=None):
        self.task_type='測試'
        print("測試開始...")

    def on_test_end(self, logs=None):
        print("測試結束.")

    def on_predict_begin(self, logs=None):
        self.task_type='預測'
        print("預測開始...")

    def on_predict_end(self, logs=None):
        print("預測結束.")

    def on_train_batch_begin(self, batch, logs=None):
        print(
            f"訓練 第 {self.epoch} 執行週期, 第 {batch} 批次開始...")

    def on_train_batch_end(self, batch, logs=None):
        print(
            f"訓練 第 {self.epoch} 執行週期, 第 {batch} 批次結束.")

    def on_test_batch_begin(self, batch, logs=None):
        print(
            f"測試 第 {self.epoch} 執行週期, 第 {batch} 批次開始...")

    def on_test_batch_end(self, batch, logs=None):
        print(
            f"測試 第 {self.epoch} 執行週期, 第 {batch} 批次結束.")

    def on_predict_batch_begin(self, batch, logs=None):
        print(
            f"預測 第 {self.epoch} 執行週期, 第 {batch} 批次開始...")

    def on_predict_batch_end(self, batch, logs=None):
        print(
            f"預測 第 {self.epoch} 執行週期, 第 {batch} 批次結束.")
```

2. 在訓練、測試、預測使用此 Callback。

5-4 回呼函數（Callbacks）

```python
1  # 訓練
2  model.fit(
3      x_train_norm, y_train, epochs=5,
4      batch_size=256, verbose=0,
5      validation_split=0.2, callbacks=[CustomCallback()]
6  )
7
8  # 測試
9  model.evaluate(
10     x_test_norm, y_test, batch_size=128,
11     verbose=0, callbacks=[CustomCallback()]
12 )
13
14 # 預測
15 model.predict(
16     x_test_norm, batch_size=128,
17     callbacks=[CustomCallback()]
18 )
```

- 訓練顯示結果：

```
訓練開始...
訓練第 0 執行週期開始...
訓練 第 0 執行週期, 第 0 批次開始...
訓練 第 0 執行週期, 第 0 批次結束.
訓練 第 0 執行週期, 第 1 批次開始...
訓練 第 0 執行週期, 第 1 批次結束.
訓練 第 0 執行週期, 第 2 批次開始...
訓練 第 0 執行週期, 第 2 批次結束.
訓練 第 0 執行週期, 第 3 批次開始...
訓練 第 0 執行週期, 第 3 批次結束.
訓練 第 0 執行週期, 第 4 批次開始...
訓練 第 0 執行週期, 第 4 批次結束.
訓練 第 0 執行週期, 第 5 批次開始...
訓練 第 0 執行週期, 第 5 批次結束.
訓練 第 0 執行週期, 第 6 批次開始...
訓練 第 0 執行週期, 第 6 批次結束.
訓練 第 0 執行週期, 第 7 批次開始...
訓練 第 0 執行週期, 第 7 批次結束.
訓練 第 0 執行週期, 第 8 批次開始...
訓練 第 0 執行週期, 第 8 批次結束.
訓練 第 0 執行週期, 第 9 批次開始...
訓練 第 0 執行週期, 第 9 批次結束.
```

- 測試顯示結果：

```
測試開始...
測試 第 0 執行週期, 第 0 批次開始...
測試 第 0 執行週期, 第 0 批次結束.
測試 第 0 執行週期, 第 1 批次開始...
測試 第 0 執行週期, 第 1 批次結束.
測試 第 0 執行週期, 第 2 批次開始...
測試 第 0 執行週期, 第 2 批次結束.
測試 第 0 執行週期, 第 3 批次開始...
測試 第 0 執行週期, 第 3 批次結束.
測試 第 0 執行週期, 第 4 批次開始...
測試 第 0 執行週期, 第 4 批次結束.
測試 第 0 執行週期, 第 5 批次開始...
測試 第 0 執行週期, 第 5 批次結束.
測試 第 0 執行週期, 第 6 批次開始...
測試 第 0 執行週期, 第 6 批次結束.
測試 第 0 執行週期, 第 7 批次開始...
測試 第 0 執行週期, 第 7 批次結束.
測試 第 0 執行週期, 第 8 批次開始...
測試 第 0 執行週期, 第 8 批次結束.
```

5-4-5 自訂 Callback

除了以上內建的 Callback 外，我們也可以自訂 Callback，觀察更多訓練過程的資訊。

範例 5. 預設的訓練過程只會顯示每個執行週期的準確率與損失，透過自訂的 callback 可以更仔細觀察每一批訓練的損失，並在整個訓練過程結束後依據收集的資料繪製線圖。

程式：**05_11_Custom_Callback_loss.ipynb**。

1. 建立模型：與 **05_02_ 模型存檔與載入.ipynb** 相似，不再贅述。
2. 自訂 Callback：在每一批的訓練結束後記錄損失至 Pandas DataFrame 中。

```python
class CustomCallback(tf.keras.callbacks.Callback):
    def __init__(self):
        self.task_type=''
        self.epoch=0
        self.batch=0
        self.df = None # pd.DataFrame(columns=['epoch', 'batch', 'metrics'])

    def on_train_begin(self, logs=None):
        self.task_type='訓練'
        print("訓練開始...")

    def on_train_end(self, logs=None):
        print(self.df.shape)
        print("訓練結束.")

    def on_epoch_begin(self, epoch, logs=None):
        self.epoch=epoch

    def on_train_batch_end(self, batch, logs=None):
        df2 = pd.DataFrame([[self.epoch, batch, logs["loss"]]], columns=['epoch', 'batch', 'metrics'])
        if self.df is None:
            self.df = df2
        else:
            self.df = pd.concat([self.df, df2], ignore_index=True)

    def save(self, file_name):
        self.df.to_excel(file_name)
```

3. 自訂 Callback 及訓練模型。

```python
custom_callback = CustomCallback()
model.fit(x_train_norm, y_train_one_hot_encoding, epochs=15, batch_size=256, verbose=0,
          validation_split=0.2, callbacks=[custom_callback])
```

4. 繪圖：圖表顯示的結果很有意思，優化的過程中損失函數並不是一路遞減，而是起起伏伏，但整體趨勢向下遞減。

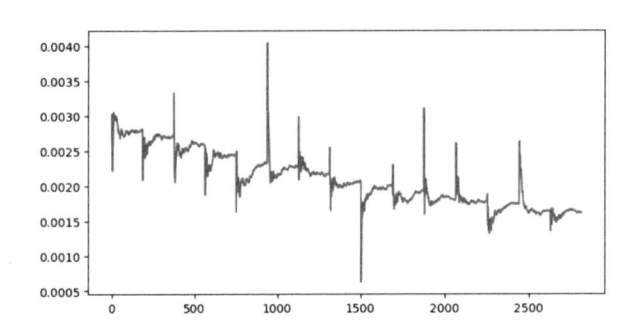

5. 依執行週期（epoch）作小計，取損失值平均值，畫出線圖。

```
1  df2 = df.groupby(by='epoch').mean()
2  plt.figure(figsize=(8, 4))
3  plt.plot(df2.index, df2.metrics);
```

- 執行結果：

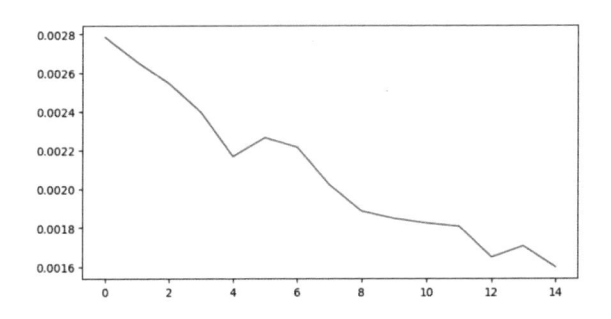

5-4-6 取得優化器的學習率變化

大部份優化器會動態調整學習率，但標準的訓練過程不會顯示學習率，藉由自訂 Callback 顯示學習率變化，讓我們能夠更深入理解優化的過程。

範例 6. 自訂 Callback，觀察學習率變化。

程式：**05_12_Custom_Callback_learning_rate.ipynb**。

1. 讀取測試資料。

第 5 章　TensorFlow 常用指令與功能

```
1  # load dataset
2  dataframe = pd.read_csv("../data/ionosphere.csv", header=None)
3  dataset = dataframe.values
4  # split into input (X) and output (Y) variables
5  X = dataset[:,0:34].astype(float)
6  Y = dataset[:,34]
7  # encode class values as integers
8  encoder = LabelEncoder()
9  encoder.fit(Y)
10 Y = encoder.transform(Y)
```

2. 建立模型：使用簡單的迴歸模型，主要是為了加大訓練週期，並能在短時間完成訓練。

```
1  model = tf.keras.models.Sequential([
2      tf.keras.layers.Input((34,)),
3      tf.keras.layers.Dense(20, activation='relu'),
4      tf.keras.layers.Dense(1, activation='sigmoid')
5  ])
```

3. 自訂 Callback：在每一訓練週期的結束後記錄學習率。

```
1  learning_rate_list = []
2  class MyCallback(tf.keras.callbacks.Callback):
3      def on_epoch_end(self, epoch, epoch_logs):
4          learning_rate_list.append(self.model.optimizer._get_current_learning_rate().numpy())
```

4. 設定優化器、動態調整的學習率：取得學習率變化的函數需有特定寫法，因為 fit 會在呼叫時，餵入相關參數。

```
1  # 設定學習率變化
2  def get_lr_metric(optimizer):
3      def lr(y_true, y_pred):
4          # return optimizer.learning_rate
5          return optimizer._get_current_learning_rate()
6      # learning_rate_list.append(lr)
7      return lr
8
9  def step_decay(epoch):
10     initial_lrate = 0.1
11     drop = 0.5
12     epochs_drop = 10.0
13     lrate = initial_lrate * math.pow(drop, math.floor((1+epoch)/epochs_drop))
14     learning_rate_list.append(lrate)
15     return lrate
16
17 learning_rate=1e-1
18 lr_schedule = tf.keras.optimizers.schedules.ExponentialDecay(
19     initial_learning_rate=learning_rate,
```

5-4 回呼函數（Callbacks）

```
20       decay_steps=10,
21       decay_rate=0.9)
22 optimizer = Adam(lr_schedule)
23 lr_metric = get_lr_metric(optimizer)
24
25 # Compile model
26 model.compile(loss='binary_crossentropy', optimizer=optimizer,
27        metrics=['accuracy', lr_metric])
```

5. 模型訓練。

```
1 history = model.fit(X, Y, epochs=50, validation_split=0.2, callbacks = [MyCallback()])
```

6. 顯示學習率變化

```
1 print(learning_rate_list)
```

- 執行結果：

```
[0.09095325, 0.08272495, 0.07524103, 0.06843417, 0.062243104, 0.05661213, 0.051490575, 0.046832357, 0.042595547, 0.03874204, 0.03523715, 0.03204933, 0.02914991, 0.026512792, 0.02411425, 0.021932695, 0.0199485, 0.018143808, 0.016502386, 0.015009457, 0.01365159, 0.012416566, 0.011293269, 0.010271597, 0.009342352, 0.008497174, 0.007728456, 0.007029281, 0.0063933604, 0.0058149695, 0.005288904, 0.004810431, 0.004375243, 0.003979426, 0.0036194175, 0.0032919776, 0.0029941616, 0.0027232869, 0.0024769187, 0.0022528379, 0.002049029, 0.001863659, 0.0016950583, 0.001541711, 0.0014022362, 0.0012753792, 0.0011599993, 0.0010550569, 0.0009596088, 0.0008727953]
```

7. 繪圖觀察學習率變化比較清楚。

```
1 import matplotlib.pyplot as plt
2
3 plt.title('Learning rate changes')
4 plt.ylabel('Learning rate')
5 plt.plot(range(1, len(learning_rate_list)+1), learning_rate_list, 5)
6 plt.ylim(learning_rate/100., learning_rate)
7 plt.show()
```

- 執行結果：學習率變化隨著執行週期逐步縮小。

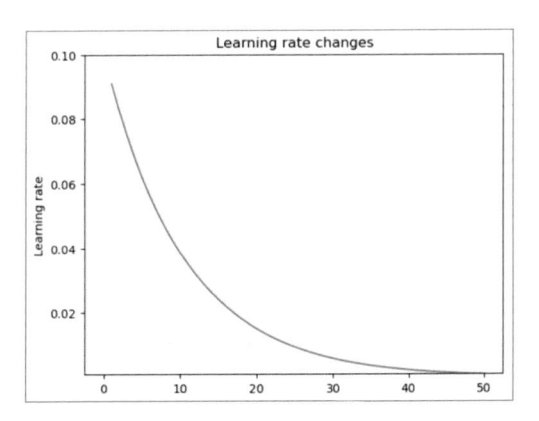

5-4-7 小結

自訂 Callback 還有個好用的功能,就是用來除錯(Debug),假如訓練出現錯誤時,像是 Nan 或優化無法收斂的情形,就可以利用 Callback 逐批檢查,還有更多 Callback 的用法,可參閱『Keras Callback API』[2]。

5-5 TensorBoard

TensorBoard 是一種視覺化的診斷工具,功能非常強大,可以顯示模型結構、訓練過程、包括圖片、文字和音訊資料。在訓練的過程中啟動 TensorBoard,能夠即時觀看訓練過程。PyTorch 也推薦 TensorBoard,表示 Tensorflow 和 PyTorch 雖然是相互競爭的死對頭,但 TensorBoard 仍然是 PyTorch 開發團隊也不得不承認的優秀工具。

5-5-1 TensorBoard 功能

TensorBoard 包含下列功能:

1. 追蹤損失和準確率等效能衡量指標(Metrics),並以視覺化呈現。

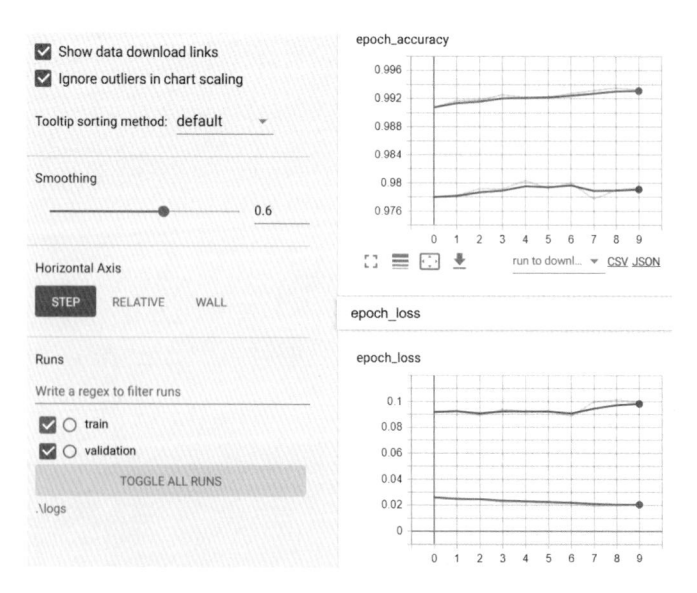

2. 顯示運算圖(Computational Graph):包括張量運算(tensor operation)和神經層(layers)。

5-5 TensorBoard

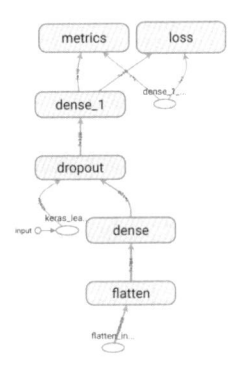

3. 直方圖（Histogram）：顯示訓練過程中的權重（weights）、偏差（bias）的機率分配。

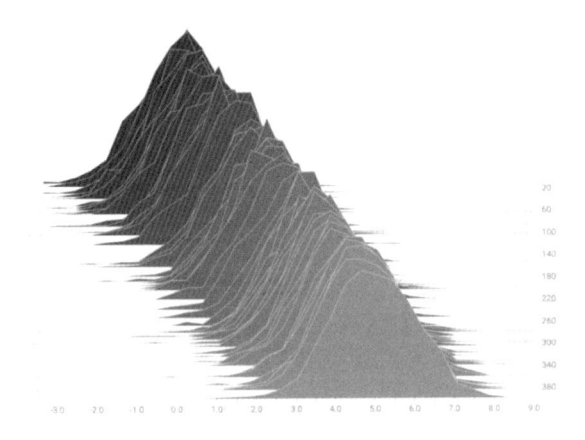

4. 詞嵌入（Word Embedding）展示：把詞嵌入向量降維，投影到三維空間來顯示。畫面右邊可輸入任意單字，例如 King，就會出現下圖，將與其相近的單字顯示出來，原理是透過詞向量（Word2Vec）將每個單字轉為向量，再利用 Cosine_Similarity 計算相似性，詳情會在後續章節介紹。

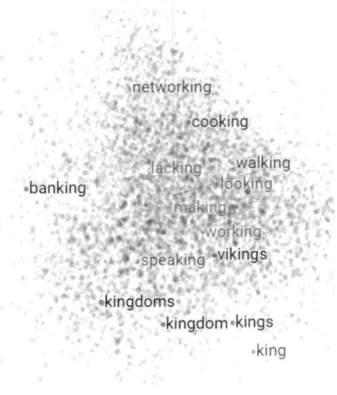

5. 顯示圖片、文字和音訊資料。

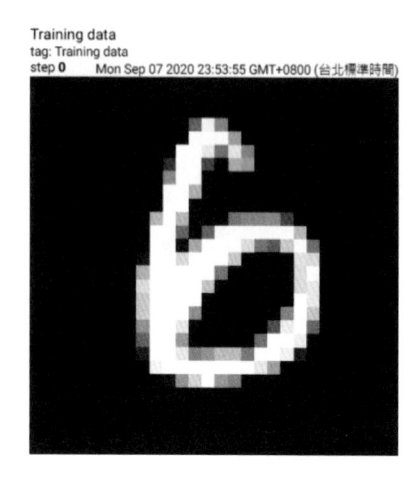

![Training data 顯示一張手寫數字 6 的圖片]

5-5-2 測試

之前在 model.fit 內指定 TensorBoard Callback，可寫入工作日誌檔（Log），內容類別是固定的，如果要自訂寫入的內容，可以直接在程式中寫入工作日誌檔，但是不能使用 model.fit，要改用自動微分（tf.GradientTape）的方式訓練模型。

範例 1. 同樣拿 MNIST 辨識作測試，前面載入資料與建立模型程式碼的流程不變，從 compile 開始作一些調整，以下僅列出關鍵的程式碼，完整的程式請參考 **05_13_TensorBoard_GradientTape.ipynb**。

1. 建立模型：與 **05_02_ 模型存檔與載入 .ipynb** 相似，不再贅述。
2. 設定優化器（optimizer）、損失函數（loss）、效能衡量指標（metrics）。

```
1  # 設定優化器(optimizer)、損失函數(Loss)、效能衡量指標(metrics)的類別
2  loss_object = tf.keras.losses.SparseCategoricalCrossentropy()
3  optimizer = tf.keras.optimizers.Adam()
4
5  # Define 訓練及測試的效能衡量指標(Metrics)
6  train_loss = tf.keras.metrics.Mean('train_loss', dtype=tf.float32)
7  train_accuracy = tf.keras.metrics.SparseCategoricalAccuracy('train_accuracy')
8  test_loss = tf.keras.metrics.Mean('test_loss', dtype=tf.float32)
9  test_accuracy = tf.keras.metrics.SparseCategoricalAccuracy('test_accuracy')
```

3. 寫入效能衡量指標：使用自動微分（tf.GradientTape）的方式訓練模型。

```
1  def train_step(model, optimizer, x_train, y_train):
2      # 自動微分
3      with tf.GradientTape() as tape:
4          predictions = model(x_train, training=True)
5          loss = loss_object(y_train, predictions)
6      grads = tape.gradient(loss, model.trainable_variables)
7      optimizer.apply_gradients(
8          zip(grads, model.trainable_variables))
9  
10     # 計算訓練的效能衡量指標
11     train_loss(loss)
12     train_accuracy(y_train, predictions)
13 
14 def test_step(model, x_test, y_test):
15     # 預測


16     predictions = model(x_test)
17     # 計算損失
18     loss = loss_object(y_test, predictions)
19 
20     # 計算測試的效能衡量指標
21     test_loss(loss)
22     test_accuracy(y_test, predictions)
```

4. 使用 tf.summary.create_file_writer 開啟 log 檔案。

```
10 train_summary_writer = tf.summary.create_file_writer(train_log_dir)
11 test_summary_writer = tf.summary.create_file_writer(test_log_dir)
```

5. 之後再使用 tf.summary.scalar 寫入變數及其儲存的內容。

```
6      with train_summary_writer.as_default():
7          tf.summary.scalar(
8              'loss', train_loss.result(), step=epoch)
9          tf.summary.scalar(
10             'accuracy', train_accuracy.result(), step=epoch)
```

6. 將訓練/測試資料轉成 Dataset：使用 Dataset 可讀取並逐批資料訓練，不必一次讀取全部資料，可節省記憶體的使用。

```python
1  # 將訓練/測試資料轉成 Dataset
2  train_dataset = tf.data.Dataset.from_tensor_slices((x_train_norm, y_train))
3  test_dataset = tf.data.Dataset.from_tensor_slices((x_test_norm, y_test))
4
5  # 每次從 60000 筆訓練資料隨機抽出 64 筆
6  # shuffle：洗牌，batch：每批 64 筆
7  train_dataset = train_dataset.shuffle(60000).batch(64)
8  # 每次從 10000 筆測試資料隨機抽出 64 筆
9  test_dataset = test_dataset.batch(64)
```

7. 模型訓練：將損失及準確率寫入 log。

```python
1  EPOCHS = 5
2
3  # 訓練 5 次
4  for epoch in range(EPOCHS):
5      # 訓練
6      for (x_train, y_train) in train_dataset:
7          train_step(model, optimizer, x_train, y_train)
8
9      # 寫入訓練 log
10     with train_summary_writer.as_default():
11         tf.summary.scalar('loss', train_loss.result(), step=epoch)
12         tf.summary.scalar('accuracy', train_accuracy.result(), step=epoch)
13
14     # 測試
15     for (x_test, y_test) in test_dataset:
16         test_step(model, x_test, y_test)
17
18     # 寫入測試 log
19     with test_summary_writer.as_default():
20         tf.summary.scalar('loss', test_loss.result(), step=epoch)
21         tf.summary.scalar('accuracy', test_accuracy.result(), step=epoch)
22
23     # 顯示結果
24     template = 'Epoch {}, Loss: {}, Accuracy: {}%, Test Loss: {}, Test Accuracy: {}%'
25     print(template.format(epoch+1,
26         train_loss.result(),
27         train_accuracy.result()*100,
28         test_loss.result(),
29         test_accuracy.result()*100))
30
31     # 重置效能衡量指標
32     train_loss.reset_state()
33     test_loss.reset_state()
34     train_accuracy.reset_state()
35     test_accuracy.reset_state()
```

5-5 TensorBoard

8. 在 jupyter notebook 內啟動，分兩步驟：

   - 先載入『TensorBoard notebook extension』擴充程式。

```
1  # 載入 TensorBoard notebook extension，
2  # 即可在 jupyter notebook 啟動 Tensorboard
3  %load_ext tensorboard
```

   - 啟動 Tensorboard 網站。

```
1  # 啟動 Tensorboard
2  %tensorboard --logdir logs/gradient_tape
```

註：% 為 jupyter notebook 的魔術方法前置符號，在終端機或 DOS 啟動，不需 %。

5-5-3 寫入圖片

範例 2. 除了訓練過程的資訊，也可以隨時把資料寫入 Log，以下示範如何將圖片寫入工作日誌檔。

程式：05_13_TensorBoard_GradientTape.ipynb 末段。

1. 設定工作日誌檔目錄，寫入圖像。

```
1  # 任意找一張圖片
2  img = x_train[0].numpy().reshape((-1, 28, 28, 1))
3  img.shape
4
5  # 指定 log 檔名
6  logdir = ".\\logs\\train_data\\" + datetime.datetime.now().strftime("%Y%m%d-%H%M%S")
7  # Creates a file writer for the log directory.
8  file_writer = tf.summary.create_file_writer(logdir)
9
10 # Using the file writer, log the reshaped image.
11 with file_writer.as_default():
12     # 將圖片寫入 log 檔
13     tf.summary.image("Training data", img, step=0)
```

   - 結果如下：

第 5 章　TensorFlow 常用指令與功能

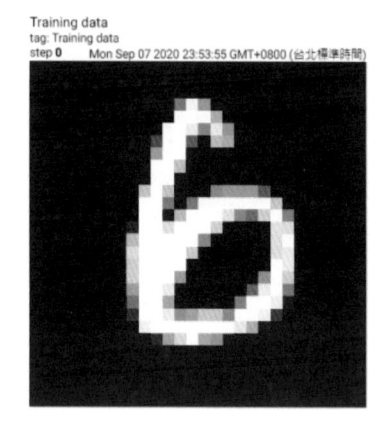

![Training data 手寫數字 6 的圖片]

2. 在後續章節介紹卷積神經網路（CNN）時，可以把卷積轉換後的圖片寫入，藉以瞭解訓練過程中圖片是如何被轉換的，有助於理解神經網路這個黑箱是如何辨識圖片的，這就是所謂的『可解釋的 AI』（Explainable AI, XAI）。

5-5-4 效能調校（Performance Tuning）

範例 3. 進行效能調校，找出最佳參數組合。

程式：**05_14_TensorBoard_Tuning.ipynb**。

1. TensorBoard 與效能調校工具 keras tuner 類似，首先設定多個調校的參數組合。

```
1  # 參數組合
2  from tensorboard.plugins.hparams import api as hp
3
4  HP_NUM_UNITS = hp.HParam('num_units', hp.Discrete([16, 32]))
5  HP_DROPOUT = hp.HParam('dropout', hp.RealInterval(0.1, 0.2))
6  HP_OPTIMIZER = hp.HParam('optimizer', hp.Discrete(['adam', 'sgd']))
```

2. 每一參數組合訓練一個模型。

```
1  # 依每一參數組合執行訓練
2  session_num = 0
3
4  for num_units in HP_NUM_UNITS.domain.values:
5      for dropout_rate in (
6          HP_DROPOUT.domain.min_value, HP_DROPOUT.domain.max_value):
7          for optimizer in HP_OPTIMIZER.domain.values:
8              hparams = {
```

```
 9                    HP_NUM_UNITS: num_units,
10                    HP_DROPOUT: dropout_rate,
11                    HP_OPTIMIZER: optimizer,
12                }
13                run_name = "run-%d" % session_num
14                print('--- Starting trial: %s' % run_name)
15                print({h.name: hparams[h] for h in hparams})
16                run('logs/hparam_tuning/' + run_name, hparams)
17                session_num += 1
```

- 啟動 Tensorboard 網站，點選『hparams』頁籤，顯示如下，第6回合準確率最佳。

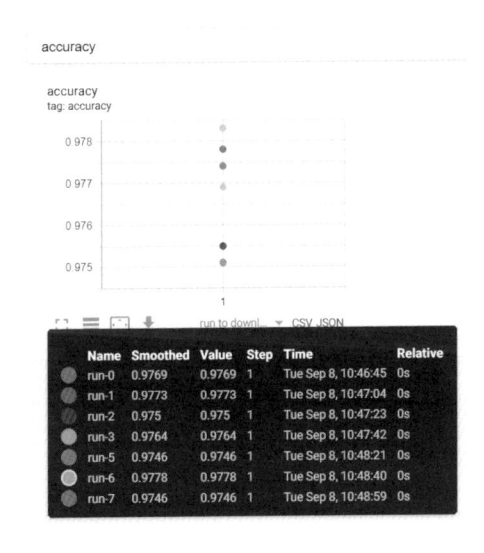

Name	Smoothed	Value	Step	Time	Relative
run-0	0.9769	0.9769	1	Tue Sep 8, 10:46:45	0s
run-1	0.9773	0.9773	1	Tue Sep 8, 10:47:04	0s
run-2	0.975	0.975	1	Tue Sep 8, 10:47:23	0s
run-3	0.9764	0.9764	1	Tue Sep 8, 10:47:42	0s
run-5	0.9746	0.9746	1	Tue Sep 8, 10:48:21	0s
run-6	0.9778	0.9778	1	Tue Sep 8, 10:48:40	0s
run-7	0.9746	0.9746	1	Tue Sep 8, 10:48:59	0s

- 詳細資訊如下：

Trial ID	Show Metrics	dropout	num_units	optimizer	accuracy
3df0d7cf35bec5a...	☐	0.20000	32.000	sgd	0.97460
3ec2aed9e07589f...	☐	0.20000	32.000	adam	0.97780
53bf5bece9190fa...	☐	0.20000	16.000	adam	0.97500
5b97f3c2967245b...	☐	0.10000	16.000	adam	0.97690
6826c7fa3322d82...	☐	0.10000	32.000	adam	0.97390
7684dcc13358fd0...	☐	0.20000	16.000	sgd	0.97640
7b29a731e3daca7...	☐	0.10000	32.000	sgd	0.97460
ae235909ec4e4d9...	☐	0.10000	16.000	sgd	0.97730

第 5 章　TensorFlow 常用指令與功能

- 從下圖的粗體線可以找到最佳參數：
    * dropout rate=0.2
    * 輸出神經元數（num_units）=32
    * 優化器（optimizer）=adam
    * 獲得最佳準確度 0.9775。

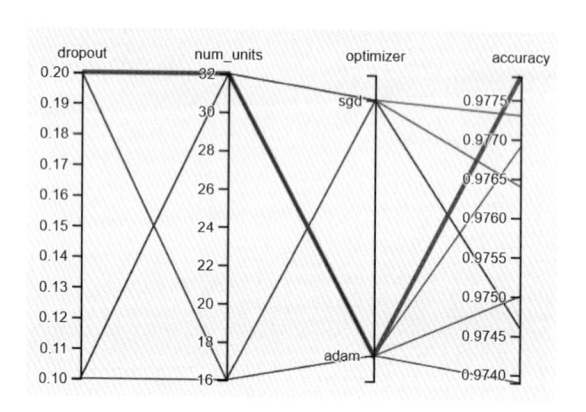

5-5-5　敏感度分析（What-If Tool, WIT）

敏感度分析能幫助我們更了解分類（classification）與迴歸（regression）模型，它擁有許多超厲害的功能，包括：

1. 在下圖左邊的欄位中修改任一個觀察值，重新預測，即可觀察變動的影響。

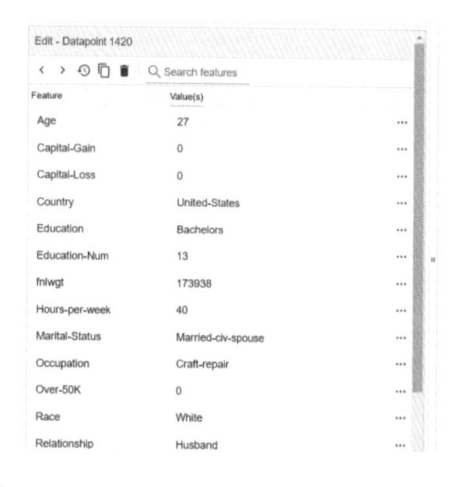

5-6 模型佈署（Deploy）

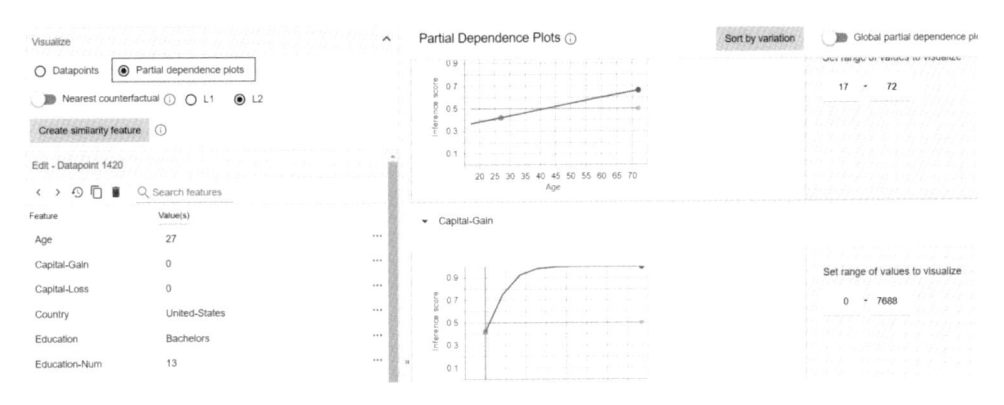

2. 點選『Partial dependence plots』，可以了解個別特徵對預測結果的影響。

3. 切割訓練資料筆數：了解測試資料預測結果的敏感度分析。

詳細操作說明可參考『A Walkthrough with UCI Census Data』[4] 及範例 [5]，可在 Colaboratoy 環境中執行，這是一個二分類的模型，由於該範例不屬於深度學習模型，筆者就不多作說明了。

5-5-6 小結

TensorBoard 隨著時間增加的功能越來越多，都快可以另外寫成一本書了，以上我們只作了很簡單的實驗，如果需要了解更多的資訊，請參閱 TensorBoard 官網指南 [6]。

5-6 模型佈署（Deploy）

一般深度學習的模型安裝的選項如下：

1. 本地伺服器（Local Server）。
2. 雲端伺服器（Cloud Server）。

3. 邊緣運算（IoT Hub）：譬如要偵測全省的溫度，我們會在各縣市安裝成千上萬個感測器，每個 IoT Hub 會安裝模型，負責多個感測器的信號接收、初步過濾和分析，分析完成後再將資料後送到資料中心。

服務呈現的方式可能是網頁、手機 App、桌面程式或 API。

5-6-1 網頁開發

要將模型安裝在本地伺服器運行，可以運用 Python 套件，例如 Django、Flask 或 Streamlit，快速開發網頁，其中以 Streamlit 最為簡單，不需要懂 HTML、CSS、Javascript，只靠 Python 一招半式就可以搞定一個美觀的網站。

範例. 建立辨識手寫阿拉伯數字的網站。

1. 安裝 Streamlit 套件：

    pip install streamlit

2. 載入相關套件。

```
1  # 載入套件
2  import streamlit as st
3  from skimage import io
4  from skimage.transform import resize
5  import numpy as np
6  import tensorflow as tf
```

3. 模型載入。

```
10  model = tf.keras.models.load_model('./model.keras')
```

4. 上傳圖檔。

```
14  # 上傳圖檔
15  uploaded_file = st.file_uploader("上傳圖片(.png)", type="png")
```

5. 檔案上傳後，執行下列工作：

    - 第 18~22 行：把圖像縮小成寬高各為（28, 28）。
    - 第 23 行：RGB 的白色為 255，但訓練資料 MNIST 的白色為 0，故需反轉顏色。
    - 第 27 行：辨識上傳檔案。

5-6 模型佈署（Deploy）

```
16  if uploaded_file is not None:
17      # 讀取上傳圖檔
18      image1 = io.imread(uploaded_file, as_gray=True)
19      # 縮小圖形為(28, 28)
20      image_resized = resize(image1, (28, 28), anti_aliasing=True)
21      # 插入第一維，代表筆數
22      X1 = image_resized.reshape(1,28,28,1)
23      # 顏色反轉
24      X1 = np.abs(1-X1)
25
26      # 預測
27      predictions = model.predict_classes(X1)[0]
28      # 顯示預測結果
29      st.write(f'預測結果:{predictions}')
30      # 顯示上傳圖檔
31      st.image(image1)
```

6. 執行：必須以 streamlit run 執行程式，而非以 python 執行。

    - streamlit run 05_15_web.py

7. 網頁顯示後，拖曳 myDigits 目錄內的任一檔案至畫面中的上傳圖檔區域，就會顯示辨識結果，也可以使用小畫家等繪圖軟體自行書寫數字。

5-6-2 桌面程式開發

我們也可以開發桌面程式提供給個人使用，以下使用 Python 內建模組 Tkinter 開發使用者介面（GUI）。

第 5 章　TensorFlow 常用指令與功能

範例. 桌面程式開發。

程式：**cnn_desktop** 資料夾。

1. 使用者介面程式：main.py。先引入套件：

```python
from tkinter import *
from tkinter import filedialog
from PIL import ImageDraw, Image, ImageGrab
import numpy as np
from skimage import color, io
import os
from cnn_class import getData, trainModel, loadModel
```

2. 使用類別定義畫面。

```python
class Paint(object):
    # 類別初始化函數
    def __init__(self):
        self.root = Tk()
        self.root.title('手寫阿拉伯數字辨識')

        #defining Canvas
        self.c = Canvas(self.root, bg='white', width=280, height=280)

        self.image1 = Image.new('RGB', (280, 280), color = 'white')
        self.draw = ImageDraw.Draw(self.image1)

        self.c.grid(row=1, columnspan=6)

        # 建立【辨識】按鈕
        self.classify_button = Button(self.root, text='辨識', command=lambda:self.classify(self.c))
        self.classify_button.grid(row=0, column=0, columnspan=2, sticky='EWNS')

        # 建立【清畫面】按鈕
        self.clear = Button(self.root, text='清畫面', command=self.clear)
        self.clear.grid(row=0, column=2, columnspan=2, sticky='EWNS')

        # 建立【存檔】按鈕
        self.savefile = Button(self.root, text='存檔', command=self.savefile)
        self.savefile.grid(row=0, column=4, columnspan=2, sticky='EWNS')

        # 建立【預測】文字框
        self.label1 = Label(self.root, height=2, width=10, text='預測：')
        self.label1.grid(row=2, column=3, columnspan=2)
        self.prediction_text = Text(self.root, height=2, width=10)
        self.prediction_text.grid(row=2, column=4, columnspan=2)

        # 定義滑鼠事件處理函數
        self.setup()

        # 監聽事件
        self.root.mainloop()
```

3. 定義按鈕及滑鼠的事件處理：特別注意 paint 函數，如果螢幕顯示設定 >100%，會造成抓取 Canvas 的筆跡範圍失準，因此，筆者在滑鼠書寫時同步寫入記憶體（第 67 行），改以記憶體內容作為模型辨識的依據。

5-6 模型佈署（Deploy）

```python
47      # 滑鼠事件處理函數
48      def setup(self):
49          self.old_x = None
50          self.old_y = None
51          self.line_width = 15
52          self.color = 'black'
53
54          # 定義滑鼠事件處理函數，包括移動滑鼠及鬆開滑鼠按鈕
55          self.c.bind('<B1-Motion>', self.paint)
56          self.c.bind('<ButtonRelease-1>', self.reset)
57
58      # 移動滑鼠 處理函數
59      def paint(self, event):
60          paint_color = self.color
61          if self.old_x and self.old_y:
62              self.c.create_line(self.old_x, self.old_y, event.x, event.y,
63                                 width=self.line_width, fill=paint_color,
64                                 capstyle=ROUND, smooth=TRUE, splinesteps=36)
65              # 顯示設定>100%, 抓到的區域會變小
66              # 畫圖同時寫到記憶體，避免螢幕字型放大，造成抓到的畫布區域不足
67              self.draw.line((self.old_x, self.old_y, event.x, event.y), fill='black', width=self.line_width)
68
69          self.old_x = event.x
70          self.old_y = event.y
71
72      # 鬆開滑鼠按鈕 處理函數
73      def reset(self, event):
74          self.old_x, self.old_y = None, None
75
76      # 【清畫面】處理函數
77      def clear(self):
78          self.c.delete("all")
79          self.image1 = Image.new('RGB', (280, 280), color = 'white')
80          self.draw = ImageDraw.Draw(self.image1)
81          self.prediction_text.delete("1.0", END)
```

```python
83      # 【存檔】處理函數
84      def savefile(self):
85          f = filedialog.asksaveasfilename( defaultextension=".png",
86              filetypes = [("png file",".png")])
87          # asksaveasfile return `None` if dialog closed with "cancel".
88          if f is None:
89              return
90          self.image1.save(f)
```

4. 辨識：與 04_02_ 手寫阿拉伯數字辨識 _ 完整版 .ipynb 類似，將筆跡轉灰階、縮小為 (28, 28)，呼叫 predict，在右下角文字框顯示預測結果。

```python
92      # 【辨識】處理函數
93      def classify(self, widget):
94          img = self.image1.resize((28, 28)).convert('L')
95          img = np.array(img)
96          img = (255 - img) / 255
97
98          img2=Image.fromarray(img)
99          img = np.reshape(img, (1, 28, 28, 1))
100
101         # Predict digit
102         pred = model.predict([img], verbose=False)
103         # Get index with highest probability
104         pred = np.argmax(pred)
105         self.prediction_text.delete("1.0", END)
106         self.prediction_text.insert(END, pred)
```

5. 模型訓練及存檔 / 載入程式：cnn_class.py。這段程式與之前開發流程相似，只是改採 CNN 演算法加上資料增補（Data Augmentation），以提高辨識準確率，下一章會有詳細討論。

6. 執行：

   python main.py

7. 操作程序如下：

   - 先在畫布中間空白處書寫阿拉伯數字 0~9。
   - 書寫完畢後，點擊【辨識】按鈕，辨識結果會出現在右下角文字框。
   - 要書寫另一個數字，先點擊【清畫面】按鈕，將畫布清為空白。
   - 點擊【存檔】按鈕可以儲存圖檔，作為其他程式測試之用。

8. 如果要重新訓練模型，可刪除 mnist_model.keras，再執行：

   python main.py

5-7 TensorFlow Dataset

Tensorflow Dataset 類似 Python Generator，可以視需要逐批讀取資料，不必一股腦把資料全部載入至記憶體，如果將龐大的資料量全部載入，記憶體可能就爆了。另外，它還有支援快取（Cache）、預取（Prefetch）、篩選（Filter）、轉換（Map）... 等功能，使用 Dataset 可有效提升模型訓練效能。

5-7-1 產生 Dataset

建立 Dataset 有很多種方式：

1. from_tensor_slices()：List 或 NumPy ndarray 資料型態轉為 Dataset。
2. from_tensors()：Tensorflow Tensor 資料型態轉為 Dataset。
3. from_generator()：Python Generator 資料型態轉。
4. TFRecordDataset()：將 TFRecord 資料型態轉。
5. TextLineDataset()：將文字檔案轉為 Dataset。

範例 1. 測試 Dataset 的相關操作。

程式：**05_17_Dataset.ipynb**。

1. 自 List 轉入 Dataset。

```
1  import tensorflow as tf
2
3  # 自 list 轉入
4  dataset = tf.data.Dataset.from_tensor_slices([8, 3, 0, 8, 2, 1])
```

2. 使用 for 迴圈自 Dataset 取出所有資料，必須以 numpy 函數轉換才能列印資料內容。

```
1  # 使用 for 迴圈可自 Dataset 取出所有資料
2  for elem in dataset:
3      print(elem.numpy())
```

3. 使用 iter 函數將 Dataset 轉成 Iterator，再使用 next 一次取一批資料。

```
1  # 轉成 iterator
2  it = iter(dataset)
3
4  # 一次取一筆
5  print(next(it).numpy())
6  print(next(it).numpy())
```

- 執行結果為前兩筆：8、3。

4. 依照維度小計（reduce），如果資料維度是一維，即為總計。

```
1  # 依照維度小計(reduce)
2  import numpy as np
3
4  # 一維資料
5  ds = tf.data.Dataset.from_tensor_slices([1, 2, 3, 4, 5])
```

第 5 章　TensorFlow 常用指令與功能

```
6
7  initial_state=0     # 起始值
8  print(ds.reduce(initial_state, lambda state, value: state + value).numpy())
```

- 執行結果：15。

5. 二維資料：依照第一維度小計（reduce）。

```
1  # 依照第一維度小計(reduce)
2  import numpy as np
3
4  # 二維資料
5  ds = tf.data.Dataset.from_tensor_slices(np.arange(1,11).reshape(2,5))
6
7  initial_state=0     # 起始值
8  print(ds.reduce(initial_state, lambda state, value: state + value).numpy())
```

- 執行結果：[7 9 11 13 15]。

6. 三維資料：依照第一維度小計（reduce）。

```
1  # 依照第一維度小計(reduce)
2  import numpy as np
3
4  # 三維資料
5  ds = tf.data.Dataset.from_tensor_slices(np.arange(1,13).reshape(2,2,3))
6
7  print('原始資料:\n', np.arange(1,13).reshape(2,2,3), '\n')
8
9  initial_state=0     # 起始值
10 print('計算結果:\n', ds.reduce(initial_state, lambda state, value: state + value).numpy())
```

- 執行結果：

```
原始資料:
[[[ 1  2  3]
  [ 4  5  6]]

 [[ 7  8  9]
  [10 11 12]]]

計算結果:
[[ 8 10 12]
 [14 16 18]]
```

7. map：以函數套用到 Dataset 內每個元素，下面程式碼將每個元素乘以 2。

```
1  # 對每個元素應用函數(map)
2  import numpy as np
3
4  # 測試資料
5  ds = tf.data.Dataset.from_tensor_slices([1, 2, 3, 4, 5])
6
```

```
7  # 對每個元素應用函數(map)
8  ds = ds.map(lambda x: x * 2)
9
10 # 轉成 iterator，再顯示
11 print(list(ds.as_numpy_iterator()))
```

- 執行結果：[2, 4, 6, 8, 10]。

8. 過濾（filter）：下面程式碼將偶數取出。

```
1  # 過濾(filter)
2  import numpy as np
3
4  # 測試資料
5  ds = tf.data.Dataset.from_tensor_slices([1, 2, 3, 4, 5])
6
7  # 對每個元素應用函數(map)
8  ds = ds.filter(lambda x: x % 2 == 0)
9
10 # 轉成 iterator，再顯示
11 print(list(ds.as_numpy_iterator()))
```

- 執行結果：[2, 4]。

9. 複製（repeat）：有時候訓練資料過少，我們會可以複製少量資料，增加訓練資料，提高模型的準確度。

```
1  # 資料複製(repeat)
2  import numpy as np
3
4  # 測試資料
5  ds = tf.data.Dataset.from_tensor_slices([1, 2, 3, 4, 5])
6
7  # 重複 3 次
8  ds = ds.repeat(3)
9
10 # 轉成 iterator，再顯示
11 print(list(ds.as_numpy_iterator()))
```

- 執行結果：[1, 2, 3, 4, 5, 1, 2, 3, 4, 5, 1, 2, 3, 4, 5]。

10. Dataset 分片（Shard）：將資料依固定間隔取樣，在分散式計算時，可利用此函數將資料分配給每一台工作站（Worker）進行運算。

```
1  # 分片(Shard)
2  import numpy as np
3
4  # 測試資料：0~10
5  ds = tf.data.Dataset.range(11)
6  print('原始資料:\n', list(ds.as_numpy_iterator()))
7
```

```
 8  # 每 3 筆間隔取樣一筆,從第一筆開始
 9  ds = ds.shard(num_shards=3, index=0)
10
11  # 轉成 iterator,再顯示
12  print('\n計算結果:\n', list(ds.as_numpy_iterator()))
```

- 執行結果:

        原始資料:
        [0, 1, 2, 3, 4, 5, 6, 7, 8, 9, 10]

        計算結果:
        [0, 3, 6, 9]

- 另外還有許多函數,譬如 take、skip、unbatch、window、zip…等,請參閱 TensorFlow 官網關於 Dataset 的說明 [7]。

11. 將 MNIST 資料轉入 Dataset。

```
 1  import tensorflow as tf
 2
 3  mnist = tf.keras.datasets.mnist
 4
 5  # 載入 MNIST 手寫阿拉伯數字資料
 6  (x_train, y_train),(x_test, y_test) = mnist.load_data()
 7
 8  # 特徵縮放,使用常態化(Normalization),公式 = (x - min) / (max - min)
 9  x_train_norm, x_test_norm = x_train / 255.0, x_test / 255.0
10
11  # 轉為 Dataset,含 X/Y 資料
12  dataset = tf.data.Dataset.from_tensor_slices((x_train_norm, y_train))
13  print(dataset)
```

- 執行結果:會顯示資料型態及維度。

```
<TensorSliceDataset shapes: ((28, 28), ()), types: (tf.float64, tf.uint8)>
```

12. 逐批取得資料:

- shuffle(10000):每次從 dataset 取出 10000 筆資料進行洗牌。
- batch(1000):隨機抽出 1000 筆資料。

```
 1  # 每次隨機抽出 1000 筆
 2  # shuffle:每次從 60000 筆訓練資料取出 10000 筆洗牌,batch:隨機抽出 1000 筆
 3  train_dataset = dataset.shuffle(10000).batch(1000)
 4  i=0
 5  for (x_train, y_train) in train_dataset:
 6      if i == 0:
 7          print(x_train.shape)
```

5-7 TensorFlow Dataset

```
 8            print(x_train[0])
 9
10        i+=1
11  print(i)
```

- 執行結果：顯示共 60 批資料，每批資料有 1000 筆。

13. 自隨機亂數產生 Dataset。

```
1  import tensorflow as tf
2
3  # 隨機亂數產生 Dataset
4  ds = tf.data.Dataset.from_tensor_slices(
5      tf.random.uniform([4, 10], minval=1, maxval=10, dtype=tf.int32))
6
7  # 轉成 iterator，再顯示
8  print(list(ds.as_numpy_iterator()))
```

- 執行結果：維度為（4, 10），每個值介於（1, 10）之間。

[array([1, 9, 1, 1, 6, 7, 5, 9, 8, 5]), array([3, 8, 1, 3, 9, 7, 1, 2, 3, 6]), array([4, 1, 5, 4, 1, 8, 5, 7, 7, 9]), array([1, 2, 7, 4, 4, 5, 2, 7, 3, 3])]

14. 從 Tensorflow Tensor 資料型態的變數轉入 Dataset。

```
1  import tensorflow as tf
2
3  # 稀疏矩陣
4  mat = tf.SparseTensor(indices=[[0, 0], [1, 2]], values=[1, 2],
5                        dense_shape=[3, 4])
6
7  # 轉入 Dataset
8  ds = tf.data.Dataset.from_tensors(mat)
9
10 # 使用迴圈自 Dataset 取出所有資料
11 for elem in ds:
12     print(tf.sparse.to_dense(elem).numpy())
```

- 執行結果：

```
[[1 0 0 0]
 [0 0 2 0]
 [0 0 0 0]]
```

第 5 章　TensorFlow 常用指令與功能

5-7-2 圖像 Dataset

通常圖檔都很大，假使一次載入所有檔案至記憶體，恐怕會發生記憶體不足的狀況，因此，TensorFlow Dataset 針對影像和文字有提供特殊類別，可以分批載入記憶體，同時提供資料增補（Data Augmentation）的功能，能夠在既有的圖像進行影像處理，產生更多的訓練資料，這些功能全都整合至 Dataset，可在訓練（fit）指令中指定資料來源為 Dataset，一氣呵成。

範例 2. 自 Python Generator 資料型態的變數轉入 Dataset，例如從網路取得壓縮檔，解壓縮後，進行資料增補（Data Augmentation），作為訓練資料。資料增補是提高圖形辨識度非常有效的方法，利用影像處理的技巧，譬如放大、縮小、偏移、旋轉、裁切…等方式，產生各種變形的訓練資料，讓訓練資料更加多樣化，進而使模型有更強的辨識能力。

程式：**05_18_Image_Dataset.ipynb**。

1. 從網路取得壓縮檔，並解壓縮。

```
untar_path = 'flower_photos'
flowers = tf.keras.utils.get_file(
    fname = untar_path,
    origin='https://storage.googleapis.com/download.tensorflow.org/example_images/flower_photos.tgz',
    extract=True)
```

2. 讀取一批檔案：ImageDataGenerator 為資料增補（Data Augmentation）類別，可生成及轉換圖片，例如圖片縮放、旋轉，後續會有更詳盡的說明。

```
# 定義參數
BATCH_SIZE = 32 # 批量
IMG_DIM = 224    # 影像寬度
NB_CLASSES = 5 # Label 類別數
flowers = flowers + '\\' + untar_path # 壓縮檔內含 flower_photos 資料夾

# 資料增補，rescale：圖片縮放，rotation_range：隨機旋轉20度以內
img_gen = tf.keras.preprocessing.image.ImageDataGenerator(rescale=1./255, rotation_range=20)

# 取一批檔案
images, labels = next(img_gen.flow_from_directory(flowers))

# 顯示圖片
import matplotlib.pyplot as plt
import numpy as np

for i in range(1, 6):
    plt.subplot(1, 5, i)
```

```
19      plt.imshow(images[i])
20      plt.title(np.argmax(labels[i]))
21      plt.axis('off')
```

- 執行結果：

Found 3670 images belonging to 5 classes.

3. 定義 generator，轉為 Dataset。

```
1  # 定義 generator 的屬性
2  gen = img_gen.flow_from_directory(
3      flowers,
4      (IMG_DIM, IMG_DIM),
5      'rgb',
6      class_mode='categorical',
7      batch_size=BATCH_SIZE,
8      shuffle=False
9  )
10
11 # 轉為 Dataset
12 ds = tf.data.Dataset.from_generator(lambda: gen,
13     output_signature=(
14         tf.TensorSpec(shape=(BATCH_SIZE, IMG_DIM, IMG_DIM, 3)),
15         tf.TensorSpec(shape=(BATCH_SIZE, NB_CLASSES))
16                 )
17     )
```

4. 自 Dataset 取一批資料。

```
1  # 取一批資料
2  item = ds.take(1)
3  for element in item.as_numpy_iterator():
4      images , label = element
5      print(np.array(images).shape, np.array(label).shape)
6      for i in range(1, 6):
7          plt.subplot(1, 5, i)
8          plt.imshow(images[i])
9          plt.title(np.argmax(labels[i]))
10         plt.axis('off')
```

第 5 章　TensorFlow 常用指令與功能

- 執行結果：

```
(32, 224, 224, 3) (32, 5)
```

可在訓練指令 fit 中指定資料來源為產生的 Dataset，在下一章會有完整範例說明。

5-7-3 TFRecord 與 Dataset

TFRecord 提供跨平台、跨語言的資料結構（record-oriented binary format），能夠儲存各種資料型態的欄位，類似 Json 格式，可序列化（serialization）為二進位的格式儲存，是由 Tensorflow 團隊所開發，遵循 Google Protocol Buffer 規範。

在操作 TFRecord 時，需要藉由 tf.train.Example 將資料封裝成 protocol message，tf.train.Example 的格式為 {"string": tf.train.Feature}，而 tf.train.Feature 可接受 BytesList、FloatList、Int64List 三種格式。BytesList 用於字串或二進位的資料，例如圖像、語音等。

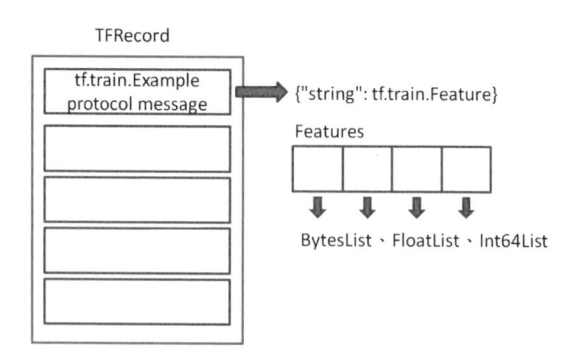

▲ 圖 5.2　TFRecord 結構

範例 3. TFRecord 相關操作測試 .。

程式：**05_19_TFRecord.ipynb**。

1. 定義 tf.train.Feature 轉換函數。

5-7 TensorFlow Dataset

```
1  # 下列函數可轉換為 tf.train.Example 的 tf.train.Feature
2  def _bytes_feature(value):
3      """Returns a bytes_list from a string / byte."""
4      if isinstance(value, type(tf.constant(0))):
5          value = value.numpy()
6      return tf.train.Feature(bytes_list=tf.train.BytesList(value=[value]))
7
8  def _float_feature(value):
9      """Returns a float_list from a float / double."""
10     return tf.train.Feature(float_list=tf.train.FloatList(value=[value]))
11
12 def _int64_feature(value):
13     """Returns an int64_list from a bool / enum / int / uint."""
14     return tf.train.Feature(int64_list=tf.train.Int64List(value=[value]))
```

2. 簡單測試。

```
1  print(_bytes_feature(b'test_string'))
2  print(_bytes_feature(u'test_bytes'.encode('utf-8')))
3
4  print(_float_feature(np.exp(1)))
5
6  print(_int64_feature(True))
7  print(_int64_feature(1))
```

- 執行結果：

```
bytes_list {
  value: "test_string"
}
bytes_list {
  value: "test_bytes"
}
float_list {
  value: 2.7182817
}
int64_list {
  value: 1
}
int64_list {
  value: 1
}
```

3. 序列化（serialization）測試。

```
1  # 序列化(serialization)
2  feature = _float_feature(np.exp(1))
3  feature.SerializeToString()
```

- 執行結果：b'\x12\x06\n\x04T\xf8-@'，為二進位的格式，有經過壓縮。

第 5 章　TensorFlow 常用指令與功能

4. 建立 tf.train.Example 訊息，含有 4 個 feature：0~3。

```
# 建立tf.train.Example訊息，含 4 個 feature

# The number of observations in the dataset.
n_observations = int(1e4)

# Boolean feature, encoded as False or True.
feature0 = np.random.choice([False, True], n_observations)

# Integer feature, random from 0 to 4.
feature1 = np.random.randint(0, 5, n_observations)

# String feature
strings = np.array([b'cat', b'dog', b'chicken', b'horse', b'goat'])
feature2 = strings[feature1]

# Float feature, from a standard normal distribution
feature3 = np.random.randn(n_observations)
```

5. 接下來要寫入 TFRecord 檔案，先定義 tf.train.Example 資料序列化函數。

```
# 序列化(serialization)
def serialize(feature0, feature1, feature2, feature3):
    """
    Creates a tf.train.Example message ready to be written to a file.
    """
    # Create a dictionary mapping the feature name to the tf.train.Example-compatible
    # data type.
    feature = {
            'feature0': _int64_feature(feature0),
            'feature1': _int64_feature(feature1),
            'feature2': _bytes_feature(feature2),
            'feature3': _float_feature(feature3),
    }

    # Create a Features message using tf.train.Example.

    example_proto = tf.train.Example(features=tf.train.Features(feature=feature))
    return example_proto.SerializeToString()
```

6. 將一筆記錄寫入 TFRecord 檔案。

```
# 將一筆記錄寫入 TFRecord 檔案
with tf.io.TFRecordWriter("test.tfrecords") as writer:
    writer.write(serialized_example)
```

7. 讀取 TFRecord 檔案。

```
# 開啟 TFRecord 檔案
filenames = ["test.tfrecords"]
raw_dataset = tf.data.TFRecordDataset(filenames)

```

```
5  ## 取得序列化的資料
6  for raw_record in raw_dataset.take(10):
7      print(repr(raw_record))
```

- 執行結果：為二進位的格式。

```
<tf.Tensor: shape=(), dtype=string, numpy=b'\nR\n\x11\n\x08feature0\x12\x05\x1a\x03\n\x01\x00\n\x11\n\x08feature1\x12\x05\x1a\x03\n\x01\x04\n\x14\n\x08feature2\x12\x08\n\x06\n\x04goat\n\x14\n\x08feature3\x12\x08\x12\x06\n\x04[\xd3|?'>
```

8. 若要取得原始資料，先反序列化（Deserialize），設定原始資料的欄位屬性，透過 parse_single_example() 來進行反序列化。

```
1  # 設定原始資料的欄位屬性
2  feature_description = {
3      'feature0': tf.io.FixedLenFeature([], tf.int64, default_value=0),
4      'feature1': tf.io.FixedLenFeature([], tf.int64, default_value=0),
5      'feature2': tf.io.FixedLenFeature([], tf.string, default_value=''),
6      'feature3': tf.io.FixedLenFeature([], tf.float32, default_value=0.0),
7  }
8
9  # 將 tf.train.Example 訊息轉為 字典(dictionary)
10 def _parse_function(example_proto):
11     return tf.io.parse_single_example(example_proto, feature_description)
```

9. 取得每一個欄位值。

```
1  # 反序列化(Deserialize)
2  parsed_dataset = raw_dataset.map(_parse_function)
3
4  # 取得每一個欄位值
5  for parsed_record in parsed_dataset.take(10):
6      print(repr(parsed_record))
```

- 執行結果：

```
{'feature0': <tf.Tensor: shape=(), dtype=int64, numpy=0>, 'feature1': <tf.Tensor: shape=(), dtype=int64, numpy=4>, 'feature2': <tf.Tensor: shape=(), dtype=string, numpy=b'goat'>, 'feature3': <tf.Tensor: shape=(), dtype=float32, numpy=0.9876>}
```

10. 從網路上取得官網的 TFRecord 檔案。

```
1  # 從網路上取的官網的 TFRecord 檔案
2  file_path = "https://storage.googleapis.com/download.tensorflow.org/" +
3              "data/fsns-20160927/testdata/fsns-00000-of-00001"
4  fsns_test_file = tf.keras.utils.get_file("fsns.tfrec", file_path)
5
6  # 顯示存檔位置
7  fsns_test_file
```

- 執行結果：顯示預設存檔位置，其中 mikec 為 Windows 登入帳號。C:\\Users\\mikec\\.keras\\datasets\\fsns.tfrec

第 5 章　TensorFlow 常用指令與功能

11. 讀取 TFRecord 檔案。

```
1  # 讀取 TFRecord 檔案
2  dataset = tf.data.TFRecordDataset(filenames = [fsns_test_file])
3
4  # 取得下一筆資料
5  raw_example = next(iter(dataset))
6  parsed = tf.train.Example.FromString(raw_example.numpy())
7  parsed.features.feature['image/text']
```

- 執行結果：該欄位為一字串。

```
bytes_list {
  value: "Rue Perreyon"
}
```

5-7-4 TextLineDataset

文字檔也可以像二進位檔案一樣儲存在 Dataset 內，並且序列化後存檔。

範例 4. TextLineDataset 相關操作測試。

程式：**05_20_TextLineDataset.ipynb**。

1. 讀取三個語料庫檔案，合併為一 TextLineDataset。

```
1  # 讀取三個檔案
2  directory_url = 'https://storage.googleapis.com/download.tensorflow.org/data/illiad/'
3  file_names = ['cowper.txt', 'derby.txt', 'butler.txt']
4
5  file_paths = [
6      tf.keras.utils.get_file(file_name, directory_url + file_name)
7      for file_name in file_names
8  ]
9
10 # 合併為一資料集
11 ds = tf.data.TextLineDataset(file_paths)
```

2. 讀取 5 筆資料。

```
1  # 讀取5筆資料
2  for line in ds.take(5):
3      print(line.numpy())
```

- 執行結果：

5-7 TensorFlow Dataset

```
b"\xef\xbb\xbfAchilles sing, O Goddess! Peleus' son;"
b'His wrath pernicious, who ten thousand woes'
b"Caused to Achaia's host, sent many a soul"
b'Illustrious into Ades premature,'
b'And Heroes gave (so stood the will of Jove)'
```

3. 輪流（interleave）：每個檔案讀 3 筆資料即換下一個檔案讀取（cycle_length=3）。

```
1  # interleave：每個檔案輪流讀取一次
2  files_ds = tf.data.Dataset.from_tensor_slices(file_paths)
3  lines_ds = files_ds.interleave(tf.data.TextLineDataset, cycle_length=3)
4
5  # 各讀 3 筆，共 9 筆
6  for i, line in enumerate(lines_ds.take(9)):
7      if i % 3 == 0:
8          print()
9      print(line.numpy())
```

- 執行結果：

```
b"\xef\xbb\xbfAchilles sing, O Goddess! Peleus' son;"
b"\xef\xbb\xbfOf Peleus' son, Achilles, sing, O Muse,"
b'\xef\xbb\xbfSing, O goddess, the anger of Achilles son of Peleus, that brought'

b'His wrath pernicious, who ten thousand woes'
b'The vengeance, deep and deadly; whence to Greece'
b'countless ills upon the Achaeans. Many a brave soul did it send'

b"Caused to Achaia's host, sent many a soul"
b'Unnumbered ills arose; which many a soul'
b'hurrying down to Hades, and many a hero did it yield a prey to dogs and'
```

範例 5. TextLineDataset 結合篩選（filter）函數。

1. 讀取鐵達尼文字檔案（.csv），匯入至 TextLineDataset。

```
1  # 讀取鐵達尼文字檔案(.csv)，匯入至TextLineDataset
2  file_path = "https://storage.googleapis.com/tf-datasets/titanic/train.csv"
3  titanic_file = tf.keras.utils.get_file("train.csv", file_path)
4  titanic_lines = tf.data.TextLineDataset(titanic_file)
```

2. 篩選生存者的資料。

```
1  # 篩選生存者的資料
2  def survived(line):
3      return tf.not_equal(tf.strings.substr(line, 0, 1), "0")
4
5  # 篩選
6  survivors = titanic_lines.skip(1).filter(survived)
```

第 5 章　TensorFlow 常用指令與功能

```
 7
 8  # 讀取10筆資料
 9  for line in survivors.take(10):
10      print(line.numpy())
```

- 執行結果：

```
b'1,female,38.0,1,0,71.2833,First,C,Cherbourg,n'
b'1,female,26.0,0,0,7.925,Third,unknown,Southampton,y'
b'1,female,35.0,1,0,53.1,First,C,Southampton,n'
b'1,female,27.0,0,2,11.1333,Third,unknown,Southampton,n'
b'1,female,14.0,1,0,30.0708,Second,unknown,Cherbourg,n'
b'1,female,4.0,1,1,16.7,Third,G,Southampton,n'
b'1,male,28.0,0,0,13.0,Second,unknown,Southampton,y'
b'1,female,28.0,0,0,7.225,Third,unknown,Cherbourg,y'
b'1,male,28.0,0,0,35.5,First,A,Southampton,y'
b'1,female,38.0,1,5,31.3875,Third,unknown,Southampton,n'
```

範例 6. TextLineDataset 結合 DataFrame。

1. 讀取鐵達尼文字檔案（.csv）。

```
1  import pandas as pd
2
3  df = pd.read_csv(titanic_file, index_col=None)
4  df.head()
```

2. 匯入 Dataset，讀取 1 筆資料。

```
1  # 匯入 Dataset
2  ds = tf.data.Dataset.from_tensor_slices(dict(df))
3
4  # 讀取1筆資料
5  for feature_batch in ds.take(1):
6      for key, value in feature_batch.items():
7          print(f"{key:20s}: {value}")
```

- 執行結果：

```
survived            : 0
sex                 : b'male'
age                 : 22.0
n_siblings_spouses  : 1
parch               : 0
fare                : 7.25
class               : b'Third'
deck                : b'unknown'
embark_town         : b'Southampton'
alone               : b'n'
```

5-7-5 Dataset 效能提升

使用 Dataset 時，可利用『預先讀取』（Prefetch）、快取（Cache）等指令，來提升資料讀取的效能。下面用時間軸的方式來展示 Prefetch 和 Cache 的用途。

1. prefetch：由於訓練時只會利用到 CPU/RAM，TensorFlow 可利用空檔先讀取下一批資料（Disk I/O），並作資料轉換。

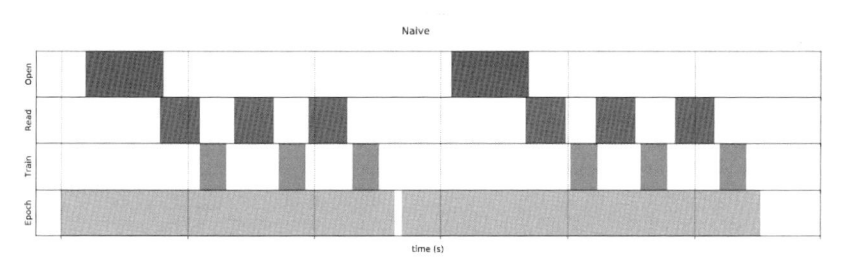

▲ 圖 5.3 不使用 prefetch 的話，開啟 Dataset、讀取資料、訓練這三個動作會依序進行，拉長執行時間

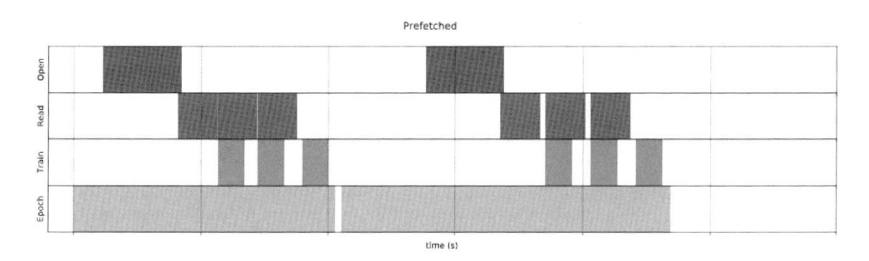

▲ 圖 5.4 使用 prefetch 的話，則會在訓練時，同時讀取下一批資料，故讀取資料和訓練一起同步進行

2. cache：可將讀出的資料留在快取記憶體裡，之後可再重複使用。

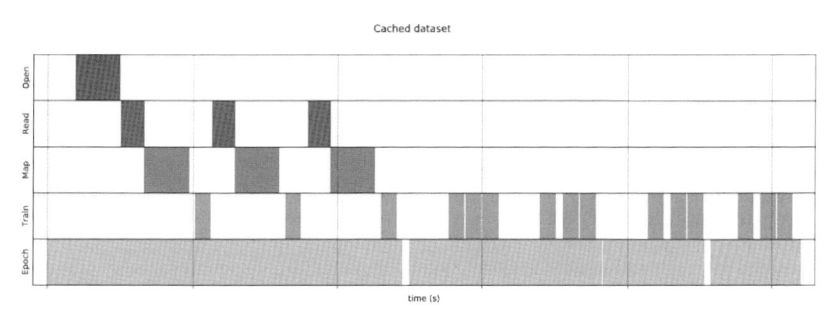

▲ 圖 5.5 使用 cache，能夠降低開啟 Dataset 讀取資料的次數，減少硬碟 IO。

詳細的情形可參考『官網 Dataset 效能說明』[8]。

▎參考資料（References）

[1] TensorFlow Save, serialize, and export models
(https://www.tensorflow.org/tutorials/keras/save_and_load)

[2] Keras Save, serialize, and export models
(https://keras.io/guides/serialization_and_saving/)

[3] Keras Callbacks API
(https://keras.io/api/callbacks/)

[4] A Walkthrough with UCI Census Data
(https://pair-code.github.io/what-if-tool/learn/tutorials/walkthrough/)

[5] 參考範例
(https://colab.research.google.com/github/pair-code/what-if-tool/blob/master/What_If_Tool_Notebook_Usage.ipynb)

[6] TensorFlow TensorBoard 指南
(https://www.tensorflow.org/tensorboard/get_started)

[7] TensorFlow Dataset 說明
(https://www.tensorflow.org/api_docs/python/tf/data/Dataset)

[8] TensorFlow Dataset 效能說明
(https://www.tensorflow.org/guide/data_performance)

第 6 章

卷積神經網路（Convolutional Neural Network）

第三波人工智慧浪潮在自然使用者介面（Natural User Interface, NUI）有突破性的進展，包括影像（Image、Video）、語音（Voice）與文字（Text）的辨識/生成/分析，電腦學會人類日常生活中所使用的溝通方式，與使用者互動不僅更具親和力，也能對週遭的環境作出更合理、更有智慧的判斷與反應，將這種能力附加到產品上，可使產品的應用發展爆發無限可能，包括自駕車（Self-Driving）、無人機（Drone）、智慧家庭（Smart Home）、製造/服務機器人（Robot）、聊天機器人（ChatBot）... 等，不勝枚舉。從這一章開始，我們逐一來探討影像（Image/Video）、語音（Voice）、文字（Text）的相關演算法。

第 6 章　卷積神經網路

▍6-1　卷積神經網路簡介

之前我們只用了十幾行程式即可辨識阿拉伯數字，令人相當興奮，但是，模型使用像素（Pixel）為特徵輸入，好像與人類辨識圖形的方式並不相同，我們應該不會逐點辨識圖形的內涵，思考一下問題的本質：

1. 寫字通常都會居中，故在中央像素的重要性應遠大於周邊的像素。
2. 像素之間應有所關聯，而非互相獨立，比如 1，為一垂直線。
3. 人類視覺應該不是逐個像素辨識，而是觀察數字的線條或輪廓，例如下圖，觀察左圖，根據輪廓會看到正面及側面照，觀察右圖，根據線條會看到女孩及人臉。

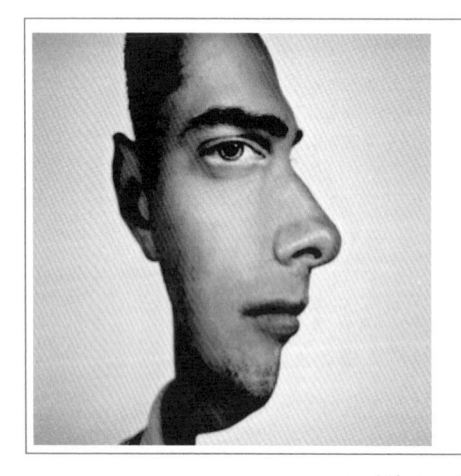

▲ 圖 6.1　線條與輪廓的觀察

因此，卷積神經網路（Convolutional Neural Network, CNN）引進了卷積層（Convolution Layer），進行『特徵萃取』（Feature Extraction），將像素轉換為各種線條特徵，再交給 Dense 層辨識，這就是圖 1.7 機器學習流程的第 3 步驟 -- 特徵工程（Feature Engineering）。

卷積（Convolution）簡單說就是將圖形逐步抽樣化（Abstraction），把不必要的資訊刪除，例如色彩、背景等，下圖經過 3 層卷積後，有些圖依稀可辨識出物體的輪廓，因而，模型就依據這些線條辨識出是人、車或其他動物。

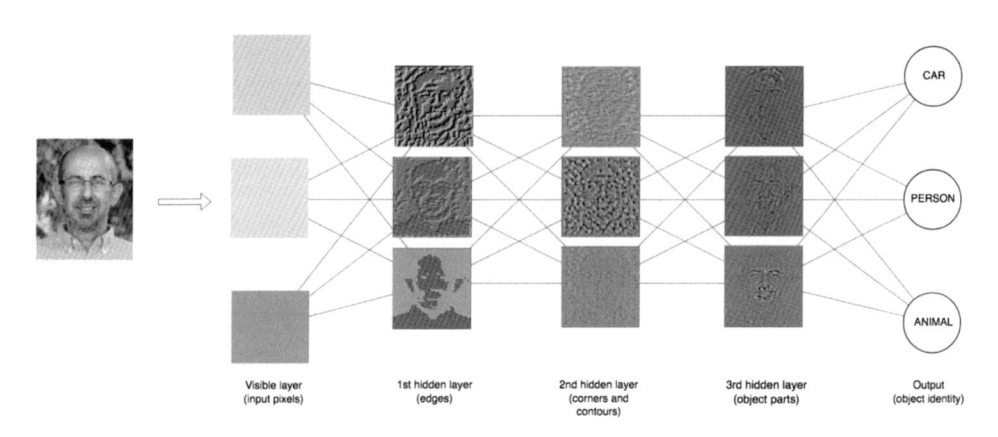

▲ 圖 6.2 卷積神經網路（Convolutional Neural Network, CNN）的特徵萃取

卷積神經網路（Convolutional Neural Network），以下簡稱 CNN，它的模型結構如下：

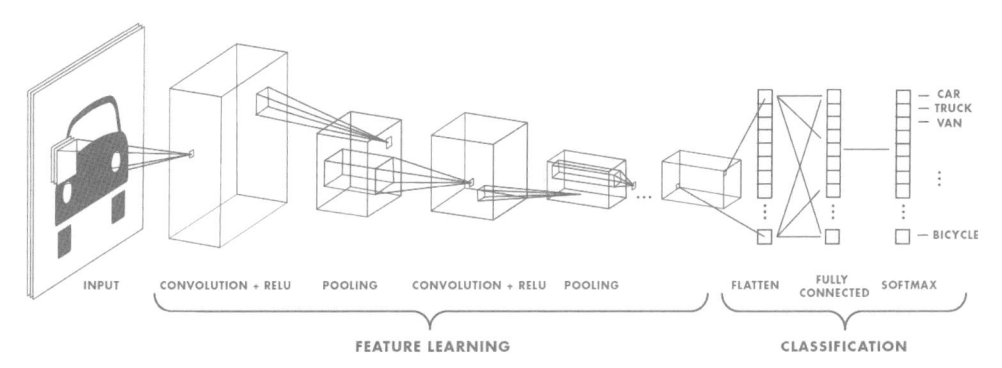

▲ 圖 6.3 卷積神經網路（Convolutional Neural Network, CNN）的模型結構

模型訓練程序如下：

1. 先輸入一張圖像，可以是彩色的，每個色彩通道（Channel）分別卷積再合併。
2. 圖像經過卷積層（Convolution Layer）運算，變成特徵圖（Feature Map），卷積可以指定很多層，注意，卷積矩陣不是固定的，而是由訓練推估出來的，與傳統的影像處理不同，卷積層通常會附加 ReLU Activation Function。
3. 卷積層後面會接一個池化層（Pooling），進行下採樣（Down Sampling），以減少模型的參數個數，避免模型過於龐大。

最後把特徵圖（Feature Map）壓扁成一維（Flatten），交給 Dense 層辨識。

6-2 卷積（Convolution）

卷積是定義一個濾波器（Filter）或稱卷積核（Kernel），對圖像進行『乘積和』運算，例如下圖所示，計算步驟如下：

1. 將輸入圖像依照濾波器裁切相同尺寸的部份圖像。
2. 裁切的圖像與濾波器相同的位置進行相乘。
3. 加總所有格的數值，即為輸出的第一格數值。
4. 逐步向右滑動視窗（如圖 6.5），回到步驟 1，計算下一格的值。
5. 滑到最右邊後，再往下滑動視窗，繼續進行。

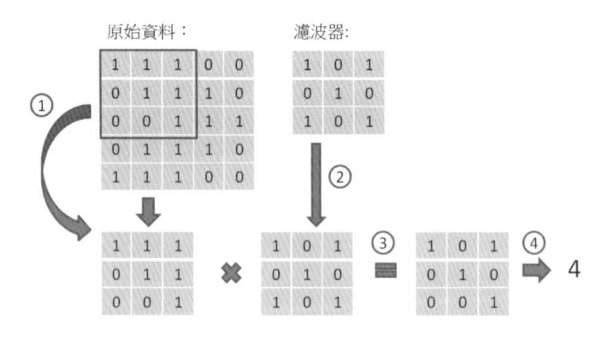

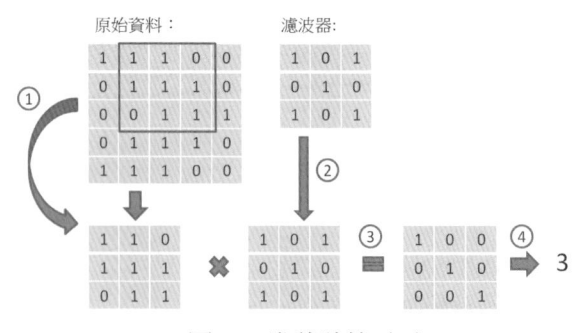

▲ 圖 6.4 卷積計算（1）

▲ 圖 6.5 卷積計算（2）

網路上有許多動畫或影片可以參考，例如『Convolutional Neural Networks—Simplified』[1] 文中卷積計算的 GIF 動畫 [2]。

6-2 卷積（Convolution）

範例. 使用程式計算卷積。

程式：06_01_convolutions.ipynb。

1. 準備資料及濾波器（Filter）：以程式驗證圖 6.4 的卷積結果。

```python
import numpy as np

# 測試資料
source_map = np.array(list('1110001110001110011001100')).astype(int)
source_map = source_map.reshape(5,5)
print('原始資料：')
print(source_map)

# 濾波器(Filter)
filter1 = np.array(list('101010101')).astype(int).reshape(3,3)
print('\n濾波器：')
print(filter1)
```

- 執行結果：

```
原始資料：
[[1 1 1 0 0]
 [0 1 1 1 0]
 [0 0 1 1 1]
 [0 0 1 1 0]
 [0 1 1 0 0]]

濾波器：
[[1 0 1]
 [0 1 0]
 [1 0 1]]
```

2. 計算卷積。

```python
# 計算卷積
# 初始化計算結果的矩陣
width = height = source_map.shape[0] - filter1.shape[0] + 1
result = np.zeros((width, height))

# 計算每一格
for i in range(width):
    for j in range(height):
        value1 =source_map[i:i+filter1.shape[0], j:j+filter1.shape[1]] * filter1
        result[i, j] = np.sum(value1)
print(result)
```

第 6 章　卷積神經網路

- 執行結果：

```
[4. 3. 4.]
[2. 4. 3.]
[2. 3. 4.]
```

3. 使用 SciPy 套件提供的卷積函數驗算，執行結果一致。

```
1  # 使用 scipy 計算卷積
2  import scipy
3
4  # convolve2d：二維卷積
5  scipy.signal.convolve2d(source_map, filter1, mode='valid')
```

4. 比較結果：完全相等。

```
1  (result == convolve2d(source_map, filter1, mode='valid')).all()
```

卷積設定有兩個重要參數：

1. 補零（Padding）：上面的卷積計算會使得圖像尺寸變小，因為，以 3x3 濾波器滑動視窗時，裁切的視窗數每列會不足 2 個，即濾波器寬度減 1，因此 Padding 有兩個選項：

   - Padding='same'：在圖像周遭補零，如圖 6.6，使計算結果的矩陣尺寸與原始圖像尺寸相同。

   - Padding='valid'：不補零，計算後圖像尺寸變小。

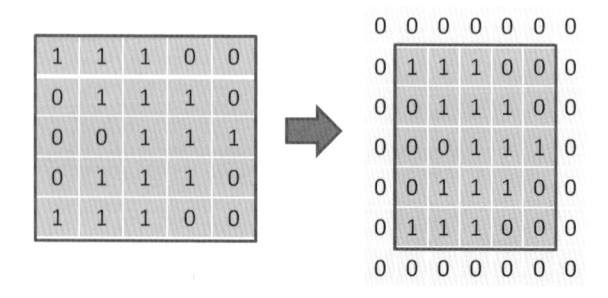

▲ 圖 6.6　Padding='same'，在圖像周遭補上不足的列與行

2. 滑動視窗的步數（Stride）：圖 6.5 的 Stride=1，圖 6.7 的 Stride=2，加大 Stride 可減少視窗數量，要估算的參數個數隨之減少。

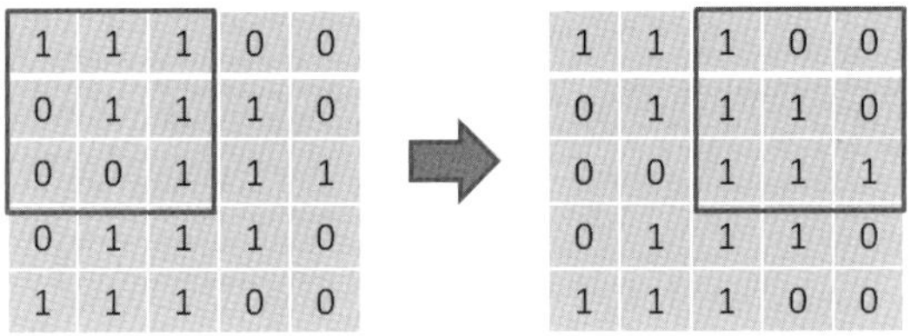

▲ 圖 6.7　Stride=2，一次滑動 2 格視窗

以上是二維的卷積（Conv2D）的運作，通常應用在圖像上。TensorFlow 還提供 Conv1D、Conv3D，其中 Conv1D 因只考慮上下文（Context Sensitive），所以可應用於語音或文字方面，Conv3D 則可應用於立體的物件。還有 Conv2DTranspose 提供反卷積（Deconvolution）或稱上採樣（Up Sampling）的功能，反向操作，以特徵圖重建圖像。同時使用卷積和反卷積，可以組合成 Encoder-Decoder 模型，它是許多生成模型的基礎演算法，還可以去除雜訊，生成乾淨的圖像。

6-3 濾波器（Filter）

不同的濾波器會產生各種影像處理的效果，以下就來看看各種濾波器的運算結果。

範例. 各種濾波器測試。

程式：**06_01_convolutions.ipynb** 下半部。

1. 先安裝 Python OpenCV 套件，它是影像處理的套件。

    pip install opencv-python

2. 定義卷積的影像轉換函數。

```
1  # 卷積的影像轉換函數，padding='same'
2  from skimage.exposure import rescale_intensity
3
4  def convolve(image, kernel):
5      # 取得圖像與濾波器的寬高
6      (iH, iW) = image.shape[:2]
7      (kH, kW) = kernel.shape[:2]
8
9      # 計算 padding='same' 單邊所需的補零行數
10     pad = int((kW - 1) / 2)
11     image = cv2.copyMakeBorder(image, pad, pad, pad, pad, cv2.BORDER_REPLICATE)
12     output = np.zeros((iH, iW), dtype="float32")
```

第 6 章　卷積神經網路

```
13
14      # 卷積
15      for y in np.arange(pad, iH + pad):
16          for x in np.arange(pad, iW + pad):
17              roi = image[y - pad:y + pad + 1, x - pad:x + pad + 1]   # 裁切圖像
18              k = (roi * kernel).sum()                                  # 卷積計算
19              output[y - pad, x - pad] = k                              # 更新計算結果的矩陣
20
21      # 調整影像色彩深淺範圍至 (0, 255)
22      output = rescale_intensity(output, in_range=(0, 255))
23      output = (output * 255).astype("uint8")
24
25      return output       # 回傳結果影像
```

3. 將影像灰階化：skimage 是 scikit-image 套件，也是一個影像處理的套件，功能較 OpenCV 簡易，Anaconda 已預先安裝，不需再安裝。

```
1   # pip install opencv-python
2   import skimage
3   import cv2
4
5   # 自 skimage 取得內建的圖像
6   image = skimage.data.chelsea()
7   cv2.imshow("original", image)
8
9   # 灰階化
10  gray = cv2.cvtColor(image, cv2.COLOR_BGR2GRAY)
11  cv2.imshow("gray", gray)
12
13  # 按 Enter 關閉視窗
14  cv2.waitKey(0)
15  cv2.destroyAllWindows()
```

- 執行結果：顏色變化請參閱範例檔。

原圖：

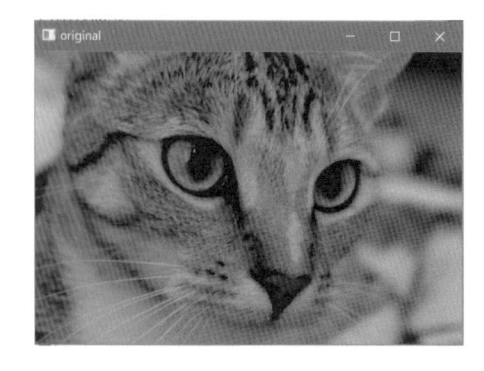

灰階化：

6-3 濾波器（Filter）

4. 模糊化（Blur）：濾波器設定為全部為 1 的矩陣，卷積後每一點數值為周圍像素的平均，會造成圖像模糊化，一般用於消除紅眼現象或是雜訊。

```
1  # 小模糊 filter
2  smallBlur = np.ones((7, 7), dtype="float") * (1.0 / (7 * 7))
3
4  # 卷積
5  convoleOutput = convolve(gray, smallBlur)
6  opencvOutput = cv2.filter2D(gray, -1, smallBlur)
7  cv2.imshow("little Blur", convoleOutput)
8
9  # 大模糊
10 largeBlur = np.ones((21, 21), dtype="float") * (1.0 / (21 * 21))
11
12 # 卷積
13 convoleOutput = convolve(gray, largeBlur)
14 opencvOutput = cv2.filter2D(gray, -1, largeBlur)
15 cv2.imshow("large Blur", convoleOutput)
16
17 # 按 Enter 關閉視窗
18 cv2.waitKey(0)
19 cv2.destroyAllWindows()
```

- 小模糊：7x7 矩陣。

- 大模糊：：21x21 矩陣，矩陣越大，影像越模糊。

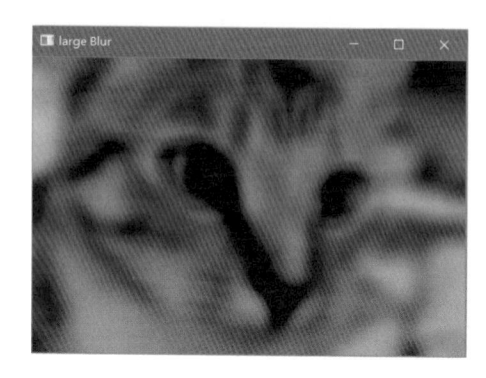

5. 銳化（sharpen）：可使圖像的對比更加明顯。

```
1  # sharpening filter
2  sharpen = np.array((
3      [0, -1, 0],
4      [-1, 5, -1],
5      [0, -1, 0]), dtype="int")
6
7  # 卷積
8  convoleOutput = convolve(gray, sharpen)
9  opencvOutput = cv2.filter2D(gray, -1, sharpen)
10 cv2.imshow("sharpen", convoleOutput)
11
12 # 按 Enter 關閉視窗
13 cv2.waitKey(0)
14 cv2.destroyAllWindows()
```

- 執行結果：卷積凸顯中間點，使圖像特徵越清晰。

6. Laplacian 邊緣偵測：可偵測圖像的輪廓。

6-3 濾波器（Filter）

```
1   # Laplacian filter
2   laplacian = np.array((
3       [0, 1, 0],
4       [1, -4, 1],
5       [0, 1, 0]), dtype="int")
6
7   # 卷積
8   convoleOutput = convolve(gray, laplacian)
9   opencvOutput = cv2.filter2D(gray, -1, laplacian)
10  cv2.imshow("laplacian edge detection", convoleOutput)
11
12  # 按 Enter 關閉視窗
13  cv2.waitKey(0)
14  cv2.destroyAllWindows()
```

- 執行結果：卷積凸顯週邊，顯現圖像週邊線條。

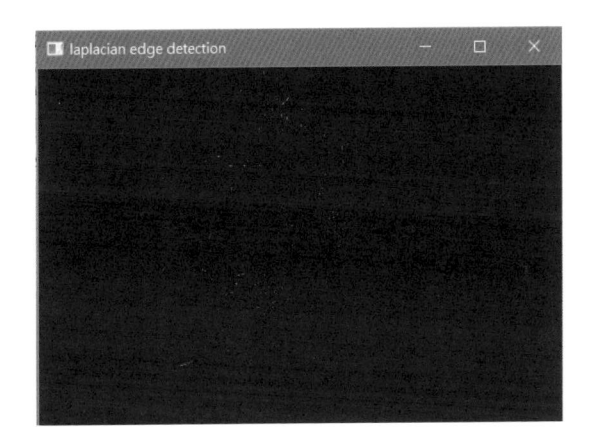

7. Sobel X 軸邊緣偵測：沿著 X 軸偵測邊緣，故可偵測垂直線特徵。

```
1   # Sobel x-axis filter
2   sobelX = np.array((
3       [-1, 0, 1],
4       [-2, 0, 2],
5       [-1, 0, 1]), dtype="int")
6
7   # 卷積
8   convoleOutput = convolve(gray, sobelX)
9   opencvOutput = cv2.filter2D(gray, -1, sobelX)
10  cv2.imshow("x-axis edge detection", convoleOutput)
11
12  # 按 Enter 關閉視窗
13  cv2.waitKey(0)
14  cv2.destroyAllWindows()
```

- 執行結果：卷積行由小至大，顯現圖像垂直線條。

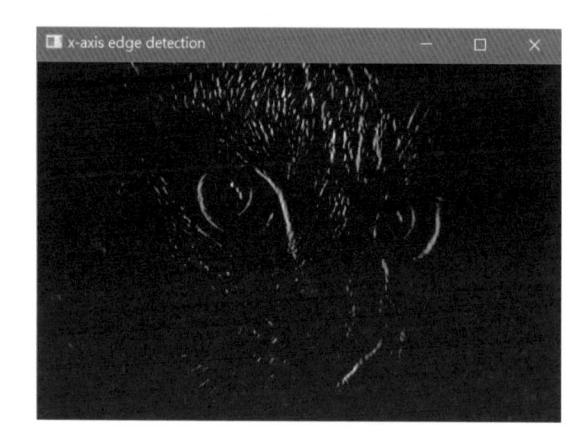

8. Sobel Y 軸邊緣偵測：沿著 Y 軸偵測邊緣，故可偵測水平線特徵。

```
1  # Sobel y-axis filter
2  sobelY = np.array((
3      [-1, -2, -1],
4      [0, 0, 0],
5      [1, 2, 1]), dtype="int")
6
7  # 卷積
8  convoleOutput = convolve(gray, sobelY)
9  opencvOutput = cv2.filter2D(gray, -1, sobelY)
10 cv2.imshow("y-axis edge detection", convoleOutput)
11
12 # 按 Enter 關閉視窗
13 cv2.waitKey(0)
14 cv2.destroyAllWindows()
```

- 執行結果：卷積列由小至大，顯現圖像水平線條。

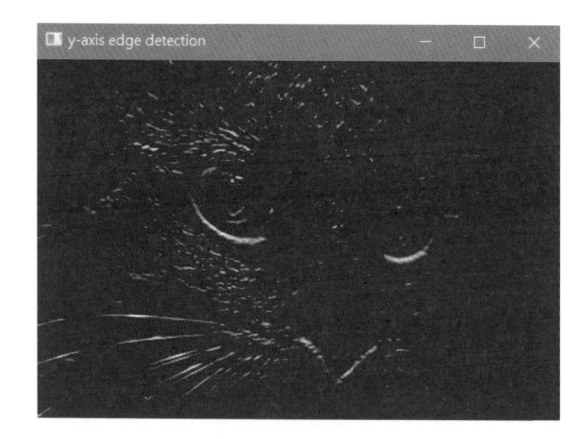

9. 使用 OpenCV 內建的 Sobel 函數,合併 X 軸邊緣與 Y 軸邊緣。

```python
1  scale = 1
2  delta = 0
3  ddepth = cv2.CV_16S
4
5  # 轉換為灰階
6  src = cv2.GaussianBlur(image, (3, 3), 0)
7  gray = cv2.cvtColor(src, cv2.COLOR_RGB2GRAY)
8
9  # X軸邊緣偵測
10 grad_x = cv2.Sobel(gray, ddepth, 1, 0, ksize=3, scale=scale, delta=delta, borderType=cv2.BORDER_DEFAULT)
11 # Y軸邊緣偵測
12 grad_y = cv2.Sobel(gray, ddepth, 0, 1, ksize=3, scale=scale, delta=delta, borderType=cv2.BORDER_DEFAULT)
13
14 # 圖像增強
15 abs_grad_x = cv2.convertScaleAbs(grad_x)
16 abs_grad_y = cv2.convertScaleAbs(grad_y)
17
18 # X軸邊緣與Y軸邊緣合成
19 grad = cv2.addWeighted(abs_grad_x, 0.5, abs_grad_y, 0.5, 0)
20
21 # 按 Enter 關閉視窗
22 cv2.imshow("Sobel testing", grad)
23 cv2.waitKey(0);
```

- 執行結果:輪廓非常明顯,較 Laplacian 演算法佳。

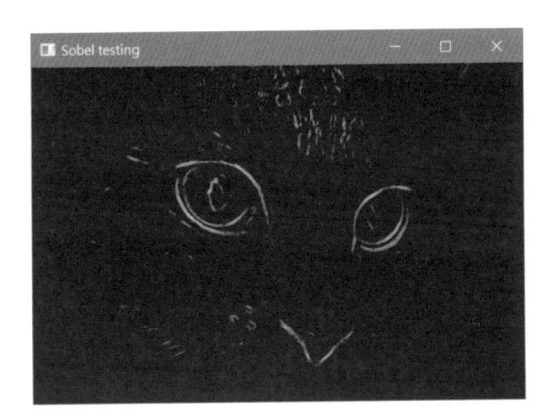

6-4 池化層(Pooling Layer)

通常我們會設定每個卷積層的濾波器個數為 4 的倍數,因此總輸出等於【筆數 x 圖像輸出寬度 x 圖像輸出高度 x 濾波器個數】,會使輸出尺寸變得很大,因此,透過池化層(Pooling Layer)進行下採樣(Down Sampling),只取滑動視窗的最大值或平均值,

第 6 章　卷積神經網路

換句話說，就是將每個滑動視窗轉化為一個點，就能有效降低每一層輸入的尺寸，同時也能保有每個視窗的特徵。我們來舉個例子說明會比較清楚。

以最大池化層（Max Pooling）為例：

1. 下圖左邊為原始圖像。
2. 假設濾波器尺寸為（2, 2）、Stride = 2。
3. 滑動視窗取（2, 2），如下圖左上角的框，取最大值 =6。
4. 接著再滑動 2 步，如圖 6.9，取最大值 =8。

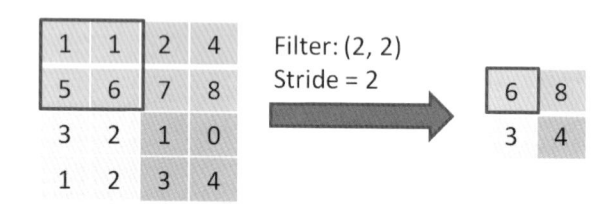

▲ 圖 6.8　最大池化層（Max Pooling）

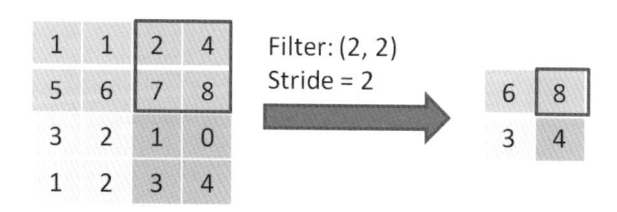

▲ 圖 6.9　最大池化層 -- 滑動 2 步

6-5　CNN 模型實作

一般卷積會採用 3x3 或 5x5 的濾波器，尺寸越大，可以萃取越大的特徵，但相對的，較小的特徵就容易被忽略。而池化層通常會採用 2x2，stride=2 的濾波器，使用越大的尺寸，會使得參數個數減少很多，但萃取到的特徵也相對減少。

範例 1. 改用 CNN 建構手寫阿拉伯數字辨識的模型。

程式：**06_02_MNIST_CNN.ipynb**。

1. 載入 MNIST 手寫阿拉伯數字資料集。

6-5 CNN 模型實作

```
1  import tensorflow as tf
2  mnist = tf.keras.datasets.mnist
3
4  # 載入 MNIST 手寫阿拉伯數字資料
5  (x_train, y_train),(x_test, y_test) = mnist.load_data()
6
7
8  ## 步驟2：資料清理，此步驟無需進行
9
10 ## 步驟3：進行特徵工程，將特徵縮放成(0, 1)之間
11
12 # 特徵縮放，使用常態化(Normalization)，公式 = (x - min) / (max - min)
13 # 顏色範圍：0~255，所以，公式簡化為 x / 255
14 # 注意，顏色0為白色，與RGB顏色不同，(0,0,0) 為黑色。
15 x_train_norm, x_test_norm = x_train / 255.0, x_test / 255.0
```

2. 改用 CNN 模型：使用兩組 Conv2D/MaxPooling2D。

```
1  # 建立模型
2  from tensorflow.keras import layers
3  import numpy as np
4
5  input_shape=(28, 28, 1)
6  # 增加一維在最後面
7  x_train_norm = np.expand_dims(x_train_norm, -1)
8  x_test_norm = np.expand_dims(x_test_norm, -1)
9
10 # CNN 模型
11 model = tf.keras.Sequential(
12     [
13         layers.Input(shape=input_shape),
14         layers.Conv2D(32, kernel_size=(3, 3), activation="relu"),
15         layers.MaxPooling2D(pool_size=(2, 2)),
16         layers.Conv2D(64, kernel_size=(3, 3), activation="relu"),
17         layers.MaxPooling2D(pool_size=(2, 2)),
18         layers.Flatten(),
19         layers.Dropout(0.5),
20         layers.Dense(10, activation="softmax"),
21     ]
22 )
```

- CNN 的卷積層（Conv2D）的輸入多一個維度，代表色彩通道（Channel），單色為 1，RGB 色彩則設為 3。

- 第 7~8 行程式：使用 np.expand_dims 增加了一維在最後面，因為 MNIST 為單色。

6-15

3. 模型訓練：與前例相同。

```
1  # 設定優化器(optimizer)、損失函數(loss)、效能衡量指標(metrics)的類別
2  model.compile(optimizer='adam',
3                loss='sparse_categorical_crossentropy',
4                metrics=['accuracy'])
5
6  # 模型訓練
7  history = model.fit(x_train_norm, y_train, epochs=5, validation_split=0.2)
8
9  # 評分(Score Model)
10 score=model.evaluate(x_test_norm, y_test, verbose=0)
11
12 for i, x in enumerate(score):
13     print(f'{model.metrics_names[i]}: {score[i]:.4f}')
```

- 執行結果：準確率為 0.9892，較之前的模型略高。

注意事項：

1. 有的模型採用連續多個卷積層（Conv2D），才接一個池化層（MaxPooling-2D），並沒有硬性規定，可依照資料多寡與實驗，調校出最佳模型及最佳參數。

2. 再強調一次，CNN 不須指定要使用何種濾波器（Filter），只要指定個數，TensorFlow 會自動配置，且會在訓練過程中找到最佳參數值。

3. 可使用 model.summary()，觀察輸出維度及參數個數。

    - 卷積層輸出的寬度/高度公式如下：

        W_out =（W−F+2P）/S+1

        其中縮寫表示：

        * W_out：輸出圖像的寬度

        * W：輸入圖像的寬度

        * F：濾波器（Filter）的寬度

        * P：單邊補零的行數（Padding）

        * S：滑動的步數（Stride）

```
Model: "sequential"
_____
Layer (type)                 Output Shape              Param #
=================================================================
conv2d (Conv2D)              (None, 26, 26, 32)        320
_____
max_pooling2d (MaxPooling2D) (None, 13, 13, 32)        0
_____
conv2d_1 (Conv2D)            (None, 11, 11, 64)        18496
_____
max_pooling2d_1 (MaxPooling2 (None, 5, 5, 64)          0
_____
flatten (Flatten)            (None, 1600)              0
_____
dropout (Dropout)            (None, 1600)              0
_____
dense (Dense)                (None, 10)                16010
=================================================================
Total params: 34,826
Trainable params: 34,826
Non-trainable params: 0
```

4. 依上述公式驗算第一層 Conv2D 輸出寬度（W_out）：

   W_out = floor（（W−F+2P）/S+1）=（28 - 3 + 2*0）/1 + 1=26

5. 驗算第一層 Conv2D 輸出參數：

   Output Filter 數量 *（Filter 寬 * Filter 高 * Input Filter 數量 + 1）= 32 *（3 * 3 * 1 + 1）= 32 * 10 = 320。其中加 1 為迴歸線的偏差項（Bias）。

6. 第一層 MaxPooling2D 輸出寬度（W_out）：

   W_out = floor（（W−F）/S+1）=（26 - 2）/ 2 + 1 = 13（無條件捨去）

7. 驗算第一層 Conv2D 輸出參數：

   Output Filter 數量 *（Filter 寬 * Filter 高 * Input Filter 數量 + 1）= 64 *（3 * 3 * 32 + 1）= 18496。

從卷積層運算觀察，CNN 模型有兩個特點：

1. 部分連接（Locally Connected or Sparse Connectivity）：Dense 每一層的神經元完全連接（Full Connected）至下一層的每個神經元，但卷積的輸出神經元則只連接滑動視窗神經元，如下圖。想像一下，假設在手臂上拍打一下，手臂以外的神經元應該不會收到訊號，既然沒收到訊號，理所當然就不必往下一層傳送訊號了，所以，下一層的神經元只會收到上一層少數神經元的訊號，接收到的範圍稱之為『感知域』（Reception Field）。

由於部分連接的關係，神經層中每條迴歸線的輸入特徵因而大幅減少，要估算的權重個數也就少了很多，於是模型即可大幅簡化。

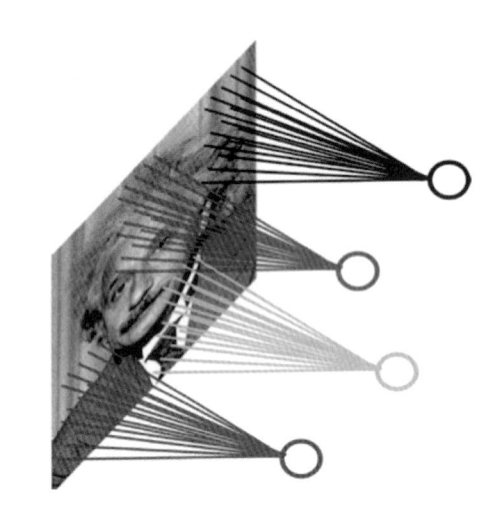

▲ 圖 6.10 部分連接（Locally Connected）

2. 權重共享（Weight Sharing）：單一濾波器應用到滑動視窗時，卷積矩陣值都是一樣的，如下圖所示，基於這個假設，要估計的權重個數就減少許多，模型複雜度因而進一步簡化了。

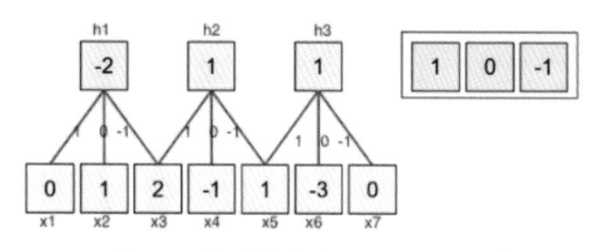

▲ 圖 6.11 權重共享（Weight Sharing）

所以，基於以上的兩個假設，CNN 可以適度縮小模型複雜度，不會因卷積計算，造成模型尺寸過大或訓練時間過久。

另外為什麼 CNN 模型輸入資料要加入色彩通道（Channel）？是因為有些情況加入色彩，會比較容易辨識，比如獅子大部份是金黃色的，又或者偵測是否有戴口罩，只要圖像上有一塊白色的矩形，我們應該就能假定有戴口罩，當然目前口罩顏色已經是五花八門，需要更多的訓練資料，才能正確辨識。

6-5 CNN 模型實作

範例 2. 使用 TensorFlow 內建的 Cifar 圖像 [3]，比較單色與彩色的圖像辨識準確率，先測試單色的圖像辨識。Cifar 資料集有 10 個類別，未去背，辨識較困難。

程式：**06_03_Cifar_gray_CNN.ipynb**。

1. 載入 Cifar10 資料。

```
1  import tensorflow as tf
2  cifar10 = tf.keras.datasets.cifar10
3
4  # 載入 cifar10 資料
5  (x_train, y_train),(x_test, y_test) = cifar10.load_data()
6
7  # 訓練/測試資料的 X/y 維度
8  print(x_train.shape, y_train.shape,x_test.shape, y_test.shape)
```

- 執行結果：訓練 / 測試資料各為 50,000 / 10,000 筆，圖像的寬和高均各為 32，為 RGB 色彩。

```
(50000, 32, 32, 3) (50000, 1) (10000, 32, 32, 3) (10000, 1)
```

2. 資料探索（EDA）：Cifar 10 資料集有 10 個類別。

```
1  # Label 名稱
2  labels_mapping = [ "airplane", "automobile", "bird", "cat", "deer", "dog", "frog", "horse", "ship", "truck" ]
```

3. 顯示單圖：

```
1  # 單圖
2  import matplotlib.pyplot as plt
3
4  plt.figure(figsize=(3,3))
5  plt.imshow(x_train[0], cmap='gray')
6  plt.title(labels_mapping[y_train[0, 0]])
7  plt.axis('off');
```

- 執行結果：因為圖像寬度 / 高度均只有 32 個像素，因此會有些模糊。

第 6 章　卷積神經網路

4. 顯示多圖：

```
1  # 多圖
2  fig = plt.figure(0)
3  fig.set_size_inches(18.5, 18.5)
4  for i in range(0,32):
5      fig.add_subplot(8, 8, i+1)
6      plt.imshow(x_train[i])
7      plt.title(labels_mapping[y_train[i, 0]])
8      plt.axis('off')
```

- 執行結果：

5. 轉換為單色：使用 TensorFlow 內建的 rgb_to_grayscale 函數，可轉換為單色，但維度不變。

```
1  # 轉成單色 : rgb_to_grayscale
2  x_train = tf.image.rgb_to_grayscale(x_train)
3  x_test = tf.image.rgb_to_grayscale(x_test)
4  print(x_train.shape, x_test.shape)
```

- 執行結果：最後一維為 1。

(50000, 32, 32, 1) (10000, 32, 32, 1)

6. 先使用完全連接層（Dense）的神經網路，與 MNIST 辨識相同，只有 Input 維度不同。

```
1  # 建立模型
2  model = tf.keras.models.Sequential([
3      tf.keras.layers.Input(x_train_norm.shape[1:]),
4      tf.keras.layers.Flatten(),
5      tf.keras.layers.Dense(128, activation='relu'),
6      tf.keras.layers.Dropout(0.2),
```

6-20

```
7     tf.keras.layers.Dense(10, activation='softmax')
8 ])
```

- 執行結果：準確率只有 32%。

7. 改用 CNN，前面加上 Conv2D、MaxPooling2D 神經層。

```
1  # 建立模型
2  model = tf.keras.models.Sequential([
3      tf.keras.layers.Input(x_train_norm.shape[1:]),
4      tf.keras.layers.Conv2D(32, (3, 3), activation='relu'),
5      tf.keras.layers.MaxPooling2D((2, 2)),
6      tf.keras.layers.Conv2D(64, (3, 3), activation='relu'),
7      tf.keras.layers.MaxPooling2D((2, 2)),
8      tf.keras.layers.Conv2D(64, (3, 3), activation='relu'),
9      tf.keras.layers.Flatten(),
10     tf.keras.layers.Dense(64, activation='relu'),
11     tf.keras.layers.Dense(10, activation='softmax')
12 ])
```

- 執行結果：準確率為 67%，提升一倍。

範例 3. 改用彩色的圖像辨識。

程式：**06_04_Cifar_RGB_CNN.ipynb**。

1. 載入 Cifar10 資料，與單色的圖像辨識相同，但不需轉換為單色。
2. CNN 模型與**範例 2** 相同，input_shape 最後一維是色彩通道，等於 3。
3. 執行結果：準確率提升為 69%，辨識效果未顯著提升，筆者認為圖像解析度不足，資料未去背，且物件拍攝角度不盡相同，光線明暗也不同，故色彩資訊幫助不大。
4. 觀察訓練結果，設定 10 個執行週期，模型訓練似乎尚未收斂，故加大執行週期為 20，測試結果如下：

    - 訓練資料準確率提升至 89%，但測試資料準確率仍只有 68%，似乎有過度擬合（Overfitting）的現象。
    - 其他改善方式：多加幾組 Conv2D、MaxPooling2D 神經層，或使用資料增補（Data Augmentation），生成更多的訓練資料，後續再討論。

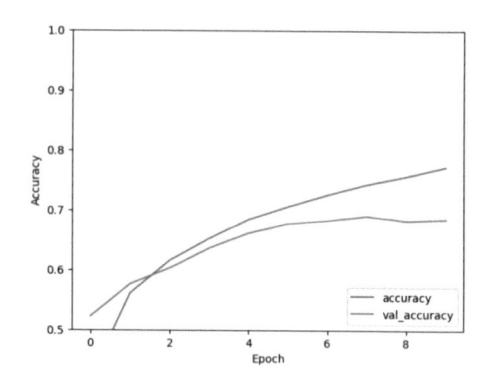

▲ 圖 6.12　10 個執行週期的模型訓練過程

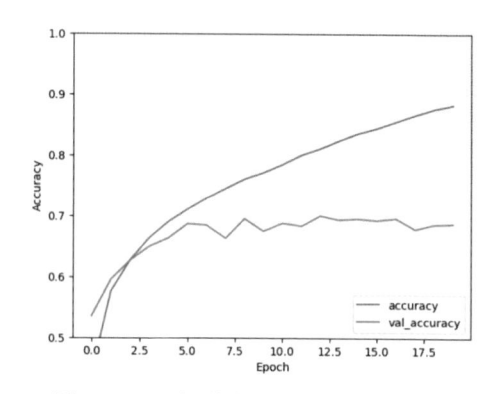

▲ 圖 6.13　20 個執行週期的模型訓練過程

6-6 資料增補（Data Augmentation）

之前我們誇讚半天的辨識手寫阿拉伯數字程式，有以下缺點：

1. 使用 MNIST 的測試資料，辨識率達 98%，但如果以在繪圖軟體裡使用滑鼠書寫的檔案測試，辨識率就差很多了。這是因為 MNIST 的訓練資料與滑鼠撰寫的樣式有所差異，MNIST 的資料應該是請受測者先寫在紙上，再掃描存檔，所以圖像會有深淺不一的灰階和鋸齒狀，與我們直接使用滑鼠在繪圖軟體內書寫的情況不太一樣，所以，假使要實際應用，應該要自行收集訓練資料，準確率才會提升。

2. 若要自行收集資料，須找上萬個測試者，可能不太容易，又加上有些人書寫可能字體歪斜、偏一邊、或字體大小不同，都會影響辨識準確度。

3. MNIST 是收集美國高中生與郵局員工的資料，美國人書寫阿拉伯數字的風格可能與我們寫法有所差異。

6-6 資料增補（Data Augmentation）

針對以上缺點，我們可以藉由『資料增補』（Data Augmentation）的方法，自動產生各種變形的訓練資料，讓模型更強健（Robust）。資料增補是將一張正常圖像轉換成各式有變形的圖像，例如旋轉、偏移、拉近/拉遠、亮度等效果，再結合這些資料當作訓練資料，訓練出來的模型就較能辨識各種風格的圖像。

TensorFlow 提供的資料增補函數 ImageDataGenerator 的參數很多元，包括：

1. width_shift_range：圖像寬度偏移的點（pixel）數或比例。
2. height_shift_range：圖像高度偏移的點（pixel）數或比例。
3. brightness_range：圖像亮度偏移的範圍。
4. shear_range：圖像順時鐘歪斜的範圍。
5. zoom_range：圖像拉近/拉遠的比例。
6. fill_mode：圖像填滿的方式，有四種方式 constant, nearest, reflect, wrap，詳見 TensorFlow ImageDataGenerator 說明 [4]。
7. horizontal_flip：圖像水平翻轉。
8. vertical_flip：圖像垂直翻轉。
9. rescale：特徵縮放。

各種效果可參見 TensorFlow Data Augmentation 說明 [5]。注意，在 TensorFlow 官方文件註明 ImageDataGenerator 已經被棄用（Deprecated），但並沒有說明新的用法，我們就繼續使用嘍。

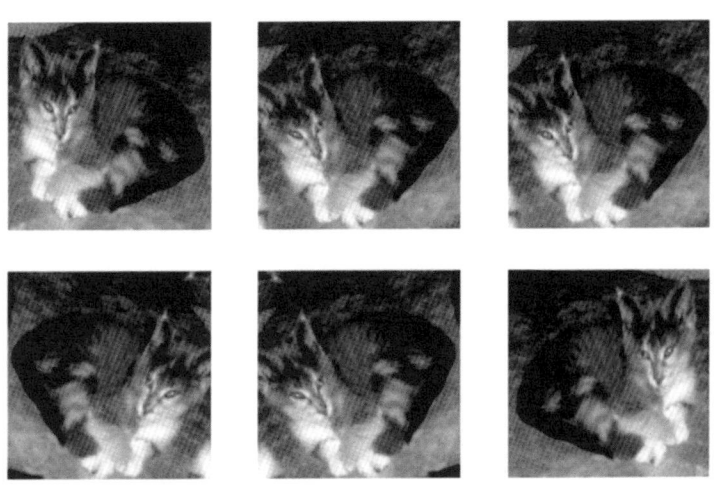

▲ 圖 6.14　左上角的原始圖像經過資料增補後，變成各種角度旋轉的圖像

第 6 章　卷積神經網路

範例 1. MNIST 加上 Data Augmentation。

程式：**06_05_Data_Augmentation_MNIST.ipynb**。

1. 載入 MNIST 資料與模型定義，與前例相同。
2. 訓練之前先進行資料增補。

```
1  # 參數設定
2  batch_size = 1000
3  epochs = 5
4
5  # 資料增補定義
6  datagen = tf.keras.preprocessing.image.ImageDataGenerator(
7          rescale=1./255,         # 特徵縮放
8          rotation_range=10,      # 旋轉 10 度
9          zoom_range=0.1,         # 拉遠/拉近 10%
10         width_shift_range=0.1,  # 寬度偏移 10%
11         height_shift_range=0.1) # 高度偏移 10%
12
13 # 增補資料，進行模型訓練
14 datagen.fit(x_train)
15 history = model.fit(datagen.flow(x_train, y_train, batch_size=batch_size), epochs=epochs,
16         validation_data=datagen.flow(x_test, y_test, batch_size=batch_size), verbose=2)
```

- 執行結果：

    * 準確度並沒有提升，但沒關係，因為我們的目的是要看自行繪製的數字是否被正確辨識。

    * 加入資料增補後，訓練時間拉長為兩倍多，以筆者的 PC 為例，由原本的 5 秒拉長至 12 秒。

3. 測試自行繪製的數字，原來的模型無法正確辨識筆者寫的 9，經過資料增補後，已經可以正確辨識了。

```
1  # 使用小畫家，繪製 0~9，實際測試看看
2  from skimage import io
3  from skimage.transform import resize
4  import numpy as np
5
6  # 讀取影像並轉為單色
7  uploaded_file = '../myDigits/9.png'
8  image1 = io.imread(uploaded_file, as_gray=True)
9
10 # 縮為 (28, 28) 大小的影像
11 image_resized = resize(image1, (28, 28))
```

6-6 資料增補（Data Augmentation）

```
12  X1 = image_resized.reshape(1,28, 28, 1) #/ 255
13
14  # 反轉顏色，顏色0 為白色，與 RGB 色碼不同，它的 0 為黑色
15  X1 = np.abs(1-X1)
16
17  # 預測
18  predictions = np.argmax(model.predict(X1, verbose=False), axis=-1)
19  print(predictions[0])
```

之前都是使用 TensorFlow/Keras 內建的資料集，這個範例使用 Kaggle 所提供的資料集，它需要做前置處理，就是進行資料清理（Data Clean），這會比較接近現實的狀況，但由於資料量較少，準確率較差，因此我們使用更複雜的 CNN 模型，再加上資料增補，以提升準確率。

範例 2. 寵物辨識模型訓練。

程式：**06_06_Data_Augmentation_Pets.ipynb**，程式修改自 Kaggle 競賽網站的範例『Keras CNN Dog or Cat Classification』[6]，Kaggle（https://www.kaggle.com/）為知名的 AI 競賽網站，也是一個很好的學習園地，這裡有很多佛心人士免費提供程式碼和資料集，各位讀者可以進去逛逛。

自寵物資料集網址（https://www.kaggle.com/c/dogs-vs-cats/data）下載檔案後解壓縮，之內還有 train.zip、test1.zip，請一併解壓縮，資料夾如下：

```
dogs-vs-cats
└ train
└ test1
```

train、test1 含 dog、cat 開頭的檔名。

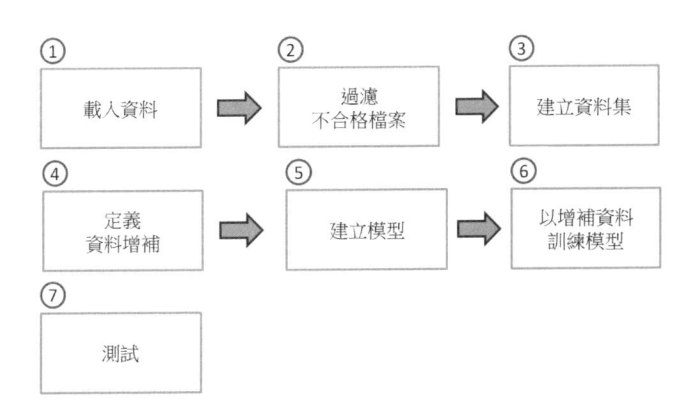

第 6 章　卷積神經網路

1. 從網路取得壓縮檔,並且解壓縮。

2. 將資料檔名整理為 data frame。

```
1  DATA_PATH = "../dogs-vs-cats"
2  filenames = os.listdir(DATA_PATH+"/train")
3  categories = []
4  for filename in filenames:
5      category = filename.split('.')[0]
6      if category == 'dog':
7          categories.append(1)
8      else:
9          categories.append(0)
10
11 df = pd.DataFrame({
12     'filename': filenames,
13     'category': categories
14 })
15 df.head()
```

- 執行結果:前 5 筆資料如下。

	filename	category
0	cat.0.jpg	0
1	cat.1.jpg	0
2	cat.10.jpg	0
3	cat.100.jpg	0
4	cat.1000.jpg	0

3. 資料探索(EDA):統計各類資料筆數。

```
1  import seaborn as sns
2
3  df["category"] = df["category"].replace({0: 'cat', 1: 'dog'})
4  chart = sns.countplot(data=df, x='category', hue='category')
```

- 執行結果:2 類資料筆數大約相等。

6-6 資料增補（Data Augmentation）

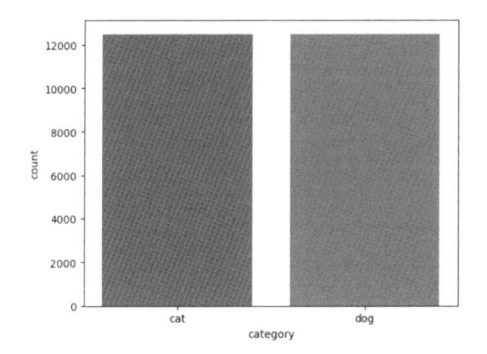

4. 隨機顯示一筆資料。

```
1  sample = random.choice(filenames)
2  image = load_img(DATA_PATH+"/train/"+sample)
3  plt.imshow(image);
```

- 執行結果：背景可能有其他物件。

5. 參數定義：FAST_RUN = True 時只訓練 3 個週期，反之訓練 50 個週期。

```
1  FAST_RUN = True
2  IMAGE_WIDTH=128
3  IMAGE_HEIGHT=128
4  IMAGE_SIZE=(IMAGE_WIDTH, IMAGE_HEIGHT)
5  IMAGE_CHANNELS=3
```

6. 資料分割：使用 Scikit-Learn 套件的 train_test_split 函數隨機切割訓練（Training）及驗證（Validation）資料。

```
1  from sklearn.model_selection import train_test_split
2
3  train_df, validate_df = train_test_split(df, test_size=0.20, random_state=42)
```

6-27

```
4  train_df = train_df.reset_index(drop=True)
5  validate_df = validate_df.reset_index(drop=True)
```

7. 資料增補：

    - 使用 flow_from_dataframe 函數自前述的 data frame 建立 Training Genera-tor。參閱 TensorFlow ImageDataGenerator 說明 [4]，還可以使用 flow_from_directory 函數自檔案目錄建立 Generator，但必須按類別名稱建立子目錄。

    - 批量（batch_size）：需視電腦的記憶體大小，設定適當值，如果模型訓練發生記憶體不足，必須縮小批量，但相對的，訓練時間就會拉長。

```
1  #定義資料增補(Data Augmentation)
2  batch_size=64
3  train_datagen = ImageDataGenerator(
4      rotation_range=15,
5      rescale=1./255,
6      shear_range=0.1,
7      zoom_range=0.2,
8      horizontal_flip=True,
9      width_shift_range=0.1,
10     height_shift_range=0.1
11 )
12
13 train_generator = train_datagen.flow_from_dataframe(
14     train_df,
15     DATA_PATH+"/train/",
16     x_col='filename',
17     y_col='category',
18     target_size=IMAGE_SIZE,
19     class_mode='categorical',
20     batch_size=batch_size
21 )
```

8. 建立 Validation Generator。

```
1  validation_datagen = ImageDataGenerator(rescale=1./255)
2  validation_generator = validation_datagen.flow_from_dataframe(
3      validate_df,
4      DATA_PATH+"/train/",
5      x_col='filename',
6      y_col='category',
7      target_size=IMAGE_SIZE,
8      class_mode='categorical',
```

```
 9      batch_size=batch_size
10  )
```

9. 測試 generator：隨機顯示訓練資料 15 筆資料增補的影像。

```
 1  # 隨機挑選一筆資料
 2  example_df = train_df.sample(n=1).reset_index(drop=True)
 3  example_generator = train_datagen.flow_from_dataframe(
 4      example_df,
 5      DATA_PATH+"/train/",
 6      x_col='filename',
 7      y_col='category',
 8      target_size=IMAGE_SIZE,
 9      class_mode='categorical'
10  )
11
12  # 顯示訓練資料前 9 筆影像
13  plt.figure(figsize=(8, 8))
14  for i in range(0, 15):
15      plt.subplot(5, 3, i+1)
16      for X_batch, Y_batch in example_generator:
17          image = X_batch[0]
18          plt.imshow(image)
19          break
20      plt.tight_layout()
21      plt.axis("off")
```

- 執行結果：含圖像旋轉、偏一邊、拉近…等效果。

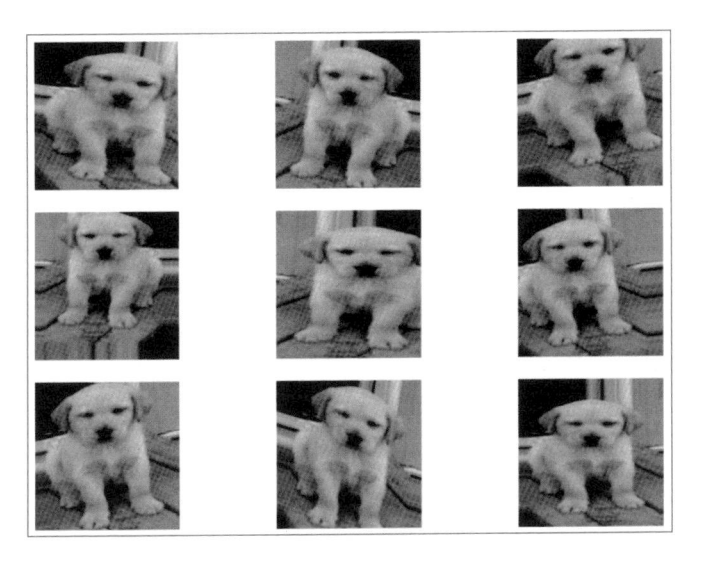

10. 定義模型：使用 CNN，特別注意，模型中額外加了 BatchNormalization 神經層，它會針對輸出進行標準化，以防止梯度爆炸或消失，後續會有較詳細的討論。

```
4  model = Sequential()
5  model.add(Input((IMAGE_WIDTH, IMAGE_HEIGHT, IMAGE_CHANNELS)))
6  model.add(Conv2D(32, (3, 3), activation='relu'))
7  model.add(BatchNormalization())
8  model.add(MaxPooling2D(pool_size=(2, 2)))
9  model.add(Dropout(0.25))
10
11 model.add(Conv2D(64, (3, 3), activation='relu'))
12 model.add(BatchNormalization())
13 model.add(MaxPooling2D(pool_size=(2, 2)))
14 model.add(Dropout(0.25))
15
16 model.add(Conv2D(128, (3, 3), activation='relu'))
17 model.add(BatchNormalization())
18 model.add(MaxPooling2D(pool_size=(2, 2)))
19 model.add(Dropout(0.25))
20
21 model.add(Flatten())
22 model.add(Dense(512, activation='relu'))
23 model.add(BatchNormalization())
24 model.add(Dropout(0.5))
25 model.add(Dense(2, activation='softmax')) # 2 because we have cat and dog classes
26
27 model.compile(loss='categorical_crossentropy', optimizer='rmsprop', metrics=['accuracy'])
```

11. 訓練模型：訓練時間有點久，只 3 個週期就可能需要幾十分鐘，最好要有 GPU 顯卡或上傳到 Colaboratory 執行。

```
1  # 有點久
2  epochs=3 if FAST_RUN else 50
3  history = model.fit(
4      train_generator,
5      epochs=epochs,
6      validation_data=validation_generator,
7  )
```

- 執行結果：準確率約 70%。

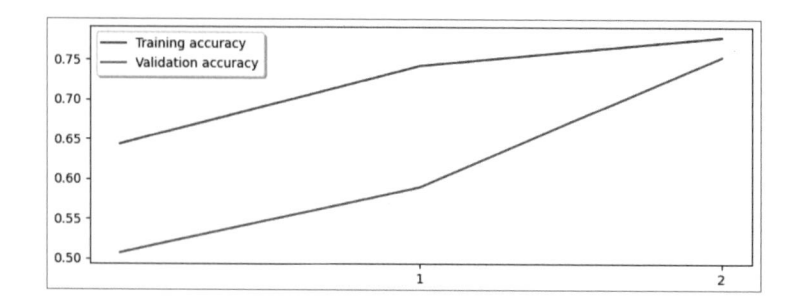

12. 模型評分：一樣對測試資料建立 data frame，並讀取圖檔合併為一個陣列，再呼叫 predict，由於合併需要很久的時間，因此只得取前 20 筆，如果要預測大量資料，可以使用 Dataset，較具效率，後續會有範例說明。

```
1  test_filenames = os.listdir(DATA_PATH+"/test1")
2  test_df = pd.DataFrame({
3      'filename': test_filenames
4  })
5  test_df.head()
```

```
1  test_data = None
2  for i, filename in enumerate(os.listdir(DATA_PATH+"/test1")):
3      if i>=20: break
4      image = load_img(DATA_PATH+"/test1/"+filename, target_size=IMAGE_SIZE)
5      if test_data is None:
6          test_data = tf.expand_dims(tf.convert_to_tensor(image), 0)
7      else:
8          test_data = tf.concat([test_data, tf.expand_dims(tf.convert_to_tensor(image), 0)], axis=0)
9  print(test_data.shape)
```

```
1  test_df = test_df.iloc[:20].copy()
2  predict = model.predict(test_data)
3  test_df['category'] = np.argmax(predict, axis=-1)
4  test_df.head()
```

- 執行結果：每一個圖檔對應預測結果。

	filename	category
0	1.jpg	1
1	10.jpg	0
2	100.jpg	0
3	1000.jpg	0
4	10000.jpg	0

13. 由於測試資料沒有答案，因此顯示前 18 筆圖像及預測結果，以比對是否正確。

```
1  sample_test = test_df.head(18)
2  plt.figure(figsize=(6, 12))
3  for index, row in sample_test.iterrows():
4      filename = row['filename']
5      category = row['category']
6      img = load_img(DATA_PATH+"/test1/"+filename, target_size=IMAGE_SIZE)
7      plt.subplot(6, 3, index+1)
8      plt.imshow(img)
```

第 6 章　卷積神經網路

```
 9      plt.xticks([], [])
10      plt.yticks([], [])
11      plt.xlabel(filename + '(' + "{}".format(category) + ')')
12  
13  plt.tight_layout()
14  plt.show()
```

- 執行結果：預測還蠻準確的。

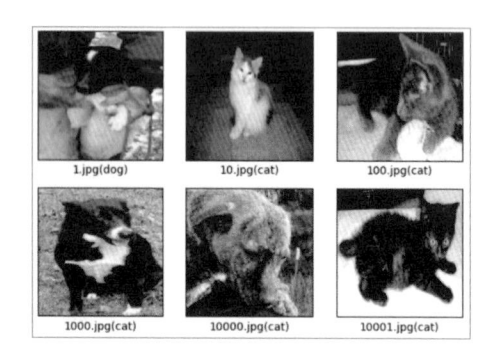

除了 TensorFlow 提供的資料增補功能之外，還有其他套件提供更多的資料增補效果，譬如 Albumentations[7]，增補的類型多達 70 種，很多都是 TensorFlow 所沒有的效果，例如下圖的顏色資料增補：

讀者也可以修改程式，辨識 Cifar 10 資料集，需更改的程式碼並不多。

6-7 可解釋的 AI（eXplainable AI, XAI）

雖然前文有說過深度學習是黑箱科學，但是，科學家們依然試圖解釋模型是如何辨識圖像，這方面的研究稱為『可解釋的 AI』（eXplainable AI, XAI），目的如下：

1. 確認模型辨識的結果是合理的：深度學習永遠不會跟你說錯，『垃圾進、垃圾出』（Garbage In, Garbage Out），確認模型推估的合理性是相當重要的。
2. 改良演算法：唯有知其所以然，才知道如何創新，光是靠參數的調校，只能有微幅的改善。目前機器學習還只能從資料中學習到知識（Knowledge Discovery from Data, KDD），要讓機器具有智慧（Wisdom）及感知（Feeling）能力，實現真正的人工智慧，勢必要有更創新的想法。

XAI 可使用視覺化的方式呈現特徵對模型的影響力，例如：

1. 使用卷積層萃取圖像的線條特徵，我們可以觀察到轉換後的結果嗎？
2. 甚至更進一步，我們可以知道哪些線條對辨識最有幫助嗎？

接下來我們以兩個實例展示相關的作法。

範例 1. 重建卷積層處理後的影像，觀察線條、輪廓，並檢視多次的卷積層 / 池化層處理後，圖像會有何種變化。

程式：**06_07_CNN_Visualization.ipynb**，程式碼修改自『Machine Learning Mastery』中的部落文 [8]。

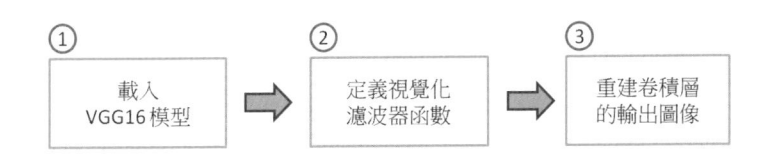

① 載入 VGG16 模型 ➡ ② 定義視覺化濾波器函數 ➡ ③ 重建卷積層的輸出圖像

1. 載入套件。

```
1  # 載入套件
2  import tensorflow as tf
3  from tensorflow.keras.applications.vgg16 import VGG16
4  import matplotlib.pyplot as plt
5  import numpy as np
```

第 6 章　卷積神經網路

2. 載入 VGG16 模型：VGG16 為知名的影像辨識模型，TensorFlow 當然沒有錯過內建此模型，包含已訓練好的模型參數，後續章節會介紹到此類預先訓練好的模型用法。

```
1  # 載入 VGG16 模型
2  model = VGG16()
3  model.summary()
```

- 執行結果：包括 16 層卷積 / 池化層。

```
Model: "vgg16"
_____
Layer (type)                 Output Shape              Param #
=================================================================
input_1 (InputLayer)         [(None, 224, 224, 3)]     0
_____
block1_conv1 (Conv2D)        (None, 224, 224, 64)      1792
_____
block1_conv2 (Conv2D)        (None, 224, 224, 64)      36928
_____
block1_pool (MaxPooling2D)   (None, 112, 112, 64)      0
_____
block2_conv1 (Conv2D)        (None, 112, 112, 128)     73856
_____
block2_conv2 (Conv2D)        (None, 112, 112, 128)     147584
_____
block2_pool (MaxPooling2D)   (None, 56, 56, 128)       0
_____
block3_conv1 (Conv2D)        (None, 56, 56, 256)       295168
_____
block3_conv2 (Conv2D)        (None, 56, 56, 256)       590080
_____
block3_conv3 (Conv2D)        (None, 56, 56, 256)       590080
_____
```

3. 定義視覺化濾波器的函數。

```
1  # 視覺化特定層的特徵圖(Feature Map)
2  def Visualize(layer_no=1, n_filters=6):
3      # 取得權重(weight)
4      filters, biases = model.layers[layer_no].get_weights()
5      # 常態化(Normalization)
6      f_min, f_max = filters.min(), filters.max()
7      filters = (filters - f_min) / (f_max - f_min)
8
9      # 繪製特徵圖
10     ix = 1
11     for i in range(n_filters):
12         f = filters[:, :, :, i]      # 取得每一個特徵圖
13         for j in range(3):           # 每列 3 張圖
14             ax = plt.subplot(n_filters, 3, ix)  # 指定子視窗
15             ax.set_xticks([])        # 無X軸刻度
```

6-7 可解釋的 AI（eXplainable AI, XAI）

```
16            ax.set_yticks([])      # 無Y軸刻度
17            plt.imshow(f[:, :, j], cmap='gray') # 以灰階繪圖
18            ix += 1
19    plt.show()
```

4. 視覺化第一層的濾波器。

```
1  Visualize(1)
```

- 執行結果：可以看出每個濾波器均不相同，表示做了不同的影像處理。

5. 視覺化第 15 層的濾波器。

```
1  Visualize(15)
```

- 執行結果：與上一張圖相對照，可以看出與第一層不同，表示又做了不同的影像處理。

6-35

第 6 章　卷積神經網路

6. 重建第一個卷積層的輸出圖像：以鳥的圖片為例，先進行卷積，接著再重建圖像。

```python
# 設定第一個卷積層的輸出為模型輸出
model2 = tf.keras.models.Model(inputs=model.inputs, outputs=model.layers[1].output)

# 載入測試的圖像
img = tf.keras.preprocessing.image.load_img('./images/bird.jpg', target_size=(224, 224))
img = tf.keras.preprocessing.image.img_to_array(img)         # 圖像轉為陣列
img = np.expand_dims(img, axis=0)                            # 加一維作為筆數
img = tf.keras.applications.vgg16.preprocess_input(img)      # 前置處理(常態化)

# 預測
feature_maps = model2.predict(img)

# 將結果以 8x8 視窗顯示
square = 8
ix = 1
plt.figure(figsize=(12,8))
for _ in range(square):
    for _ in range(square):
        ax = plt.subplot(square, square, ix)
        ax.set_xticks([])
        ax.set_yticks([])
        plt.imshow(feature_maps[0, :, :, ix-1], cmap='gray')
        ix += 1
plt.show()
```

- 執行結果：可以看見第一層影像處理結果，有的濾波器可以抓到線條，有的則是漆黑一片。

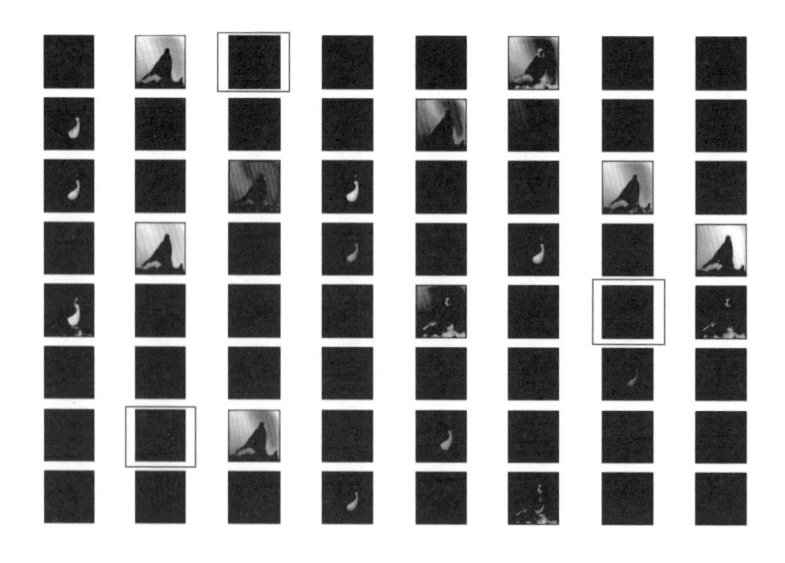

6-7 可解釋的 AI（eXplainable AI, XAI）

7. 重建 2, 5, 9, 13, 17 多層卷積層的輸出圖像。

```
1  # 取得 2, 5, 9, 13, 17 卷積層輸出
2  ixs = [2, 5, 9, 13, 17]
3  outputs = [model.layers[i].output for i in ixs]
4  model2 = tf.keras.models.Model(inputs=model.inputs, outputs=outputs)
5
6  # 載入測試的圖像
7  img = tf.keras.preprocessing.image.load_img('./images/bird.jpg', target_size=(224, 224))
8  img = tf.keras.preprocessing.image.img_to_array(img)      # 圖像轉為陣列
9  img = np.expand_dims(img, axis=0)                          # 加一維作為筆數
10 img = tf.keras.applications.vgg16.preprocess_input(img)   # 前置處理(常態化)
11
12 # 預測
13 feature_maps = model2.predict(img)
14
15 # 將結果以 8x8 視窗顯示
16 square = 8
17 for fmap in feature_maps:
18     ix = 1
19     plt.figure(figsize=(12,8))
20     for _ in range(square):
21         for _ in range(square):
22             ax = plt.subplot(square, square, ix)
23             ax.set_xticks([])
24             ax.set_yticks([])
25             plt.imshow(fmap[0, :, :, ix-1], cmap='gray')
26             ix += 1
27     plt.show()
```

- 執行結果：在第 9 層影像處理結果中，還能夠明顯看到線條，但到第 17 層影像處理結果，已經是抽象到認不出來是鳥的地步了。

第 9 層影像處理結果：

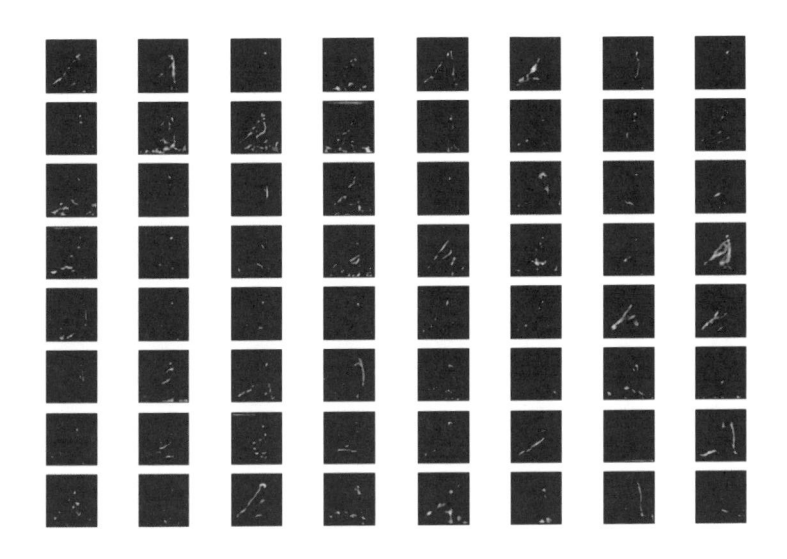

第 6 章　卷積神經網路

第 17 層影像處理結果：

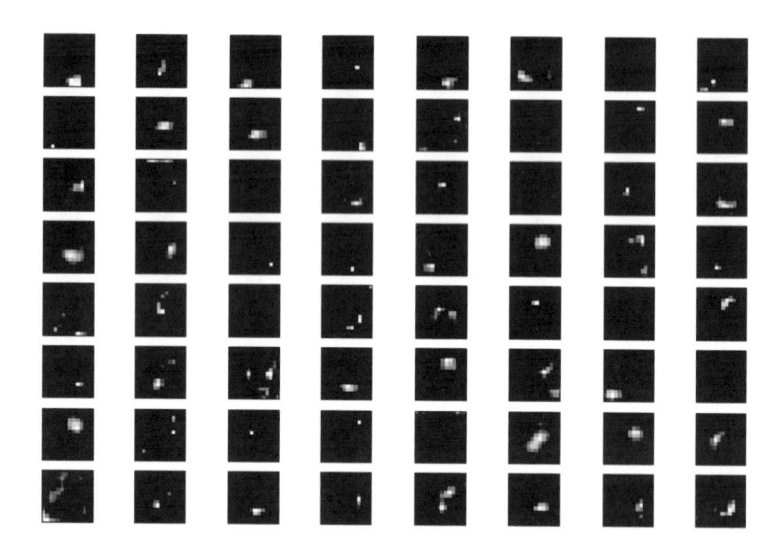

從以上的實驗，可以很清楚看到 CNN 的處理過程，我們雖然不明白辨識的邏輯，但是至少能夠觀察到整個模型處理的過程。

SHAP（SHapley Additive exPlanations）套件是另一種視覺化工具，可觀察特徵的影響力，它是由 Scott Lundberg 及 Su-In Lee 所開發的，提供 Shapley value 的計算，並具備多種視覺化的圖形，藉以提供模型的解釋能力，使用說明可參考 SHAP GitHub [9]，以下僅說明神經網路的應用。

Shapley value 是由多人賽局理論（Game Theory）而發展出來的，原本是用來公平分配利益給團隊中的每個人，學者將概念套用到機器學習領域，解釋個別特徵對預測結果的影響力，更詳細的介紹可參考維基百科 [10]。

範例 2. 使用 Shap 套件觀察圖像的那些位置對辨識最有幫助。

程式：**06_08_Shap_MNIST.ipynb**。

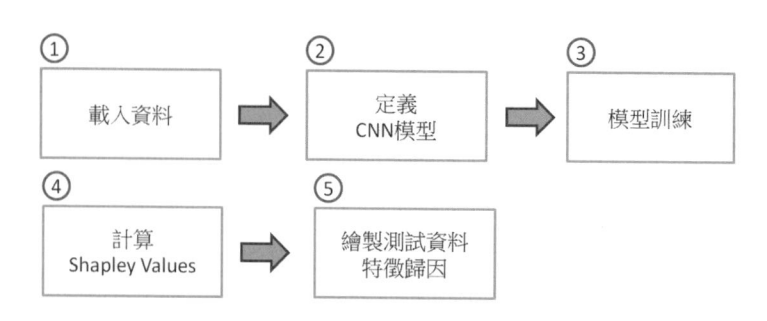

① 載入資料 → ② 定義 CNN模型 → ③ 模型訓練
④ 計算 Shapley Values → ⑤ 繪製測試資料 特徵歸因

6-7 可解釋的 AI（eXplainable AI, XAI）

1. 載入 MNIST 資料集：請注意，目前 TensorFlow 2.x 版執行 shap 有 bug，所以務必使用 tf.compat.v1.disable_v2_behavior()，切換回 TensorFlow 1.x 版。

```
1  import tensorflow as tf
2
3  # 目前 tensorflow 2.x 版執行 shap 有 bug
4  tf.compat.v1.disable_v2_behavior()
5
6  # 載入 MNIST 手寫阿拉伯數字資料
7  mnist = tf.keras.datasets.mnist
8  (x_train, y_train),(x_test, y_test) = mnist.load_data()
```

2. 定義 CNN 模型：與前面模型相同，也可使用其他模型做測試。

```
1  # 建立模型
2  from tensorflow.keras import layers
3  import numpy as np
4
5  # 增加一維在最後面
6  x_train = np.expand_dims(x_train, -1)
7  x_test = np.expand_dims(x_test, -1)
8  x_train_norm, x_test_norm = x_train / 255.0, x_test / 255.0
9
10 # CNN 模型
11 input_shape=(28, 28, 1)
12 model = tf.keras.Sequential(
13     [
14         tf.keras.Input(shape=input_shape),
15         layers.Conv2D(32, kernel_size=(3, 3), activation="relu"),
16         layers.MaxPooling2D(pool_size=(2, 2)),
17         layers.Conv2D(64, kernel_size=(3, 3), activation="relu"),
18         layers.MaxPooling2D(pool_size=(2, 2)),
19         layers.Flatten(),
20         layers.Dropout(0.5),
21         layers.Dense(10, activation="softmax"),
22     ]
23 )
24
25 # 設定優化器(optimizer)、損失函數(loss)、效能衡量指標(metrics)的類別
26 model.compile(optimizer='adam',
27               loss='sparse_categorical_crossentropy',
28               metrics=['accuracy'])
```

3. 模型訓練：與前面相同。

```
1  # 模型訓練
2  history = model.fit(x_train_norm, y_train, epochs=5, validation_split=0.2)
3
4  # 評分(Score Model)
5  score=model.evaluate(x_test_norm, y_test, verbose=0)
6
7  for i, x in enumerate(score):
8      print(f'{model.metrics_names[i]}: {score[i]:.4f}')
```

第 6 章　卷積神經網路

4. Shapley Values 計算：測試第 1 筆資料。

```
1  import shap
2  import numpy as np
3
4  # 計算 Shap value 的 base
5  # 目前 tensorflow 2.x 版執行 shap 有 bug
6  # background = x_train_norm[np.random.choice(x_train_norm.shape[0], 100, replace=False)]
7  # e = shap.DeepExplainer(model, background)        # shap values 不明顯
8  e = shap.DeepExplainer(model, x_train_norm[:100])
9
10 # 測試第 1 筆
11 shap_values = e.shap_values(x_test_norm[:1])
```

5. 繪製 1 筆測試資料的特徵歸因。

```
1  # 繪製特徵的歸因(feature attribution)
2  # 一次只能顯示一列
3  shap.image_plot(shap_values, -x_test_norm[:1])
```

- 執行結果：請參看程式執行結果，紅色的區塊代表貢獻率較大的區域，藍色表負的貢獻率。

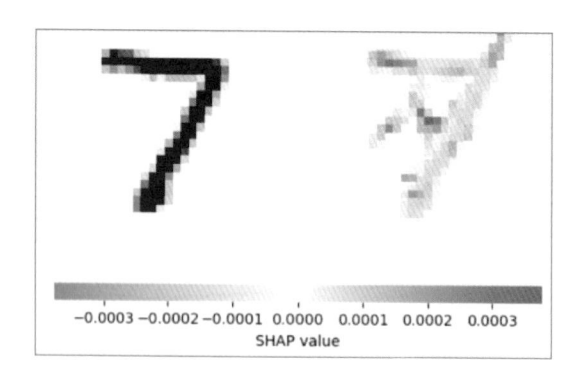

從 Shap 套件的功能，我們很容易判斷出中央位置是辨識的重點區域，這與我們認知是一致的。另一個名為 LIME[11] 的套件，與 Shap 套件齊名，讀者如果對這領域有興趣，可以由此深入研究，筆者就偷懶一下嘍。

還有一篇論文[12] 提出了 Class Activation Mapping 概念，可以描繪辨識的熱區，如下圖所示。Kaggle 也有一篇超讚的實作[13]，值得大家好好欣賞一番。

6-40

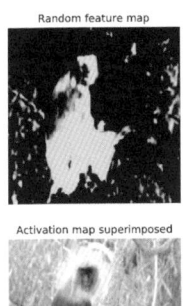

▲ 圖 6.15 左上角的圖像為原圖，左下角的圖像顯示了辨識熱區，即猴子的頭和頸部都是辨識的主要關鍵區域

透過以上視覺化的輔助，不只可以幫助我們更瞭解 CNN 模型的運作，也能夠讓我們在收集資料時，有較明確的方向知道重點應該要放在哪裡，當然，如果未來能有更創新的想法，來改良演算法，那就可以開香檳慶祝了。

6-8 卷積神經網路的缺點

CNN 利用卷積萃取特徵，以線條及輪廓取代像素作為神經網路的輸入，有效提高辨識的準確度，但是它仍存在一些缺點：

1. 卷積不管特徵在圖像的所在位置，只針對局部視窗進行特徵萃取，因此，下列兩張圖，辨識結果是相同的，這種現象稱為『位置無差異性』（Position Invariant）。

▲ 圖 6.16 左圖是正常的人臉，右圖五官移位，兩者對 CNN 來說是無差異的，圖片來源：Disadvantages of CNN models[14]

6-41

2. 圖像中的物件如果經過旋轉或傾斜，CNN 就無法辨識了，如下圖：

▲ 圖 6.17 右圖為左圖旋轉近 180 度，CNN 就無法辨識了，
圖片來源：Disadvantages of CNN models

3. 圖像座標轉換，人眼可以辨識不同的物件特徵，但對於 CNN 來說卻難以理解，如下圖：

▲ 圖 9.16 右圖為上下顛倒的左圖，人眼可以看出年輕人與老年人，
然而 CNN 就很難理解，圖片來源：Disadvantages of CNN models

因此，Geoffrey Hinton 等學者就提出了『膠囊神經網路』（Capsule Networks）[15]，用來改善以上的問題，不過該演算法並沒有引發太多關注。

參考資料（References）

[1]　Prateek Karkare,《Convolutional Neural Networks—Simplified》, 2019
    (https://medium.com/x8-the-ai-community/cnn-9c5e63703c3f)

參考資料（References）

[2] 《Convolutional Neural Networks—Simplified》文中卷積計算的 GIF 動畫
(https://miro.medium.com/max/963/1*wpbLgTW_lopZ6JtDqVByuA.gif)

[3] TensorFlow 內建的 Cifar 圖像
(https://www.tensorflow.org/datasets/catalog/cifar10?hl=zh-tw)

[4] TensorFlow ImageDataGenerator 說明
(https://www.tensorflow.org/api_docs/python/tf/keras/preprocessing/image/ImageDataGenerator)

[5] TensorFlow Data Augmentation 說明
(https://www.tensorflow.org/tutorials/images/data_augmentation)

[6] Keras CNN Dog or Cat Classification 範例
(https://www.kaggle.com/code/uysimty/keras-cnn-dog-or-cat-classification)

[7] Albumentations
(https://github.com/albumentations-team/albumentations)

[8] Jason Brownlee,《How to Visualize Filters and Feature Maps in Convolutional Neural Networks》, 2019
(https://machinelearningmastery.com/how-to-visualize-filters-and-feature-maps-in-convolutional-neural-networks/)

[9] SHAP GitHub
(https://github.com/slundberg/shap)

[10] 維基百科中關於 Shapley value 的介紹
(https://en.wikipedia.org/wiki/Shapley_value)

[11] LIME 套件的安裝與介紹說明
(https://github.com/marcotcr/lime)

[12] Bolei Zhou, Aditya Khosla, Agata Lapedriza et al,《Learning Deep Features for Discriminative Localization》, 2015
(https://arxiv.org/pdf/1512.04150.pdf)

[13] Kaggle 中介紹的實作
(https://www.kaggle.com/aakashnain/what-does-a-cnn-see)

[14] Disadvantages of CNN models
(https://iq.opengenus.org/disadvantages-of-cnn/)

[15] Capsule Networks
(https://cedar.buffalo.edu/~srihari/CSE676/9.12%20CapsuleNets.pdf)

第 7 章

預先訓練的模型
(Pre-trained Model)

透過 CNN 模型和資料增補的強化,我們已經能夠建立準確度還不錯的模型,然而,與近幾年影像辨識競賽中的冠、亞軍模型相較,只能算是小巫見大巫了,有些冠、亞軍模型的神經層數量高達 100 多層,訓練資料有 100 多萬筆,若要自行訓練這些模型就需要花上幾個星期甚至幾個月的時間,難道縮短訓練時間的辦法,只剩購置企業級伺服器這個選項嗎?

第 7 章 預先訓練的模型（Pre-trained Model）

幸好 TensorFlow/Keras、PyTorch 等深度學習框架早已為我們這些中小企業設想好了，套件提供許多預先訓練好的模型，可以直接套用，也可以只採用部份模型，再接上自訂的神經層，進行其他物件的辨識，這些預先訓練好的模型就稱為『Pre-trained Model』或『Keras Applications』，也稱為『基礎模型』（Foundation Model）。

■ 7-1 預先訓練模型的簡介

在 ImageNet 歷年舉辦的影像辨識競賽（ILSVRC），近幾年產生的冠亞軍，大都是 CNN 模型的變型，整個演進過程非常精彩，簡述如下：

1. 2012 年冠軍 AlexNet 一舉將錯誤率減少 10% 以上，且首度導入 Dropout 層。
2. 2014 年亞軍 VGGNet 承襲 AlexNet 思路，建立更多層的模型，例如 VGG 16/19 分別包括 16 及 19 層卷積層及池化層。
3. 2014 年圖像分類冠軍 GoogLeNet 導入多種不同尺寸的 Kernel，讓系統決定最佳 Kernel 尺寸，並引進 Inception、Batch Normalization… 等觀念，參見『Batch Normalization: Accelerating Deep Network Training by Reducing Internal Covariate Shift』[1]。
4. 2015 年冠軍 ResNets 發現到 20 層以上的模型其前面幾層會發生退化（degradation）的狀況，因而提出以『殘差』（Residual）方法來解決問題，參見『Deep Residual Learning for Image Recognition』[2]。

Keras 收錄許多預先訓練的模型，稱為 Keras Applications[3]，隨著版本的更新，提供的模型愈來愈多，目前（2024 年）包括：

Model	Size (MB)	Top-1 Accuracy	Top-5 Accuracy	Parameters	Depth	Time (ms) per inference step (CPU)	Time (ms) per inference step (GPU)
Xception	88	79.0%	94.5%	22.9M	81	109.4	8.1
VGG16	528	71.3%	90.1%	138.4M	16	69.5	4.2
VGG19	549	71.3%	90.0%	143.7M	19	84.8	4.4
ResNet50	98	74.9%	92.1%	25.6M	107	58.2	4.6
ResNet50V2	98	76.0%	93.0%	25.6M	103	45.6	4.4
ResNet101	171	76.4%	92.8%	44.7M	209	89.6	5.2
ResNet101V2	171	77.2%	93.8%	44.7M	205	72.7	5.2
ResNet152	232	76.6%	93.1%	60.4M	311	127.4	6.5

7-1 預先訓練模型的簡介

Model	Size (MB)	Top-1 Accuracy	Top-5 Accuracy	Parameters	Depth	Time (ms) per inference step (CPU)	Time (ms) per inference step (GPU)
ResNet152V2	232	78.0%	94.2%	60.4M	307	107.5	6.6
InceptionV3	92	77.9%	93.7%	23.9M	189	42.2	6.9
InceptionResNetV2	215	80.3%	95.3%	55.9M	449	130.2	10.0
MobileNet	16	70.4%	89.5%	4.3M	55	22.6	3.4
MobileNetV2	14	71.3%	90.1%	3.5M	105	25.9	3.8
DenseNet121	33	75.0%	92.3%	8.1M	242	77.1	5.4
DenseNet169	57	76.2%	93.2%	14.3M	338	96.4	6.3
DenseNet201	80	77.3%	93.6%	20.2M	402	127.2	6.7
NASNetMobile	23	74.4%	91.9%	5.3M	389	27.0	6.7
NASNetLarge	343	82.5%	96.0%	88.9M	533	344.5	20.0
EfficientNetB0	29	77.1%	93.3%	5.3M	132	46.0	4.9
EfficientNetB1	31	79.1%	94.4%	7.9M	186	60.2	5.6
EfficientNetB2	36	80.1%	94.9%	9.2M	186	80.8	6.5
EfficientNetB3	48	81.6%	95.7%	12.3M	210	140.0	8.8
EfficientNetB4	75	82.9%	96.4%	19.5M	258	308.3	15.1
EfficientNetB5	118	83.6%	96.7%	30.6M	312	579.2	25.3
EfficientNetB6	166	84.0%	96.8%	43.3M	360	958.1	40.4
EfficientNetB7	256	84.3%	97.0%	66.7M	438	1578.9	61.6
EfficientNetV2B0	29	78.7%	94.3%	7.2M	-	-	-
EfficientNetV2B1	34	79.8%	95.0%	8.2M	-	-	-
EfficientNetV2B2	42	80.5%	95.1%	10.2M	-	-	-
EfficientNetV2B3	59	82.0%	95.8%	14.5M	-	-	-
EfficientNetV2S	88	83.9%	96.7%	21.6M	-	-	-
EfficientNetV2M	220	85.3%	97.4%	54.4M	-	-	-
EfficientNetV2L	479	85.7%	97.5%	119.0M	-	-	-
ConvNeXtTiny	109.42	81.3%	-	28.6M	-	-	-
ConvNeXtSmall	192.29	82.3%	-	50.2M	-	-	-
ConvNeXtBase	338.58	85.3%	-	88.5M	-	-	-
ConvNeXtLarge	755.07	86.3%	-	197.7M	-	-	-
ConvNeXtXLarge	1310	86.7%	-	350.1M	-	-	-

▲ 圖 7.1 Keras 提供的預先訓練模型（Pre-trained Model）

第 7 章 預先訓練的模型（Pre-trained Model）

上述表格的欄位說明如下：

1. Size：模型檔案大小。
2. Top-1 Accuracy：預測機率最大的類別為正確答案。
3. Top-5 Accuracy：預測機率前 5 名含正確答案即算預測正確。。
4. Parameters：模型參數（權重、偏差）的數目。
5. Depth：模型層數。

Keras 研發團隊將這些模型先進行訓練與參數調校，並且存檔，使用者就不用自行訓練，直接套用即可，故稱為預先訓練的模型（Pre-trained Model）。

這些預先訓練的模型主要應用在圖像辨識，各模型結構的複雜度和準確率有所差異，下圖是各模型的比較，這裡提供各位一個簡單的選用原則，如果是注重準確率，可選擇準確率較高的模型，例如 ResNet 152，反之，如果要佈署在手機上，就可考慮使用檔案較小的模型，例如 MobileNet。

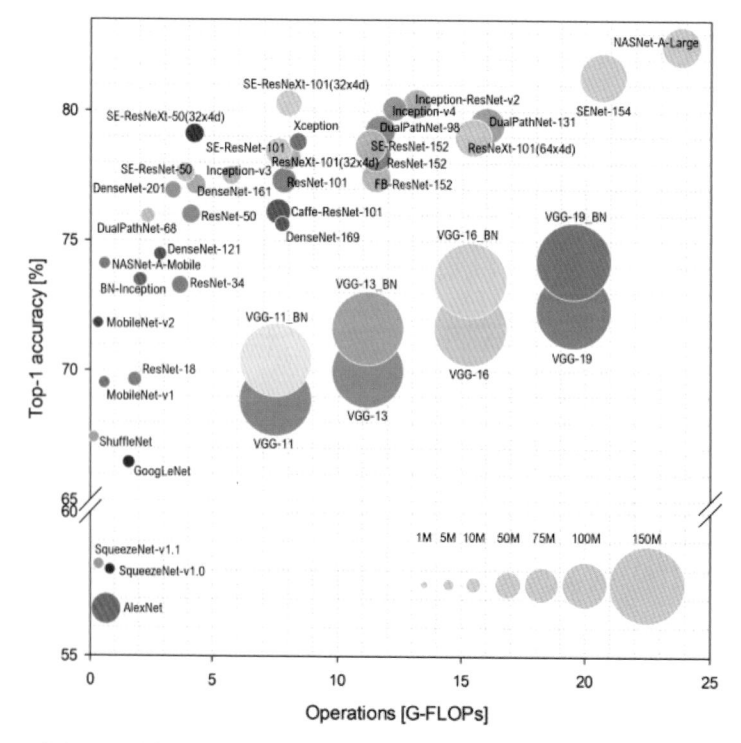

▲ 圖 7.2 預先訓練模型的準確率與計算速度之比較，圖形來源：
How to Choose the Best Keras Pre-Trained Model for Image Classification[4]

這些模型使用 ImageNet 100 多萬張圖片作為訓練資料集，內含 1,000 種類別，詳情請參考 yrevar GitHub[5]，類別範圍幾乎涵蓋了日常生活中會看到的物件，例如動物、植物、交通工具 ... 等，如果要辨識的物件屬於這 1000 種，可以直接套用模型，反之，如果要辨識這 1000 種以外的物件，就需要接上自訂的輸入層及辨識層（Dense），只利用預先訓練模型的中間層萃取特徵。

因此應用這些預先訓練的模型，有 3 種方式：

1. 採用完整的模型，可辨識 ImageNet 所提供 1000 種物件類別。
2. 採用部分的模型，只萃取特徵，通常是進行特徵比對，而非辨識。
3. 採用部分的模型，並接上自訂的輸入層和辨識層（Dense），即可辨識這 1000 種以外的物件類別。

以下就依序介紹這 3 種方式的實作。

7-2 採用完整模型

預先訓練的模型的第一種用法，是採用完整模型來辨識 1000 種物件類別。

範例. 使用 VGG16 模型進行物件的辨識。

程式：**07_01_Keras_applications.ipynb**。

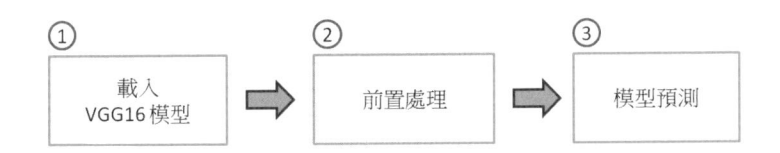

1. 載入套件：

```
1  import tensorflow as tf
2  from tensorflow.keras.applications.vgg16 import VGG16
3  from tensorflow.keras.preprocessing import image
4  from tensorflow.keras.applications.vgg16 import preprocess_input
5  from tensorflow.keras.applications.vgg16 import decode_predictions
6  import numpy as np
```

第 7 章　預先訓練的模型（Pre-trained Model）

2. 載入 VGG16 模型：顯示和繪製模型結構。

```
1  model = VGG16(weights='imagenet')
2  print(model.summary())
```

- 執行 VGG16 時，系統會先下載模型檔案至使用者 Home 資料夾下的 /.keras/models/。

    * Linux/Mac：~/.keras/models/

    * Windows：%HomePath%/.keras/models/

- 檔案名稱為 vgg16_weights_tf_dim_ordering_tf_kernels.h5。

- VGG16 類別的參數 weight 有 3 種選項：

    * None：表示此模型還未經訓練，只有模型結構。

    * imagenet：已使用 ImageNet 圖片完成訓練，載入該模型權重。

    * 檔案路徑：使用自訂的權重檔。

- include_top：

    * True：預設值，表示採用完整的模型。

    * False：不包含最上面的三層，一層是 Flatten、另外兩層則是 Dense。注意，最上面是指最後面的神經層，檔案名稱會含有 notop，譬如 vgg16_weights_tf_dim_ordering_tf_kernels_notop.h5。

- 執行結果：VGG 16 使用多組的卷積/池化層，共有 16 層的卷積/池化層。

```
Model: "vgg16"
_____
Layer (type)                 Output Shape              Param #
=================================================================
input_2 (InputLayer)         [(None, 224, 224, 3)]     0
_____
block1_conv1 (Conv2D)        (None, 224, 224, 64)      1792
_____
block1_conv2 (Conv2D)        (None, 224, 224, 64)      36928
_____
block1_pool (MaxPooling2D)   (None, 112, 112, 64)      0
_____
block2_conv1 (Conv2D)        (None, 112, 112, 128)     73856
_____
block2_conv2 (Conv2D)        (None, 112, 112, 128)     147584
_____
block2_pool (MaxPooling2D)   (None, 56, 56, 128)       0
_____
block3_conv1 (Conv2D)        (None, 56, 56, 256)       295168
_____
block3_conv2 (Conv2D)        (None, 56, 56, 256)       590080
_____
block3_conv3 (Conv2D)        (None, 56, 56, 256)       590080
_____
block3_pool (MaxPooling2D)   (None, 28, 28, 256)       0
_____
block4_conv1 (Conv2D)        (None, 28, 28, 512)       1180160
_____
block4_conv2 (Conv2D)        (None, 28, 28, 512)       2359808
_____
block4_conv3 (Conv2D)        (None, 28, 28, 512)       2359808
_____
block4_pool (MaxPooling2D)   (None, 14, 14, 512)       0
_____
block5_conv1 (Conv2D)        (None, 14, 14, 512)       2359808
_____
block5_conv2 (Conv2D)        (None, 14, 14, 512)       2359808
_____
block5_conv3 (Conv2D)        (None, 14, 14, 512)       2359808
_____
block5_pool (MaxPooling2D)   (None, 7, 7, 512)         0
_____
flatten (Flatten)            (None, 25088)             0
_____
fc1 (Dense)                  (None, 4096)              102764544
_____
fc2 (Dense)                  (None, 4096)              16781312
_____
predictions (Dense)          (None, 1000)              4097000
=================================================================
```

第 7 章　預先訓練的模型（Pre-trained Model）

模型結構圖：為單純的順序型模型，後三層為 Dense。

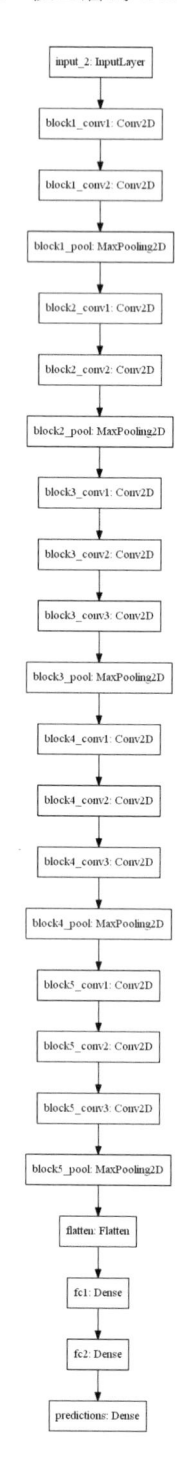

```
input_2: InputLayer
        ↓
block1_conv1: Conv2D
        ↓
block1_conv2: Conv2D
        ↓
block1_pool: MaxPooling2D
        ↓
block2_conv1: Conv2D
        ↓
block2_conv2: Conv2D
        ↓
block2_pool: MaxPooling2D
        ↓
block3_conv1: Conv2D
        ↓
block3_conv2: Conv2D
        ↓
block3_conv3: Conv2D
        ↓
block3_pool: MaxPooling2D
        ↓
block4_conv1: Conv2D
        ↓
block4_conv2: Conv2D
        ↓
block4_conv3: Conv2D
        ↓
block4_pool: MaxPooling2D
        ↓
block5_conv1: Conv2D
        ↓
block5_conv2: Conv2D
        ↓
block5_conv3: Conv2D
        ↓
block5_pool: MaxPooling2D
        ↓
flatten: Flatten
        ↓
fc1: Dense
        ↓
fc2: Dense
        ↓
predictions: Dense
```

3. 任選一張圖片，比如大象的側面照，進行模型預測。

```
1  IMAGE_PATH = '../images/'
2
3  # 任選一張圖片，例如大象側面照
4  img_path = IMAGE_PATH + 'elephant.jpg'
5  # 載入圖檔，並縮放寬高為 (224, 224)
6  img = image.load_img(img_path, target_size=(224, 224))
7
8  # 加一維，變成 (1, 224, 224)
9  x = image.img_to_array(img)
10 x = np.expand_dims(x, axis=0)
11 x = preprocess_input(x)
12
13 # 預測
14 preds = model.predict(x)
15 # decode_predictions：取得前 3 名的物件，每個物件屬性包括（類別代碼，名稱，機率）
16 print('Predicted:', decode_predictions(preds, top=3)[0])
```

- 執行結果：前三名的結果分別是印度象、非洲象、圖斯克象。

  [（'n02504013', 'Indian_elephant', 0.71942127），（'n02504458', 'African_elephant', 0.24141161），（'n01871265', 'tusker', 0.03627622）]

- decode_predictions：會根據模型取得前 N 名的物件資訊，包括類別代碼、名稱及機率。

4. 再換一張圖片，比方大象的正面照，進行模型預測。

```
1  # 任選一張圖片，例如大象正面照
2  img_path = IMAGE_PATH + 'elephant2.jpg'
3  # 載入圖檔，並縮放寬高為 (224, 224)
4  img = image.load_img(img_path, target_size=(224, 224))
5
6  # 加一維，變成 (1, 224, 224)
7  x = image.img_to_array(img)
8  x = np.expand_dims(x, axis=0)
9  x = preprocess_input(x)
10
11 # 預測
12 preds = model.predict(x)
13 # decode_predictions：取得前 3 名的物件，每個物件屬性包括（類別代碼，名稱，機率）
14 print('Predicted:', decode_predictions(preds, top=3)[0])
```

第 7 章　預先訓練的模型（Pre-trained Model）

- 執行結果：前三名的結果分別是圖斯克象、非洲象、印度象。

  [（'n01871265', 'tusker', 0.6267539）,（'n02504458', 'African_elephant', 0.3303416）,（'n02504013', 'Indian_elephant', 0.04290244）]

- 不論正面或是側面，都可以正確辨識，也不用另外去背。

5. 改用 ResNet 50 模型測試：程式碼完全一樣，只是換了命名空間及模型名稱。

```
1  from tensorflow.keras.applications.resnet50 import ResNet50
2  from tensorflow.keras.preprocessing import image
3  from tensorflow.keras.applications.resnet50 import preprocess_input
4  from tensorflow.keras.applications.resnet50 import decode_predictions
5  import numpy as np
6
7  # 預先訓練好的模型 -- ResNet50
8  model = ResNet50(weights='imagenet')
```

6. 任選一張圖片，例如老虎的大頭照，進行模型預測。

```
1  # 任意一張圖片，例如老虎大頭照
2  img_path = IMAGE_PATH + 'tiger3.jpg'
3  # 載入圖檔，並縮放寬高為 (224, 224)
4  img = image.load_img(img_path, target_size=(224, 224))
5
6  # 加一維，變成 (1, 224, 224)
7  x = image.img_to_array(img)
8  x = np.expand_dims(x, axis=0)
9  x = preprocess_input(x)
10
11 # 預測
12 preds = model.predict(x)
13 # decode_predictions: 取得前 3 名的物件，每個物件屬性包括 (類別代碼, 名稱, 機率)
14 print('Predicted:', decode_predictions(preds, top=3)[0])
```

- 執行結果：前三名的結果分別是老虎、虎貓、美洲虎。

  [（'n02129604', 'tiger', 0.8657895）,（'n02123159', 'tiger_cat', 0.13371062）,（'n02128925', 'jaguar', 0.00046872292）]

- 可以改用 tiger1.jpg、tiger2.jpg 再嘗試看看，結果應該相去不遠。

7. 改用 MobileNetV2 模型測試：程式碼完全一樣，但 preprocess_input、decode_predictions 的命名空間須使用其他模型，MobileNetV2 本身沒有自己的函數。

```python
1  from tensorflow.keras.applications import MobileNetV2
2  from tensorflow.keras.preprocessing import image
3  from tensorflow.keras.applications.vgg16 import preprocess_input
4  from tensorflow.keras.applications.vgg16 import decode_predictions
5  import numpy as np
6
7  # 載入預先訓練模型
8  model = MobileNetV2(weights='imagenet')
```

7-3 採用部分模型

預先訓練模型的第二種用法是採用部分模型，只萃取特徵，不作辨識。例如，一個 3D 模型的網站，提供模型搜尋功能，使用者可上傳要搜尋的圖檔，網站即時比對出相似的圖檔，顯示在網頁上讓使用者勾選下載，操作請參考 Sketchfab 網站（https://sketchfab.com/），類似的功能應可適用到許多場域，譬如比對嫌疑犯、商品推薦…等。

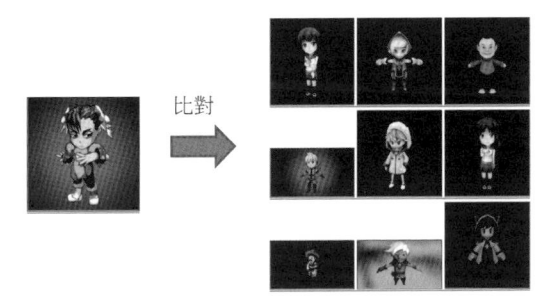

▲ 圖 7.3 3D 模型搜尋，資料來源：
Using Keras' Pretrained Neural Networks for Visual Similarity Recommendations [6]

範例. 採用 VGG16 部分模型進行圖像相似度比較。

程式：**07_02_ 圖像相似度比較 .ipynb**。

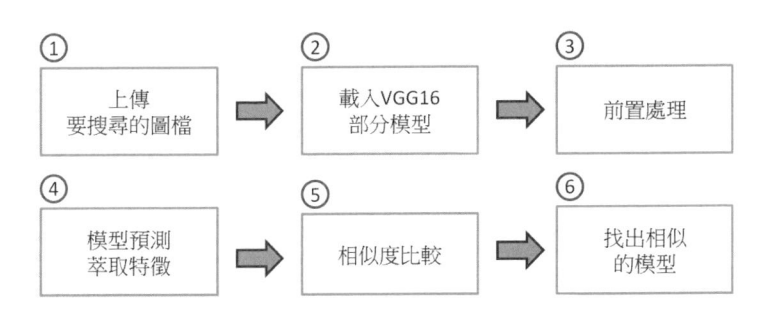

第 7 章　預先訓練的模型（Pre-trained Model）

1. 載入套件。

```python
from tensorflow.keras.applications.vgg16 import VGG16
from tensorflow.keras.preprocessing import image
from tensorflow.keras.applications.vgg16 import preprocess_input
import numpy as np
from os import listdir
from os.path import isfile, join
```

2. 載入 VGG 16 模型：include_top=False 表示不包含最上面的三層（辨識層）。

```python
# 載入VGG 16 模型, 不含最上面的三層(辨識層)
model = VGG16(weights='imagenet', include_top=False)
model.summary()
```

- 執行結果：模型不包含 Dense。

```
Model: "vgg16"
_____
Layer (type)                 Output Shape              Param #
=================================================================
input_1 (InputLayer)         [(None, None, None, 3)]   0
_____
block1_conv1 (Conv2D)        (None, None, None, 64)    1792
_____
block1_conv2 (Conv2D)        (None, None, None, 64)    36928
_____
block1_pool (MaxPooling2D)   (None, None, None, 64)    0
_____
block2_conv1 (Conv2D)        (None, None, None, 128)   73856
_____
block2_conv2 (Conv2D)        (None, None, None, 128)   147584
_____
block2_pool (MaxPooling2D)   (None, None, None, 128)   0
_____
block3_conv1 (Conv2D)        (None, None, None, 256)   295168
_____
block3_conv2 (Conv2D)        (None, None, None, 256)   590080
_____
block3_conv3 (Conv2D)        (None, None, None, 256)   590080
_____
block3_pool (MaxPooling2D)   (None, None, None, 256)   0
_____
block4_conv1 (Conv2D)        (None, None, None, 512)   1180160
_____
block4_conv2 (Conv2D)        (None, None, None, 512)   2359808
_____
block4_conv3 (Conv2D)        (None, None, None, 512)   2359808
_____
block4_pool (MaxPooling2D)   (None, None, None, 512)   0
_____
block5_conv1 (Conv2D)        (None, None, None, 512)   2359808
_____
block5_conv2 (Conv2D)        (None, None, None, 512)   2359808
_____
block5_conv3 (Conv2D)        (None, None, None, 512)   2359808
_____
block5_pool (MaxPooling2D)   (None, None, None, 512)   0
=================================================================
```

7-3 採用部分模型

3. 萃取特徵：任選一張圖片，例如大象的側面照，取得圖檔的特徵向量。

```
1  # 任選一張圖片，例如大象側面照，取得圖檔的特徵向量
2  IMAGE_PATH = '../images/'
3  img_path = IMAGE_PATH + 'elephant.jpg'
4
5  # 載入圖檔，並縮放寬高為 (224, 224)
6  img = image.load_img(img_path, target_size=(224, 224))
7
8  # 加一維，變成 (1, 224, 224)
9  x = image.img_to_array(img)
10 x = np.expand_dims(x, axis=0)
11 x = preprocess_input(x)
12
13 # 取得圖檔的特徵向量
14 features = model.predict(x)
15 print(features[0])
```

- 執行結果：得到圖檔的特徵向量如下。

```
[[[ 0.         0.         0.         ...  0.         0.         0.        ]
  [ 0.         0.        41.877056   ...  0.         0.         0.        ]
  [ 1.0921738  0.        22.865002   ...  0.         0.         0.        ]
  ...
  [ 0.         0.         0.         ...  0.         0.         0.        ]
  [ 0.         0.         0.         ...  0.         0.         0.        ]
  [ 0.         0.         0.         ...  0.         0.         0.        ]]

 [[ 0.         0.        36.385143   ...  0.         0.         3.2606328]
  [ 0.         0.        80.49929    ...  8.425463   0.         0.        ]
  [ 0.         0.        48.48268    ...  0.         0.         0.        ]
  ...
  [ 0.         0.         0.         ...  4.342996   0.         0.        ]
  [ 0.         0.         0.         ...  0.         0.         0.        ]
  [ 0.         0.         0.         ...  0.         0.         0.        ]]
```

4. 相似度比較：使用 cosine_similarity 比較特徵向量，先取得 images 資料夾下所有 .jpg 檔案名稱。

```
1  # 取得 images 目錄下所有 .jpg 檔案名稱
2  image_files = np.array([f for f in listdir(IMAGE_PATH)
3          if isfile(join(IMAGE_PATH, f)) and f[-3:] == 'jpg'])
4  image_files
```

- 執行結果：

```
array(['bird.jpg', 'bird2.jpg', 'deer.jpg', 'elephant.jpg',
       'elephant2.jpg', 'lion1.jpg', 'lion2.jpg', 'panda1.jpg',
       'panda2.jpg', 'panda3.jpg', 'tiger1.jpg', 'tiger2.jpg',
       'tiger3.jpg'], dtype='<U13')
```

5. 取得 images 資料夾下所有 .jpg 檔案的像素。

```
1  # 合併所有圖檔的像素
2  X = None
3  for f in image_files:
4      image_file = join(IMAGE_PATH, f)
5      # 載入圖檔，並縮放寬高為 (224, 224)
6      img = image.load_img(image_file, target_size=(224, 224))
7      img2 = image.img_to_array(img)
8      img2 = np.expand_dims(img2, axis=0)
9      if X is None:
10         X = img2
11     else:
12         X = np.concatenate((X, img2), axis=0)
13
14 X = preprocess_input(X)
```

6. 取得所有圖檔的特徵向量。

```
1  # 取得所有圖檔的特徵向量
2  features = model.predict(X)
3
4  features.shape, X.shape
```

- 執行結果：輸出與輸入的維度比較。

    ((13, 7, 7, 512), (13, 224, 224, 3))

7. 使用 cosine_similarity 函數比較特徵向量相似度。Cosine Similarity 計算兩個向量的夾角，如下圖，判斷兩個向量的方向是否近似，Cosine 介於（-1, 1）之間，越接近 1，表示方向越相近。

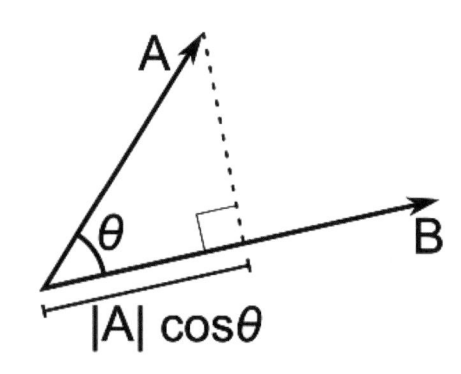

▲ 圖 7.4　夾角與 Cosine 函數

Cosine（θ）計算公式：

$$\cos(\theta) = \frac{\mathbf{A} \cdot \mathbf{B}}{\|\mathbf{A}\|\|\mathbf{B}\|}$$

```
1  # 使用 cosine_similarity 比較 Tiger2.jpg 與其他圖檔特徵向量
2  from sklearn.metrics.pairwise import cosine_similarity
3
4  no=-2
5  print(image_files[no])
6
7  # 轉為二維向量，類似扁平層(Flatten)
8  features2 = features.reshape((features.shape[0], -1))
9
10 # 排除 Tiger2.jpg 的其他圖檔特徵向量
11 other_features = np.concatenate((features2[:no], features2[no+1:]))
12
13 # 使用 cosine_similarity 計算 Cosine 函數
14 similar_list = cosine_similarity(features2[no:no+1], other_features,
15                                  dense_output=False)
16
17 # 顯示相似度，由大排到小
18 print(np.sort(similar_list[0])[::-1])
19
20 # 依相似度，由大排到小，顯示檔名
21 image_files2 = np.delete(image_files, no)
22 image_files2[np.argsort(similar_list[0])[::-1]]
```

- 執行結果：與 tiger2.jpg 比較的相似度。

  [0.35117537 0.26661643 0.19401284 0.19142228 0.1704499 0.14298241

  0.10661671 0.10612212 0.09741708 0.09370482 0.08440351 0.08097083]

- 對應的檔名：

  ['tiger1.jpg', 'tiger3.jpg', 'lion1.jpg', 'elephant.jpg',
   'elephant2.jpg', 'lion2.jpg', 'panda2.jpg', 'panda3.jpg',
   'bird.jpg', 'panda1.jpg', 'bird2.jpg', 'deer.jpg'], dtype='<U13')

觀察比對的結果，如預期一樣是正確的。利用這種方式，不只能夠比較 Image-Net 1000 類中的物件，也可以比較其他的物件，比如 3D 模型圖檔，讀者可以在網路上自行下載一些圖檔測試看看囉。

第 7 章　預先訓練的模型（Pre-trained Model）

7-4 轉移學習（Transfer Learning）

預先訓練模型的第三種用法是採用部分的模型再加上自訂的輸入層和辨識層（Dense），能夠不受限於模型原先辨識的物件，也就是所謂的『轉移學習』（Transfer Learning）或者翻譯為『遷移學習』。其實不使用預先訓練模型，直接建構 CNN 模型，也是可以辨識出任何物件的，為什麼要使用預先訓練的模型呢？優點如下：

1. 使用大量高品質的資料：ImageNet 為普林斯頓大學與史丹福大學所主導的專案，有名校掛保證！☺，加上設計較複雜的模型結構，例如 ResNet 高達 150 層，準確率因此大大提高。
2. 使用較少的訓練資料：因為模型前半段已經預先訓練好了，只需訓練自訂的辨識層。
3. 訓練速度比較快：只需要重新訓練自訂的辨識層，預先訓練模型的神經層設定 trainable=False 即可。

一般的轉移學習分為兩階段：

1. 建立預先訓練的模型（Pre-trained Model）：包括之前的 Keras Applications，以及後面章節會談到的自然語言模型 -- Transformer，目前最夯的 GPT 模型也屬於此類的演算法，均是利用大量的訓練資料和複雜的模型結構，取得圖像與自然語言特徵向量。
2. 微調（Fine Tuning）：依照特定任務需求，個別建模並訓練，例如上面所述，利用預先訓練模型的前半段，再加入自訂的神經層，進行特殊模型的建構。

範例. 使用 ResNet152V2 模型，辨識花朵資料集，程式源自 Tensorflow 官網所提供的範例『Load images』[7]，筆者進行一些修改和註解。

程式：**07_03_Flower_ResNet.ipynb**。

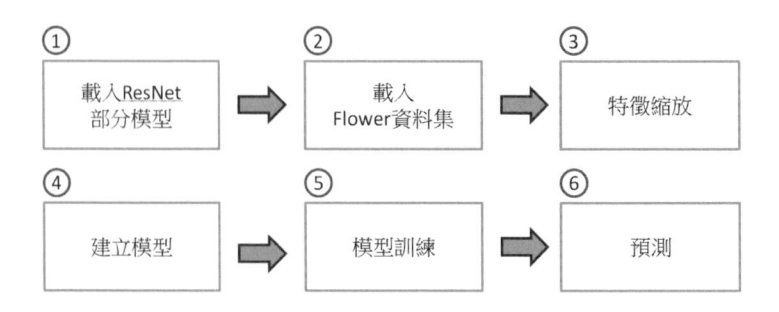

① 載入 ResNet 部分模型 ➡ ② 載入 Flower 資料集 ➡ ③ 特徵縮放

④ 建立模型 ➡ ⑤ 模型訓練 ➡ ⑥ 預測

7-4 轉移學習（Transfer Learning）

1. 載入套件：引進 ResNet152V2 模型。

```
1  import tensorflow as tf
2  from tensorflow.keras.applications.resnet_v2 import ResNet152V2
3  from tensorflow.keras.preprocessing import image
4  from tensorflow.keras.applications.resnet_v2 import preprocess_input
5  from tensorflow.keras.applications.resnet_v2 import decode_predictions
6  from tensorflow.keras.layers import Dense, GlobalAveragePooling2D
7  from tensorflow.keras.models import Model
8  import numpy as np
```

2. 下載 Flower 資料集。

```
1  # 下載資料集來源：https://www.tensorflow.org/tutorials/load_data/images
2  import pathlib
3  dataset_url = "https://storage.googleapis.com/download.tensorflow.org/example_images/flower_photos.tgz"
4  archive = tf.keras.utils.get_file(origin=dataset_url, untar=True)
5  data_dir = pathlib.Path(archive).with_suffix('')
```

- 執行結果：共 3670 個檔案、5 種類別（class），其中 2936 個檔案作為訓練之用，734 個檔案作為驗證。

3. 轉換為 Dataset：使用 image_dataset_from_directory，將資料夾轉為 Dataset。

```
1   # 參數設定
2   batch_size = 64
3   img_height = 224
4   img_width = 224
5   
6   # 載入 Flower 訓練資料
7   train_ds = tf.keras.preprocessing.image_dataset_from_directory(
8     data_dir,
9     validation_split=0.2,
10    subset="training",
11    seed=123,
12    image_size=(img_height, img_width),
13    batch_size=batch_size)
14  
15  # 載入 Flower 驗證資料
16  val_ds = tf.keras.preprocessing.image_dataset_from_directory(
17    data_dir,
18    validation_split=0.2,
19    subset="validation",
20    seed=123,
21    image_size=(img_height, img_width),
22    batch_size=batch_size)
```

第 7 章　預先訓練的模型（Pre-trained Model）

4. 進行特徵工程，將特徵縮放成（0, 1）之間。

```
1  from tensorflow.keras import layers
2
3  normalization_layer = tf.keras.layers.experimental.preprocessing.Rescaling(1./255)
4  normalized_ds = train_ds.map(lambda x, y: (normalization_layer(x), y))
5  normalized_val_ds = val_ds.map(lambda x, y: (normalization_layer(x), y))
```

5. 顯示 ResNet152V2 完整的模型結構。

```
1  base_model = ResNet152V2(weights='imagenet')
2  print(base_model.summary())
```

- 執行結果：

    * 共有 152 層卷積／池化層，再加上其他類型的神經層，總共有 566 層。

    * 輸入層（InputLayer）維度為 （224, 224, 3）。

    ```
    Model: "resnet152v2"
    _____
    Layer (type)                    Output Shape
    =================================================================
    input_5 (InputLayer)            [(None, 224, 224, 3)
    ```

    * 最後兩層為 GlobalAveragePooling、Dense，若加上 include_top= False，這兩層則會被移除。

    ```
    post_bn (BatchNormalization)     (None, 7, 7, 2048)
    _____
    post_relu (Activation)           (None, 7, 7, 2048)
    _____
    avg_pool (GlobalAveragePooling2  (None, 2048)
    _____
    predictions (Dense)              (None, 1000)
    =================================================================
    ```

6. 建立模型結構：使用 Function API 加上自訂的辨識層（GlobalAveragePooling、Dense），再指定 Model 的輸入／輸出。

```
1  # 預先訓練好的模型 -- ResNet152V2
2  base_model = ResNet152V2(weights='imagenet', include_top=False)
3  print(base_model.summary())
4
5  # 加上自訂的辨識層(Dense)
6  x = base_model.output
7  x = GlobalAveragePooling2D()(x)
8  predictions = Dense(10, activation='softmax')(x)
```

7-4 轉移學習（Transfer Learning）

```
 9
10  # 指定自訂的輸入層及辨識層(Dense)
11  model = Model(inputs=base_model.input, outputs=predictions)
12
13  # 模型前段不需訓練了
14  for layer in base_model.layers:
15      layer.trainable = False
16
17  # 設定優化器(optimizer)、損失函數(loss)、效能衡量指標(metrics)的類別
18  model.compile(optimizer='rmsprop', loss='sparse_categorical_crossentropy',
19                metrics=['accuracy'])
```

7. 模型訓練：設定快取（cache）、prefetch，提升訓練效率，時間還是有點久，建議使用 GPU 顯卡訓練，如本機無 GPU 顯卡，可使用 Colaboratory。由於在 Windows 作業系統下，TensorFlow 只有舊版 v2.10 才支援 GPU，因此，若要使用 GPU 加速訓練速度，也可以在 WSL（Windows Subsystem for Linux）內訓練，訓練速度還不錯。

```
1  # 設定快取(cache)、prefetch，以增進訓練效率
2  AUTOTUNE = tf.data.AUTOTUNE
3  normalized_ds = normalized_ds.cache().prefetch(buffer_size=AUTOTUNE)
4  normalized_val_ds = normalized_val_ds.cache().prefetch(buffer_size=AUTOTUNE)
5
6  # 模型訓練
7  history = model.fit(normalized_ds, validation_data = normalized_val_ds, epochs=5)
```

- 執行結果：非常誇張，訓練及驗證準確率均達 100%，但雲端執行只有 93% 左右。

8. 繪製訓練過程的準確率 / 損失函數。

```
1  # 對訓練過程的準確率繪圖
2  import matplotlib.pyplot as plt
3
4  plt.figure(figsize=(8, 6))
5  plt.plot(history.history['accuracy'], 'r', label='Train')
6  plt.plot(history.history['val_accuracy'], 'g', label='Validation')
7  plt.xlabel('Epoch')
8  plt.ylabel('Accuracy')
9  plt.legend();
```

- 執行結果：下方圖表可見，隨著訓練週期的增長，驗證準確率並沒有提高，這是因為預先訓練的模型已將大部份的神經層訓練過了。

7-19

第 7 章　預先訓練的模型（Pre-trained Model）

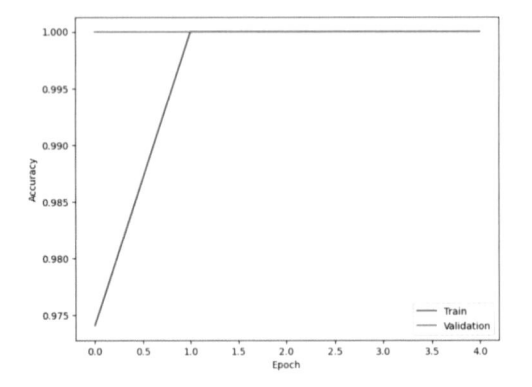

9. 顯示辨識的類別。

```
1  # 辨識的類別
2  class_names = train_ds.class_names
3  print(class_names)
```

- 執行結果：['daisy', 'dandelion', 'roses', 'sunflowers', 'tulips']。

10. 預測：任選一張圖片，譬如玫瑰花，預測結果還蠻正確的。

```
1   # 任選一張圖片，例如玫瑰
2   IMAGE_PATH = '../images/'
3   img_path = IMAGE_PATH + 'rose.png'
4   # 載入圖檔，並縮放寬高為 (224, 224)
5   img = image.load_img(img_path, target_size=(224, 224))
6
7   # 加一維，變成 (1, 224, 224, 3)
8   x = image.img_to_array(img)
9   x = np.expand_dims(x, axis=0)
10  x = preprocess_input(x)
11
12  # 預測
13  preds = model.predict(x)
14
15  # 顯示預測結果
16  y_pred = [round(i * 100, 2) for i in preds[0]]
17  print(f'預測機率(%):{y_pred}')
18  print(f'預測類別:{class_names[np.argmax(preds)]}')
```

- 執行結果：

預測機率(%):[0.03, 0.0, 99.78, 0.04, 0.15, 0.0, 0.0, 0.0, 0.0, 0.0]
預測類別：roses

- 再任選一張圖片，例如雛菊（daisy），執行結果也是 OK。

注意，筆者一開始使用 cifar 10 內建資料集，它的圖片寬高只有（28, 28），而 ResNet 的訓練模型的輸入維度則為（224, 224），雖然還是可以訓練，因為 ResNet 會自動將 cifar 10 圖像放大，但也因此造成圖像模糊，模型辨識能力變差。提醒讀者在使用預先訓練模型時要特別留意，大部份模型的輸入維度都是（224, 224）以上。

筆者同時將此程式放到 Colaboratory，並指定使用 GPU 顯卡，訓練時間相差 10~20 倍，工欲善其事，必先利其器，花錢買張好一點顯卡，會省很多時間。要在 Colaboratory 上傳資料檔，可直接點擊【資料夾圖示】，再自檔案總管拖曳檔案至 Colaboratory notebook 顯示的資料夾上即可。

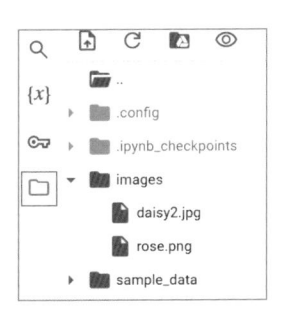

也可以使用程式上傳：

7-5 Batch Normalization 說明

上一節我們使用複雜的 ResNet152V2 模型，其中內含許多的 Batch Normalization 神經層，它在神經網路的反向傳導時可消除梯度消失（Gradient Vanishing）或梯度爆炸（Gradient Exploding）現象，所以，我們花點時間研究其原理與應用時機。

當神經網路包含很多神經層時，經常會在其中放置一些 Batch Normalization 層，顧名思義，它的用途應該是特徵縮放，然而，究竟內部是如何運作的？有哪些好處？運用的時機？擺放的位置？

第 7 章　預先訓練的模型（Pre-trained Model）

Sergey Ioffe 與 Christian Szegedy 在 2015 年首次提出 Batch Normalization，論文標題為『Batch Normalization: Accelerating Deep Network Training by Reducing Internal Covariate Shift』[8]。簡單來說，Batch Normalization 即為特徵縮放，將前一層的輸出標準化後，再轉至下一層，標準化公式如下：

$$\frac{x - \mu}{\delta}$$

標準化的好處就是讓收斂速度快一點，假如沒有標準化的話，模型通常會針對梯度較大的變數先優化，進而造成收斂路線曲折前進，如下圖，左圖是特徵未標準化的優化路徑，右圖則是標準化後的優化路徑。

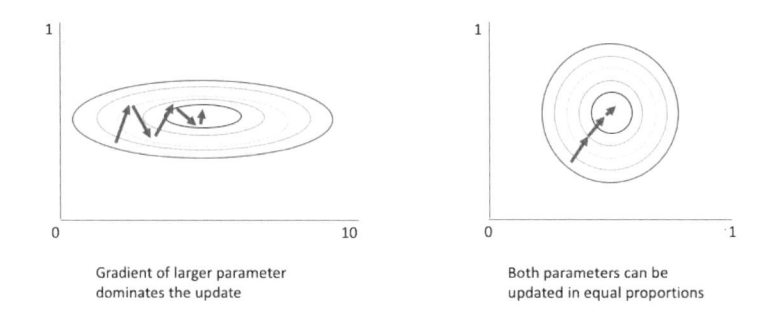

▲ 圖 7.5 未標準化 vs. 標準化優化過程的示意圖，
　　圖片來源：Why Batch Normalization Matters? [9]

Batch Normalization 另外再引進兩個變數 γ、β，分別控制規模縮放（Scale）和偏移（Shift）。

▲ 圖 7.6 Batch Normalization 公式，圖片來源：Why Batch Normalization Matters? [9]

7-5 Batch Normalization 說明

補充說明：

1. 標準化是在訓練時『逐批』處理的，而非同時所有資料一起標準化，通常加在 Activation Function 之前。
2. ε 是為了避免分母為 0 而加上的一個微小正數。
3. γ、β 值是由訓練過程中計算出來的，並不是事先設定好的。

假設我們要建立小狗的辨識模型，收集資料全部是黃色小狗的圖片進行訓練，模型完成後，拿花狗的圖片來辨識，效果想當然會變差，要改善的話必須重新收集資料再訓練一次，這種現象就稱為『Covariate Shift』，正式的定義是『假設我們要使用 X 預測 Y 時，當 X 的分配隨著時間有所變化時，模型就會逐漸失效』。股價預測也有類似的狀況，當股價長期趨勢上漲時，原來的模型就慢慢失準了，除非納入最新的資料重新訓練模型。

由於神經網路的權重會隨著反向傳導不斷更新，每一層的輸出都會受到上一層的輸出影響，這是一種迴歸的關係，隨著神經層越多，整個神經網路的輸出有可能會逐漸偏移，此種現象稱之為『Internal Covariate Shift』。

而 Batch Normalization 就可以矯正『Internal Covariate Shift』現象，它在輸出至下一層的神經層時，每批資料都會先被標準化，這使得輸入資料的分佈為 N（0, 1），即標準常態分配，因此，不管有多少層神經層，都不用擔心發生輸出逐漸偏移的問題。

至於什麼是梯度消失或爆炸？這是由於 CNN 模型共享權值（Shared Weights）的關係，使得梯度逐漸消失或爆炸，原因如下，相同的 W 值經過很多層的傳導：

- 如果 W<1 ➔ 對於輸出的影響力，離輸出越遠的神經層 n 愈大，W^n 會趨近於 0，影響力逐漸遞減，即梯度消失（Gradient Vanishing）。
- 反之若 W>1 ➔ W^n 會趨近於 ∞，則造成模型求解無法收斂，即梯度爆炸（Gradient Explosion）。

只要經過 Batch Normalization，將每一批標準化後，梯度都會重新計算，這樣就不會有梯度消失和梯度爆炸的狀況發生了。除此之外，根據原作者的說法，Batch Normalization 還有以下優點：

- 優化收斂速度快（Train faster）。
- 可使用較大的學習率（Use higher learning rates），加速訓練過程。
- 權重初始化較容易（Parameter initialization is easier）。

第 7 章　預先訓練的模型（Pre-trained Model）

- 不使用 Batch Normalization 時，Activation function 容易在訓練過程中消失或提早停止學習，但如果經過 Batch Normalization 則又會再復活（Makes activation functions viable by regulating the inputs to them）。

- 準確率全面性提升（Better results overall）。

- 類似 Dropout 的效果，可防止過度擬合（It adds noise which reduces over-fitting with a regularization effect），所以，當使用 Batch Normalization 時，就**不需要加 Dropout 層了**，為避免效果加乘過強，反而造成低度擬合（Under-fitting）。

有一篇文章『On The Perils of Batch Norm』[10] 做了一個很有趣的實驗，使用兩個資料集模擬『Internal Covariate Shift』現象，一個是 MNIST 資料集，背景是單純白色，另一個則是 SVHN 資料集，有複雜的背景，實驗過程如下：

1. 首先合併兩個資料集來訓練第一種模型，如下圖所示：

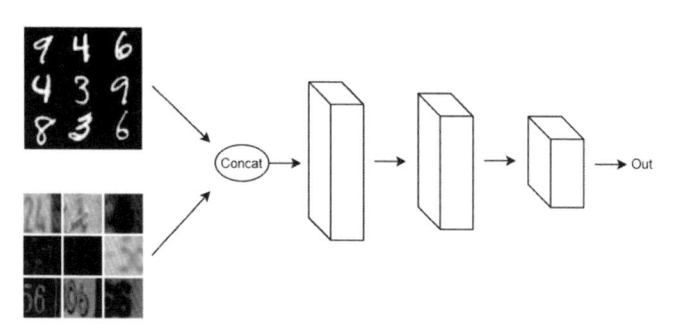

▲ 圖 7.7　合併兩個資料集來訓練一個模型

2. 再使用兩個資料集各自分別訓練模型，但共享權值，為第二種模型，如下圖：

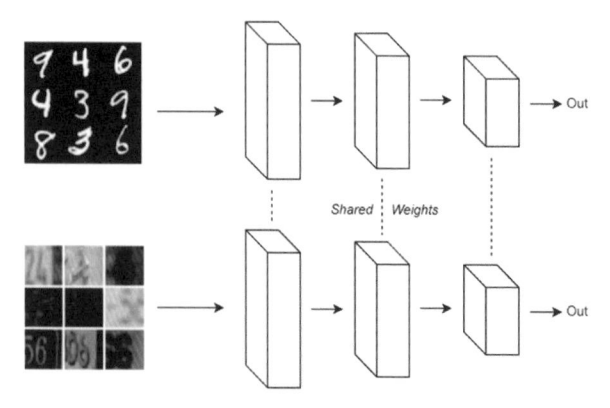

▲ 圖 7.8　使用兩個資料集個別訓練模型，但共享權值

兩種模型都有插入 Batch Normalization，比較結果，前者即單一模型準確度較高，因為 Batch Normalization 可以矯正『Internal Covariate Shift』現象。後者則由於資料集內容的不同，兩個模型共享權值本來就不合理。

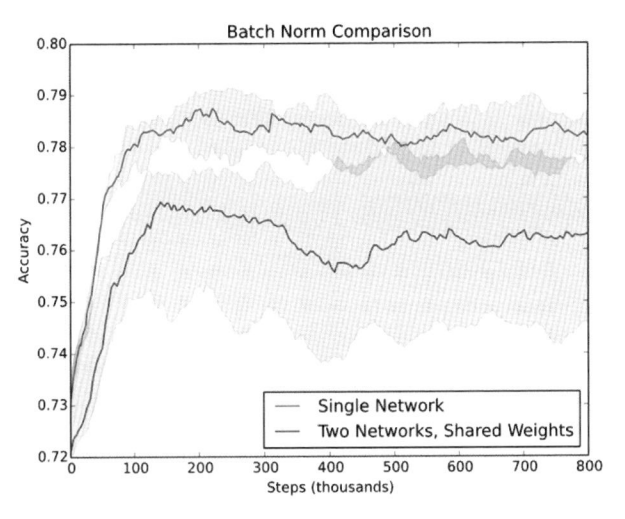

▲ 圖 7.9 兩種模型準確率比較

3. 第三種模型：使用兩個資料集訓練兩個模型，個別作 Batch Normalization，但不共享權值。比較結果，第三種模型效果最好。

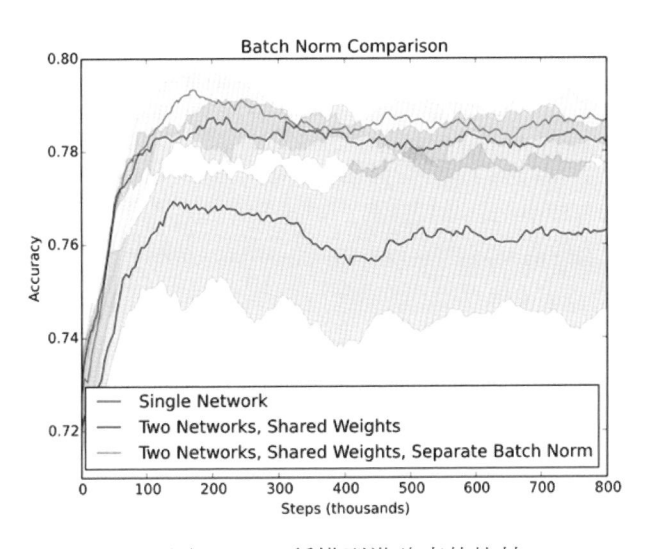

▲ 圖 7.10 三種模型準確率的比較

第 7 章 預先訓練的模型（Pre-trained Model）

參考資料（References）

[1] Sergey Ioffe、Christian Szegedy,《Batch Normalization: Accelerating Deep Network Training by Reducing Internal Covariate Shift》, 2015
(http://proceedings.mlr.press/v37/ioffe15.pdf)

[2] Kaiming He、Xiangyu Zhang、Shaoqing Ren、Jian Sun,《Deep Residual Learning for Image Recognition》, 2015
(https://arxiv.org/abs/1512.03385)

[3] Keras 官網關於 Keras Applications 的介紹
(https://keras.io/api/applications/)

[4] Marie Stephen Leo,《How to Choose the Best Keras Pre-Trained Model for Image Classification》, 2020
(https://towardsdatascience.com/how-to-choose-the-best-keras-pre-trained-model-for-image-classification-b850ca4428d4)

[5] yrevar GitHub
(https://gist.github.com/yrevar/942d3a0ac09ec9e5eb3a)

[6] Ethan Rosenthal,《Using Keras' Pretrained Neural Networks for Visual Similarity Recommendations》, 2016
(https://www.ethanrosenthal.com/2016/12/05/recasketch-keras/)

[7] ensorflow 官網範例『Load images』
(https://www.tensorflow.org/tutorials/load_data/images)

[8] Sergey Ioffe、Christian Szegedy,《Batch Normalization: Accelerating Deep Network Training by Reducing Internal Covariate Shift》, 2015
(https://arxiv.org/pdf/1502.03167.pdf)

[9] Aman Sawarn, Why Batch Normalization Matters?, 2020
(https://medium.com/towards-artificial-intelligence/why-batch-normalization-matters-4a6d753ba309)

[10] alexirpan, On The Perils of Batch Norm, 2017
(https://www.alexirpan.com/2017/04/26/perils-batch-norm.html)

第三篇

進階的影像應用

恭喜各位勇士們通過卷積神經網路（CNN）關卡，越過一座高山，本篇就來好好秀一下努力的成果，展現 CNN 在各種領域應用上有哪些厲害的功能吧！

本篇的菜色超澎湃，包括下列主題：

- 物件偵測（Object Detection）。
- 語義分割（Semantic Segmentation）。
- 人臉辨識（Facial Recognition）。
- 風格轉換（Style Transfer）。
- 光學文字辨識（Optical Character Recognition, OCR）。

第 8 章

物件偵測
(Object Detection)

前面介紹的圖像辨識模型,一張圖片中僅能辨識一個物件,接下來要登場的物件偵測可以在一張圖片中同時偵測多個物件,並且標示出物件的位置。但是標示位置有什麼用處呢?現今最熱門的物件偵測演算法 YOLO,發明人 Joseph Redmon 提出一張有趣的照片:

第 8 章　物件偵測（Object Detection）

▲ 圖 8.1 機器人煎餅，
圖片來源：Real-Time Grasp Detection Using Convolutional Neural Networks [1]

機器人要能完成煎餅的任務，它必須知道煎餅的所在位置，才能夠將煎餅翻面，如果有兩張以上的餅，還需知道要翻哪一張。不只機器人工作時需要電腦視覺，其他領域也會用到物件偵測，譬如：

1. 自駕車（Self-driving Car）：需要即時掌握前方路況及閃避障礙物。
2. 智慧交通：車輛偵測，利用一輛車在兩個時間點的位置，計算車速，進而可以推算道路壅塞的狀況，也可以用來偵測違規車輛。
3. 機器人、玩具、無人機、飛彈…等都可以作類似的應用。
4. 異常偵測（Anomaly Detection）：可以在生產線上架設攝影機，即時偵測異常的瑕疵，像是印刷電路板、產品外觀…等。
5. 無人商店的購物籃掃描，自動結帳。

8-1 圖像辨識模型的發展

綜觀歷年 ImageNet ILSVRC 挑戰賽（Large Scale Visual Recognition Challenge）的競賽題目，從 2011 年的影像分類（Classification）與定位（Classification with Localization）[2]，到 2017 年，題目擴展至物體定位（Object Localization）、物體偵測（Object Detection）、影片物體偵測（Object Detection from Video）[3]。我們可從中

8-1 圖像辨識模型的發展

觀察到圖像辨識模型的發展史，了解到整個技術的演進。目前圖像辨識大概分為下列四大類型，如下圖所示：

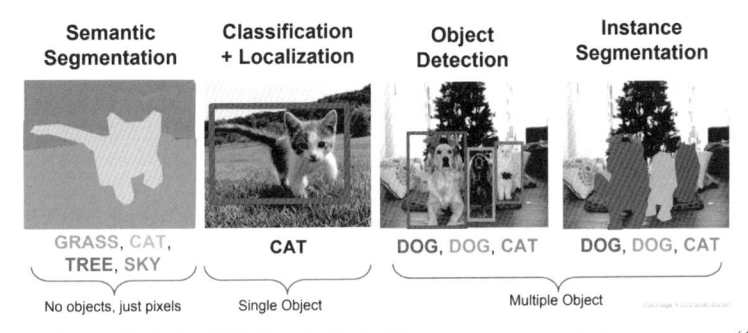

▲ 圖 8.2 物件偵測類型，圖片來源：Detection and Segmentation[4]

1. 語義分割（Semantic Segmentation）：按照物件類別來劃分像素區域，但不區分實例（Instance）。拿上圖的第 4 張照片為例，照片中有 2 隻狗，都使用同一種顏色表達，即是語義分割，2 隻狗使用不同顏色來表示，區分實例，則稱為實例分割。

2. 定位（Classification + Localization）：標記單一物件（Single Object）的類別與所在的位置。

3. 物件偵測（Object Detection）：標記多個物件（Multiple Object）的類別與所在的位置。

4. 實例分割（Instance Segmentation）：標記實例（Instance），同一類的物件使用不同顏色區分，並標示個別的位置，尤其是物件之間有重疊時。

另外，YOLO v11 還提供額外功能：

1. 姿態辨識（Pose Estimation）：偵測人體的關鍵點（Key Points），藉以判斷姿勢正確與否或是狀態（站立／坐下）。

2. 物件追蹤（Object Tracking）：追蹤物件移動的軌跡。

3. 物件方向偵測（Oriented Object Detection, OOB）：偵測物件面對的方向，標記的邊界框（Bounding Box）是旋轉的矩形，與一般的物件偵測不同，後者永遠是垂直的矩形，OOB 標記的位置更精準。

在本書初版時，我們看到許多演算法只解決上述類別的單一功能，發展至今（2024），YOLO 演算法已經可以處理所有類別的功能，因此這一版就聚焦在 YOLO 及前期演算法的發展歷程，讀者可以欣賞創意發想的過程，研發團隊前仆後繼，不斷努力，最終 YOLO 成為經典（State of the Art，簡稱 SOTA）模型。

第 8 章　物件偵測（Object Detection）

8-2 影像金字塔與滑動視窗

物件偵測要同時偵測物件類別與位置，直覺上就是拆分為兩項任務（Task）：

1. 分類（Classification）：辨識物件的類別。
2. 迴歸（Regression）：找到物件的位置，包括物件左上角的座標和寬度/高度。

最初的作法是結合影像金字塔（Image Pyramid）與滑動視窗（Sliding Win-dow），步驟如下：

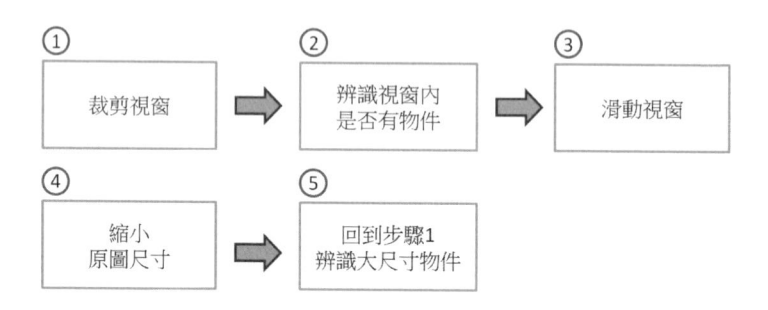

① 裁剪視窗 ➡ ② 辨識視窗內是否有物件 ➡ ③ 滑動視窗

④ 縮小原圖尺寸 ➡ ⑤ 回到步驟1辨識大尺寸物件

1. 設定某一尺寸的視窗，譬如寬高各為 128 像素，由原圖左上角擷取視窗。
2. 辨識視窗內是否有物件存在。
3. 再滑動視窗一格，與前面介紹的卷積作法類似，再次擷取，並回到步驟 2，直到全圖掃描完為止。
4. 縮小原圖尺寸後，再重新回到步驟 1，尋找更大尺寸的物件。

這種將原圖縮小成各種尺寸的方式稱為影像金字塔，詳情請參閱『Image Pyramids with Python and OpenCV』一文 [5]，如下圖所示：

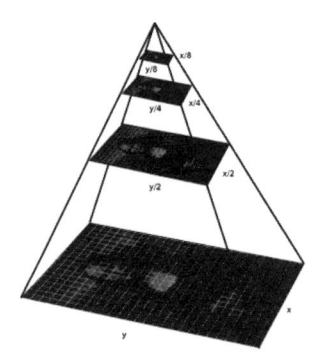

▲ 圖 8.3 影像金字塔（Image Pyramid），最下層為原圖，往上逐步縮小原圖尺寸，圖片來源：IIP Image [6]

8-2 影像金字塔與滑動視窗

範例. 對圖片滑動視窗並作影像金字塔（Image Pyramid）。

程式：08_01_Sliding_Window_And_Image_Pyramid.py，程式修改自『Sliding Windows for Object Detection with Python and OpenCV』[7]。

1. 先安裝套件 OpenCV、imutils：

   pip install opencv-python imutils

2. 載入套件。

```
1  # 載入套件
2  import cv2
3  import time
4  import imutils
```

3. 定義影像金字塔操作函數：逐步縮小原圖尺寸，以便找到較大尺寸的物件。

```
1   # 影像金字塔操作
2   # image：原圖，scale：每次縮小倍數，minSize：最小尺寸
3   def pyramid(image, scale=1.5, minSize=(30, 30)):
4       # 第一次傳回原圖
5       yield image
6
7       while True:
8           # 計算縮小後的尺寸
9           w = int(image.shape[1] / scale)
10          # 縮小
11          image = imutils.resize(image, width=w)
12          # 直到最小尺寸為止
13          if image.shape[0] < minSize[1] or image.shape[1] < minSize[0]:
14              break
15          # 傳回縮小後的圖像
16          yield image
```

4. 定義滑動視窗函數。

```
1  # 滑動視窗
2  def sliding_window(image, stepSize, windowSize):
3      for y in range(0, image.shape[0], stepSize):      # 向下滑動 stepSize 格
4          for x in range(0, image.shape[1], stepSize):  # 向右滑動 stepSize 格
5              # 傳回裁剪後的視窗
6              yield (x, y, image[y:y + windowSize[1], x:x + windowSize[0]])
```

5. 測試。

```
1  # 讀取一個圖檔
2  image = cv2.imread('./lena.jpg')
```

第 8 章　物件偵測（Object Detection）

```
 3
 4  # 視窗尺寸
 5  (winW, winH) = (128, 128)
 6
 7  # 取得影像金字塔各種尺寸
 8  for resized in pyramid(image, scale=1.5):
 9      # 滑動視窗
10      for (x, y, window) in sliding_window(resized, stepSize=32,
11                                  windowSize=(winW, winH)):
12          # 視窗尺寸不合即放棄,滑動至邊緣時,尺寸過小
13          if window.shape[0] != winH or window.shape[1] != winW:
14              continue
15          # 標示滑動的視窗
16          clone = resized.copy()
17          cv2.rectangle(clone, (x, y), (x + winW, y + winH), (0, 255, 0), 2)
18          cv2.imshow("Window", clone)
19          cv2.waitKey(1)
20          # 暫停
21          time.sleep(0.025)
22
23  # 結束時關閉視窗
24  cv2.destroyAllWindows()
```

6. 執行指令：

   python 08_01_Sliding_Window_And_Image_Pyramid.py

7. 執行結果：

這種方式非常簡單，但有以下缺點：

1. 以滑動視窗比對，偵測時間較長，再加上影像金字塔操作增加數倍的掃描時間。

2. 如果使用像素比對，物件的背景必須單純，否則比對會很困難。

3. 如果物件只有局部影像,而非完整物件,應該偵測不到。
4. 若偵測人物或動物,姿勢不同也很困難比對。

8-3 方向梯度直方圖(HOG)

像素比對顯然不是好方法,與卷積一樣,應萃取線條特徵或輪廓,因此,有學者研發出『方向梯度直方圖』(Histogram of oriented gradient, HOG),抓取圖像輪廓線條,類似前面所提的 Sobel、Laplacian 演算法,HOG 詳細內容可參閱『CV17 HOG 特徵提取算法』[8],由於後面有更好的作法,因此,本文不在此詳細說明 HOG,只著重在實作。

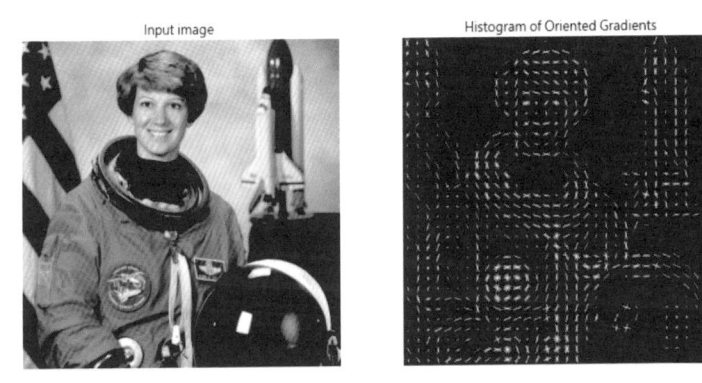

▲ 圖 8.4 HOG 處理:左圖為原圖,右圖為 HOG 處理後的輸出

根據『Histogram of Oriented Gradients and Object Detection』[9] 一文的介紹,結合了 HOG 的物件偵測,流程如下:

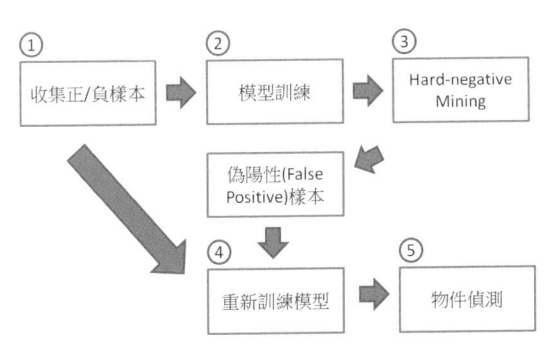

▲ 圖 8.5 結合 HOG 的物件偵測之流程圖

第 8 章　物件偵測（Object Detection）

1. 收集正樣本（Positive set）：收集目標物件的各式圖像樣本，包括不同視角、尺寸、背景的圖像。
2. 收集負樣本（Negative set）：收集無目標物件的各式圖像樣本，若有找到相近的物件則更好，可增強辨識準確度。
3. 使用以上正 / 負樣本與分類演算法訓練二分類模型，判斷是否包含目標物件，一般使用『支援向量機』（SVM）演算法。
4. Hard-negative Mining：掃描負樣本，使用滑動視窗的技巧，將每個視窗餵入模型來預測，如果有偵測到目標物件，即是偽陽性（False Positive），接著將這些圖像加到訓練資料集中重新進行訓練，這個步驟可以重複很多次，能夠有效地提高模型準確率，類似整體學習（Ensemble Learning）的 Boosting 演算法。
5. 使用最後的模型進行物件偵測：使用滑動視窗與影像金字塔技巧，對測試圖像進行裁切，再餵入模型進行辨識，找出含有物件的視窗。
6. 篩選合格的視窗：使用 Non-Maximum Suppression（NMS）演算法，從所有合格的視窗中篩選出最適合的視窗。

範例 1. 使用 HOG、滑動視窗及 SVM 進行物件偵測。

程式：**08_02_HOG-Face-Detection.ipynb**，修改自 Scikit-Image 的範例。

1. 載入套件：本例使用 scikit-image 套件，OpenCV 也有支援類似的函數。

```
1  # Scikit-Image 的範例
2  # 載入套件
3  import numpy as np
4  import matplotlib.pyplot as plt
5  from skimage.feature import hog
6  from skimage import data, exposure
```

2. HOG 測試：使用 Scikit-Image 內建的女太空人圖像來測試 HOG 的效果。

```
1  # 測試圖片
2  image = data.astronaut()
3
4  # 取得圖片的 hog
5  fd, hog_image = hog(image, orientations=8, pixels_per_cell=(16, 16),
6                      cells_per_block=(1, 1), visualize=True, multichannel=True)
7
8  # 原圖與 hog 圖比較
9  fig, (ax1, ax2) = plt.subplots(1, 2, figsize=(12, 6), sharex=True, sharey=True)
10
11 ax1.axis('off')
12 ax1.imshow(image, cmap=plt.cm.gray)
```

8-3 方向梯度直方圖（HOG）

```
13  ax1.set_title('Input image')
14
15  # 調整對比，讓顯示比較清楚
16  hog_image_rescaled = exposure.rescale_intensity(hog_image, in_range=(0, 10))
17
18  ax2.axis('off')
19  ax2.imshow(hog_image_rescaled, cmap=plt.cm.gray)
20  ax2.set_title('Histogram of Oriented Gradients')
21  plt.show()
```

- 執行結果：原圖與 HOG 處理過後的圖比較。

3. 收集正樣本（positive set）：使用 scikit-learn 內建的人臉資料集作為正樣本，共有 13233 筆。

```
1  # 收集正樣本 (positive set)
2  # 使用 scikit-learn 的人臉資料集
3  from sklearn.datasets import fetch_lfw_people
4  faces = fetch_lfw_people()
5  positive_patches = faces.images
6  positive_patches.shape
```

4. 觀察正樣本中部份的圖片。

```
1  # 顯示正樣本部份圖片
2  fig, ax = plt.subplots(4,6)
3  for i, axi in enumerate(ax.flat):
4      axi.imshow(positive_patches[500 * i], cmap='gray')
5      axi.axis('off')
```

8-9

第 8 章 物件偵測（Object Detection）

- 執行結果：每張圖片寬高為（62, 47）。

5. 收集負樣本（negative set）：任意找些非人臉的資料集作為負樣本，這裡使用 Scikit-Image 內建的資料集。

```
1  # 收集負樣本 (negative set)：使用 Scikit-Image 的非人臉資料
2  from skimage import data, transform, color
3
4  imgs_to_use = ['hubble_deep_field', 'coffee','chelsea']
5  images = [color.rgb2gray(getattr(data, name)())
6            for name in imgs_to_use]
7  len(images)
```

6. 增加負樣本筆數：將負樣本轉換為不同的尺寸，也可以使用資料增補技術。

```
1  # 將負樣本轉換為不同的尺寸
2  from sklearn.feature_extraction.image import PatchExtractor
3
4  # 轉換為不同的尺寸
5  def extract_patches(img, N, scale=1.0, patch_size=positive_patches[0].shape):
6      extracted_patch_size = tuple((scale * np.array(patch_size)).astype(int))
7      # PatchExtractor：產生不同尺寸的圖像
8      extractor = PatchExtractor(patch_size=extracted_patch_size,
9                                 max_patches=N, random_state=0)
10     patches = extractor.transform(img[np.newaxis])
11     if scale != 1:
12         patches = np.array([transform.resize(patch, patch_size)
13                             for patch in patches])
14     return patches
```

8-3 方向梯度直方圖（HOG）

```
15
16  # 產生 27000 筆圖像
17  negative_patches = np.vstack([extract_patches(im, 1000, scale)
18                                  for im in images for scale in [0.5, 1.0, 2.0]])
19  negative_patches.shape
```

- 執行結果：產生 27000 筆圖像。

7. 觀察負樣本中部份的圖片。

```
1  # 顯示部份負樣本
2  fig, ax = plt.subplots(4,6)
3  for i, axi in enumerate(ax.flat):
4      axi.imshow(negative_patches[600 * i], cmap='gray')
5      axi.axis('off')
```

- 執行結果：

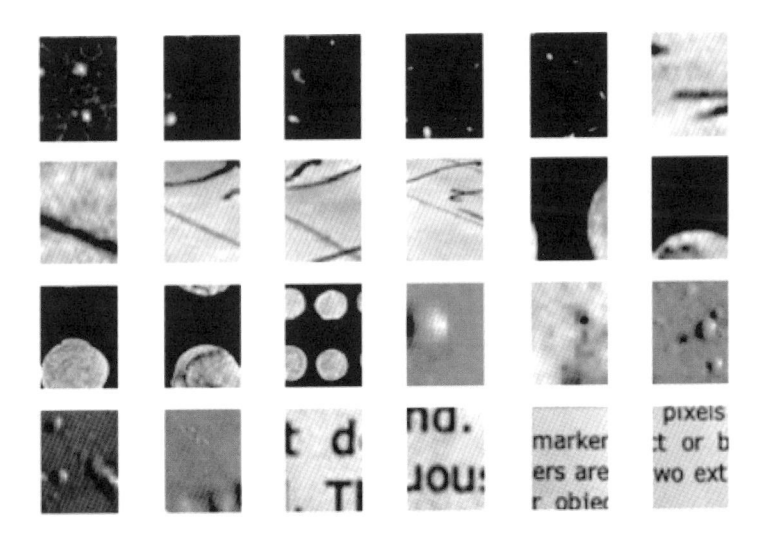

8. 合併正 / 負樣本。

```
1  # 合併正樣本與負樣本
2  from skimage import feature      # To use skimage.feature.hog()
3  from itertools import chain
4
5  X_train = np.array([feature.hog(im)
6                      for im in chain(positive_patches,
7                                      negative_patches)])
8  y_train = np.zeros(X_train.shape[0])
9  y_train[:positive_patches.shape[0]] = 1
```

第 8 章 物件偵測（Object Detection）

9. 使用 SVM 進行二分類的訓練：使用 GridSearchCV 尋求最佳參數值。

```
1  # 使用 SVM 作二分類的訓練
2  from sklearn.svm import LinearSVC
3  from sklearn.model_selection import GridSearchCV
4
5  # C為矯正過度擬合強度的倒數，使用 GridSearchCV 尋求最佳參數值
6  grid = GridSearchCV(LinearSVC(dual=False), {'C': [1.0, 2.0, 4.0, 8.0]},cv=3)
7  grid.fit(X_train, y_train)
8  grid.best_score_
```

- 執行結果：最佳模型準確率為 98.77%。

10. 取得最佳參數值。

```
1  # C 最佳參數值
2  grid.best_params_
```

11. 依最佳參數值再訓練一次，取得最終模型。

```
1  # 依最佳參數值再訓練一次
2  model = grid.best_estimator_
3  model.fit(X_train, y_train)
```

12. 新圖像測試：需先轉為灰階圖像。

```
1  # 取新圖像測試
2  test_img = data.astronaut()
3  test_img = color.rgb2gray(test_img)
4  test_img = transform.rescale(test_img, 0.5)
5  test_img = test_img[:120, 60:160]
6
7
8  plt.imshow(test_img, cmap='gray')
9  plt.axis('off');
```

- 執行結果：

8-3 方向梯度直方圖（HOG）

13. 定義滑動視窗函數。

```
# 滑動視窗函數
def sliding_window(img, patch_size=positive_patches[0].shape,
                   istep=2, jstep=2, scale=1.0):
    Ni, Nj = (int(scale * s) for s in patch_size)
    for i in range(0, img.shape[0] - Ni, istep):
        for j in range(0, img.shape[1] - Ni, jstep):
            patch = img[i:i + Ni, j:j + Nj]
            if scale != 1:
                patch = transform.resize(patch, patch_size)
            yield (i, j), patch
```

14. 計算 Hog：使用滑動視窗來計算每一滑動視窗的 Hog，餵入模型辨識。

```
# 使用滑動視窗計算每一視窗的 Hog
indices, patches = zip(*sliding_window(test_img))
patches_hog = np.array([feature.hog(patch) for patch in patches])

# 辨識每一視窗
labels = model.predict(patches_hog)
labels.sum() # 偵測到的總數
```

- 執行結果：共有 55 個合格視窗。

15. 顯示這 55 個合格視窗。

```
# 將每一個偵測到的視窗顯示出來
fig, ax = plt.subplots()
ax.imshow(test_img, cmap='gray')
ax.axis('off')

# 取得左上角座標
Ni, Nj = positive_patches[0].shape
indices = np.array(indices)

# 顯示
for i, j in indices[labels == 1]:
    ax.add_patch(plt.Rectangle((j, i), Nj, Ni, edgecolor='red',
                               alpha=0.3, lw=2, facecolor='none'))
```

- 執行結果：

第 8 章 物件偵測（Object Detection）

16. 篩選合格視窗：使用 Non-Maximum Suppression（NMS）演算法，從所有合格的視窗中篩選出最適合的視窗。以下採用『Non-Maximum Suppression for Object Detection in Python』[10] 一文的程式碼。

    - 定義 NMS 演算法函數：這是由 Pedro Felipe Felzenszwalb 等學者發明的演算法，執行速度較慢，Tomasz Malisiewicz[11] 因而提出改善的演算法。函數的重疊比例門檻（overlapThresh）參數一般設為 0.3~0.5 之間。

```
1   # Non-Maximum Suppression演算法 by Felzenszwalb et al.
2   # boxes：所有候選的視窗，overlapThresh：視窗重疊的比例門檻
3   def non_max_suppression_slow(boxes, overlapThresh=0.5):
4       if len(boxes) == 0:
5           return []
6   
7       pick = []          # 儲存篩選的結果
8       x1 = boxes[:,0]    # 取得候選的視窗的左/上/右/下 座標
9       y1 = boxes[:,1]
10      x2 = boxes[:,2]
11      y2 = boxes[:,3]
12  
13      # 計算候選視窗的面積
14      area = (x2 - x1 + 1) * (y2 - y1 + 1)
15      idxs = np.argsort(y2)    # 依視窗的底Y座標排序
16  
17      # 比對重疊比例
18      while len(idxs) > 0:
19          # 最後一筆
20          last = len(idxs) - 1
21          i = idxs[last]
22          pick.append(i)
23          suppress = [last]
24  
25          # 比對最後一筆與其他視窗重疊的比例
26          for pos in range(0, last):
27              j = idxs[pos]
28  
29              # 取得所有視窗的涵蓋範圍
```

8-3 方向梯度直方圖（HOG）

```
30              xx1 = max(x1[i], x1[j])
31              yy1 = max(y1[i], y1[j])
32              xx2 = min(x2[i], x2[j])
33              yy2 = min(y2[i], y2[j])
34              w = max(0, xx2 - xx1 + 1)
35              h = max(0, yy2 - yy1 + 1)
36
37              # 計算重疊比例
38              overlap = float(w * h) / area[j]
39
40              # 如果大於門檻值，則儲存起來
41              if overlap > overlapThresh:
42                  suppress.append(pos)
43
44          # 刪除合格的視窗，繼續比對
45          idxs = np.delete(idxs, suppress)
46
47      # 傳回合格的視窗
48      return boxes[pick]
```

17. 呼叫 non_max_suppression_slow 函數，篩選出最適合的邊界框（Bounding box）。

```
1  # 使用 Non-Maximum Suppression演算法，剔除多餘的視窗。
2  candidate_boxes = []
3  for i, j in indices[labels == 1]:
4      candidate_boxes.append([j, i, Nj, Ni])
5  final_boxes = non_max_suppression_slow(np.array(candidate_boxes).reshape(-1, 4))
6
7  # 將每一個合格的視窗顯示出來
8  fig, ax = plt.subplots()
9  ax.imshow(test_img, cmap='gray')
10 ax.axis('off')
11
12 # 顯示
13 for i, j, Ni, Nj in final_boxes:
14     ax.add_patch(plt.Rectangle((i, j), Ni, Nj, edgecolor='red',
15                                alpha=0.3, lw=2, facecolor='none'))
```

- 執行結果：得到 3 個合格邊界框。

第 8 章　物件偵測（Object Detection）

這個範例類似 CNN，先使用 HOG 萃取輪廓特徵，再以輪廓特徵為 X，以 SVM 進行分類，找出合格邊界框，過程中省略了一些細節，譬如 Hard-negative mining、影像金字塔，這個例子無法偵測多個實體（Instance）。所以我們再來看一個範例，可使用任何 CNN 模型結合影像金字塔，進行多物件、多實體的偵測。

範例 2. 使用 ResNet50 進行物件偵測。

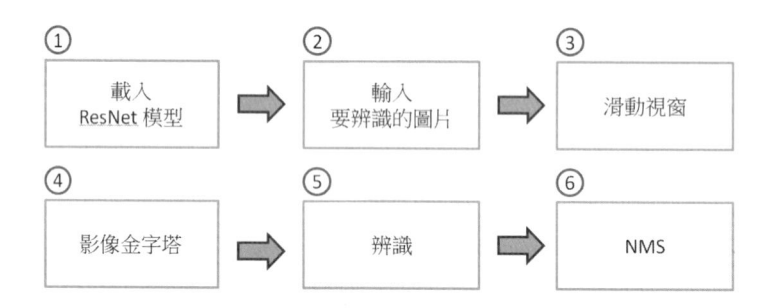

```
①載入 ResNet 模型 → ②輸入要辨識的圖片 → ③滑動視窗
④影像金字塔 → ⑤辨識 → ⑥NMS
```

程式：**08_03_Object_Detection.ipynb**。

1. 載入套件。

```
1  # 載入套件，需額外安裝 imutils 套件
2  from tensorflow.keras.applications import ResNet50
3  from tensorflow.keras.applications.resnet import preprocess_input
4  from tensorflow.keras.preprocessing.image import img_to_array
5  from tensorflow.keras.applications import imagenet_utils
6  from imutils.object_detection import non_max_suppression
7  import numpy as np
8  import imutils
9  import time
10 import cv2
```

2. 參數設定：此範例是辨識三隻斑馬同時存在的圖像，另外也可以辨識騎自行車的圖像（bike.jpg）。

```
1  # 參數設定
2  image_path = './images_test/zebra.jpg'  # 要辨識的圖檔
3  WIDTH = 600                # 圖像縮放為 (600, 600)
4  PYR_SCALE = 1.5            # 影像金字塔縮放比例
5  WIN_STEP = 16              # 視窗滑動步數
6  ROI_SIZE = (250, 250)      # 視窗大小
7  INPUT_SIZE = (224, 224)    # CNN的輸入尺寸
```

8-3 方向梯度直方圖(HOG)

3. 載入 ResNet50 模型。

```
1  # 載入 ResNet50 模型
2  model = ResNet50(weights="imagenet", include_top=True)
```

4. 讀取要辨識的圖片。

```
1  # 讀取要辨識的圖片
2  orig = cv2.imread(image_path)
3  orig = imutils.resize(orig, width=WIDTH)
4  (H, W) = orig.shape[:2]
```

5. 定義滑動視窗和影像金字塔函數,這部分與範例 2 的流程相同。

```
1   # 定義滑動視窗與影像金字塔函數
2
3   # 滑動視窗
4   def sliding_window(image, step, ws):
5       for y in range(0, image.shape[0] - ws[1], step):    # 向下滑動 stepSize 格
6           for x in range(0, image.shape[1] - ws[0], step): # 向右滑動 stepSize 格
7               # 傳回裁剪後的視窗
8               yield (x, y, image[y:y + ws[1], x:x + ws[0]])
9
10  # 影像金字塔操作
11  # image : 原圖, scale : 每次縮小倍數, minSize : 最小尺寸
12  def image_pyramid(image, scale=1.5, minSize=(224, 224)):
13      # 第一次傳回原圖
14      yield image
15
16      # keep looping over the image pyramid
17      while True:
18          # 計算縮小後的尺寸
19          w = int(image.shape[1] / scale)
20          image = imutils.resize(image, width=w)
21
22          # 直到最小尺寸為止
23          if image.shape[0] < minSize[1] or image.shape[1] < minSize[0]:
24              break
25
26          # 傳回縮小後的圖像
27          yield image
```

6. 產生影像金字塔,並逐一進行視窗辨識。

```
1  # 輸出候選框
2  rois = []              # 候選框
3  locs = []              # 位置
4  SHOW_BOX = False       # 是否顯示要找的框
5
6  # 產生影像金字塔
```

8-17

第 8 章　物件偵測（Object Detection）

```
7  pyramid = image_pyramid(orig, scale=PYR_SCALE, minSize=ROI_SIZE)
8  # 逐一視窗辨識
9  for image in pyramid:
10     # 框與原圖的比例
11     scale = W / float(image.shape[1])
12
13     # 滑動視窗
14     for (x, y, roiOrig) in sliding_window(image, WIN_STEP, ROI_SIZE):
15         # 取得候選框
16         x = int(x * scale)
17         y = int(y * scale)
18         w = int(ROI_SIZE[0] * scale)
19         h = int(ROI_SIZE[1] * scale)
20
21         # 縮放圖形以符合模型輸入規格
22         roi = cv2.resize(roiOrig, INPUT_SIZE)
23         roi = img_to_array(roi)
24         roi = preprocess_input(roi)
25
26         # 加入輸出變數中
27         rois.append(roi)
28         locs.append((x, y, x + w, y + h))
29
30         # 是否顯示要找的框
31         if SHOW_BOX:
32             clone = orig.copy()
33             cv2.rectangle(clone, (x, y), (x + w, y + h),
34                 (0, 255, 0), 2)
35
36             # 顯示正在找的框
37             cv2.imshow("Visualization", clone)
38             cv2.imshow("ROI", roiOrig)
39             cv2.waitKey(0)
40
41 cv2.destroyAllWindows()
```

7. 預測：辨識機率必須大於設定值，並進行 NMS。

```
1  # 預測
2  MIN_CONFIDENCE = 0.9  # 辨識機率門檻值
3
4  rois = np.array(rois, dtype="float32")
5  preds = model.predict(rois)
6  preds = imagenet_utils.decode_predictions(preds, top=1)
7  labels = {}
8
9  # 檢查預測結果，機率須大於設定值
10 for (i, p) in enumerate(preds):
11     # grab the prediction information for the current ROI
12     (imagenetID, label, prob) = p[0]
13
14     # 機率大於設定值，則放入候選名單
15     if prob >= MIN_CONFIDENCE:
16         # 放入候選名單
17         box = locs[i]
```

8-3 方向梯度直方圖（HOG）

```
18              L = labels.get(label, [])
19              L.append((box, prob))
20              labels[label] = L
22  # 掃描每一個類別
23  for label in labels.keys():
24      # 複製原圖
25      clone = orig.copy()
26
27      # 畫框
28      for (box, prob) in labels[label]:
29          (startX, startY, endX, endY) = box
30          cv2.rectangle(clone, (startX, startY), (endX, endY),
31              (0, 255, 0), 2)
32
33      # 顯示 NMS(non-maxima suppression) 前的框
34      cv2.imshow("Before NMS", clone)
35      clone = orig.copy()
36
37      # NMS
38      boxes = np.array([p[0] for p in labels[label]])
39      proba = np.array([p[1] for p in labels[label]])
40      boxes = non_max_suppression(boxes, proba)
41
42      for (startX, startY, endX, endY) in boxes:
43          # 畫框及類別
44          cv2.rectangle(clone, (startX, startY), (endX, endY), (0, 255, 0), 2)
45          y = startY - 10 if startY - 10 > 10 else startY + 10
46          cv2.putText(clone, label, (startX, y),
47              cv2.FONT_HERSHEY_SIMPLEX, 0.45, (0, 255, 0), 2)
48
49      # 顯示
50      cv2.imshow("After NMS", clone)
51      cv2.waitKey(0)
52
53  cv2.destroyAllWindows()     # 關閉所有視窗
```

- 執行結果：因為圖中的斑馬有重疊，沒有偵測到。

- 試試看騎自行車的圖像，images_test/bike.jpg，結果也只抓到兩輛，後續的演算法可以改善這個缺點。

由於物件偵測的應用範圍很廣，因此，許多學者紛紛投入研究，提出各種改良的演算法，試圖提高準確率並加快辨識速度，接下來我們就遵循前人的研究軌跡，逐步深入探討。

8-3 R-CNN 系列演算法

滑動視窗並結合 HOG 的演算法雖然很好用，但是它還是有以下的缺點：

1. 滑動視窗加上影像金字塔，需要檢查的視窗個數太多了，耗時過久。
2. 一個 SVM 分類器只能偵測一個物件。
3. 簡單的 CNN 模型辨識並不準確，尤其是重疊的物件。

由於 CNN 以線條及輪廓為特徵辨識圖片效果極為顯著，因此，從 2014 年開始改良 CNN 的演算法陸續出現，用以偵測單張圖片內多個物件，演進過程如下：

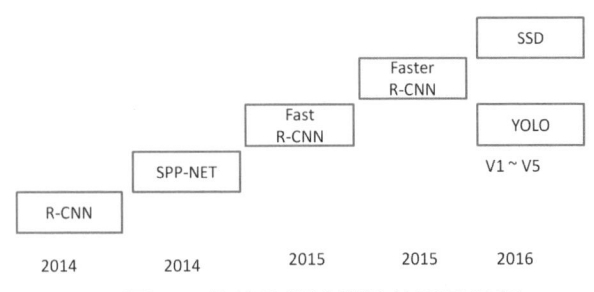

▲ 圖 8.6 物件偵測演算法的發展過程

註：至今（2024）YOLO 成為主流，已發展至 v11。

8-3 R-CNN 系列演算法

R-CNN 系列演算法主要是由 Ross B. Girshick 研發團隊於 2014 年提出，他們提出的第一個演算法為 Regions with CNN，簡稱 R-CNN，論文標題為『Rich feature hierarchies for accurate object detection and semantic segmentation』[12]。

步驟如下：

1. 讀取要辨識的圖片。
2. 使用區域推薦（Region Proposal）演算法，找到 2000 個候選區域（Regions of interest）。
3. 使用 CNN 對每一個候選區域萃取特徵。
4. 使用 SVM 辨識，以 CNN 萃取的特徵為 Input。

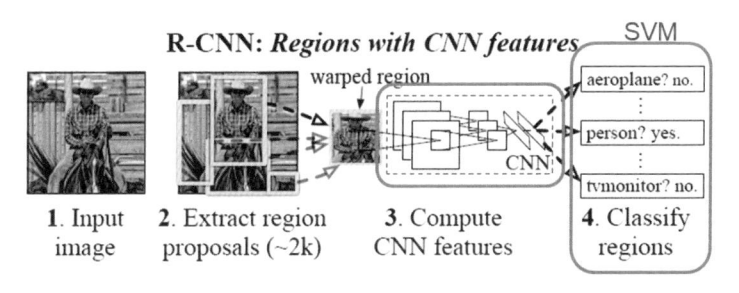

▲ 圖 8.7 R-CNN 架構，圖片來源：
Rich feature hierarchies for accurate object detection and semantic segmentation

完整的架構如下：

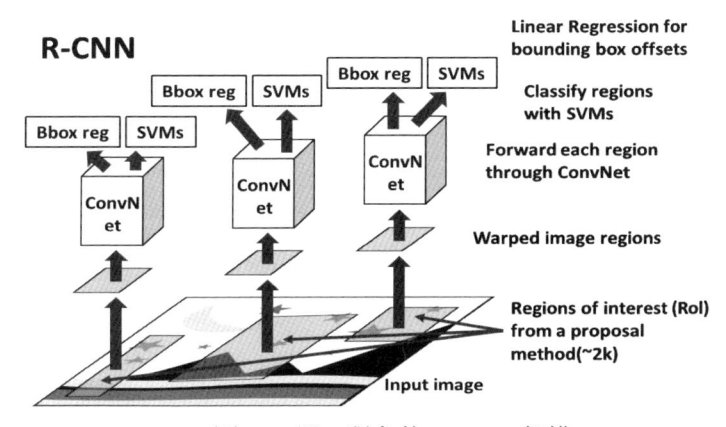

▲ 圖 8.8 另一視角的 R-CNN 架構

第 8 章 物件偵測（Object Detection）

處理流程如下：

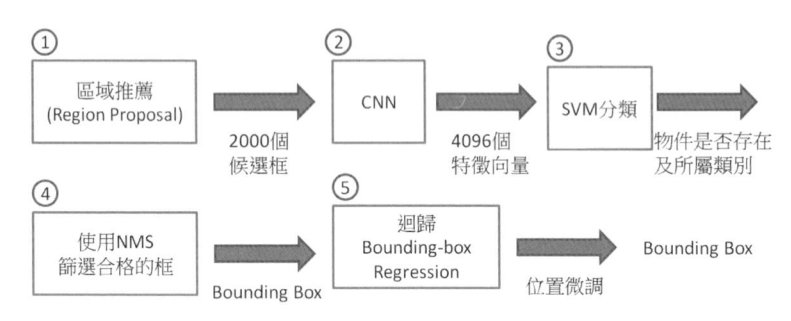

▲ 圖 8.9　R-CNN 處理流程

1. 區域推薦（Region Proposal）：改善滑動視窗需要檢查過多視窗，改用區域推薦演算法，只找出 2000 個候選區域（Regions of interest）輸入到模型。

    區域推薦（Region Proposal）也有多種演算法，R-CNN 所採用的是 Selective Search，它會依據顏色（color）、紋理（texture）、規模（Scale）、空間關係（Enclosure）來進行合併，接著再選取 2000 個最有可能包含物件的區域。

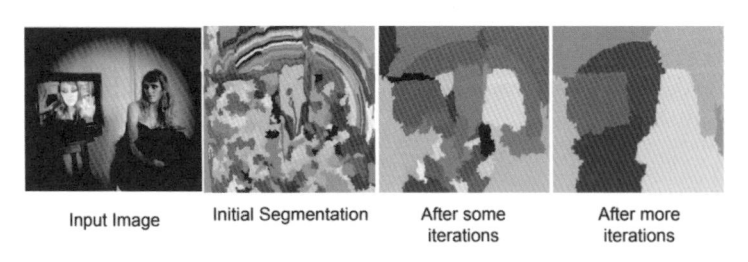

    ▲ 圖 8.10　區域推薦（Region Proposal）：最左邊的圖為原圖，將顏色、紋理、規模、空間關係相近的區域合併，最後變成最右邊圖的區域。

2. 特徵萃取（Feature Extractor）：將 2000 個候選區域使用影像變形轉換（Image Warping），轉成幾個固定尺寸的圖像，餵入不同的 CNN 進行特徵萃取，每個候選區域轉換成 4096 個元素的特徵向量。

3. SVM 分類器：輸入特徵向量，偵測物件是否存在與所屬的類別，注意，每一個類別使用不同的二分類 SVM。

4. 使用 Non-Maximum Suppression（NMS）篩選合格的框：上個步驟會找到許多的邊界框（Bounding Box），再利用 NMS 篩選可信度較高的邊界框，作法是計算與基準框的 IoU（Intersection-over Union）重疊的比例，IoU 值越高表示重疊度高，就可以把它們過濾掉，類似上一節的作法。

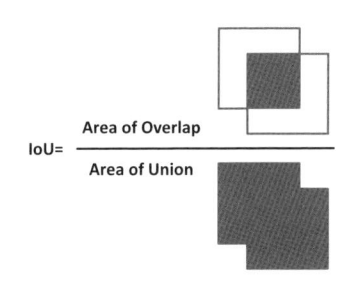

▲ 圖 8.11 IoU：分母為邊界框與目標框聯集的面積，分子邊界框為與目標框交集的面積

5. 位置微調：利用迴歸（Bounding-box Regression）微調邊界框的位置。

利用迴歸計算邊界框的四個變數：中心點（P_x, P_y）與寬高（P_w, P_h），其微調公式如下，G 為預估值。推論過程有點複雜，詳情可參考原文附錄 C。

$$\hat{G}_x = P_w d_x(P) + P_x \quad (1)$$
$$\hat{G}_y = P_h d_y(P) + P_y \quad (2)$$
$$\hat{G}_w = P_w \exp(d_w(P)) \quad (3)$$
$$\hat{G}_h = P_h \exp(d_h(P)). \quad (4)$$

損失函數如下，採用 Ridge Regression，以最小平方法估算出來的權重：

$$\mathbf{w}_\star = \operatorname*{argmin}_{\hat{\mathbf{w}}_\star} \sum_i^N (t_\star^i - \hat{\mathbf{w}}_\star^\mathsf{T} \phi_5(P^i))^2 + \lambda \|\hat{\mathbf{w}}_\star\|^2. \quad (5)$$

微調後的目標值 $t_\star$：

$$t_x = (G_x - P_x)/P_w \quad (6)$$
$$t_y = (G_y - P_y)/P_h \quad (7)$$
$$t_w = \log(G_w/P_w) \quad (8)$$
$$t_h = \log(G_h/P_h). \quad (9)$$

第 8 章　物件偵測（Object Detection）

整個 R-CNN 處理流程涉及相當多的演算法，包括：

1. 區域推薦（Region Proposal）：Selective Search。
2. 特徵萃取（Feature Extractor）：採用 AlexNet，也可使用其他 CNN 模型，如 VGG。
3. SVM 分類器：分類，判斷邊界框屬於哪一類的物件。
4. Non-Maximum Suppression（NMS）：去除冗餘的邊界框。
5. 迴歸（Bounding-box Regression）：微調邊界框。

範例. 使用 R-CNN 偵測空照圖中的飛機。

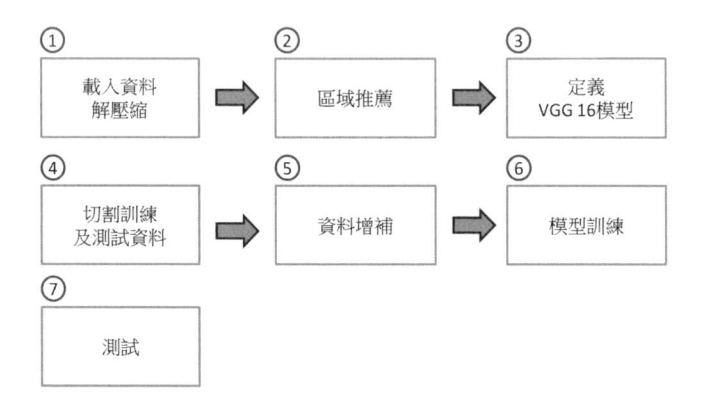

```
① 載入資料解壓縮  →  ② 區域推薦  →  ③ 定義 VGG 16 模型
④ 切割訓練及測試資料  →  ⑤ 資料增補  →  ⑥ 模型訓練
⑦ 測試
```

程式：**08_04_RCNN.ipynb**。程式執行有點久，若本機無 GPU 顯卡，可在 Colaboratory 執行。

1. 需安裝 OpenCV 擴展版：先解除安裝 OpenCV，再安裝擴展版，一般版與擴展版只能擇其一。

    pip uninstall opencv-contrib-python opencv-python -y

    pip install opencv-contrib-python

2. 自 RCNN GitHub[13] 下載訓練資料集（Images.zip、Airplanes_Annotations.zip）：可利用程式解壓縮圖像。

```
1  import zipfile
2  import os
3
4  # 圖像訓練資料
5  path_to_zip_file = './images_Object_Detection/Images.zip'
```

8-24

8-3 R-CNN 系列演算法

```
6  directory_to_extract_to = './images_Object_Detection/'
7
8  # 檢查目錄是否存在
9  if not os.path.isdir(directory_to_extract_to):
10     # 解壓縮
11     with zipfile.ZipFile(path_to_zip_file, 'r') as zip_ref:
12         zip_ref.extractall(directory_to_extract_to)
```

3. 解壓縮標記訓練資料。

```
1  # 標註訓練資料
2  path_to_zip_file = './images_Object_Detection/Airplanes_Annotations.zip'
3  directory_to_extract_to = './images_Object_Detection/'
4
5  # 檢查目錄是否存在
6  if not os.path.isdir(directory_to_extract_to):
7      # 解壓縮
8      with zipfile.ZipFile(path_to_zip_file, 'r') as zip_ref:
9          zip_ref.extractall(directory_to_extract_to)
```

4. 載入套件。

```
1  # 載入套件
2  import os,cv2
3  import pandas as pd
4  import matplotlib.pyplot as plt
5  import numpy as np
6  import tensorflow as tf
```

5. 顯示 1 張圖像訓練資料並包含標記。

```
1  # 設定圖像及標註目錄
2  path = "./images_Object_Detection/Images"
3  annot = "./images_Object_Detection/Airplanes_Annotations"
4
5  # 顯示1張圖像訓練資料含標註
6  for e,i in enumerate(os.listdir(annot)):
7      if e < 10:
8          # 讀取圖像
9          filename = i.split(".")[0]+".jpg"
10         print(filename)
11         img = cv2.imread(os.path.join(path,filename))
12         df = pd.read_csv(os.path.join(annot,i))
13         plt.axis('off')
14         plt.imshow(img)
15         for row in df.iterrows():
16             x1 = int(row[1][0].split(" ")[0])
17             y1 = int(row[1][0].split(" ")[1])
18             x2 = int(row[1][0].split(" ")[2])
19             y2 = int(row[1][0].split(" ")[3])
20             cv2.rectangle(img,(x1,y1),(x2,y2),(255,0,0), 2)
21         plt.figure()
```

8-25

第 8 章　物件偵測（Object Detection）

```
22          plt.axis('off')
23          plt.imshow(img)
24          break
```

- 執行結果：

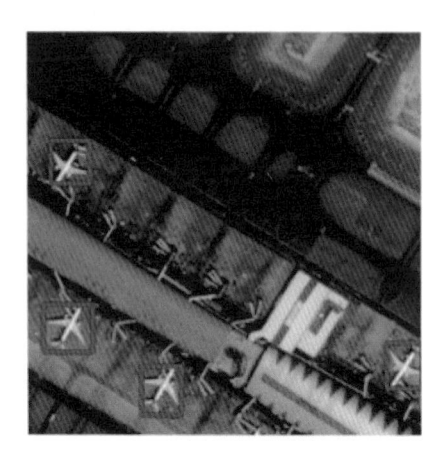

6. 區域推薦（Region Proposal）：使用 Selective Search 演算法，OpenCV 擴展版提供現成的函數 createSelectiveSearchSegmentation（假如要自行開發，可以參照『Lung-Ying Ling, R-CNN 學習筆記』[14] 一文的內容。

```
1   # 區域推薦(Region Proposal) :Selective Search
2   # 讀取圖像
3   im = cv2.imread(os.path.join(path,"42850.jpg"))
4
5   # Selective Search
6   cv2.setUseOptimized(True);
7   ss = cv2.ximgproc.segmentation.createSelectiveSearchSegmentation()
8   ss.setBaseImage(im)
9   ss.switchToSelectiveSearchFast()
10  rects = ss.process()
11
12  # 輸出
13  imOut = im.copy()
14  for i, rect in (enumerate(rects)):
15      x, y, w, h = rect
16  #     print(x,y,w,h)
17  #     imOut = imOut[x:x+w,y:y+h]
18      cv2.rectangle(imOut, (x, y), (x+w, y+h), (0, 255, 0), 1, cv2.LINE_AA)
19
20  plt.imshow(imOut)
```

- 執行結果：會依顏色、紋理等萃取出 2000 個候選區域（綠色）。

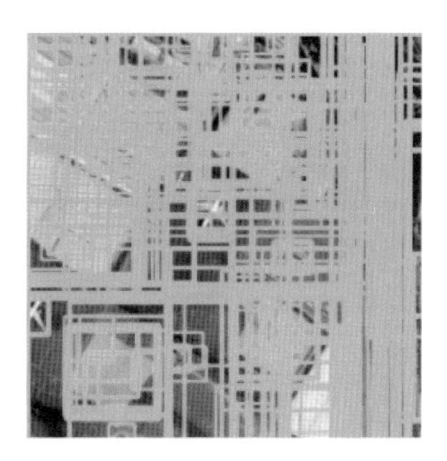

7. 定義 IoU 計算函數：計算兩個框的 IoU。

```
1   # 定義 IoU 計算函數
2   def get_iou(bb1, bb2):
3       assert bb1['x1'] < bb1['x2']
4       assert bb1['y1'] < bb1['y2']
5       assert bb2['x1'] < bb2['x2']
6       assert bb2['y1'] < bb2['y2']
7
8       x_left = max(bb1['x1'], bb2['x1'])
9       y_top = max(bb1['y1'], bb2['y1'])
10      x_right = min(bb1['x2'], bb2['x2'])
11      y_bottom = min(bb1['y2'], bb2['y2'])
12
13      if x_right < x_left or y_bottom < y_top:
14          return 0.0
15
16      intersection_area = (x_right - x_left) * (y_bottom - y_top)
17
18      bb1_area = (bb1['x2'] - bb1['x1']) * (bb1['y2'] - bb1['y1'])
19      bb2_area = (bb2['x2'] - bb2['x1']) * (bb2['y2'] - bb2['y1'])
20
21      iou = intersection_area / float(bb1_area + bb2_area - intersection_area)
22      assert iou >= 0.0
23      assert iou <= 1.0
24      return iou
```

8. 篩選訓練資料：找出檔名為 airplane 開頭的檔案，並使用區域推薦將每個檔案各取出 2000 個候選區域。要留意的是，每個檔案必須包含 30 個以上的正樣本（IoU>70%）與 30 個以上的負樣本（IoU<30%），才能被列為訓練資料。

第 8 章　物件偵測（Object Detection）

```
1   # 篩選訓練資料
2
3   # 儲存正樣本及負樣本的候選框
4   train_images=[]
5   train_labels=[]
6
7   # 掃描每一個標註
8   for e,i in enumerate(os.listdir(annot)):
9       try:
10          # 取得飛機的圖像
11          if i.startswith("airplane"):
12              filename = i.split(".")[0]+".jpg"
13              print(e,filename)
14
15              # 讀取標註檔案
16              image = cv2.imread(os.path.join(path,filename))
17              df = pd.read_csv(os.path.join(annot,i))
18
19              # 取得所有標註的座標
20              gtvalues=[]
21              for no, row in df.iterrows():
22                  x1 = int(row.iloc[0].split(" ")[0])
23                  y1 = int(row.iloc[0].split(" ")[1])
24                  x2 = int(row.iloc[0].split(" ")[2])
25                  y2 = int(row.iloc[0].split(" ")[3])
26                  gtvalues.append({"x1":x1,"x2":x2,"y1":y1,"y2":y2})
```

```
28              # 區域推薦
29              ss.setBaseImage(image)
30              ss.switchToSelectiveSearchFast()
31              ssresults = ss.process()
32              imout = image.copy()
33
34              # 初始化
35              counter = 0          # 正樣本筆數
36              falsecounter = 0     # 負樣本筆數
37              flag = 0             # 1:正負樣本筆數均 >= 30
38              fflag = 0            # 1:正樣本筆數 >= 30
39              bflag = 0            # 1:負樣本筆數 >= 30
40
41              # 掃描每一個候選框
42              for e,result in enumerate(ssresults):
43                  if e < 2000 and flag == 0:
44                      for gtval in gtvalues:
45                          x,y,w,h = result
46                          # 比較區域推薦區域與標註的 IoU
47                          iou = get_iou(gtval,{"x1":x,"x2":x+w,"y1":y,"y2":y+h})
48
49                          # 收集30筆正樣本
50                          if counter < 30:
51                              if iou > 0.70:
52                                  timage = imout[y:y+h,x:x+w]
53                                  resized = cv2.resize(timage, (224,224),
54                                              interpolation = cv2.INTER_AREA)
```

8-28

```
55                        train_images.append(resized)
56                        train_labels.append(1)
57                        counter += 1
58              else :
59                  fflag =1
60
61              # 收集30筆負樣本
62              if falsecounter <30:
63                  if iou < 0.3:
64                      timage = imout[y:y+h,x:x+w]
65                      resized = cv2.resize(timage, (224,224),
66                                        interpolation = cv2.INTER_AREA)
67                      train_images.append(resized)
68                      train_labels.append(0)
69                      falsecounter += 1
70              else :
71                  bflag = 1
72
73              # 超過30筆正樣本及負樣本，表有物件在框裡面
74              if fflag == 1 and bflag == 1:
75                  print("inside")
76                  flag = 1
77      except Exception as e:
78          print(e)
79          print("error in "+filename)
80          continue
```

9. 定義模型：使用 VGG 16，加上自訂的神經層。

```
1  from tensorflow.keras.layers import Dense
2  from tensorflow.keras import Model
3  from tensorflow.keras import optimizers
4  from tensorflow.keras.preprocessing.image import ImageDataGenerator
5  from tensorflow.keras.applications.vgg16 import VGG16
6
7  vggmodel = VGG16(weights='imagenet', include_top=True)
8
9  # VGG16 前端的神經層不重作訓練
10 for layers in (vggmodel.layers)[:15]:
11     print(layers)
12     layers.trainable = False
13
14 # 接自訂神經層作辨識
15 X= vggmodel.layers[-2].output
16 predictions = Dense(2, activation="softmax")(X)
17 model_final = Model(inputs = vggmodel.input, outputs = predictions)
18
19 # 訂定損失函數、優化器、效能衡量指標
20 from tensorflow.keras.optimizers import Adam
21 opt = Adam(lr=0.0001)
22 model_final.compile(loss = tf.keras.losses.categorical_crossentropy,
23                     optimizer = opt, metrics=["accuracy"])
24 model_final.summary()
```

第 8 章 物件偵測（Object Detection）

10. 定義轉換函數：將標記 Y 轉為二個變數。

```
1  # 定義函數，將標記 Y 轉為二個變數，
2  from sklearn.preprocessing import LabelBinarizer
3
4  class MyLabelBinarizer(LabelBinarizer):
5      def transform(self, y):
6          Y = super().transform(y)
7          if self.y_type_ == 'binary':
8              return np.hstack((Y, 1-Y))
9          else:
10             return Y
11     def inverse_transform(self, Y, threshold=None):
12         if self.y_type_ == 'binary':
13             return super().inverse_transform(Y[:, 0], threshold)
14         else:
15             return super().inverse_transform(Y, threshold)
```

11. 前置處理及訓練資料 / 測試資料分割。

```
1  # 資料前置處理，切割訓練及測試資料
2  from sklearn.model_selection import train_test_split
3
4  # 筆者 PC 記憶體不足，只取 10000
5  X_new = np.array(train_images[:10000])
6  y_new = np.array(train_labels[:10000])
7
8  # 標記 Y 轉為二個變數，
9  lenc = MyLabelBinarizer()
10 Y =  lenc.fit_transform(y_new)
11
12 # 切割訓練及測試資料
13 X_train, X_test , y_train, y_test = train_test_split(X_new, Y, test_size=0.10)
14 print(X_train.shape,X_test.shape,y_train.shape,y_test.shape)
```

12. 進行資料增補（Data Augmentation），以提高模型準確率，因為飛機停放的方向可能會有偏斜。

```
1  # 資料增補(Data Augmentation)
2  trdata = ImageDataGenerator(horizontal_flip=True,
3                              vertical_flip=True, rotation_range=90)
4  traindata = trdata.flow(x=X_train, y=y_train)
5  tsdata = ImageDataGenerator(horizontal_flip=True,
6                              vertical_flip=True, rotation_range=90)
7  testdata = tsdata.flow(x=X_test, y=y_test)
```

13. 模型訓練：原作者訓練週期（epoch）達 1000 次之多，故設定檢查點與提前結束的 Callback，以縮短訓練時間，筆者只測試 20 epochs。

8-30

8-3 R-CNN 系列演算法

```
1  # 模型訓練
2  from tensorflow.keras.callbacks import ModelCheckpoint, EarlyStopping
3  # 定義模型存檔及提早結束的 Callback
4  checkpoint = ModelCheckpoint("ieeercnn_vgg16_1.keras", monitor='val_loss',
5                               verbose=1, save_best_only=True,
6                               save_weights_only=False, mode='auto')
7  early = EarlyStopping(monitor='val_loss', min_delta=0, patience=100,
8                        verbose=1, mode='auto')
9
10 # 模型訓練,節省時間,只訓練 20 epochs,正式專案還是要訓練較多週期
11 # hist = model_final.fit_generator(generator= traindata, steps_per_epoch= 10,
12 #         epochs= 1000, validation_data= testdata, validation_steps=2,
13 #         callbacks=[checkpoint,early])
14 hist = model_final.fit(traindata, steps_per_epoch= 10,
15         epochs= 20, validation_data= testdata, validation_steps=2,
16         callbacks=[checkpoint,early])
```

14. 繪製模型訓練過程的準確率。

```
1  # 繪製模型訓練過程的準確率
2  import matplotlib.pyplot as plt
3  plt.plot(hist.history['accuracy'])
4  plt.plot(hist.history['val_accuracy'])
5  plt.ylabel("Accuracy")
6  plt.xlabel("Epoch")
7  plt.legend(["Accuracy","Validation Accuracy"])
8  plt.show()
```

- 執行結果:準確率並未穩定上升,這表示訓練週期不足,由於筆者只著重在演算法的研究,所以沒有繼續訓練下去,如果用於正式專案務必多訓練幾個週期比較妥當。

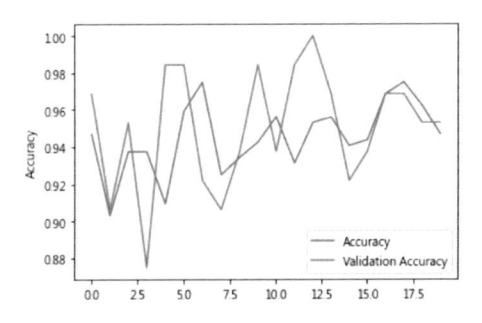

15. 任選一張圖片測試。

```
1  # 任選一張圖片測試
2  im = X_test[100]
```

第 8 章　物件偵測（Object Detection）

```python
3  plt.imshow(im)
4  img = np.expand_dims(im, axis=0)
5  out= model_final.predict(img)
6
7  # 顯示預測結果
8  if out[0][0] > out[0][1]:
9      print("有飛機")
10 else:
11     print("沒有飛機")
```

- 執行結果：圖片有偵測到飛機。

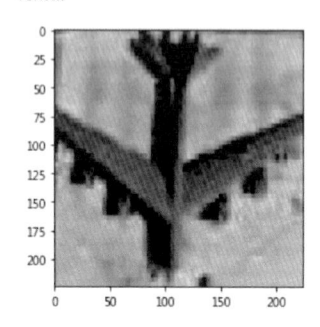

有飛機

16. 測試所有檔名為 4 開頭的檔案。

```python
1  # 測試所有檔名為 4 開頭的檔案
2  z=0
3  for e,i in enumerate(os.listdir(path)):
4      if i.startswith("4"):
5          z += 1
6          img = cv2.imread(os.path.join(path,i))
7          # 區域推薦
8          ss.setBaseImage(img)
9          ss.switchToSelectiveSearchFast()
10         ssresults = ss.process()
11         imout = img.copy()
12
13         # 物件偵測
14         for e,result in enumerate(ssresults):
15             if e < 2000:
16                 x,y,w,h = result
17                 timage = imout[y:y+h,x:x+w]
18                 resized = cv2.resize(timage, (224,224), interpolation = cv2.INTER_AREA)
19                 img = np.expand_dims(resized, axis=0)
20                 out= model_final.predict(img)
21
22                 # 機率 > 0.65 才算偵測到飛機
23                 if out[0][0] > 0.65:
24                     cv2.rectangle(imout, (x, y), (x+w, y+h), (0, 255, 0), 1, cv2.LINE_AA)
25         plt.figure()
26         plt.imshow(imout)
```

8-32

8-3 R-CNN 系列演算法

- 執行結果：下面這張圖片有偵測到飛機，但其他部分的圖片並沒有正確偵測到，是因為筆者設定的訓練週期不足。

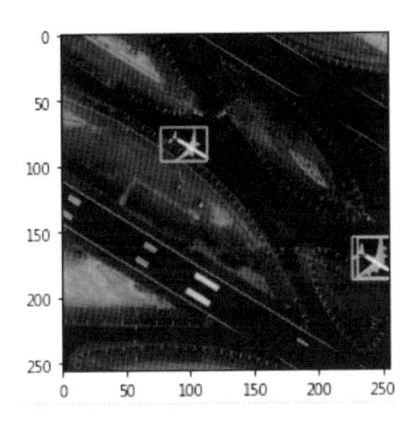

以上的範例並未使用 Non-Maximum Suppression（NMS）、Bounding-box Regression。

R-CNN 依然不盡理想的原因如下：

1. 每張圖經由區域推薦處理過後，各會產生出 2000 個候選區域，然後每個框都需經過辨識，執行時間還是過長。
2. 區域推薦不具備自我學習能力，不會隨著模型訓練越來越精準。
3. 透過 CNN 模型萃取 4096 個特徵向量，合計有 2000 x 4096 = 8,192,000 個特徵向量，記憶體消耗也很大。
4. 每筆資料都要經過 CNN、SVM、迴歸三個模型的訓練與預測，過於複雜。
5. 演算分兩階段執行：區域推薦及物件辨識，並非一貫化（End-to-End），可能只找到區域最佳解。

總體而論，物件偵測不只追求高準確率，更要求能夠即時偵測，像是自駕車，總不能等撞到障礙物後才偵測到，那就悲劇了。原作者雖然以 Caffe（C++）開發 R-CNN，企圖縮短偵測時間，但仍需要 40 多秒才能偵測一張圖像，因此引發一波演算法的改良浪潮，參閱圖 8.6。

8-33

8-4 R-CNN 改良

接著 Kaiming He 等學者提出 SPP-Net（Spatial Pyramid Pooling in Deep Convo-lutional Networks for Visual Recognition）演算法 [15]，針對 R-CNN 把每個候選區域都需要變形轉換（Warp）成一致的尺寸，才能輸入 CNN 的缺點進行改良，作法如下：

1. R-CNN 每一個尺寸候選區域都需轉換為固定尺寸才能輸入 CNN 模型，各個區域的長寬不一，非等比例的轉換會造成準確度降低，SPP-Net 作者所提出『Spatial pyramid pooling』（SPP）神經層，各種尺寸的輸入圖像都能產生一個固定長度的輸出，將之放在最後一個卷積層的後面，負責轉換成固定長度的特徵萃取。
2. 其他的處理與 R-CNN 類似。

R-CNN 與 SPP-Net 的模型結構比較，示意圖如下：

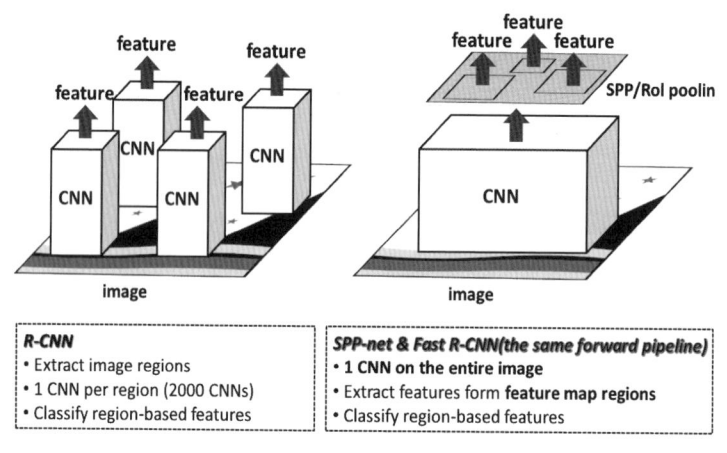

▲ 圖 8.12 R-CNN 與 SPP-Net 模型結構比較

SPP 還是有缺點：

1. 雖然解決了 CNN 計算過多的狀況，但沒有處理分類（SVM）與迴歸過慢的問題。
2. 特徵向量太占記憶體空間。

SPP 詳細處理流程可參閱論文，中文說明可參閱『SPP-Net 論文詳解』[16]。接著 Ross B. Girshick 團隊陸續再提出 Fast R-CNN、Faster R-CNN 等演算法。

8-4 R-CNN 改良

Fast R-CNN 作法：

1. 將整個圖像直接經由 CNN 轉成特徵向量，不再使用 2000 個候選區域輸入 CNN。
2. 自訂一個 RoI（Region of Interest）池化層，透過候選區域在整個圖像的所在位置，換算出每個候選區域的特徵向量。
3. 其他流程與 R-CNN 類似。

Fast R-CNN 模型結構如下：

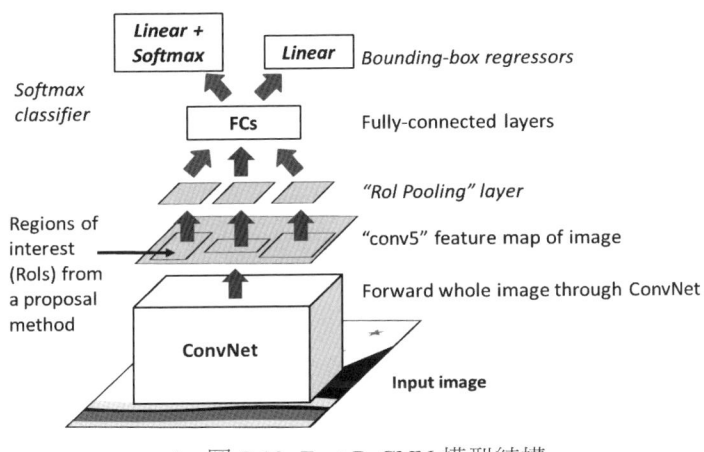

▲ 圖 8.13 Fast R-CNN 模型結構

優點：

1. CNN 模型只需訓練原圖就好，不用訓練 2000 個候選區域。
2. 透過 ROI pooling 得到固定尺寸的特徵後，只要連接一個 Dense 進行分類即可。

由於使用區域推薦演算法找 2000 張候選區域，還是太耗時，Ross B. Girshick 決定放棄使用 selective search，引進 RPN（Region Proposal Network）神經層，開發 Faster R-CNN 模型，CNN 輸出的特徵圖（Feature Map）同時提供 RPN 及分類器使用，可以同步處理，大大提高執行速度，可參照下圖說明。

第 8 章　物件偵測（Object Detection）

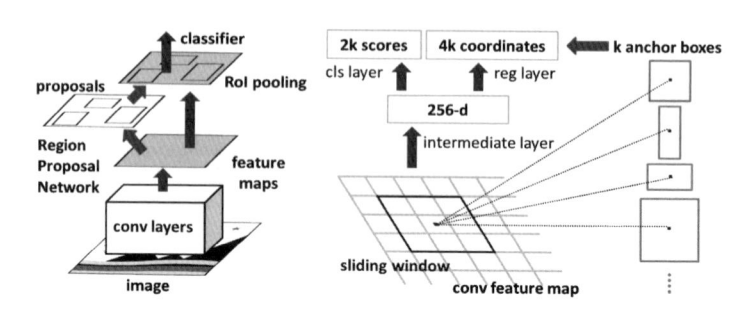

▲ 圖 8.14　Faster R-CNN 模型結構

其中 RPN 會依據 CNN 輸出的特徵圖產生固定幾種尺寸的 Anchor Box，作為候選視窗，不再使用 Selective Search 費力的找尋 2000 個候選區域。

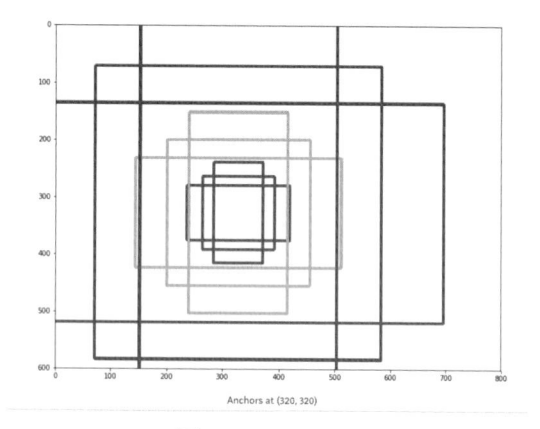

▲ 圖 8.15　Anchor Box

雖然 Ross B. Girshick 在 GitHub 放上 Faster R-CNN 程式碼[17]，但安裝不僅超複雜，執行環境的要求也很高（Caffe/C++），建議讀者改用 Detectron 套件測試，它也是 Ross B. Girshick 開發的套件，目前已開發至第二版（Detectron 2），它使用 PyTorch 框架，只能安裝在 Linux/Mac 環境，Windows 使用者可以在 Colaboratory 上進行測試。詳細使用說明請參閱『Getting Started with Detectron2』[18]，可直接測試內文的範例[19]。

範例. 使用 Detectron2 套件進行物件偵測。

程式：**Detectron2_Tutorial.ipynb**，請在 **Colaboratory** 上執行，記得要在選單『執行階段』選取 **GPU**。

1. 安裝 Detectron2 套件。

```
1  !python -m pip install pyyaml==5.1
2  import sys, os, distutils.core
3  # Note: This is a faster way to install detectron2 in Colab, but it does not include all functionalities
4  # See https://detectron2.readthedocs.io/tutorials/install.html for full installation instructions
5  !git clone 'https://github.com/facebookresearch/detectron2'
6  dist = distutils.core.run_setup("./detectron2/setup.py")
7  !python -m pip install {' '.join([f"'{x}'" for x in dist.install_requires])}
8  sys.path.insert(0, os.path.abspath('./detectron2'))
9
10 # Properly install detectron2. (Please do not install twice in both ways)
11 # !python -m pip install 'git+https://github.com/facebookresearch/detectron2.git'
```

2. 自 Model Zoo 下載 Detectron2 預先訓練的模型,進行物件偵測。

```
cfg = get_cfg()
# add project-specific config (e.g., TensorMask) here if you're not running a model in detectron2's core library
cfg.merge_from_file(model_zoo.get_config_file("COCO-InstanceSegmentation/mask_rcnn_R_50_FPN_3x.yaml"))
cfg.MODEL.ROI_HEADS.SCORE_THRESH_TEST = 0.5  # set threshold for this model
# Find a model from detectron2's model zoo. You can use the https://dl.fbaipublicfiles... url as well
cfg.MODEL.WEIGHTS = model_zoo.get_checkpoint_url("COCO-InstanceSegmentation/mask_rcnn_R_50_FPN_3x.yaml")
predictor = DefaultPredictor(cfg)
outputs = predictor(im)
```

3. 顯示物件偵測結果。

第 8 章 物件偵測（Object Detection）

- 執行結果：效果超好，非常厲害，就連背景中旁觀的人群都可以被正確偵測。

4. 上傳之前用 ResNet50 偵測失敗的斑馬照片測試看看。

```
# Upload the results
from google.colab import files
files.upload()
```

5. 讀取檔案，進行物件偵測。

```
# 讀取檔案，進行物件偵測
im = cv2.imread("./zebra.jpg")
cv2_imshow(im)
predictor = DefaultPredictor(cfg)
outputs = predictor(im)
outputs
```

- 執行結果：偵測到 3 個物件及 3 個信賴度，數值都相當高。

    [46.5412, 94.6141, 234.9006, 258.9107],

    [180.8245, 86.8508, 418.6142, 261.7740],

    [342.8438, 103.8605, 563.8304, 266.2300]

    信賴度：[0.9992, 0.9986, 0.9983]。

6. 顯示物件偵測結果。

```
v = Visualizer(im[:, :, ::-1], MetadataCatalog.get(cfg.DATASETS.TRAIN[0]), scale=1.2)
out = v.draw_instance_predictions(outputs["instances"].to("cpu"))
cv2_imshow(out.get_image()[:, :, ::-1])
```

- 執行結果：

這個套件真的超強,除了成功偵測到所有物件外,也可以進行實例分割(Instance Segmentation),掃描到的物件不僅有邊界框(Bounding Box),還有準確的遮罩(Mask)。

檔案後面還示範了以下功能:

1. 使用自訂的資料集,偵測自己有興趣的物件。在 Colaboratory 上訓練只需幾分鐘的時間就可以完成。
2. 人體骨架的偵測。
3. 全景視訊的物件偵測(筆者測試時有出現錯誤)。

Meta 公司除了開發這個套件外,還有一個更新的專案 Segment Anything Model 2(SAM 2)[20],功能更強大,值得一試。專案內的 notebooks 資料夾含有圖片、視訊的物件偵測範例,建議要在 A100 GPU 卡執行,免費的 Colaboratory 上執行可能會發生記憶體不足的現象,需使用付費的 Colaboratory 才能使用 A100 GPU。

8-5 YOLO 演算法簡介

由於 R-CNN 系列的演算法分為兩階段(Two Stage),第一階段先利用區域推薦找出候選區域,第二階段才進行物件偵測,所以在偵測速度上始終是一個瓶頸,難以滿足即時偵測的要求,後續學者就提出一階段(Single Shot)的演算法,主要有兩類,YOLO(You Only Look Once)及 SSD(Single Shot MultiBox Detector)。

R-CNN 經過一連串的改良後,物件偵測的速度比較如下表,最新版速度比原版增快了 250 倍。

	R-CNN	Fast R-CNN	Faster R-CNN
Test Time per Image	50 Seconds	2 Seconds	0.2 Seconds
Speed Up	1x	25x	250x

▲ 圖 8.16 R-CNN 各演算法之物件偵測的速度

看似很好了,然而 YOLO 發明人 Joseph Redmon 在 2016 年的 CVPR 研討會(You Only Look Once: Unified, Real-Time Object Detection)[21] 中有兩張投影片非常有意思,一輛轎車平均車身長約 8 英呎(Feet),假如使用 Faster R-CNN 偵測下一個路況的話,車

第 8 章　物件偵測（Object Detection）

子早已行駛了 12 英呎，也就是車子又開了 1 又 1/2 個車身的距離，相對的，如果使用 YOLO 偵測下一個路況，車子則只行駛了 2 英呎，即 1/4 個車身的距離，安全性是否會提高許多？相信答案已不言而喻，非常有說服力。

	Pascal 2007 mAP	Speed	
DPM v5	33.7	.07 FPS	14 s/img
R-CNN	66.0	.05 FPS	20 s/img
Fast R-CNN	70.0	.5 FPS	2 s/img
Faster R-CNN	73.2	7 FPS	140 ms/img

▲ 圖 8.17　Faster R-CNN 演算法物件偵測的速度

	Pascal 2007 mAP	Speed	
DPM v5	33.7	.07 FPS	14 s/img
R-CNN	66.0	.05 FPS	20 s/img
Fast R-CNN	70.0	.5 FPS	2 s/img
Faster R-CNN	73.2	7 FPS	140 ms/img
YOLO	63.4	45 FPS	22 ms/img

▲ 圖 8.18　YOLO 演算法物件偵測的速度

YOLO（You Only Look Once）是現在最夯的物件偵測演算法，於 2016 年由 Joseph Redmon 提出，他本人開發至第三版，但因某些因素離開此研究領域，其他學者繼續接手，直至 2024 年已開發到第 11 版了。其中台灣中研院王建堯博士也開發好幾個版本，可謂台灣之光，相關新聞可參見『人工智慧技術 YOLOv4 出自中研院團隊』[22] 報導。

- v1 2015 Joseph Redmon 提出。
- v2 2016 加入 batch normalization, anchor boxes, and dimension clusters.
- v3 2018 使用 more efficient backbone network, multiple anchors and spatial pyramid pooling 加強模型效能。
- v4 2020 中研院，引進 Mosaic data augmentation, a new anchor-free detection head, and a new loss function.
- v5 2020 Ultralytics, 改善 hyperparameter optimization, integrated experiment tracking and automatic export to popular export formats.
- v6 2022 美團提出。
- v7 2022 中研院，增加額外應用(tasks)：pose estimation on the COCO keypoints dataset.
- v8 2023 Ultralytics, 支援更多應用(tasks).
- v9 2023 中研院,引進 Programmable Gradient Information (PGI) and the Generalized Efficient Layer Aggregation Network (GELAN).
- v10 2023 中國清華大學使用 Ultralytics 套件加強即時偵測，並免除 NMS 處理。
- v11 2024 Ultralytics 加強模型在各種應用的效能。

▲ 圖 8.19　YOLO 版本演進，資料來源：Ultralytics 網站首頁 [23]

8-5 YOLO 演算法簡介

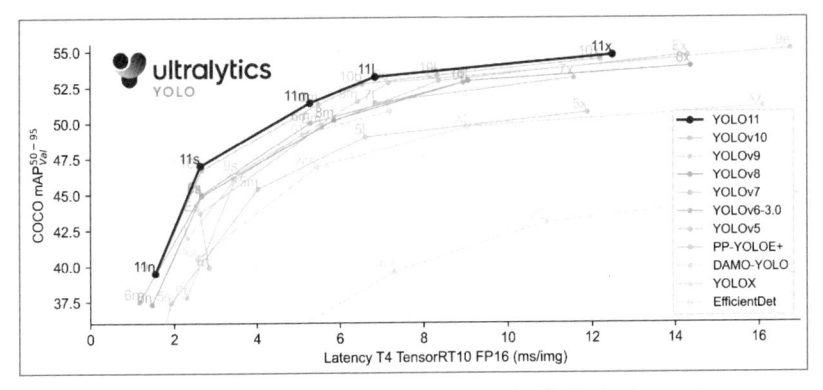

YOLO 各版本的平均準確度（mAP）與速度的比較，如下圖所示：

▲ 圖 8.20 YOLO 各版本（v5~v11）準確度（mAP）與速度的的比較，圖片來源：Ultralytics GitHub [24]

YOLO 的速度部份是犧牲準確率換來的，作法如下：

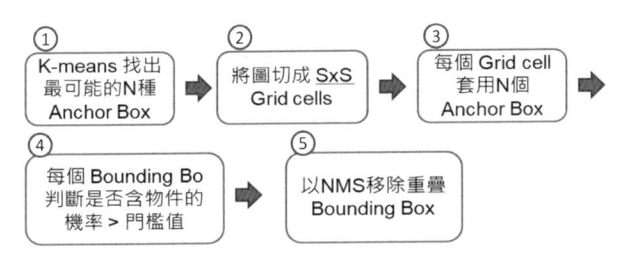

▲ 圖 8.21 YOLO 的處理流程

放棄區域推薦，以集群演算法 K-Means，從訓練資料中找出最常見的 N 種尺寸的 Anchor Box。

1. 直接將圖像劃分成（s, s）個網格（Grid）：每個網格只檢查多種不同尺寸的 Anchor Box 是否含有物件而已。
2. 輸入 CNN 模型，計算每個 Anchor Box 含有物件的機率。
3. 同時計算每一個網格可能有各種物件的機率，假設每一網格最多只含一個物件。
4. 合併步驟 3、4 的資訊，並找出合格的候選區域。
5. 以 NMS 移除重疊 Bounding Box。

第 8 章　物件偵測（Object Detection）

觀察下面示意圖，有助於 YOLO 的理解。

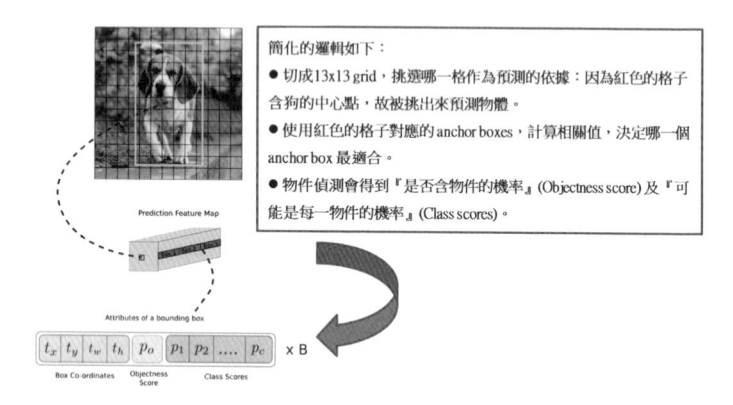

▲ 圖 8.22　YOLO 物件偵測的簡單流程

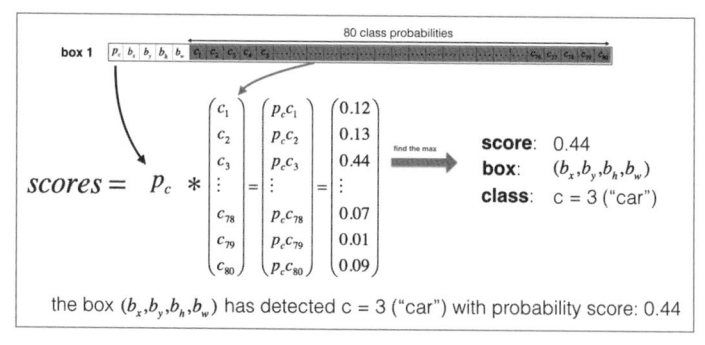

▲ 圖 8.23　每個 anchor box 的預測結果

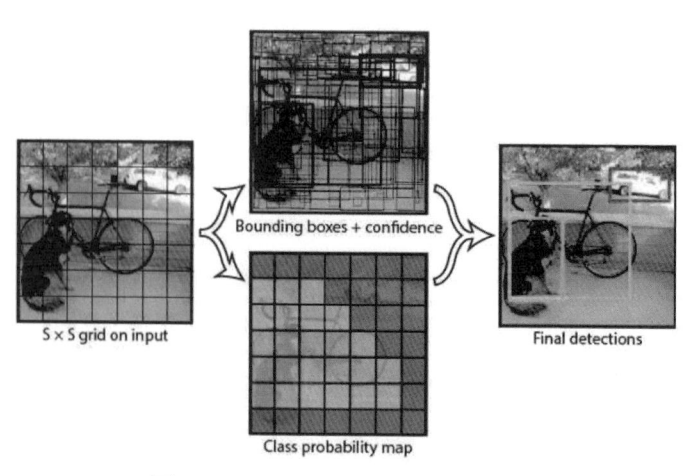

▲ 圖 8.24　YOLO 整體處理流程示意圖，
圖片來源：You Only Look Once: Unified, Real-Time Object Detection [25]

8-6 YOLO 訓練與推論

YOLO 前 4 版使用 C/C++ 建構 Darknet 基礎平台，再提供 Python API（或稱 Python binding）訓練模型及推論，程式建置較為複雜，第 5 版之後就全部改用 Python，以 PyTorch 作為開發工具，安裝及訓練就簡單許多，不過，就無法使用 TensorFlow/Keras 呼叫了，因此，以下章節要請讀者先安裝 PyTorch，安裝說明請參閱 PyTorch 官網首頁 [26]，按選單操作，即可產生安裝指令，注意，產生的安裝指令中的 pip3 應改為 pip。

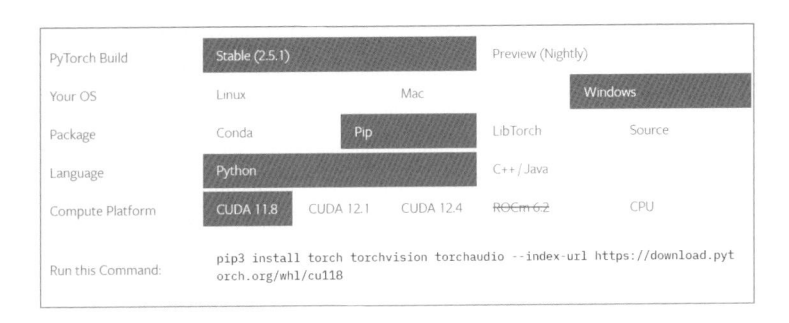

PyTorch 安裝後，即可開始測試 YOLO，以下採用 Ultralytics 開發的 v11 版本，文件說明較完整，安裝程序也較簡單。

YOLO v11 安裝指令如下：

- pip install ultralytics

註：可將 OpenCV 還原為正常版本：

- pip uninstall opencv-contrib-python opencv-python -y
- pip install opencv-python

8-7 YOLO 各項功能

依照 Ultralytics 說明文件 [27]，YOLO 支援的任務（Tasks）包括：

1. 分類（Classification）：偵測單一物件（Single Object）的類別。
2. 物件偵測（Object Detection）：偵測多個物件（Multiple Object）的類別與所在的位置。
3. 語義分割（Semantic Segmentation）：按照物件類別標示不同顏色的像素。

第 8 章　物件偵測（Object Detection）

4. 姿態辨識（Pose Estimation）：偵測人體重要的關鍵點（Keypoints），藉以判斷姿勢正確與否，或是站立 / 坐下。
5. 物件方向偵測（Oriented Object Detection, OBB）：偵測物件面對的方向，標記的邊界框（Bounding Box）是旋轉的矩形，與一般的物件偵測不同，後者永遠是垂直的矩形，OBB 標記的位置更精準。

依據以上任務類別，可衍生出各種解決方案：

- Object Counting: Learn to perform real-time object counting with YOLO11. Gain the expertise to accurately count objects in live video streams.
- Object Cropping: Master object cropping with YOLO11 for precise extraction of objects from images and videos.
- Object Blurring: Apply object blurring using YOLO11 to protect privacy in image and video processing.
- Workouts Monitoring: Discover how to monitor workouts using YOLO11. Learn to track and analyze various fitness routines in real time.
- Objects Counting in Regions: Count objects in specific regions using YOLO11 for accurate detection in varied areas.
- Security Alarm System: Create a security alarm system with YOLO11 that triggers alerts upon detecting new objects. Customize the system to fit your specific needs.
- Heatmaps: Utilize detection heatmaps to visualize data intensity across a matrix, providing clear insights in computer vision tasks.
- Instance Segmentation with Object Tracking NEW: Implement instance segmentation and object tracking with YOLO11 to achieve precise object boundaries and continuous monitoring.
- VisionEye View Objects Mapping: Develop systems that mimic human eye focus on specific objects, enhancing the computer's ability to discern and prioritize details.
- Speed Estimation: Estimate object speed using YOLO11 and object tracking techniques, crucial for applications like autonomous vehicles and traffic monitoring.
- Distance Calculation: Calculate distances between objects using bounding box centroids in YOLO11, essential for spatial analysis.
- Queue Management: Implement efficient queue management systems to minimize wait times and improve productivity using YOLO11.
- Parking Management: Organize and direct vehicle flow in parking areas with YOLO11, optimizing space utilization and user experience.
- Analytics: Conduct comprehensive data analysis to discover patterns and make informed decisions, leveraging YOLO11 for descriptive, predictive, and prescriptive analytics.
- Live Inference with Streamlit: Leverage the power of YOLO11 for real-time object detection directly through your web browser with a user-friendly Streamlit interface.
- Track Objects in Zone NEW: Learn how to track objects within specific zones of video frames using YOLO11 for precise and efficient monitoring.

▲ 圖 8.25　YOLO 解決方案 [28]

所有功能的操作程序都很類似：

1. 下載 YOLO 權重檔：每種任務的權重檔均不相同，且分為特大 / 大 / 中 / 小 / 微（x/l/m/s/n）等不同尺寸的檔案。
2. 建立 YOLO 模型物件，並載入 YOLO 權重檔。
3. 輸入圖檔或視訊檔，進行推論。

4. 顯示輸出結果：包括類別、邊界框、關鍵點⋯等。

8-8 圖像分類（Image Classification）

分類是最簡單的任務，一張圖片只有一個物件，預設分類模型是以 ImageNet 資料集訓練而成的，可辨識 1000 種物件，與預先訓練模型類似，請參考上一章說明，其他任務則是以 COCO 資料集訓練而成的，可辨識 80 種物件，類別可參閱 coco.yaml[29]。

1	人	21	大象	41	紅酒杯	61	餐桌
2	自行車	22	熊	42	杯子	62	廁所
3	汽車	23	斑馬	43	叉子	63	電視
4	機車	24	長頸鹿	44	刀子	64	筆電
5	飛機	25	背包	45	湯匙	65	滑鼠
6	公車	26	傘	46	碗	66	遙控器
7	火車	27	手提包	47	香蕉	67	鍵盤
8	卡車	28	領帶	48	蘋果	68	手機
9	船	29	手提箱	49	三明治	69	微波爐
10	紅綠燈	30	飛盤	50	柳丁	70	烤箱
11	消防栓	31	雙板滑雪板	51	花椰菜	71	烤麵包機
12	停止標誌	32	單板滑雪板	52	紅蘿蔔	72	水槽
13	停車收費碼錶	33	運動類用球	53	熱狗	73	冰箱
14	長椅	34	風箏	54	比薩	74	書
15	鳥	35	棒球棒	55	甜甜圈	75	時鐘
16	貓	36	棒球手套	56	蛋糕	76	花瓶
17	狗	37	滑板	57	椅子	77	剪刀
18	馬	38	衝浪板	58	沙發	78	泰迪熊
19	羊	39	網球拍	59	植物盆栽	79	吹風機
20	牛	40	瓶子	60	床	80	牙刷

▲ 表．COCO 資料集 80 個類別

第 8 章　物件偵測（Object Detection）

所有任務均可以使用指令（CLI）或程式進行訓練及推論，詳細說明可參閱圖像分類文件 [30]。

先使用指令測試。

1. 使用最小的模型 YOLO11n-cls 測試，指令如下。

   - yolo classify predict model=yolo11n-cls.pt source='https://ultralytics.com/images/bus.jpg'

   - 結果會顯示在螢幕上，並將圖片存至 runs\classify\predict\bus.jpg。

   - 模型及測試圖檔會自動下載至執行資料夾內，分類的模型名稱以【-cls】結尾。

2. 原圖如下：

3. 辨識結果：會找到最大的物件 -- 車輛，各種車輛的預測機率如下。

   minibus 0.57

   police_van 0.34

   trolleybus 0.04

   recreational_vehicle 0.01

   streetcar 0.01

範例 1. 採用程式進行分類推論。

程式：**08_05_YOLO_Classification.ipynb**。

8-8 圖像分類（Image Classification）

1. 載入相關套件。

```
1  from ultralytics import YOLO
```

2. 建立 YOLO 物件並載入 YOLO 權重檔。

```
1  model = YOLO("yolo11n-cls.pt")
```

3. 辨識。

```
1  results = model("https://ultralytics.com/images/bus.jpg")
2  results
```

4. 執行結果：內容超豐富，包括 1000 種類別名稱。

```
Found https://ultralytics.com/images/bus.jpg locally at bus.jpg
image 1/1 F:\0_AI\Books\Tensorflow\00_V2\src\08\bus.jpg: 224x224 minibus 0.57, police_van 0.34, trolleybus 0.04, recreational_vehicle 0.01, streetcar 0.01, 5.0ms
Speed: 11.0ms preprocess, 5.0ms inference, 0.0ms postprocess per image at shape (1, 3, 224, 224)

[ultralytics.engine.results.Results object with attributes:

boxes: None
keypoints: None
masks: None
names: {0: 'tench', 1: 'goldfish', 2: 'great_white_shark', 3: 'tiger_shark', 4: 'hammerhead', 5: 'electric_ray', 6: 'stingray', 7: 'cock', 8: 'hen', 9: 'ostrich', 10: 'brambling', 11: 'goldfinch', 12: 'house_finch', 13: 'junco', 14: 'indigo_bunting', 15: 'robin', 16: 'bulbul', 17: 'jay', 18: 'magpi
```

範例 2. 採用指令進行分類訓練。

1. 使用內建資料集 MNIST160 訓練模型，內有訓練及測試資料各 80 筆。指令如下：

    yolo classify train data=mnist160 model=yolo11n-cls.yaml epochs=100 imgsz=64

    - 若發生 worker error，上述指令可加『workers=0』，降低執行緒數目。
    - 觀察訓練及測試資料下載的資料夾，每個子資料夾存放同一類別的檔案。
    - 下載的資料夾定義在 C:\Users\< 登入帳號 >\AppData\Roaming\Ultralytics\ 資料夾中的 settings.json，可自行修改，注意，路徑的最下層資料夾須為 datasets。也可以使用 yolo settings 指令查詢，以 yolo settings dataset_dir='< 路徑 >' 指令修改。

2. 執行結果：會產生 last.pt 及 best.pt，存於 runs\classify\train5\weights 資料夾內，前者是最新模型，後者是最佳模型。

3. 採用最佳模型評分：使用測試資料。

    yolo classify val model=runs/classify/train/weights/best.pt data=mnist160 split=test

第 8 章　物件偵測（Object Detection）

4. 執行結果：只有 80 筆訓練資料，準確率可以達到 75%，也算厲害。
   - top1 準確率：75%。
   - top5 準確率：95%，前 5 名含正確答案即算預測正確。

5. 任意取一個圖檔測試：筆者取測試資料的【9】辨識。

   yolo classify predict model=runs/classify/train/weights/best.pt source='./content/datasets/mnist160/test/9/214.png'

6. 執行結果：各種數字的預測機率如下，預測正確。

   9 1.00

   8 0.00

   7 0.00

   6 0.00

   3 0.00

7. 再使用筆者手寫的數字測試：路徑使用【/】，才不會出現警告訊息。

   yolo classify predict model=runs/classify/train/weights/best.pt source='../myDigits/9.png'

8. 執行結果：辨識錯誤，預測最大機率者為 4，其他數字辨識結果也不盡理想，訓練資料還是需要多一點。

   4 0.32

   9 0.26

   7 0.19

   5 0.10

   2 0.03

9. 筆者同時在 Colaboratory 測試，可參閱 **08_06_YOLO_Classification_Training_colab.ipynb**。

10. 如果要使用自訂的資料集訓練模型,只要資料夾內含 train 及 test 子資料夾,其下子資料夾再存放同一類別的檔案即可,如下圖。

```
data
└ train
    └ cat
    └ dog
└ test
    └ cat
    └ dog
```

11. 使用自訂的資料集訓練模型指令如下:

    yolo classify train data='f:/data' model=yolo11n-cls.yaml epochs=20 imgsz=256

    - 出現【OMP: Error #15: Initializing libiomp5md.dll, but found libiomp5md.dll already initialized.】錯誤時,須先執行 set KMP_DUPLICATE_LIB_OK=TRUE,Linux/Mac 作業系統須將 set 改為 export。

12. 筆者自 cat-or-not GitHub[31],下載 data.zip,解壓縮並更改資料夾名稱為 train 及 test,即可依照上一步驟的指令訓練模型。

13. 執行結果:
    - top1 準確率:72.9%。
    - top5 準確率:100%。

使用 YOLO 進行圖像分類,其實與自訂 CNN 模型結構差不多,只是 YOLO 模型結構較複雜,且可使用指令直接執行,但是,資料轉換及清理還是需要事先處理好。

8-9 物件偵測（Object Detection）

物件偵測可辨識單張圖像中的多個物件,並指出所在位置（Location）。與圖像分類一樣有不同尺寸的模型可選用,可詳閱『YOLO Object detection』文件說明[32]。

本節除了介紹 YOLO 物件偵測外,也介紹 TensorFlow Object Detection API,讓 TensorFlow 開發者也可以實踐物件偵測。

第 8 章　物件偵測（Object Detection）

8-9-1　YOLO 物件偵測（Object Detection）

可以使用指令（CLI）或程式進行推論，先使用指令測試。

1. 使用最小的模型 YOLO11n 測試，指令如下。

    - yolo detect predict model=yolo11n.pt source='https://ultralytics.com/images/bus.jpg'
    - 結果會顯示在螢幕上，並將圖片存至 runs\detect\predict\bus.jpg。
    - 模型及測試圖檔會自動下載至執行資料夾內。

2. 原圖如下：

3. 辨識結果：查看 runs\detect\predict\bus.jpg，可以看到只有露出一條腿或一隻手仍可被正確辨識。

8-9 物件偵測（Object Detection）

範例 1. 採用程式進行物件偵測。

程式：**08_07_YOLO_Object_Detection.ipynb**。

1. 載入相關套件。

```
from ultralytics import YOLO
```

2. 建立 YOLO 物件並載入 YOLO 權重檔。

```
model = YOLO("yolo11n.pt")
```

3. 辨識。

```
results = model("https://ultralytics.com/images/bus.jpg")
results
```

4. 執行結果：內容超豐富，包括 80 種類別名稱，並辨識出 4 個人及一輛巴士。

```
Found https://ultralytics.com/images/bus.jpg locally at bus.jpg
image 1/1 F:\0_AI\Books\Tensorflow\00_V2\src\08\bus.jpg: 640x480 4 persons, 1 bus, 40.6ms
Speed: 3.0ms preprocess, 40.6ms inference, 94.2ms postprocess per image at shape (1, 3, 640, 480)
[ultralytics.engine.results.Results object with attributes:

boxes: ultralytics.engine.results.Boxes object
keypoints: None
masks: None
names: {0: 'person', 1: 'bicycle', 2: 'car', 3: 'motorcycle', 4: 'airplane', 5: 'bus', 6: 'train', 7: 'truck', 8: 'boat', 9: 'traffic light', 10: 'fire hydrant', 11: 'stop sign', 12: 'parking meter', 13: 'bench', 14: 'bird', 15: 'cat', 16: 'dog', 17: 'horse', 18: 'sheep', 19: 'cow', 20: 'elephant', 21: 'bear', 22: 'zebra', 23: 'giraffe', 24: 'backpack', 25: 'umbrella', 26: 'handbag', 27: 'tie', 28: 'suitcase', 29: 'frisbee', 30: 'skis', 31: 'snowboard', 32: 'sports ball', 33: 'kite', 34: 'baseball bat', 35: 'baseball glove', 36: 'skateboard', 37: 'surfboard', 38: 'tennis racket', 39: 'bottle', 40: 'wine glass', 41: 'cup', 42: 'fork', 43: 'knife', 44: 'spoon', 45: 'bowl', 46: 'banana', 47: 'apple', 48: 'sandwich', 49: 'orange', 50: 'broccoli', 51: 'carrot', 52: 'hot dog', 53: 'pizza', 54: 'donut', 55: 'cake', 56: 'chair', 57: 'couch', 58: 'potted plant', 59: 'bed', 60: 'dining table', 61: 'toilet', 62: 'tv', 63: 'laptop', 64: 'mouse', 65: 'remote', 66: 'keyboard', 67: 'cell phone', 68: 'microwave', 69: 'oven', 70: 'toaster', 71: 'sink', 72: 'refrigerator', 73: 'book', 74: 'clock', 75: 'vase', 76: 'scissors', 77: 'teddy bear', 78: 'hair drier', 79: 'toothbrush'}
```

5. 解析結果：找出邊界框（Bounding box），或顯示物件偵測圖片。

```
for result in results:
    boxes = result.boxes  # Boxes object for bounding box outputs
    print(boxes)
    masks = result.masks  # Masks object for segmentation masks outputs
    keypoints = result.keypoints  # Keypoints object for pose outputs
    probs = result.probs  # Probs object for classification outputs
    obb = result.obb  # Oriented boxes object for OBB outputs
    result.show()  # display to screen
    result.save(filename="result.jpg")   # save to disk
```

第 8 章　物件偵測（Object Detection）

6. 執行結果：

   - 找到 4 個人（0）及一輛巴士（5），數字代表類別代碼。

     ```
     cls: tensor([5., 0., 0., 0., 0.], device='cuda:0')
     ```

   - 邊界框（Bounding box）座標：4 種格式，（x,y）為左上角或右下角座標，（w,h）為寬高，n 為百分比。

   ```
   xywh: tensor([[400.0136, 478.8883, 792.3620, 499.0480],
           [740.4135, 636.7728, 138.7925, 483.8794],
           [143.3527, 651.8801, 191.8959, 504.6299],
           [283.7633, 634.5622, 121.4086, 451.7472],
           [ 34.4536, 714.2139,  68.8638, 316.2908]], device='cuda:0')
   xywhn: tensor([[0.4938, 0.4434, 0.9782, 0.4621],
           [0.9141, 0.5896, 0.1713, 0.4480],
           [0.1770, 0.6036, 0.2369, 0.4672],
           [0.3503, 0.5876, 0.1499, 0.4183],
           [0.0425, 0.6613, 0.0850, 0.2929]], device='cuda:0')
   xyxy: tensor([[3.8327e+00, 2.2936e+02, 7.9619e+02, 7.2841e+02],
           [6.7102e+02, 3.9483e+02, 8.0981e+02, 8.7871e+02],
           [4.7405e+01, 3.9957e+02, 2.3930e+02, 9.0420e+02],
           [2.2306e+02, 4.0869e+02, 3.4447e+02, 8.6044e+02],
           [2.1726e-02, 5.5607e+02, 6.8886e+01, 8.7236e+02]], device='cuda:0')
   xyxyn: tensor([[4.7317e-03, 2.1237e-01, 9.8296e-01, 6.7446e-01],
           [8.2842e-01, 3.6559e-01, 9.9977e-01, 8.1362e-01],
           [5.8524e-02, 3.6997e-01, 2.9543e-01, 8.3722e-01],
           [2.7538e-01, 3.7842e-01, 4.2527e-01, 7.9670e-01],
           [2.6822e-05, 5.1488e-01, 8.5044e-02, 8.0774e-01]], device='cuda:0')
   ```

範例 2. 採用指令進行物件偵測模型之訓練。

1. 使用內建資料集 COCO8 訓練模型，每一類別只有訓練及驗證資料各 4 筆，主要用於測試及偵錯。指令如下：

   yolo detect train data=coco8.yaml model=yolo11n.yaml epochs=3 imgsz=640 workers=0

   - YOLO11 會記住第一次下載資料集的資料夾，之後都會下載至該資料夾，要改變資料夾，請執行 yolo settings，依據資訊修改設定檔內容。

8-9 物件偵測（Object Detection）

```
D:\0_AI\Books\Tensorflow\00_V2>yolo settings
JSONDict("C:\Users\mikec\AppData\Roaming\Ultralytics\settings.json"):
{
  "settings_version": "0.0.6",
  "datasets_dir": "D:\\0_AI\\Books\\Tensorflow\\00_V2\\src\\08\\datasets",
```

- 出現【OMP: Error #15: Initializing libiomp5md.dll, but found libiomp5md.dll already initialized.】錯誤時，須執行 set KMP_DUPLICATE_LIB_OK= TRUE，Linux/Mac 作業系統須將 set 改為 export。

- 出現【存取被拒…】訊息，可能是 GPU 顯卡版本過舊，請在擁有較新的顯卡的 PC 上測試或使用 Colaboratory 測試。

- 要尋找物件偵測的資料集可參閱『Object Detection Datasets Overview』[33]，其中 Roboflow 提供許多資料集，可供測試。

2. 執行結果：會產生 last.pt 及 best.pt，存於 runs\classify\train\weights\ 資料夾內，後者是最佳模型。

3. 採用最佳模型評分：使用驗證資料評分。

   yolo detect val model=runs/detect/train/weights/best.pt data=coco8.yaml

4. 執行結果：會存在 runs\detect\val 資料夾。

   - 原圖：

第 8 章　物件偵測（Object Detection）

- 執行結果：左上角毛毯錯認為狗，其他辨識結果均正確。

5. 任意取一個圖檔測試：辨識不到任何物件，應該是訓練資料太少了。

    yolo detect predict model=runs/detect/train/weights/best.pt source='./bus.jpg'

8-9-2 TensorFlow Object Detection API

由於物件偵測應用廣泛，TensorFlow 與 PyTorch 套件都有特別提供 API，可直接呼叫相關模型，TensorFlow Object Detection API 就是 TensorFlow 所支援的版本。API 支援的演算法包含 FasterRCNN+InceptionResNet V2、SSD+mobilenet V2，前者準確率較高，後者模型檔較輕巧。

範例. TensorFlow Object Detection API 測試。

程式：**08_08_TF_Object_Detection_API.ipynb**，可自 TensorFlow Object Detection[34] 下載範例。由於範例程式碼已久未更新，有些警告訊息，故以下僅說明重要程式碼。

1. 安裝 TensorFlow Hub 套件：

    pip install tensorflow_hub

8-9 物件偵測（Object Detection）

2. 下載圖檔，並設定解析度。

```
1  # By Heiko Gorski, Source: https://commons.wikimedia.org/wiki/File:Naxos_Taverna.jpg
2  image_url = "https://upload.wikimedia.org/wikipedia/commons/6/60/Naxos_Taverna.jpg"
3  downloaded_image_path = download_and_resize_image(image_url, 1280, 856, True)
```

3. 下載模型：有多個模型可供選擇。

```
1  module_handle = "https://tfhub.dev/google/faster_rcnn/openimages_v4/inception_resnet_v2/1"
2  # @param ["https://tfhub.dev/google/openimages_v4/ssd/mobilenet_v2/1",
3  # "https://tfhub.dev/google/faster_rcnn/openimages_v4/inception_resnet_v2/1"]
4
5  detector = hub.load(module_handle).signatures['default']
```

4. 物件偵測：

```
1  def run_detector(detector, path):
2      img = load_img(path)
3
4      converted_img  = tf.image.convert_image_dtype(img, tf.float32)[tf.newaxis, ...]
5      start_time = time.time()
6      result = detector(converted_img)
7      end_time = time.time()
8
9      result = {key:value.numpy() for key,value in result.items()}
10
11     print("Found %d objects." % len(result["detection_scores"]))
12     print("Inference time: ", end_time-start_time)
13
14     image_with_boxes = draw_boxes(
15         img.numpy(), result["detection_boxes"],
16         result["detection_class_entities"], result["detection_scores"])
17
18     display_image(image_with_boxes)
19
20 run_detector(detector, downloaded_image_path)
```

5. 執行結果：找到 100 個物件。

8-55

第 8 章　物件偵測（Object Detection）

如果要自訂資料集，偵測其他物件的話，Tensorflow Object Detection API 的官網文件[35] 有非常詳盡的解說，讀者可依指示自行測試。

8-10　資料標記（Data Annotation）

如果是自行蒐集訓練資料，必須在每一張圖片標記物件的類別及邊界框，標記檔的格式有 Pascal VOC、COCO、YOLO…等，YOLO v11 有 API 支援 COCO 轉換為 YOLO 格式（convert_cococ 函數），同時還支援各種任務的格式轉換，請參閱『YOLO v11 Simple Utilities』[36]。簡單的講，COCO 標記檔為 JSON 格式，一個圖檔會以 List 定義多個物件類別及邊界框，而 YOLO 則是一般的文字檔，一個圖檔對應一個標記檔，每一行只定義一個物件類別及邊界框，如下，欄位依序為類別代碼、x、y、w、h，以比例表示座標及寬/高度：

```
45 0.479492 0.688771 0.955609 0.5955
45 0.736516 0.247188 0.498875 0.476417
50 0.637063 0.732938 0.494125 0.510583
45 0.339438 0.418896 0.678875 0.7815
```

要從頭標記物件或修改標記檔，可安裝標記工具軟體，例如 LabelImg[37] 或 Label Studio[38]…等，LabelImg 雖已不再更新，不過使用比較直覺，而 Label Studio 功能比較多元，但需要較長的學習時間，本文說明 LabelImg 使用方法。

1. 安裝指令：會安裝執行檔 labelimg。
   - pip install labelimg
2. 執行：可單獨輸入 labelimg，或後面帶參數指定圖檔資料夾及類別列表檔。
   - labelimg <image 資料夾 > <class file>
   - 筆者將圖像及標記資料夾內所有檔案複製到 images_labels 資料夾，並建立 classes.txt，內容含所有類別名稱。然後執行 labelimg images_labels，出現畫面如下：

▲ 圖 8.26 LabelImg 標記工具

3. 操作說明如下：
   - 出現第一張圖，而且顯示邊界框，右半部指出每一邊界框對應的類別名稱，要修改可點擊『Edit Label』。
   - 點擊『Next Image』，可修改下一張圖。點擊『Prev Image』，可修改上一張圖。
   - 修改完記得要點擊『Save』存檔。
   - 存檔格式有 Pascal VOC、COCO、YOLO，可點擊『Save』下面的按鈕切換，優先選擇 YOLO。
   - 要增加標記可點擊『Create RectBox』，再以滑鼠拖曳一個區域，並輸入類別名稱。

8-11 物件偵測的效能衡量指標

物件偵測的效能衡量指標是採『平均精確度均值』（Mean Average Precision，mAP），YOLO v3 官網展示的圖表針對各種模型比較 mAP。

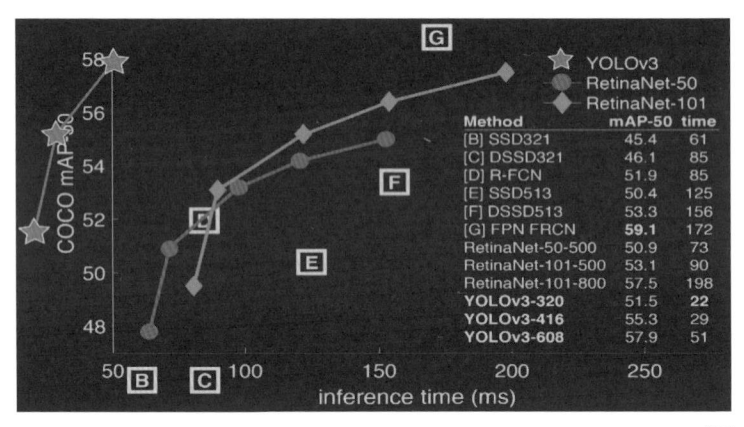

▲ 圖 8.27 YOLO 與其他模型比較，圖片來源：YOLO 官網 [17]

第四章介紹過的 ROC/AUC 效能衡量指標，是以預測機率為基準，計算各種閾值（門檻值）下的真陽率與偽陽率，以偽陽率為 X 軸，真陽率為 Y 軸，繪製出 ROC 曲線。而 mAP 也類似 ROC/AUC，以 IoU 為基準，計算各種閾值（門檻值）下的精確率（Precision）與召回率（Recall），以召回率為 X 軸，精確率為 Y 軸，繪製出 mAP 曲線。

不過，物件偵測模型通常是多分類，不是二分類，因此，採取計算各個種類的平均精確度，繪製後如下左方圖表，通常會調整成右方圖表的粗線，因為，在閾值低的精確率一定比閾值高的精確率更好，所以作此調整。

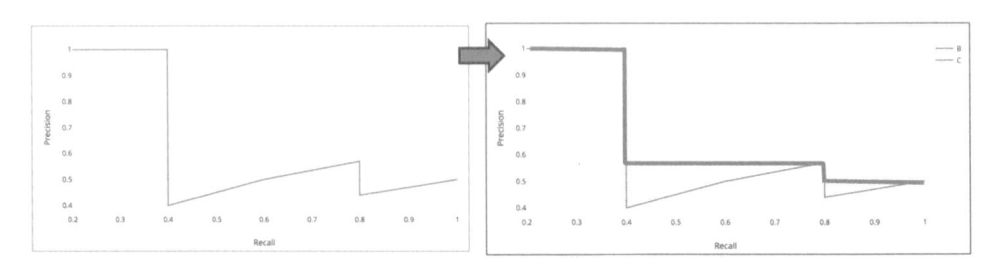

▲ 圖 8.30 mAP 曲線，左圖是實際計算的結果，右圖是調整後的結果

8-12 實例分割（Instance Segmentation）

實例分割（Instance Segmentation）或語義分割（Semantics Segmentation）是更精準的物件偵測，以不同顏色的像素遮罩（Mask）物件，而非以矩形邊界框標示，可以讓電腦或自駕車辨識物件的形狀或大小。

以下依然以 YOLO v11 為例說明，可詳閱『YOLO v11 Instance Segmentation』[39]。

可以使用指令（CLI）或程式進行推論，先使用指令測試。

1. 使用最小的模型 YOLO11n-seg 測試，指令如下，實例分割模型一律以【-seg】結尾。
   - yolo segment predict model=yolo11n-seg.pt source='https://ultralytics.com/images/bus.jpg'
   - 結果會顯示在螢幕如下，並將圖片存至 runs\segment\predict\bus.jpg。
     4 persons, 1 bus, 1 stop sign
   - 模型及測試圖檔會自動下載至執行資料夾內。

2. 原圖如下：

第 8 章　物件偵測（Object Detection）

3. 辨識結果：查看 runs\segment\predict\bus.jpg，除了邊界框，還有不同顏色的遮罩（Mask）。

範例 1. 採用程式進行實例分割。

程式：**08_09_Instance_Segmentation.ipynb**。

1. 載入相關套件。

```
1  from ultralytics import YOLO
```

2. 建立 YOLO 物件並載入 YOLO 權重檔。

```
1  model = YOLO("yolo11n-seg.pt")
```

3. 辨識。

```
1  results = model("https://ultralytics.com/images/bus.jpg")
2  results
```

4. 解析結果。

```
1  for result in results:
2      boxes = result.boxes    # Boxes object for bounding box outputs
3      masks = result.masks    # Masks object for segmentation masks outputs
```

8-12 實例分割（Instance Segmentation）

```
4    print(masks)
5    keypoints = result.keypoints  # Keypoints object for pose outputs
6    probs = result.probs  # Probs object for classification outputs
7    obb = result.obb  # Oriented boxes object for OBB outputs
8    result.show()  # display to screen
9    result.save(filename="result.jpg")  # save to disk
```

5. 執行結果：顯示遮罩（Mask）的每個像素。

```
xy: [array([[     185.62,         232.
            [     185.62,         237.94],
            [     183.94,         239.62],
            ...,
            [     410.06,         239.62],
            [     408.38,         237.94],
            [     408.38,         232.88]],
            [     803.25,         398.25],
            [     801.56,         398.25],
            [     799.88,         399.94],
            [     798.19,         399.94],
            [     796.5,          401.62],
            [     796.5,          403.31],
```

範例 2. 採用指令進行實例分割模型之訓練。

1. 使用內建資料集 COCO8 訓練模型，每一類別只有訓練及驗證資料各 4 筆，主要適用於測試及偵錯。指令如下：

    yolo segment train data=coco8-seg.yaml model=yolo11n-seg.yaml epochs=3 imgsz=640 workers=0

    - 出　現【OMP: Error #15: Initializing libiomp5md.dll, but found libiomp5md.dll already initialized.】錯誤時，須執行 set KMP_DUPLICATE_LIB_OK= TRUE，Linux/Mac 作業系統須將 set 改為 export。

    - YOLO11 支援的實例分割資料集有許多種，可參閱『Instance Segmentation Datasets Overview』[40]。

    - 執行結果：會產生 last.pt 及 best.pt，存於 runs\segment\train\weights\ 資料夾內，後者是最佳模型。

2. 採用最佳模型評分：使用驗證資料評分。

3. yolo segment val model=runs/segment/train/weights/best.pt

4. 執行結果儲存在 runs\segment\val 資料夾。

第 8 章 物件偵測（Object Detection）

- 原圖：

- 執行結果：左上角毛毯錯認為狗，其他辨識結果均正確。

5. 任意取一個圖檔測試：辨識不到任何物件，應該是訓練資料太少了。

    - yolo segment predict model=runs/segment/train/weights/best.pt source='./bus.jpg'

8-13 姿態辨識（Pose Estimation）

姿態辨識（Pose Estimation）偵測物件的關鍵點（Key Point），通常是人體的關節，藉以辨識人體的姿勢，例如健身動作是否正確。

以下依然以 YOLO v11 為例說明，可詳閱『YOLO v11 Pose Estimation』[41]。

可以使用指令（CLI）或程式進行推論，先使用指令測試。

1. 使用最小的模型 YOLO11n-pose 測試，指令如下，實例分割模型一律以【-pose】結尾。

    - yolo pose predict model=yolo11n-pose.pt source='https://ultralytics.com/images/bus.jpg'
    - 結果會顯示在螢幕如下，並將圖片存至 runs\pose\predict\bus.jpg。4 persons
    - 模型及測試圖檔會自動下載至執行資料夾內。

2. 原圖如下：

3. 辨識結果：從臉部到手腳的關鍵點均被標示出來，要辨識姿態，可進一步計算關鍵點之間的距離與關聯。

8-63

第 8 章 物件偵測（Object Detection）

範例 1. 採用程式進行姿態辨識。

程式：**08_10_Pose_Estimation.ipynb**。

1. 載入相關套件。

```
1  from ultralytics import YOLO
```

2. 建立 YOLO 物件並載入 YOLO 權重檔。

```
1  model = YOLO("yolo11n-pose.pt")
```

3. 辨識。

```
1  results = model("https://ultralytics.com/images/bus.jpg")
2  results
```

4. 解析結果：。

```
1  for result in results:
2      boxes = result.boxes      # Boxes object for bounding box outputs
3      masks = result.masks      # Masks object for segmentation masks outputs
4      keypoints = result.keypoints  # Keypoints object for pose outputs
5      print(keypoints)
6      probs = result.probs      # Probs object for classification outputs
7      obb = result.obb  # Oriented boxes object for OBB outputs
```

8-64

8-13 姿態辨識（Pose Estimation）

```
8    result.show()    # display to screen
9    result.save(filename="result.jpg")    # save to disk
```

5. 執行結果：顯示關鍵點。

```
xy: tensor([[[142.3645, 441.8607],
             [147.9897, 431.4179],
             [130.5441, 433.3708],
             [  0.0000,   0.0000],
             [107.1834, 440.6561],
             [157.4519, 493.1139],
             [ 94.2637, 499.2470],
             [176.4613, 550.9833],
             [110.6623, 567.5746],
             [174.2181, 532.3488],
             [162.0030, 534.3918],
```

範例 2. 採用指令進行姿態辨識模型之訓練。

1. 使用內建資料集 COCO8 訓練模型，每一類別只有訓練及驗證資料各 4 筆，主要用於測試及偵錯。指令如下：

    - yolo pose train data=coco8-pose.yaml model=yolo11n-pose.yaml epochs=3 imgsz=640 workers=0

    - 出　現【OMP: Error #15: Initializing libiomp5md.dll, but found libiomp5md.dll already initialized.】錯誤時，須執行 set KMP_DUPLICATE_LIB_OK= TRUE，Linux/Mac 作業系統須將 set 改為 export。

    - YOLO v11 支援的姿態辨識資料集有許多種，可參閱『Pose Estimation Datasets Overview』[42]。

2. 執行結果：會產生 last.pt 及 best.pt，存於 runs\pose\train\weights\ 資料夾內，後者是最佳模型。

3. 採用最佳模型評分：使用驗證資料評分。

    yolo pose val model=runs/pose/train/weights/best.pt

4. 執行結果：會存在 runs\pose\val 資料夾。

8-65

第 8 章　物件偵測（Object Detection）

- 原圖：

- 執行結果：只能辨識人體，其他動物無法辨識，主要是訓練資料只標記人體關節。

5. 任意取一個圖檔測試：辨識不到任何物件，應該是訓練資料太少了。

    yolo pose predict model=runs/pose/train/weights/best.pt source='./bus.jpg'

8-14 旋轉邊界框物件偵測（Oriented Bounding Boxes Object Detection）

姿態辨識還可用來辨識體育運動姿勢是否標準，協助運動員提升成績，另外還有手勢偵測[43]、體感遊戲、製作皮影戲[44]等，也很好玩。

8-14 旋轉邊界框物件偵測（Oriented Bounding Boxes Object Detection）

旋轉邊界框物件偵測（Oriented Bounding Boxes Object Detection, OBB）除了偵測物件外，還可以偵測物件的方向，更精準指出物件的位置。

以下依然以 YOLO v11 為例說明，可詳閱『YOLO v11 Oriented Bounding Boxes Object Detection』[45]。

可以使用指令（CLI）或程式進行推論，先使用指令測試。

1. 使用最小的模型 YOLO11n-obb 測試，指令如下，實例分割模型一律以【-obb】結尾。

    - yolo obb predict model=yolo11n-obb.pt source='https://ultralytics.com/images/boats.jpg'
    - 執行結果會將圖片存至 runs\obb\predict\boats.jpg。
    - 模型及測試圖檔會自動下載至執行資料夾內。

2. 原圖如下：有許多不同方向的小船。

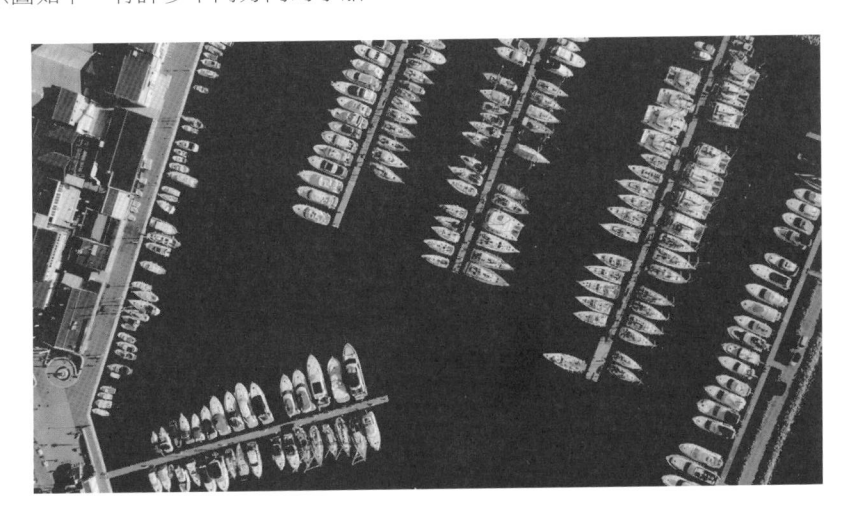

3. 辨識結果：畫面顯示有點擁擠。

範例 1. 採用程式進行姿態辨識。

程式：**08_11_YOLO_OBB.ipynb**。

1. 載入相關套件。

```
from ultralytics import YOLO
```

2. 建立 YOLO 物件並載入 YOLO 權重檔。

```
model = YOLO("yolo11n-obb.pt")
```

3. 辨識。

```
results = model("https://ultralytics.com/images/boats.jpg")
results
```

4. 解析結果：。

```
for result in results:
    boxes = result.boxes  # Boxes object for bounding box outputs
    masks = result.masks  # Masks object for segmentation masks outputs
    keypoints = result.keypoints  # Keypoints object for pose outputs
    probs = result.probs  # Probs object for classification outputs
    obb = result.obb  # Oriented boxes object for OBB outputs
    print(obb)
```

8-14 旋轉邊界框物件偵測（Oriented Bounding Boxes Object Detection）

```
8    result.show()   # display to screen
9    result.save(filename="result.jpg")   # save to disk
```

5. 執行結果：顯示類別、可信度（Confidence）及邊界框含旋轉角度（xywhr）。

```
cls: tensor([1., 1., 1., 1., 1., 1., 1., 1., 1., 1., 1., 1., 1., 1., 1., 1., 1., 1., 1., 1., 1., 1., 1., 1., 1.,
        1., 1., 1., 7., 1., 7., 1., 1., 1., 1., 1., 1., 1., 1., 1., 1., 1., 1., 1., 1., 1., 1., 1., 1.,
        1., 1., 1., 1.,
            1., 1., 1., 1., 1., 1., 1., 1., 1., 1., 1., 1., 7., 1., 1., 1., 1., 1., 1., 1., 1., 1., 1.,
        1., 1., 1., 1., 1., 1., 1., 1., 1., 1., 1., 1., 1., 7., 1., 1., 1., 1., 1., 1., 1., 1., 1., 1.,
        1., 1., 1., 1.,
            1., 1., 1., 1., 1., 1., 1., 1., 1., 1., 1., 1., 1., 7., 1., 1., 1., 1., 1.], device='cuda:0')
conf: tensor([0.8425, 0.8336, 0.8318, 0.8297, 0.8263, 0.8244, 0.8243, 0.8218, 0.8197, 0.8181, 0.8177, 0.8162, 0.8147, 0.8135, 0.8132, 0.8122, 0.8106,
        0.8105, 0.8101, 0.8097, 0.8094, 0.8076, 0.8074, 0.8067, 0.8067, 0.8067, 0.8065, 0.8061, 0.8059, 0.8055, 0.8052, 0.8035, 0.8026, 0.8014, 0.7996, 0.7991,
        0.7979, 0.7965, 0.7965,
            0.7909, 0.7904, 0.7895, 0.7891, 0.7887, 0.7887, 0.7884, 0.7861, 0.7851, 0.7846, 0.7845, 0.7845, 0.7843, 0.7841, 0.7840, 0.7838, 0.7835, 0.7807,
        0.7806, 0.7804, 0.7777, 0.7772, 0.7754, 0.7744, 0.7742, 0.7736, 0.7720, 0.7711, 0.7696, 0.7692, 0.7669, 0.7664, 0.7652, 0.7638, 0.7634, 0.7625, 0.7595,
        0.7591, 0.7493,
            0.7490, 0.7480, 0.7475, 0.7473, 0.7471, 0.7468, 0.7462, 0.7456, 0.7430, 0.7369, 0.7368, 0.7358, 0.7353, 0.7345, 0.7341, 0.7340, 0.7278, 0.7273,
        0.7270, 0.7266, 0.7253, 0.7252, 0.7224, 0.7213, 0.7201, 0.7196, 0.7160, 0.7131, 0.7102, 0.7079, 0.7063, 0.7025, 0.7014, 0.7007, 0.6997, 0.6957, 0.6889]
```

```
xywhr: tensor([[3.2304e+02, 4.4219e+02, 7.6033e+01, 2.7323e+01, 3.7826e-01],
        [1.7504e+03, 6.9368e+02, 1.0658e+02, 3.2217e+01, 4.1058e-01],
        [1.8187e+03, 5.4377e+02, 9.5871e+01, 2.9535e+01, 4.4453e-01],
        [1.6765e+03, 8.6366e+02, 1.1517e+02, 3.3924e+01, 4.2344e-01],
        [1.6191e+03, 1.0020e+03, 1.1964e+02, 3.5574e+01, 4.0201e-01],
```

範例 2. 採用指令進行姿態辨識模型之訓練。

1. 使用內建資料集 DOTA8 訓練模型，每一類別只有訓練及驗證資料各 4 筆，主要用於測試及偵錯。指令如下：

   yolo obb train data=dota8.yaml model=yolo11n-obb.yaml epochs=3 imgsz=640 workers=0

   - 出　現【OMP: Error #15: Initializing libiomp5md.dll, but found libiomp5md.dll already initialized.】錯誤時，須執行 set KMP_DUPLICATE_LIB_OK= TRUE，Linux/Mac 作業系統須將 set 改為 export。

   - YOLO11 支援的 OBB 資料集有許多種，可參閱『Oriented Bounding Box（OBB）Datasets Overview』[46]。

2. 執行結果：會產生 last.pt 及 best.pt，存於 runs\obb\train\weights\ 資料夾內，後者是最佳模型。

3. 採用最佳模型評分：使用驗證資料評分。

   yolo obb val model=runs/obb/train/weights/best.pt

4. 執行結果：會存在 runs\obb\val 資料夾。

第 8 章 物件偵測（Object Detection）

- 原圖：

- 執行結果：辨識出足球 / 棒球體育場。

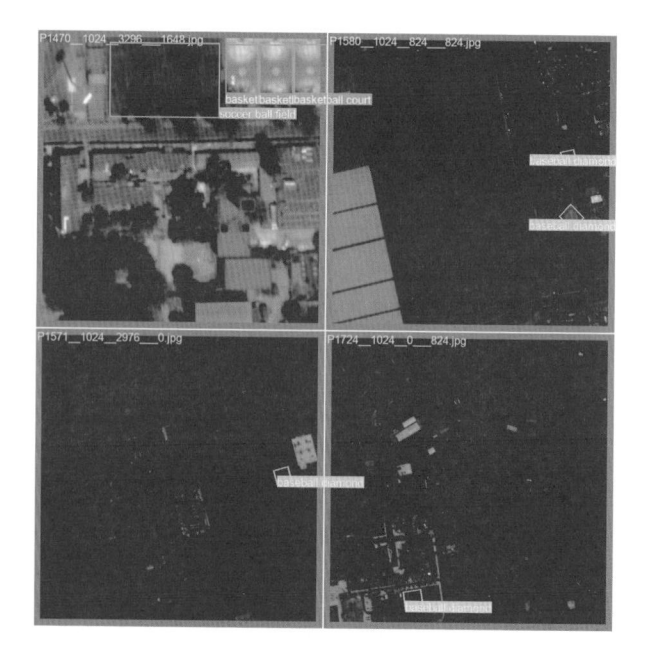

8-15 物件追蹤（Object Tracking）

5. 任意取一個圖檔測試：辨識不到任何物件，應該是訓練資料太少了或類別不符。

   yolo obb predict model=runs/obb/train/weights/best.pt source='./boats.jpg'

8-15 物件追蹤（Object Tracking）

YOLO v11 不僅提供基礎任務，也提供許多解決方案，請參閱 8.7 節，本節以物件追蹤（Object Tracking）為例，物件追蹤可以賦予偵測物件唯一的識別碼，在視訊中追蹤物件的移動，實際的應用包括無人搬運車導引、車輛行駛速度計算、罪犯的追蹤…等，可參閱『清大獲得 2018 NVIDIA 機器人冠軍』[47]。

▲ 圖 8.28 物件追蹤（Object Tracking）的應用案例，圖片來源：『Multi-Object Tracking with Ultralytics YOLO』[48]

YOLO v11 採用 2 種演算法，可以在組態檔（yaml）指定：

1. BoT-SORT[49]：組態檔為 botsort.yaml，未指定組態檔時，BoT-SORT 為預設值，此演算法準確率較高。
2. ByteTrack[50]：組態檔為 bytetrack.yaml，此演算法較簡單快速。

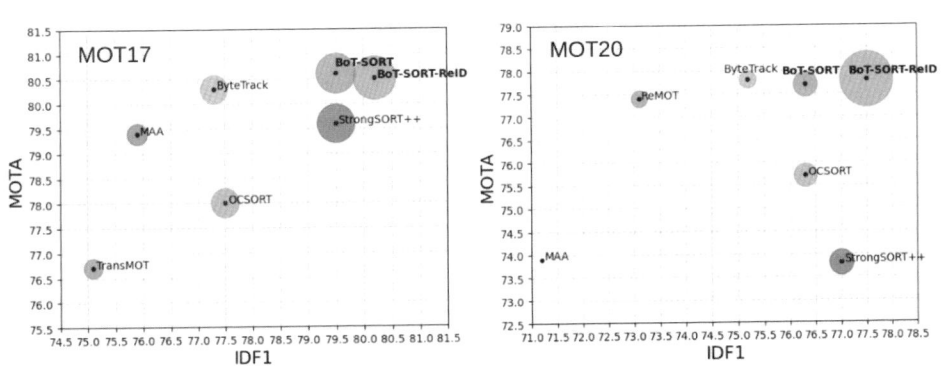

▲ 圖 8.29 BoT-SORT 的效能指標（Benchmark）

第 8 章　物件偵測（Object Detection）

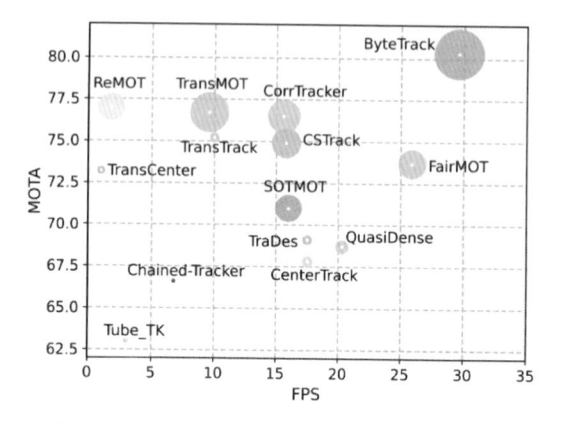

▲ 圖 8.30　ByteTrack 的效能指標（Benchmark）

相關專有名詞說明如下：

- MOTA（Multiple Object Tracking Accuracy）：多個物件追蹤準確率。
- IDF1（Identification F1-Score）：依物件識別碼（ID）的 F1 效能指標。
- FPS（Frame Per Second）：每秒影格數。

公式可參閱『多目標跟蹤的性能評估：指標與計算方法』[51]，相關說明可詳閱『Multi-Object Tracking with Ultralytics YOLO』[52]。

物件偵測、實例分割、姿態辨識模型均支援物件追蹤，可以使用指令（CLI）或程式進行推論，先使用指令測試。

1. 使用最小的模型 YOLO11n 測試，指令如下。

    - yolo track model=yolo11n.pt source="https://youtu.be/LNwODJXcvt4"
    - 出現【OMP: Error #15: Initializing libiomp5md.dll, but found libiomp5md.dll already initialized.】錯誤時，須執行 set KMP_DUPLICATE_LIB_OK= TRUE，Linux/Mac 作業系統須將 set 改為 export。
    - 執行結果會將圖片存至 runs\detect\track\LNwODJXcvt4.avi。
    - 模型會自動下載至執行資料夾內。

2. 辨識結果：每個物件額外多一個識別碼（id）。

8-15 物件追蹤（Object Tracking）

3. 實例分割指令：
   - yolo track model=yolo11n-seg.pt source="https://youtu.be/LNwODJXcvt4"

4. 姿態辨識指令：
   - yolo track model=yolo11n-pose.pt source="https://youtu.be/LNwODJXcvt4"

5. 可指定自訂模型及組態檔：
   - yolo track model=path/to/best.pt tracker="bytetrack.yaml" source="https://youtu.be/LNwODJXcvt4"

範例 1. 採用程式進行物件追蹤。

程式：**08_12_YOLO_Object_Tracking.ipynb**。

1. 載入相關套件。

```
1 from ultralytics import YOLO
```

2. 建立 YOLO 物件並載入 YOLO 權重檔。

```
1 model = YOLO("yolo11n.pt")
```

3. 物件追蹤：不加【stream=True】參數，會出現錯誤訊息，應該是記憶體不足，筆者在本機及 Colaboratory 均有相同情形。

```
1 results = model.track("https://youtu.be/LNwODJXcvt4", stream=True)
2 results
```

第 8 章　物件偵測（Object Detection）

4. 建立影格輸出資料夾。

```
1  import os
2  os.makedirs('./result', exist_ok=True)
```

5. 解析結果：將每一幀影格逐一存檔，影格過多會造成錯誤。

```
1  for i, result in enumerate(results):
2      if i>20: break
3      boxes = result.boxes          # Boxes object for bounding box outputs
4      masks = result.masks          # Masks object for segmentation masks outputs
5      keypoints = result.keypoints  # Keypoints object for pose outputs
6      probs = result.probs          # Probs object for classification outputs
7      obb = result.obb              # Oriented boxes object for OBB outputs
8      # result.show()               # display to screen
9      result.save(filename=f"result/{i:04d}.jpg")  # save to disk
```

6. 客製化：

- 可以加參數過濾辨識結果，例如可信度（conf）、iou 門檻。

- 修改組態檔：2 個演算法的組態檔可參 Ultralytics GitHub[53]，程式碼如下：
  results = model.track（source="https://youtu.be/LNwODJXcvt4", tracker="custom_tracker.yaml"）

- 即時追蹤（Persisting Tracks Loop）：以攝影機或視訊檔為輸入，持續追蹤，程式碼如下，可參閱 08_13_YOLO_Object_Tracking_with_File.py：

```
1   import cv2
2   from ultralytics import YOLO
3
4   # Load the YOLO11 model
5   model = YOLO("yolo11n.pt")
6
7   # Open the video file
8   video_path = "./walk.mp4"
9   cap = cv2.VideoCapture(video_path)
10
11  # Loop through the video frames
12  while cap.isOpened():
13      # Read a frame from the video
14      success, frame = cap.read()
15
16      if success:
```

8-74

8-15 物件追蹤（Object Tracking）

```
17              # persisting tracks between frames
18              results = model.track(frame, persist=True)
19
20              # Visualize the results on the frame
21              annotated_frame = results[0].plot()
22
23              # Display the annotated frame
24              cv2.imshow("YOLO11 Tracking", annotated_frame)
```

- 可針對物件繪出移動的軌跡，重要程式碼如下，詳細可參閱 08_14_YOLO_Object_Tracking_by_ID.py：

```
29              # Visualize the results on the frame
30              annotated_frame = results[0].plot()
31
32              # Plot the tracks
33              for box, track_id in zip(boxes, track_ids):
34                  x, y, w, h = box
35                  track = track_history[track_id]
36                  track.append((float(x), float(y)))  # x, y center point
37                  if len(track) > 30:  # retain 90 tracks for 90 frames
38                      track.pop(0)
39
40                  # Draw the tracking lines
41                  points = np.hstack(track).astype(np.int32).reshape((-1, 1, 2))
42                  cv2.polylines(annotated_frame, [points], isClosed=False,
43                      color=(230, 230, 230), thickness=10)
```

- 執行結果：有兩條白色彎曲的線條，分別代表小孩及娃娃車的移動軌跡。

- 可以使用多執行緒（Multithreaded Tracking）加快處理速度，請參閱官網。

第 8 章　物件偵測（Object Detection）

8-16　YOLO 測試心得

Ultralytics 已經將 YOLO 產品化，不僅可載入自家模型，也可載入其他版本的模型，同時也擴充任務變成解決方案，使 YOLO 功能極大化，儼然已是 AI 影像辨識的經典演算法（SOTA），企業應可大膽應用在各項業務上，包括銷售、結帳、服務、行銷…等，軍事上也有許多應用，例如無人機、機器狗、飛彈…等，另外，人流計算（Object Counter）、物件邊緣偵測、標誌（Logo）/ 交通號誌偵測、手勢偵測…等，各種整合式的解決方案可參閱『Ultralytics Solutions: Harness YOLO11 to Solve Real-World Problems』[54]，均附帶範例程式，可直接執行、測試。

另外有幾點要特別提醒：

1. 筆者在測試時，執行指令時誤用任務及模型，致使資料與模型不一致，發生以下錯誤，讀者如有類似情形，可將資料資料夾刪除，再修正指令重新執行。

```
File "C:\Users\mikec\anaconda3\Lib\site-packages\ultralytics\data\dataset.py", line 64, in __init__
    super().__init__(*args, **kwargs)
File "C:\Users\mikec\anaconda3\Lib\site-packages\ultralytics\data\base.py", line 74, in __init__
    self.labels = self.get_labels()
                  ^^^^^^^^^^^^^^^^^
File "C:\Users\mikec\anaconda3\Lib\site-packages\ultralytics\data\dataset.py", line 161, in get_labels
    len_cls, len_boxes, len_segments = (sum(x) for x in zip(*lengths))
                                        ^^^^^^^^^^^^^^^^^^^^^^^^^^^^^
ValueError: not enough values to unpack (expected 3, got 0)
```

2. 除了上述的資料集外，YOLO v11 還提供各種任務的資料集列表[55]，讓使用者可自行下載並訓練模型。當然我們也可以自行蒐集資料，訓練模型。

8-17　總結

這一章我們認識了許多物件偵測的演算法，包括 HOG、R-CNN、YOLO，同時也實作許多範例，像是傳統的影像金字塔、R-CNN、PyTorch Detectron2、YOLO、TensorFlow Object Detection API，還包含圖像和視訊偵測，也可自訂資料集訓練模型，證明我們的確有能力，將物件偵測技術導入到專案中使用。

演算法各有優劣，Faster R-CNN 雖然較慢，但準確度高，儘管 YOLO 早期為了提升執行速度犧牲了準確度，但經過幾個版本升級後，準確度也已大幅提高。所以建議讀者在實際應用時，還是應該多方嘗試，找出最適合的模型，譬如在邊緣運算的場域使用較輕量的模型，不只辨識速度快，更要節省記憶體的使用。

參考資料（References）

在完稿的同時，YOLO v12 已經發表了，真是學無止境，AI 浪潮一波接著一波，不過，功能完整性還是以 Ultralytics 的 YOLO v11 為首，不必擔心所學又過時了。

參考資料（References）

[1] Joseph Redmon、Anelia Angelova,《Real-Time Grasp Detection Using Convolutional Neural Networks》, 2015
(https://docs.google.com/presentation/d/1Zc9-iR1eVz-zysinwb7bzLGC2no2ZiaD897_14dGbhw/edit?usp=sharing)

[2] 2011 年 ImageNet ILSVRC 挑戰賽比賽說明
(http://image-net.org/challenges/LSVRC/2011/index)

[3] 2017 年 ImageNet ILSVRC 挑戰賽比賽說明
(http://image-net.org/challenges/LSVRC/2017/)

[4] Fei-Fei Li、Justin Johnson、Serena Yeung,《Lecture 11: Detection and Segmentation》, 2017
(http://cs231n.stanford.edu/slides/2017/cs231n_2017_lecture11.pdf)

[5] Adrian Rosebrock,《Image Pyramids with Python and OpenCV》, 2015
(https://www.pyimagesearch.com/2015/03/16/image-pyramids-with-python-and-opencv/)

[6] IIP Image
(https://iipimage.sourceforge.io/documentation/images/)

[7] Adrian Rosebrock,《Sliding Windows for Object Detection with Python and OpenCV》, 2015
(https://www.pyimagesearch.com/2015/03/23/sliding-windows-for-object-detection-with-python-and-opencv/)

[8] CV17 HOG 特徵提取算法
(https://blog.csdn.net/henghuizan2771/article/details/124186797)

[9] Adrian Rosebrock,《Histogram of Oriented Gradients and Object Detection》, 2014
(https://www.pyimagesearch.com/2014/11/10/histogram-oriented-gradients-object-detection/)

第 8 章　物件偵測（Object Detection）

[10] Adrian Rosebrock,《Non-Maximum Suppression for Object Detection in Python》, 2014

(https://www.pyimagesearch.com/2014/11/17/non-maximum-suppression-object-detection-python/)

[11] Tomasz Malisiewicz,《Ensemble of Exemplar-SVMs for Object Detection and Beyond》

(http://www.cs.cmu.edu/~tmalisie/projects/iccv11/index.html)

[12] Ross Girshick、Jeff Donahue、Trevor Darrell、Jitendra Malik,《Rich feature hierarchies for accurate object detection and semantic segmentation》, 2014

(https://arxiv.org/pdf/1311.2524.pdf)

[13] RCNN GitHub

(https://github.com/1297rohit/RCNN)

[14] Lung-Ying Ling, R-CNN 學習筆記, 2019

(https://laptrinhx.com/r-cnn-xue-xi-bi-ji-1145354539/)

[15] Kaiming He、Xiangyu Zhang、Shaoqing Ren、Jian Sun,《Spatial Pyramid Pooling in Deep Convolutional Networks for Visual Recognition》, 2015

(https://arxiv.org/abs/1406.4729)

[16] v1_vivian,《SPP-Net 論文詳解》, 2017

(https://www.itread01.com/content/1542334444.html)

[17] Ross B. Girshick 於 GitHub 上放置的 Faster R-CNN 程式碼

(https://github.com/rbgirshick/py-faster-rcnn)

[18] Getting Started with Detectron2

(https://detectron2.readthedocs.io/en/latest/tutorials/getting_started.html)

[19] Detectron2 範例檔

(https://colab.research.google.com/drive/16jcaJoc6bCFAQ96jDe2HwtXj7BMD_-m5)

[20] Segment Anything Model 2 (SAM 2)

(https://github.com/facebookresearch/sam2)

參考資料（References）

[21] You Only Look Once: Unified, Real-Time Object Detection
(https://docs.google.com/presentation/d/1kAa7NOamBt4calBU9iHgT8a86RRHz9Yz2oh4-GTdX6M/edit#slide=id.p)

[22] 人工智慧技術 YOLOv4 出自中研院團隊
(https://news.ltn.com.tw/news/life/breakingnews/3215570)

[23] Ultralytics 網站首頁
(https://docs.ultralytics.com/)

[24] Ultralytics GitHub
(https://github.com/ultralytics/ultralytics)

[25] Joseph Redmon、Santosh Divvala、Ross Girshick、Ali Farhadi,《You Only Look Once: Unified, Real-Time Object Detection》, 2016
(https://docs.google.com/presentation/d/1kAa7NOamBt4calBU9iHgT8a86RRHz9Yz2oh4-GTdX6M/edit?usp=sharing)

[26] PyTorch 官網首頁
(https://pytorch.org/)

[27] Ultralytics 說明文件
(https://docs.ultralytics.com/quickstart/)

[28] YOLO 解決方案
(https://docs.ultralytics.com/solutions/#solutions)

[29] COCO 類別
(https://github.com/ultralytics/ultralytics/blob/main/ultralytics/cfg/datasets/coco.yaml)

[30] YOLO 圖像分類文件說明
(https://docs.ultralytics.com/tasks/classify/)

[31] Cat-or-not GitHub
(https://github.com/rpeden/cat-or-not/releases)

[32] YOLO Object detection 文件說明
(https://docs.ultralytics.com/tasks/detect/)

第 8 章　物件偵測（Object Detection）

[33] Object Detection Datasets Overview
(https://docs.ultralytics.com/datasets/detect/)

[34] TensorFlow Object Detection
(https://www.tensorflow.org/hub/tutorials/object_detection)

[35] Tensorflow Object Detection API 官網文件
(https://tensorflow-object-detection-api-tutorial.readthedocs.io/en/latest/training.html)

[36] YOLO v11 Simple Utilities
(https://docs.ultralytics.com/usage/simple-utilities/)

[37] LabelImg
(https://github.com/tzutalin/labelImg)

[38] Label Studio
(https://labelstud.io/)

[39] YOLO v11 Instance Segmentation
(https://docs.ultralytics.com/tasks/segment/)

[40] Instance Segmentation Datasets Overview
(https://docs.ultralytics.com/datasets/segment/#supported-datasets)

[41] YOLO v11 Pose Estimation
(https://docs.ultralytics.com/tasks/pose/)

[42] Pose Estimation Datasets Overview
(https://docs.ultralytics.com/datasets/pose/#dataset-yaml-format)

[43] Oz Ramos,《Introducing Handsfree.js - Integrate hand, face, and pose gestures to your frontend》
(https://dev.to/midiblocks/introducing-handsfree-js-integrate-hand-face-and-pose-gestures-to-your-frontend-4g3p)

[44] Jen Looper,《Ombromanie: Creating Hand Shadow stories with Azure Speech and TensorFlow.js Handposes》
(https://dev.to/azure/ombromanie-creating-hand-shadow-stories-with-azure-speech-and-tensorflow-js-handposes-3cln)

參考資料（References）

[45] YOLO v11 Oriented Bounding Boxes Object Detection
(https://docs.ultralytics.com/tasks/obb/)

[46] Oriented Bounding Box (OBB) Datasets Overview
(https://docs.ultralytics.com/datasets/obb/)

[47] 清大獲得 2018 NVIDIA 機器人冠軍
(https://www.youtube.com/watch?v=_OqdnG4AII8)

[48] YOLO v11, Multi-Object Tracking with Ultralytics YOLO
(https://docs.ultralytics.com/modes/track/)

[49+]BoT-SORT GitHub
(https://github.com/NirAharon/BoT-SORT)

[50] ByteTrack GitHub
(https://github.com/ifzhang/ByteTrack)

[51] 多目標跟蹤的性能評估：指標與計算方法
(https://cloud.baidu.com/article/3271389)

[52] Multi-Object Tracking with Ultralytics YOLO
(https://docs.ultralytics.com/modes/track/#available-trackers)

[53] Ultralytics GitHub
(https://github.com/ultralytics/ultralytics/tree/main/ultralytics/cfg/trackers)

[54] Ultralytics Solutions: Harness YOLO11 to Solve Real-World Problems
(https://docs.ultralytics.com/solutions/)

[55] YOLO 各種任務的資料集列表
(https://docs.ultralytics.com/datasets/#ultralytics-explorer)

第 8 章　物件偵測（Object Detection）

MEMO

第 9 章

生成式 AI
(Generative AI)

生成式 AI (Generative AI),簡稱 Gen AI 或 GAI,在 2022 年引爆一股熱潮,先有 MidJourney,輸入提示可生成精美圖像,有人使用 MidJourney 生成的圖像參加數位藝術競賽,竟然獲得首獎,之後又有攝影大師在獲得 Sony 世界攝影大獎優勝後,宣稱他的作品是以 MidJourney 生成的,圖像生成工具就此如雨後春筍般推陳出新,演變至今,生成的人像品質已直逼真人的照片,例如 Flux、Stable Diffusion…等工具。

第 9 章　生成式 AI（Generative AI）

▲ 圖 9.1 使用 MidJourney 生成的圖像獲得數位藝術競賽首獎，
圖片來源：An AI-Generated Painting Wins Art Contest, Annoys Artists [1]

▲ 圖 9.2 使用 MidJourney 獲得 Sony 世界攝影大獎優勝，
圖片來源：Sony 世界攝影大獎優勝作品「竟是 AI 畫的」[2]

緊接著 2022 年 11 月 OpenAI 推出 ChatGPT，輸入提示可生成論文、程式碼、翻譯、摘要、關鍵詞（Keywords）、宣傳口號（Slogan）…等，生成式 AI 就此佔據新聞版面，在『主權 AI』（Sovereign AI）的訴求下，各國也紛紛投入大語言模型（LLM）的訓練，針對本國語言 / 文化特色建立特有的基礎模型（Foundation Model），各行各業也針對產業特性進行模型微調（Fine Tuning）或 RAG，希望建構企業專屬的知識庫，提升企業運作效率。

本章先就圖像進行討論，後續章節談到自然語言處理時，再說明文字及語音生成，本章內容包括：

- 編碼器與解碼器（Encoder-decoder）網路模型。

- 生成對抗網路（GAN）。

- 擴散模型（Diffusion Model）。

- Stable Diffusion 及 Dalle-E 實作。

9-1 編碼器與解碼器（Encoder-decoder）

編碼器與解碼器（Encoder-decoder）原理是先利用 CNN 進行特徵萃取（Feature Extraction），萃取的特徵向量再透過反卷積神經網路重建影像，如下圖：

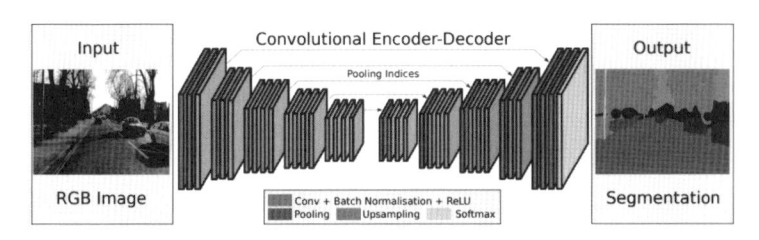

▲ 圖 9.3 編碼器與解碼器的示意圖，圖片來源：SegNet: A Deep Convolutional Encoder-Decoder Architecture for Image Segmentation [3]

許多生成式 AI 的演算法都以 Encoder-decoder 架構為框架，建構出更複雜的模型，以下我們先來探究一個較簡單的模型 -- AutoEncoder。

9-2 自動編碼器（AutoEncoder）

自動編碼器（AutoEncoder, AE）是編碼器與解碼器『對稱』的網路模型，透過編碼器萃取訓練資料的共同特徵，雜訊會被過濾掉，接著解碼器再依據特徵向量重建影像，達到『去雜訊』（Denosing）的功用，模型也可以再擴展至語義分割（Semantic segmentation）、風格轉換（Style Transfer）、U-net、生成對抗網路（GAN）…等演算法。

AutoEncoder 由 Encoder 與 Decoder 兩個子網路組合而成：

- 編碼器（Encoder）：即為萃取特徵的過程，類似於 CNN 模型，但不含最後的分類層（Dense），萃取的特徵向量也稱為『隱含表徵』（Latent Representation）。
- 解碼器（Decoder）：根據萃取的特徵來重建影像。

▲ 圖 9.4 自動編碼器（AutoEncoder）示意圖

第 9 章　生成式 AI（Generative AI）

接下來，我們實作 AutoEncoder，使用 MNIST 資料集，示範如何將雜訊去除。

範例. 實作 AutoEncoder，進行雜訊去除。

程式：**09_01_MNIST_Autoencoder.ipynb**。

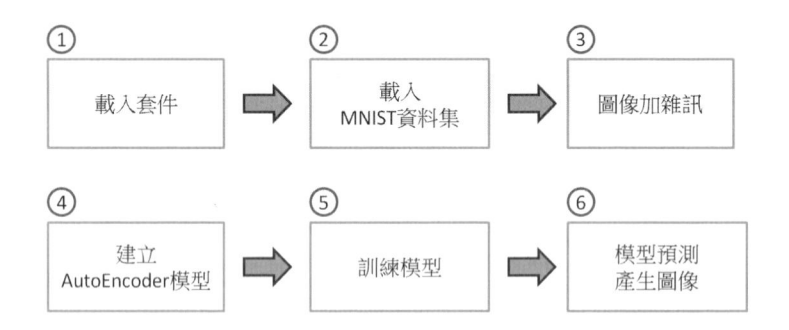

```
①載入套件  →  ②載入MNIST資料集  →  ③圖像加雜訊
④建立AutoEncoder模型  →  ⑤訓練模型  →  ⑥模型預測產生圖像
```

1. 載入套件。

```
1  # 載入相關套件
2  import numpy as np
3  import tensorflow as tf
4  import tensorflow.keras as K
5  import matplotlib.pyplot as plt
6  from tensorflow.keras.layers import Dense, Conv2D, MaxPooling2D, UpSampling2D
```

2. 超參數設定。

```
1  # 超參數設定
2  batch_size = 128      # 訓練批量
3  max_epochs = 50       # 訓練執行週期
4  filters = [32,32,16]  # 三層卷積層的輸出個數
```

3. 取得 MNIST 訓練資料，只取圖像 (x)，不需要 Label（Y），因為程式只要進行特徵萃取，不用辨識。

```
1   # 只取 X ，不需 Y
2   (x_train, _), (x_test, _) = K.datasets.mnist.load_data()
3
4   # 常態化
5   x_train = x_train / 255.
6   x_test = x_test / 255.
7
8   # 加一維：色彩
9   x_train = np.reshape(x_train, (len(x_train),28, 28, 1))
10  x_test = np.reshape(x_test, (len(x_test), 28, 28, 1))
```

9-2 自動編碼器（AutoEncoder）

4. 在圖像中加入隨機雜訊，以利後續實驗，觀察 AutoEncoder 是否能去除雜訊。

```python
1  # 在既有圖像加雜訊
2  noise = 0.5
3
4  # 固定隨機亂數
5  np.random.seed(11)
6  tf.random.set_seed(11)
7
8  # 隨機加雜訊
9  x_train_noisy = x_train + noise * np.random.normal(loc=0.0,
10                                    scale=1.0, size=x_train.shape)
11 x_test_noisy = x_test + noise * np.random.normal(loc=0.0,
12                                    scale=1.0, size=x_test.shape)
13
14 # 加完裁切數值，避免大於 1
15 x_train_noisy = np.clip(x_train_noisy, 0, 1)
16 x_test_noisy = np.clip(x_test_noisy, 0, 1)
17
18 # 轉換為浮點數
19 x_train_noisy = x_train_noisy.astype('float32')
20 x_test_noisy = x_test_noisy.astype('float32')
```

5. 建立編碼器（Encoder）模型：使用卷積層與池化層，call 函數會在物件建立後自動被呼叫，串連各個神經層，這是以類別(Class)建構神經網路的方式，與 PyTorch 很類似。

```python
1  # 編碼器(Encoder)
2  class Encoder(K.layers.Layer):
3      def __init__(self, filters):
4          super(Encoder, self).__init__()
5          self.conv1 = Conv2D(filters=filters[0], kernel_size=3, strides=1,
6                              activation='relu', padding='same')
7          self.conv2 = Conv2D(filters=filters[1], kernel_size=3, strides=1,
8                              activation='relu', padding='same')
9          self.conv3 = Conv2D(filters=filters[2], kernel_size=3, strides=1,
10                             activation='relu', padding='same')
11         self.pool = MaxPooling2D((2, 2), padding='same')
12
13
14     def call(self, input_features):
15         x = self.conv1(input_features)
16         #print("Ex1", x.shape)
17         x = self.pool(x)
18         #print("Ex2", x.shape)
19         x = self.conv2(x)
20         x = self.pool(x)
21         x = self.conv3(x)
22         x = self.pool(x)
23         return x
```

第 9 章　生成式 AI（Generative AI）

6. 建立解碼器（Decoder）模型：使用卷積層和上採樣層（UpDampling2D），卷積層的輸出個數與 Encoder 相反，以便將圖像還原成原圖尺寸，上採樣層與池化層作用相反，是將圖像放大。

```
1  # 解碼器(Decoder)
2  class Decoder(K.layers.Layer):
3      def __init__(self, filters):
4          super(Decoder, self).__init__()
5          self.conv1 = Conv2D(filters=filters[2], kernel_size=3, strides=1,
6                              activation='relu', padding='same')
7          self.conv2 = Conv2D(filters=filters[1], kernel_size=3, strides=1,
8                              activation='relu', padding='same')
9          self.conv3 = Conv2D(filters=filters[0], kernel_size=3, strides=1,
10                             activation='relu', padding='valid')
11         self.conv4 = Conv2D(1, 3, 1, activation='sigmoid', padding='same')
12         self.upsample = UpSampling2D((2, 2))
13
14     def call(self, encoded):
15         x = self.conv1(encoded)
16         # 上採樣
17         x = self.upsample(x)
18
19         x = self.conv2(x)
20         x = self.upsample(x)
21
22         x = self.conv3(x)
23         x = self.upsample(x)
24
25         return self.conv4(x)
```

7. 結合編碼器（Encoder）、解碼器（Decoder），建立 AutoEncoder 模型。

```
1  # 建立 Autoencoder 模型
2  class Autoencoder(K.Model):
3      def __init__(self, filters):
4          super(Autoencoder, self).__init__()
5          self.loss = []
6          self.encoder = Encoder(filters)
7          self.decoder = Decoder(filters)
8
9      def call(self, input_features):
10         #print(input_features.shape)
11         encoded = self.encoder(input_features)
12         #print(encoded.shape)
13         reconstructed = self.decoder(encoded)
14         #print(reconstructed.shape)
15         return reconstructed
```

8. 訓練模型。

9-2 自動編碼器（AutoEncoder）

```
1  model = Autoencoder(filters)
2
3  model.compile(loss='binary_crossentropy', optimizer='adam')
4
5  loss = model.fit(x_train_noisy,
6                   x_train,
7                   validation_data=(x_test_noisy, x_test),
8                   epochs=max_epochs,
9                   batch_size=batch_size)
```

9. 繪製損失函數。

```
1  # 繪製損失函數
2  plt.plot(range(max_epochs), loss.history['loss'])
3  plt.xlabel('Epochs')
4  plt.ylabel('Loss')
5  plt.show()
```

- 執行結果：損失隨著訓練次數越來越小，且趨於收斂。

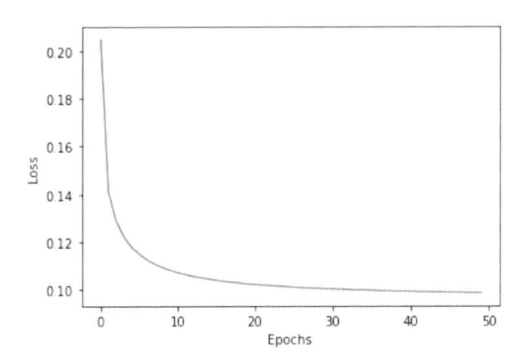

10. 比較含雜訊的圖像與去除雜訊後的圖像。

```
1  number = 10   # how many digits we will display
2  plt.figure(figsize=(20, 4))
3  for index in range(number):
4      # 加了雜訊的圖像
5      ax = plt.subplot(2, number, index + 1)
6      plt.imshow(x_test_noisy[index].reshape(28, 28), cmap='gray')
7      ax.get_xaxis().set_visible(False)
8      ax.get_yaxis().set_visible(False)
9
10     # 重建的圖像
11     ax = plt.subplot(2, number, index + 1 + number)
12     plt.imshow(tf.reshape(model(x_test_noisy)[index], (28, 28)), cmap='gray')
13     ax.get_xaxis().set_visible(False)
14     ax.get_yaxis().set_visible(False)
15 plt.show()
```

第 9 章　生成式 AI（Generative AI）

- 執行結果：效果相當好，雜訊被有效剔除。

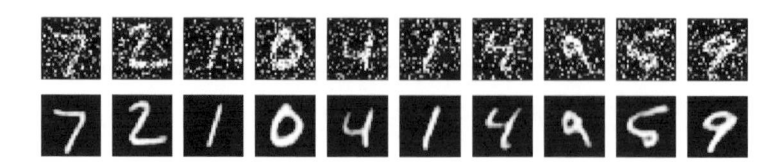

自動編碼器（AutoEncoder, AE）除了影像生成外，還有許多應用，包括：

1. 降維（Dimensionality Reduction）：減少輸入特徵的維度。
2. 異常檢測（Anomaly Detection）：AutoEncoder 預測值與實際值若有顯著差異，即可被視為離群值（Outlier）。
3. 去躁（Denosing）：如上例，如果輸入的影像有雜訊，經過 AutoEncoder 推論後，會得到乾淨的影像。
4. 影像修復（Image Inpainting）：影像若有缺陷，例如照片存放過久，會產生斑點，可利用 AutoEncoder 修復，如下圖。

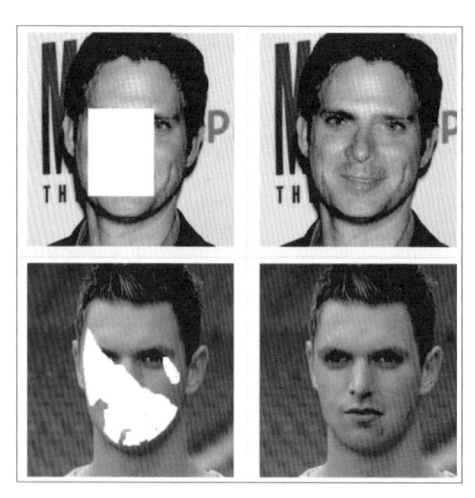

▲ 圖 9.5　影像修復（Image Inpainting）

5. 推薦系統（Recommender System）：可將使用者與商品轉換為『隱含表徵』（Latent Representation），再據以推測使用者偏好。
6. 語義分割（Semantic segmentation）：下面有範例說明。

想瞭解詳細內容，可參閱『PyImageSearch, Introduction to Autoencoders』[4]。

9-3 變分自編碼器（Variational AutoEncoder）

AutoEncoder 屬於非監督式學習算法，不需要標記（Labeling），透過特徵萃取得到每一類別資料的共同特徵，如果，將萃取的特徵由固定的向量改為常態機率分配，在訓練時估計其平均數與變異數，解碼時依據機率分配進行隨機抽樣，每次輸出都不一樣，生成的影像就具有多樣性，這種演算法稱為 Variational AutoEncoder（VAE），屬於 AutoEncoder 的變形（Variants），VAE 常與生成對抗網路（GAN）相提並論，也是現今最夯的擴散模型（Diffusion Model）的起源。

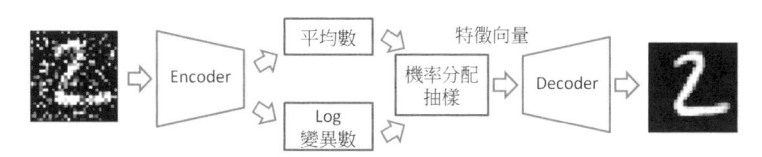

▲ 圖 9.6 Variational AutoEncoder（VAE）的架構

範例. 建立 VAE 模型，使用 MNIST 資料集，生成影像。

程式：**09_02_MNIST_VAE.ipynb**，程式修改自 Convolutional Variational Autoencoder（CAE）[5]。

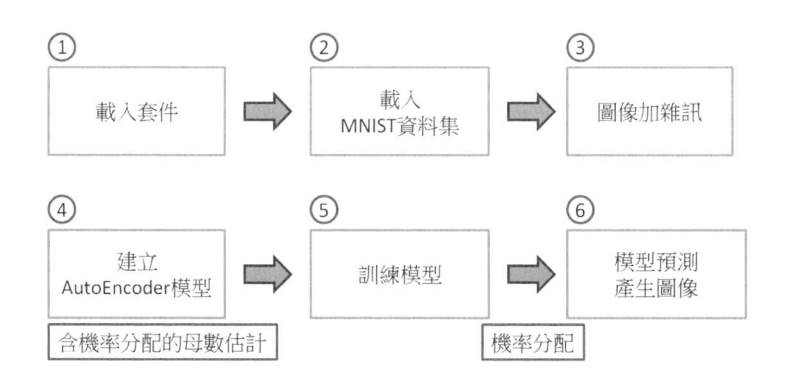

* VAE 的編碼器輸出不是特徵向量，而是機率分配的母數 μ 和 $\log(\delta)$。

1. 安裝 TensorFlow-Probability 套件：TensorFlow-Probability 套件提供機率分配的相關函數，與 TensorFlow 的資料結構相容。

```
1  # tensorflow 機率套件
2  !pip install tensorflow-probability
```

第 9 章　生成式 AI (Generative AI)

```
3
4  # 生成動畫(GIF)的套件
5  !pip install imageio
```

2. 載入套件。

```
1  from IPython import display
2  import glob
3  import imageio
4  import matplotlib.pyplot as plt
5  import numpy as np
6  import PIL
7  import tensorflow as tf
8  import tensorflow_probability as tfp
9  import time
```

3. 取得 MNIST 訓練資料。

```
1  (train_images, _), (test_images, _) = tf.keras.datasets.mnist.load_data()
```

4. 前置處理，像素轉換為 0 或 1。

```
1  def preprocess_images(images):
2    images = images.reshape((images.shape[0], 28, 28, 1)) / 255.
3    return np.where(images > .5, 1.0, 0.0).astype('float32')
4
5  train_images = preprocess_images(train_images)
6  test_images = preprocess_images(test_images)
```

5. 超參數設定。

```
1  train_size = 60000
2  batch_size = 32
3  test_size = 10000
```

6. 轉換為 dataset，隨機抽樣。

```
1  train_dataset = (tf.data.Dataset.from_tensor_slices(train_images)
2                   .shuffle(train_size).batch(batch_size))
3  test_dataset = (tf.data.Dataset.from_tensor_slices(test_images)
4                  .shuffle(test_size).batch(batch_size))
```

9-10

9-3 變分自編碼器（Variational AutoEncoder）

7. 定義 VAE 模型：與上一節範例類似，除了 encoder 最後一層，decoder 採反卷積（Conv2DTranspose）。

```python
class CVAE(tf.keras.Model):
    def __init__(self, latent_dim):
        super(CVAE, self).__init__()
        self.latent_dim = latent_dim
        self.encoder = tf.keras.Sequential(
            [
                tf.keras.layers.Input((28, 28, 1)),
                tf.keras.layers.Conv2D(
                    filters=32, kernel_size=3, strides=(2, 2), activation='relu'),
                tf.keras.layers.Conv2D(
                    filters=64, kernel_size=3, strides=(2, 2), activation='relu'),
                tf.keras.layers.Flatten(),
                # No activation
                tf.keras.layers.Dense(latent_dim + latent_dim),
            ]
        )

        self.decoder = tf.keras.Sequential(
            [
                tf.keras.layers.Input((latent_dim,)),
                tf.keras.layers.Dense(units=7*7*32, activation=tf.nn.relu),
                tf.keras.layers.Reshape(target_shape=(7, 7, 32)),
                tf.keras.layers.Conv2DTranspose(
                    filters=64, kernel_size=3, strides=2, padding='same',
                    activation='relu'),
                tf.keras.layers.Conv2DTranspose(
                    filters=32, kernel_size=3, strides=2, padding='same',
                    activation='relu'),
                # No activation
                tf.keras.layers.Conv2DTranspose(
                    filters=1, kernel_size=3, strides=1, padding='same'),
            ]
        )
```

8. 定義上述類別的抽樣函數、估計平均數（mu）和 Log 變異數（logvar）、產生雜訊與解碼。由於反向傳導不能流經隨機神經層，因此定義 reparameterize 函數讓『隱含表徵』（Latent Representation）輸出等於平均數 +（Log 變異數 *0.5）平方 * 雜訊（ε）。

$$z = \mu + \sigma \odot \epsilon$$

第 9 章　生成式 AI（Generative AI）

```
35    @tf.function
36    def sample(self, eps=None):
37      if eps is None:
38        eps = tf.random.normal(shape=(100, self.latent_dim))
39      return self.decode(eps, apply_sigmoid=True)
40
41    def encode(self, x):
42      mean, logvar = tf.split(self.encoder(x), num_or_size_splits=2, axis=1)
43      return mean, logvar
44
45    def reparameterize(self, mean, logvar):
46      eps = tf.random.normal(shape=mean.shape)
47      return eps * tf.exp(logvar * .5) + mean
48
49    def decode(self, z, apply_sigmoid=False):
50      logits = self.decoder(z)
51      if apply_sigmoid:
52        probs = tf.sigmoid(logits)
53        return probs
54      return logits
```

9. 定義特殊的損失函數（loss）：主要用於最小化實際與理論機率分配之差。

```
1  optimizer = tf.keras.optimizers.Adam(1e-4)
2
3  def log_normal_pdf(sample, mean, logvar, raxis=1):
4    log2pi = tf.math.log(2. * np.pi)
5    return tf.reduce_sum(
6        -.5 * ((sample - mean) ** 2. * tf.exp(-logvar) + logvar + log2pi),
7        axis=raxis)
8
9  def compute_loss(model, x):
10   mean, logvar = model.encode(x)
11   z = model.reparameterize(mean, logvar)
12   x_logit = model.decode(z)
13   cross_ent = tf.nn.sigmoid_cross_entropy_with_logits(logits=x_logit, labels=x)
14   logpx_z = -tf.reduce_sum(cross_ent, axis=[1, 2, 3])
15   logpz = log_normal_pdf(z, 0., 0.)
16   logqz_x = log_normal_pdf(z, mean, logvar)
17   return -tf.reduce_mean(logpx_z + logpz - logqz_x)
18
19 @tf.function
20 def train_step(model, x, optimizer):
21   with tf.GradientTape() as tape:
22     loss = compute_loss(model, x)
23   gradients = tape.gradient(loss, model.trainable_variables)
24   optimizer.apply_gradients(zip(gradients, model.trainable_variables))
```

9-3 變分自編碼器（Variational AutoEncoder）

10. 定義重建影像的函數。

```
1  def generate_and_save_images(model, epoch, test_sample):
2    mean, logvar = model.encode(test_sample)
3    z = model.reparameterize(mean, logvar)
4    predictions = model.sample(z)
5    fig = plt.figure(figsize=(4, 4))
6
7    for i in range(predictions.shape[0]):
8      plt.subplot(4, 4, i + 1)
9      plt.imshow(predictions[i, :, :, 0], cmap='gray')
10     plt.axis('off')
11
12   # tight_layout minimizes the overlap between 2 sub-plots
13   plt.savefig('image_at_epoch_{:04d}.png'.format(epoch))
14   plt.show()
```

11. 訓練模型：訓練同時生成影像。

```
1  epochs = 10
2  # set the dimensionality of the latent space to a plane for visualization later
3  latent_dim = 2
4  num_examples_to_generate = 16
5
6  # keeping the random vector constant for generation (prediction) so
7  # it will be easier to see the improvement.
8  random_vector_for_generation = tf.random.normal(
9      shape=[num_examples_to_generate, latent_dim])
10 model = CVAE(latent_dim)
11
12 # Pick a sample of the test set for generating output images
13 assert batch_size >= num_examples_to_generate
14 for test_batch in test_dataset.take(1):
15   test_sample = test_batch[0:num_examples_to_generate, :, :, :]
16
17 generate_and_save_images(model, 0, test_sample)
18
19 for epoch in range(1, epochs + 1):
20   start_time = time.time()
21   for train_x in train_dataset:
22     train_step(model, train_x, optimizer)
23   end_time = time.time()
24
25   loss = tf.keras.metrics.Mean()
26   for test_x in test_dataset:
27     loss(compute_loss(model, test_x))
28   elbo = -loss.result()
29   display.clear_output(wait=False)
```

第 9 章　生成式 AI（Generative AI）

```
30    print('Epoch: {}, Test set ELBO: {}, time elapse for current epoch: {}'
31          .format(epoch, elbo, end_time - start_time))
32    generate_and_save_images(model, epoch, test_sample)
```

12. 顯示最後結果。

```
1  def display_image(epoch_no):
2    return PIL.Image.open('image_at_epoch_{:04d}.png'.format(epoch_no))
3
4  plt.imshow(display_image(epoch))
5  plt.axis('off')   # Display images
```

13. 執行結果：剛開始是一片雜訊，之後不斷更新，愈來愈清晰。

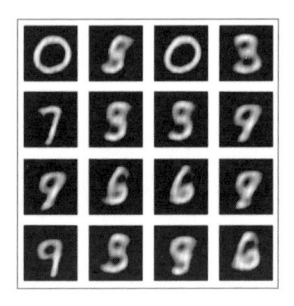

14. 將所有執行週期的輸出圖檔合成動畫檔（GIF）。

```
1  anim_file = 'cvae.gif'
2
3  with imageio.get_writer(anim_file, mode='I') as writer:
4    filenames = glob.glob('image*.png')
5    filenames = sorted(filenames)
6    for filename in filenames:
7      image = imageio.imread(filename)
8      writer.append_data(image)
9    image = imageio.imread(filename)
10   writer.append_data(image)
```

15. 從『隱含表徵』（Latent Representation）產生圖像：

```
1  def plot_latent_images(model, n, digit_size=28):
2    """Plots n x n digit images decoded from the latent space."""
3
4    norm = tfp.distributions.Normal(0, 1)
5    grid_x = norm.quantile(np.linspace(0.05, 0.95, n))
6    grid_y = norm.quantile(np.linspace(0.05, 0.95, n))
```

9-14

9-3 變分自編碼器（Variational AutoEncoder）

```
7    image_width = digit_size*n
8    image_height = image_width
9    image = np.zeros((image_height, image_width))
10
11   for i, yi in enumerate(grid_x):
12     for j, xi in enumerate(grid_y):
13       z = np.array([[xi, yi]])
14       x_decoded = model.sample(z)
15       digit = tf.reshape(x_decoded[0], (digit_size, digit_size))
16       image[i * digit_size: (i + 1) * digit_size,
17             j * digit_size: (j + 1) * digit_size] = digit.numpy()
18
19   plt.figure(figsize=(10, 10))
20   plt.imshow(image, cmap='Greys_r')
21   plt.axis('Off')
22   plt.show()
23
24 plot_latent_images(model, 20)
```

- 執行結果：

第 9 章　生成式 AI（Generative AI）

9-4 Conditional VAE

以上的 AutoEncoder、Variational AutoEncoder（VAE）都是非監督式學習（Unsupervised Learning），沒有使用標籤（Label），即 Y，因此，生成影像時無法指定特定類別，生成內容無法控制，若要能指定類別，須將 Y 也輸入到模型中，此演算法稱為 Conditional VAE，模型結構如下圖，先將 Y 進行 One-hot encoding 轉換，再分別輸入至 Encoder、Decoder 中即可。

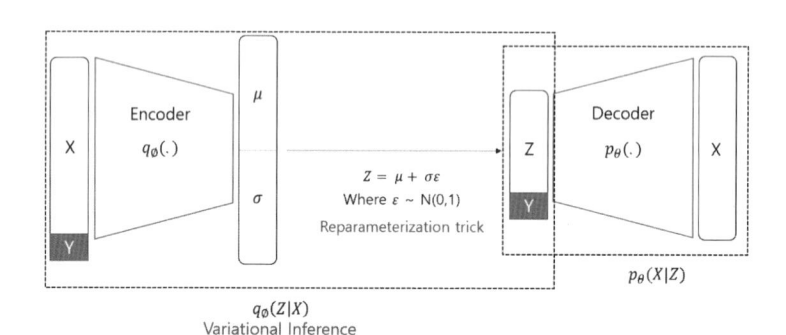

▲ 圖 9.7　Conditional VAE 模型結構，圖片來源：mingukkang, CVAE [6]

範例. 以 Conditional VAE 生成指定的阿拉伯數字。

程式：CVAE 資料夾，程式修改自 gozsoy, conditional-vae [7]。以下說明主程式 main.py 邏輯。

1. 資料前置處理：包括 X 的特徵縮放 / 標準化、Y 的 one-hot encoding，並轉換為 Dataset，依照上圖，我們要同時輸入 X/Y。

```
10  def prepare_data():
11      (x_train, y_train), (x_test, y_test) = tf.keras.datasets.mnist.load_data()
12
13      # 特徵縮放
14      x_train = x_train/255.
15      dataset_mean,dataset_std = np.mean(x_train),np.std(x_train)
16      # 標準化(standardization)
17      x_train = (x_train - dataset_mean) / (dataset_std)
18      x_train = np.expand_dims(x_train,axis=3)
19      x_train = tf.cast(x_train,dtype=tf.float32)
20
21      # convert labels into one-hot encoding
22      def convert_onehot(idx):
23          arr = np.zeros((10))
24          arr[idx] = 1.0
```

9-4 Conditional VAE

```
25            return arr
26      y_train_onehot = np.zeros((y_train.shape[0],10))
27      for idx,temp_y in enumerate(y_train):
28          y_train_onehot[idx] = convert_onehot(temp_y)
29
30      # 轉換為 Dataset
31      train_ds = tf.data.Dataset.from_tensor_slices((x_train,y_train_onehot))
32      train_ds = train_ds.shuffle(1000).batch(64)
33      train_ds = train_ds.prefetch(tf.data.AUTOTUNE)
34      return train_ds, dataset_mean, dataset_std
```

2. 定義 KL 損失：與上個範例類似。

```
36  # KL loss
37  def kl_loss(z_mu,z_rho):
38      sigma_squared = tf.math.softplus(z_rho) ** 2
39      kl_1d = -0.5 * (1 + tf.math.log(sigma_squared) - z_mu ** 2 - sigma_squared)
40      # sum over sample dim, average over batch dim
41      kl_batch = tf.reduce_mean(tf.reduce_sum(kl_1d,axis=1))
42      return kl_batch
43
44  @tf.function
45  def elbo(z_mu,z_rho,decoded_img,original_img):
46      # reconstruction loss
47      mse = tf.reduce_mean(tf.reduce_sum(tf.square(original_img - decoded_img),axis=1))
48      # kl loss
49      kl = kl_loss(z_mu, z_rho)
50      return mse, kl
```

3. 定義梯度下降函數：

- 模型定義在 model.py，採用 Functional API 定義模型，而非之前的順序型（Sequential）模型，因為模型有多個輸入（X、Y）。

- call 函數可在物件建立後呼叫並再給予參數，串連各個子神經網路。

- 模型結構與 VAE 類似，只是多了 Y 的輸入。

- 這一段程式在訓練時會產生警告訊息【是否忘記定義損失函數】（forget to provide a `loss` argument），但是作者明明有利用 elbo 定義損失函數，並與輸入的變數串連，因此筆者使用 tf.get_logger().setLevel（'ERROR'）隱藏警告訊息，讀者如果有發現解決方法，歡迎提供給筆者。

```
52  @tf.function
53  def train_step(imgs, labels):
54      # training loop
55      with tf.GradientTape() as tape:
56          # forward pass
```

```
57          z_mu, z_rho, decoded_imgs = model(imgs, labels)
58
59          # compute loss
60          mse, kl = elbo(z_mu,z_rho,decoded_imgs,imgs)
61          loss = mse + beta * kl
62
63      # compute gradients
64      gradients = tape.gradient(loss, model.trainable_variables)
65
66      # update weights
67      optimizer.apply_gradients(zip(gradients, model.trainable_variables))
68      kl_loss_tracker(loss)
69      mse_loss_tracker(loss)
70
71      return z_mu, z_rho, mse, kl
```

4. 訓練模型,並生成圖像:生成圖像的函數定義在 utils.py。

```
73  def train(latent_dim,beta,epochs,train_ds,dataset_mean,dataset_std):
74      for epoch in range(epochs):
75          label_list = []
76          z_mu_list = []
77
78          for _,(imgs, labels) in train_ds.enumerate():
79              z_mu, z_rho, mse, kl = train_step(imgs, labels)
80
81              # save encoded means and labels for latent space visualization
82              if len(label_list) == 0:
83                  label_list = labels
84              else:
85                  label_list = np.concatenate((label_list, labels))
86
87              if len(z_mu_list) == 0:
88                  z_mu_list = z_mu
89              else:
90                  z_mu_list = np.concatenate((z_mu_list, z_mu),axis=0)
93          # update metrics
94          kl_loss_tracker.update_state(beta * kl)
95          mse_loss_tracker.update_state(mse)
96
97          # generate new samples
98          generate_conditioned_digits(model,dataset_mean,dataset_std)
99
100         # display metrics at the end of each epoch.
101         epoch_kl,epoch_mse = kl_loss_tracker.result(),mse_loss_tracker.result()
102         print(f'epoch: {epoch}, mse: {epoch_mse:.4f}, kl_div: {epoch_kl:.4f}')
```

```
103
104        # reset metric states
105        kl_loss_tracker.reset_state()
106        mse_loss_tracker.reset_state()
107
108    return model,z_mu_list,label_list
```

5. 執行：

   python main.py

   - 結果儲存在 results/generated_conditoned_digits.png，每個數字生成 20 個，生成的圖像有些微差異。

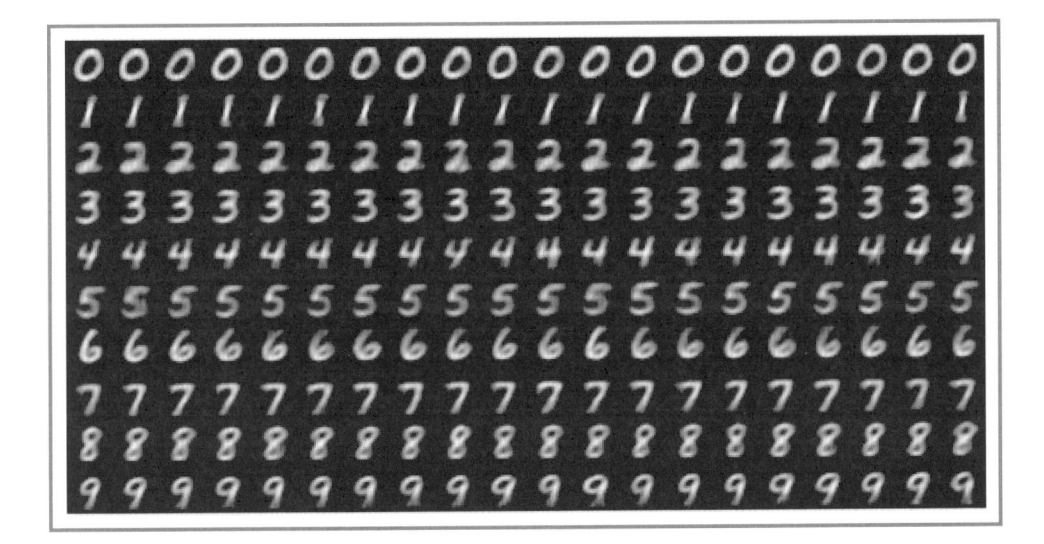

9-5 U-Net

U-Net 也是一種 AutoEncoder 的變形（Variant），由於它的模型結構為 U 型而得名，常用於語義分割（Semantic Segmentation）或稱影像分割（Image Segmen-ation），可作為醫療疾病的診斷，如黃斑部病變或腫瘤的偵測。

第 9 章　生成式 AI（Generative AI）

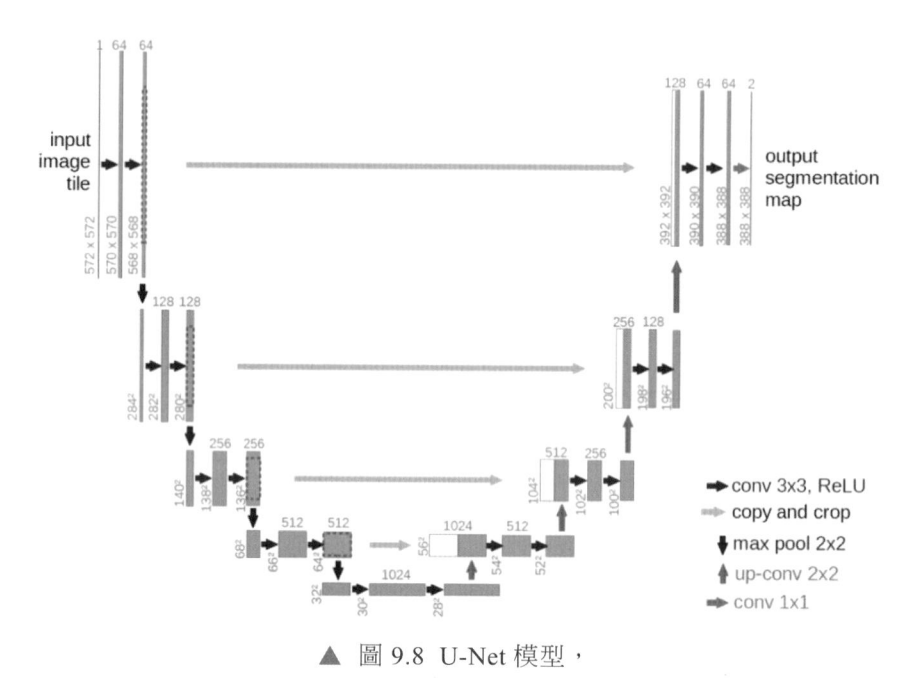

▲ 圖 9.8　U-Net 模型，

圖片來源：U-Net: Convolutional Networks for Biomedical Image Segmentation [6]

傳統 AutoEncoder 的問題點發生在前半段的編碼器（Encoder），由於它萃取特徵的過程，會使輸出的尺寸（Size）越變越小，接著解碼器（Decoder）再透過這些變小的特徵，重建出一個與原圖同樣大小的新圖像，因此原圖的很多資訊，像是前文所提到的雜訊，就沒辦法傳遞到解碼器了，這個特點應用在去除雜訊上是十分恰當的，但假若是要偵測異常點（如檢測黃斑部病變）的話，那就糟糕了，經過模型過濾後，異常點通通都不見了。

所以，U-Net 在原有編碼器與解碼器的聯繫上，增加了一些連結，每一段編碼器的輸出都與其對面的解碼器相連接，使得編碼器每一層的資訊，都會額外輸入到一樣尺寸的解碼器，如上圖橫跨 U 型兩側的中間長箭頭，這樣在重建的過程中就比較不會遺失重要資訊。

範例. 以 U-Net 實作語義分割。

程式：09_03_Image_segmentation.ipynb，修改自 PyImageSearch 的『U-Net Image Segmentation in Keras』[8]，它又是修改自 TensorFlow 官網的範例 [9]，故兩篇文章均可參閱。

9-5 U-Net

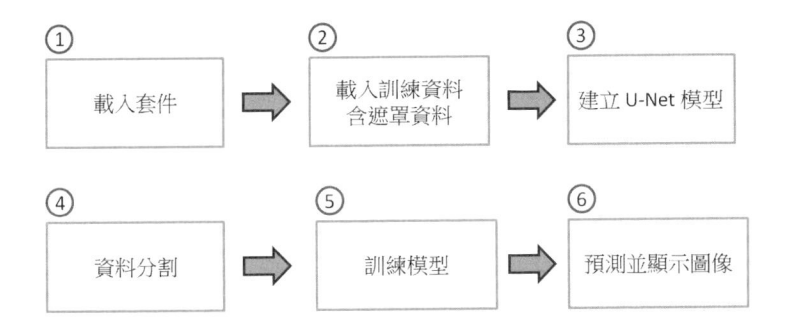

```
① 載入套件  →  ② 載入訓練資料含遮罩資料  →  ③ 建立 U-Net 模型
④ 資料分割  →  ⑤ 訓練模型  →  ⑥ 預測並顯示圖像
```

1. 安裝 tensorflow_datasets 套件：它提供許多資料集，包括本範例的寵物資料集（Pets），檔案為 TensorFlow 特有的 TFDS（TensorFlow Datasets）格式，把圖片及標籤（Label）合併為一個檔案，讀取較方便、迅速，也可以自 Oxford 大學資訊工程學系網址 [10] 下載原始圖檔及標籤，自行處理。

    - 原圖：http://www.robots.ox.ac.uk/~vgg/data/pets/data/images.tar.gz
    - 註解：http://www.robots.ox.ac.uk/~vgg/data/pets/data/annotations.tar.gz

2. 載入套件。

```
1  import tensorflow as tf
2  from tensorflow import keras
3  from tensorflow.keras import layers
4  import tensorflow_datasets as tfds
5  import matplotlib.pyplot as plt
6  import numpy as np
```

3. 下載資料：資料預設會存在 C:\Users\< 登入帳號 >\tensorflow_datasets \ox-ford_iiit_pet\4.0.0 資料夾。

```
1  dataset, info = tfds.load('oxford_iiit_pet:4.*.*', with_info=True)
2  print(info)
```

- 可能會出現過多警告訊息，超過 Jupyter Notebook 的限制，可以下列方式處理。
- 產生 Jupyter 組態檔：jupyter server --generate-config。
- 修改 C:\Users\< 使用者名稱 >\.jupyter\jupyter_server_config.py，尋找 c.ServerApp.iopub_msg_rate_limit，解除註解並改為：

    c.ServerApp.iopub_msg_rate_limit = 1000000

第 9 章　生成式 AI（Generative AI）

- 筆者在 WSL 及 Colaboratory 上執行，都出現以下錯誤，只有在 Windows 作業系統才能正常執行：

    NotImplementedError: `.as_dataset()` not implemented for ArrayRecord files. Please, use `.as_data_source()`.

4. 定義前置處理（Data preprocessing）函數。

```
# 圖像縮放
def resize(input_image, input_mask):
    input_image = tf.image.resize(input_image, (128, 128), method="nearest")
    input_mask = tf.image.resize(input_mask, (128, 128), method="nearest")
    return input_image, input_mask

# 資料增補
def augment(input_image, input_mask):
    if tf.random.uniform(()) > 0.5:
        # Random flipping of the image and mask
        input_image = tf.image.flip_left_right(input_image)
        input_mask = tf.image.flip_left_right(input_mask)
    return input_image, input_mask

# 特徵縮放
def normalize(input_image, input_mask):
    input_image = tf.cast(input_image, tf.float32) / 255.0
    input_mask -= 1
    return input_image, input_mask

def load_image_train(datapoint):
    input_image = datapoint["image"]
    input_mask = datapoint["segmentation_mask"]
    input_image, input_mask = resize(input_image, input_mask)
    input_image, input_mask = augment(input_image, input_mask)
    input_image, input_mask = normalize(input_image, input_mask)
    return input_image, input_mask

def load_image_test(datapoint):
    input_image = datapoint["image"]
    input_mask = datapoint["segmentation_mask"]
    input_image, input_mask = resize(input_image, input_mask)
    input_image, input_mask = normalize(input_image, input_mask)
    return input_image, input_mask
```

9-5 U-Net

5. 前置處理（Data preprocessing）：準備訓練、驗證及測試資料。

```
1  # 參數定義
2  BATCH_SIZE = 64
3  BUFFER_SIZE = 1000
4
5  # 訓練、驗證及測試資料準備
6  train_dataset = dataset["train"].map(load_image_train, num_parallel_calls=tf.data.AUTOTUNE)
7  test_dataset = dataset["test"].map(load_image_test, num_parallel_calls=tf.data.AUTOTUNE)
8  print(train_dataset)
9
10 train_batches = train_dataset.cache().shuffle(BUFFER_SIZE).batch(BATCH_SIZE).repeat()
11 train_batches = train_batches.prefetch(buffer_size=tf.data.experimental.AUTOTUNE)
12 validation_batches = test_dataset.take(3000).batch(BATCH_SIZE)
13 test_batches = test_dataset.skip(3000).take(669).batch(BATCH_SIZE)
14 print(train_batches)
```

6. 資料視覺化（Data Visualization）：顯示一批圖像。

```
1  def display(display_list):
2    plt.figure(figsize=(15, 15))
3
4    title = ["Input Image", "True Mask", "Predicted Mask"]
5
6    for i in range(len(display_list)):
7      plt.subplot(1, len(display_list), i+1)
8      plt.title(title[i])
9      plt.imshow(tf.keras.utils.array_to_img(display_list[i]))
10     plt.axis("off")
11   plt.show()
12
13 # 顯示一批資料
14 sample_batch = next(iter(test_batches))
15 random_index = np.random.choice(sample_batch[0].shape[0])
16 sample_image, sample_mask = sample_batch[0][random_index], sample_batch[1][random_index]
17 display([sample_image, sample_mask])
```

- 執行結果：左為原圖，右為遮罩圖（Mask），將貓的形狀標示出來。

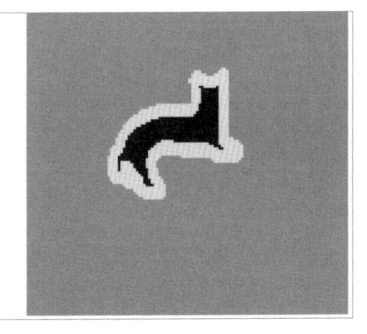

第 9 章 生成式 AI（Generative AI）

7. 建立 U-Net 模型模塊（Blocks）。

```python
def double_conv_block(x, n_filters):

    # Conv2D then ReLU activation
    x = layers.Conv2D(n_filters, 3, padding = "same", activation = "relu", kernel_initializer = "he_normal")(x)
    # Conv2D then ReLU activation
    x = layers.Conv2D(n_filters, 3, padding = "same", activation = "relu", kernel_initializer = "he_normal")(x)
    return x


def downsample_block(x, n_filters):
    f = double_conv_block(x, n_filters)
    p = layers.MaxPool2D(2)(f)
    p = layers.Dropout(0.3)(p)
    return f, p

def upsample_block(x, conv_features, n_filters):
    # upsample
    x = layers.Conv2DTranspose(n_filters, 3, 2, padding="same")(x)
    # concatenate
    x = layers.concatenate([x, conv_features])
    # dropout
    x = layers.Dropout(0.3)(x)
    # Conv2D twice with ReLU activation
    x = double_conv_block(x, n_filters)
    return x
```

8. 建立 U-Net 模型。

- 分為編碼器（encoder）及解碼器（decoder），中間連接處稱為 bottle-neck。
- 編碼器（encoder）及解碼器（decoder）是對稱的。
- 使用 tf.keras.utils.plot_model（unet_model, show_shapes=True），無法繪製漂亮的 U 型結構圖，會出現一長串的區塊。

```python
def build_unet_model():
    inputs = layers.Input(shape=(128,128,3))

    # encoder: contracting path - downsample
    # 1 - downsample
    f1, p1 = downsample_block(inputs, 64)
    # 2 - downsample
    f2, p2 = downsample_block(p1, 128)
    # 3 - downsample
    f3, p3 = downsample_block(p2, 256)
    # 4 - downsample
    f4, p4 = downsample_block(p3, 512)

    # 5 - bottleneck
    bottleneck = double_conv_block(p4, 1024)

    # decoder: expanding path - upsample
```

```
18      # 6 - upsample
19      u6 = upsample_block(bottleneck, f4, 512)
20      # 7 - upsample
21      u7 = upsample_block(u6, f3, 256)
22      # 8 - upsample
23      u8 = upsample_block(u7, f2, 128)
24      # 9 - upsample
25      u9 = upsample_block(u8, f1, 64)
26
27      # outputs
28      outputs = layers.Conv2D(3, 1, padding="same", activation = "softmax")(u9)
29      # unet model with Keras Functional API
30      unet_model = tf.keras.Model(inputs, outputs, name="U-Net")
31      return unet_model
```

9. 訓練模型：訓練需要數小時，可以先去散步、喝咖啡，再回來看結果。

```
1  NUM_EPOCHS = 20
2  TRAIN_LENGTH = info.splits["train"].num_examples
3  STEPS_PER_EPOCH = TRAIN_LENGTH // BATCH_SIZE
4  VAL_SUBSPLITS = 5
5  TEST_LENTH = info.splits["test"].num_examples
6  VALIDATION_STEPS = TEST_LENTH // BATCH_SIZE // VAL_SUBSPLITS
7
8  unet_model.compile(optimizer=tf.keras.optimizers.Adam(),
9                     loss="sparse_categorical_crossentropy",
10                    metrics=["accuracy"])
11 model_history = unet_model.fit(train_batches,
12                                epochs=NUM_EPOCHS,
13                                steps_per_epoch=STEPS_PER_EPOCH,
14                                validation_steps=VALIDATION_STEPS,
15                                validation_data=validation_batches)
```

10. 預測並顯示圖像。

```
1  def create_mask(pred_mask):
2    pred_mask = tf.argmax(pred_mask, axis=-1)
3    pred_mask = pred_mask[..., tf.newaxis]
4    return pred_mask[0]
5
6  def show_predictions(dataset=None, num=1):
7    if dataset:
8      for image, mask in dataset.take(num):
9        pred_mask = unet_model.predict(image)
10       display([image[0], mask[0], create_mask(pred_mask)])
```

第 9 章　生成式 AI（Generative AI）

```
11    else:
12        display([sample_image, sample_mask,
13                 create_mask(model.predict(sample_image[tf.newaxis, ...]))])
14
15 count = 0
16 for i in test_batches:
17     count +=1
18 print("number of batches:", count)
19
20 show_predictions(test_batches.skip(5), 3)
```

- 執行結果：效果還不錯。

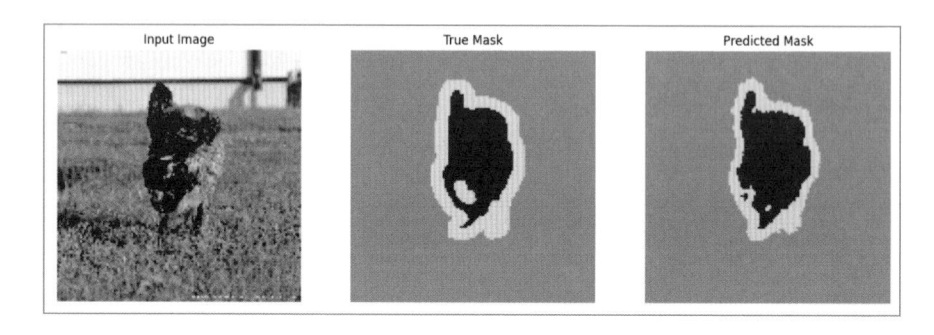

醫療診斷如黃斑部病變或腫瘤，我們可以使用大量的 X 光片標記異常位置，餵入 U-NET 模型訓練，之後我們就可以為新的 X 光片找到異常位置。

U-NET 演算法有許多變形（Variants），包括 U-NET++、3D U-Net、Ternaus-Net、Res-UNet、Attention Res-UNet、Dense U-Net…等，可參閱『U-Net and its variants for medical image segmentation: theory and applications』[11]，效能評估可參閱『Performance Analysis of UNet and Variants for Medical Image Segmenta-tion』[12]。

9-6 風格轉換（Style Transfer）-- 人人都可以是畢卡索

繼續來認識另一個有趣的 AutoEncoder 變形，稱為『風格轉換』（Neural Style Transfer），把一張照片轉換成某一幅畫的風格，如下圖。讀者可以在手機下載『Prisma』App 來玩玩，它能夠在拍照後，將照片風格即時轉換，內建近二十種的大師畫風可供選擇，只是轉換速度有點慢。

9-6 風格轉換（Style Transfer）-- 人人都可以是畢卡索

▲ 圖 9.9 風格轉換（Style Transfer），
原圖 + 風格圖像 = 生成圖像，圖片來源：fast-style-transfer GitHub [13]

之前有一則關於美圖影像實驗室（MTlab）的新聞，『催生全球首位 AI 繪師 Andy，美圖搶攻人工智慧卻面臨一大挑戰』[14]，該公司號稱投資了 1.99 億元人民幣，研發團隊超過 60 人，將風格轉換速度縮短到 3 秒鐘，開發成『美圖秀秀』App，大受歡迎，之後更趁勢推出專屬手機，狂銷 100 多萬台，算得上少數成功的 AI 商業模式。

風格轉換演算法由 Leon A. Gatys 等學者於 2015 年提出[15]，主要作法是重新定義損失函數，分為『內容損失』（Content Loss）與『風格損失』（Style Loss），並利用 AutoEncoder 的解碼器合成圖像，隨著訓練週期，損失逐漸變小，亦即生成的圖像會越接近於原圖與風格圖的合成。

內容損失函數比較單純，即原圖與生成圖像的像素差異平方和，定義如下：

$$J_{content}(C, G) = \frac{1}{4 \times n_H \times n_W \times n_C} \sum (a^{(C)} - a^{(G)})^2$$

n_H、n_W：原圖的寬、高。

n_C：色彩通道數。

$a^{(C)}$：原圖的像素。

$a^{(G)}$：生成圖像的像素。

風格損失函數為該演算法的重點，如何量化抽象的畫風是一大挑戰，Gatys 等學者想到的方法是，先定義 Gram 矩陣（Matrix）後，再利用 Gram 矩陣來定義風格損失。

Gram Matrix：兩個特徵向量進行點積，代表特徵的關聯性，顯現那些特徵是同時出現的，亦即風格。因此，風格損失就是風格圖像與生成圖像的 Gram 差異平方和，如下：

第 9 章　生成式 AI（Generative AI）

$$J_{style}(S,G) = \frac{1}{4 \times n_c^2 \times (n_H \times n_W)^2} \sum_{i=l}^{n_c} \sum_{j=1}^{n_c} (G^{(S)} - G^{(G)})^2$$

$G^{(S)}$：風格圖像的 Gram。

$G^{(G)}$：生成圖像的 Gram。

上式只是單一神經層的風格損失，結合所有神經層的風格損失，定義如下：

$$J_{style}(S,G) = \sum_{l} \lambda^{(l)} J_{style}^{(l)}(S,G)$$

λ：每一層的權重。

總損失函數：

$$J(G) = \alpha J_{content}(C,G) + \beta J_{style}(S,G)$$

α、β：控制內容與風格的比重，可以控制生成圖像要偏重風格的比例。

接下來，我們就來進行實作。

範例. 使用風格轉換演算法進行圖檔的轉換。提醒一下，範例中的內容圖即是原圖的意思。

程式：09_04_Neural_Style_Transfer.ipynb，修改自 TensorFlow 官網提供的範例『Neural Style Transfer』[16]，若本機無 GPU 顯卡，建議使用 09_05_Neural_Style_Transfer_colab.ipynb 在 Colaboratory 上執行，注意圖片資料夾的設定，在設定的資料夾上傳圖片。

9-28

9-6 風格轉換（Style Transfer） -- 人人都可以是畢卡索

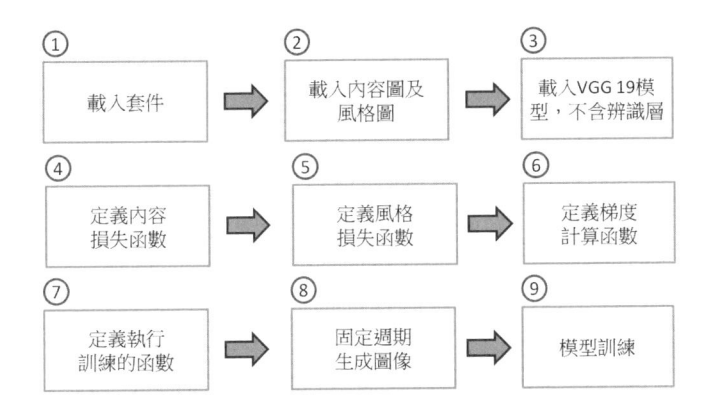

```
① 載入套件 → ② 載入內容圖及風格圖 → ③ 載入VGG 19模型，不含辨識層
④ 定義內容損失函數 → ⑤ 定義風格損失函數 → ⑥ 定義梯度計算函數
⑦ 定義執行訓練的函數 → ⑧ 固定週期生成圖像 → ⑨ 模型訓練
```

1. 載入套件。

```
1  # 載入相關套件
2  import os
3  import time
4  import sys
5  import matplotlib.pyplot as plt
6  from PIL import Image
7  import numpy as np
8  import tensorflow as tf
9  from tensorflow.keras.applications.vgg19 import VGG19
10 from tensorflow.keras.preprocessing.image import load_img
11 from tensorflow.keras.preprocessing.image import img_to_array
```

2. 載入內容圖檔。

```
1  # 載入內容圖檔
2  content_path = "./style_transfer/chicago.jpg"
3  content_image = load_img(content_path)
4  plt.imshow(content_image)
5  plt.axis('off')
6  plt.show()
```

- 執行結果：

9-29

3. 載入風格圖檔。

```
1  # 載入風格圖檔
2  style_path = "./style_transfer/wave.jpg"
3  style_image = load_img(style_path)
4  plt.imshow(style_image)
5  plt.axis('off')
6  plt.show()
```

- 執行結果：

4. 定義圖像前置處理的函數。

```
1  # 載入圖像並進行前置處理
2  def load_and_process_img(path_to_img):
3      img = load_img(path_to_img)
4      img = img_to_array(img)
5      img = np.expand_dims(img, axis=0)
6      img = tf.keras.applications.vgg19.preprocess_input(img)
7      # print(img.shape)
8  
9      # 回傳影像陣列
10     return img
```

5. 定義由陣列還原成圖像的函數。

```
1  # 由陣列還原影像
2  def deprocess_img(processed_img):
3      x = processed_img.copy()
4      if len(x.shape) == 4:
5          x = np.squeeze(x, 0)
6  
7      # 前置處理的還原
8      x[:, :, 0] += 103.939
9      x[:, :, 1] += 116.779
10     x[:, :, 2] += 123.68
11     x = x[:, :, ::-1]
12  
13     # 裁切元素值在(0, 255)之間
14     x = np.clip(x, 0, 255).astype('uint8')
15     return x
```

9-6 風格轉換（Style Transfer）-- 人人都可以是畢卡索

6. 定義內容圖和風格圖輸出的神經層名稱。

```
1  # 定義內容圖輸出的神經層名稱
2  content_layers = ['block5_conv2']
3
4  # 定義風格圖輸出的神經層名稱
5  style_layers = ['block1_conv1',
6                  'block2_conv1',
7                  'block3_conv1',
8                  'block4_conv1',
9                  'block5_conv1'
10                 ]
11
12 num_content_layers = len(content_layers)
13 num_style_layers = len(style_layers)
```

7. 定義模型：載入 VGG 19 模型，不含辨識層，加上自訂的輸出。

```
1  from tensorflow.python.keras import models
2
3  def get_model():
4      # 載入 VGG19，不含辨識層
5      vgg = tf.keras.applications.vgg19.VGG19(include_top=False, weights='imagenet')
6      vgg.trainable = False   # 不重新訓練
7      for layer in vgg.layers:
8          layer.trainable = False
9
10     # 以之前定義的內容圖及風格圖神經層為輸出
11     style_outputs = [vgg.get_layer(name).output for name in style_layers]
12     content_outputs = [vgg.get_layer(name).output for name in content_layers]
13     model_outputs = style_outputs + content_outputs
14
15     # 建立模型
16     return models.Model(vgg.input, model_outputs)
```

8. 定義內容損失函數：為內容圖與生成圖特徵向量之差的平方和，也可以採用上述理論的公式。

```
1  # 內容損失函數
2  def get_content_loss(base_content, target):
3      # 下面可附加 『/ (4. * (channels ** 2) * (width * height) ** 2)』
4      return tf.reduce_mean(tf.square(base_content - target))
```

9. 定義風格損失函數：先定義 Gram Matrix 計算函數，再定義風格損失函數。

```
1  # 計算 Gram Matrix 函數
2  def gram_matrix(input_tensor):
3      # We make the image channels first
4      channels = int(input_tensor.shape[-1])
5      a = tf.reshape(input_tensor, [-1, channels])
6      n = tf.shape(a)[0]
7      gram = tf.matmul(a, a, transpose_a=True)
```

```
 8       return gram / tf.cast(n, tf.float32)
 9
10  # 風格損失函數
11  def get_style_loss(base_style, gram_target):
12      # 取得風格圖的高、寬、色彩數
13      height, width, channels = base_style.get_shape().as_list()
14
15      # 計算 Gram Matrix
16      gram_style = gram_matrix(base_style)
17
18      # 計算風格損失
19      return tf.reduce_mean(tf.square(gram_style - gram_target))
```

10. 定義圖像的特徵向量計算函數。

```
 1  # 計算內容圖及風格圖的特徵向量
 2  def get_feature_representations(model, content_path, style_path):
 3      # 載入圖檔
 4      content_image = load_and_process_img(content_path)
 5      style_image = load_and_process_img(style_path)
 6
 7      # 設定模型
 8      style_outputs = model(style_image)
 9      content_outputs = model(content_image)
10
11
12      # 取得特徵向量
13      style_features = [style_layer[0] for style_layer
14                        in style_outputs[:num_style_layers]]
15      content_features = [content_layer[0] for content_layer
16                          in content_outputs[num_style_layers:]]
17      return style_features, content_features
```

11. 定義梯度計算函數。

```
 1  # 計算梯度
 2  def compute_grads(cfg):
 3      with tf.GradientTape() as tape:
 4          # 累計損失
 5          all_loss = compute_loss(**cfg)
 6
 7      # 取得梯度
 8      total_loss = all_loss[0]
 9      # cfg['init_image']：內容圖影像陣列
10      return tape.gradient(total_loss, cfg['init_image']), all_loss
```

12. 定義所有層的損失計算函數：按照內容圖與風格圖的權重比例，計算總損失。

```
 1  # 計算所有層的損失
 2  def compute_loss(model, loss_weights, init_image, gram_style_features, content_features):
 3      # 內容圖及風格圖的權重比例
 4      style_weight, content_weight = loss_weights
 5
 6      # 取得模型輸出
```

9-6 風格轉換（Style Transfer）-- 人人都可以是畢卡索

```
7      model_outputs = model(init_image)
8      style_output_features = model_outputs[:num_style_layers]
9      content_output_features = model_outputs[num_style_layers:]
10
11     # 累計風格分數
12     style_score = 0
13     weight_per_style_layer = 1.0 / float(num_style_layers)
14     for target_style, comb_style in zip(gram_style_features, style_output_features):
15         style_score += weight_per_style_layer * get_style_loss(comb_style[0], target_style)
16
17     # 累計內容分數
18     content_score = 0
19     weight_per_content_layer = 1.0 / float(num_content_layers)
20     for target_content, comb_content in zip(content_features, content_output_features):
21         content_score += weight_per_content_layer* get_content_loss(comb_content[0], target_content)
22
23     # 乘以權重比例
24     style_score *= style_weight
25     content_score *= content_weight
26
27     # 總損失
28     loss = style_score + content_score
29
30     return loss, style_score, content_score
```

13. 定義執行訓練的函數：這是程式的核心，在訓練的過程中，在固定週期生成圖像，可以看到圖像逐漸轉變的過程。

```
1  # 執行訓練的函數
2  import IPython.display
3
4  def run_style_transfer(content_path, style_path, num_iterations=1000,
5                         content_weight=1e3, style_weight=1e-2):
6      # 取得模型
7      model = get_model()
8
9      # 取得內容圖及風格圖的神經層輸出
10     style_features, content_features = get_feature_representations(model, content_path, style_path)
11     gram_style_features = [gram_matrix(style_feature) for style_feature in style_features]
12
13     # 載入內容圖
14     init_image = load_and_process_img(content_path)
15     init_image = tf.Variable(init_image, dtype=tf.float32)
16
17     # 指定優化器
18     opt = tf.optimizers.Adam(learning_rate=5, beta_1=0.99, epsilon=1e-1)
19
20     # 初始化變數
21     iter_count = 1
22     best_loss, best_img = float('inf'), None
23     loss_weights = (style_weight, content_weight)
24     cfg = {  # 組態
25             'model': model,
26             'loss_weights': loss_weights,
27             'init_image': init_image,
28             'gram_style_features': gram_style_features,
29             'content_features': content_features
30     }
31
32     # 參數設定
33     num_rows = 2   # 輸出小圖以 2 列顯示
34     num_cols = 5   # 輸出小圖以 5 行顯示
35     # 每N個週期數生成圖像，計算 N : display_interval
36     display_interval = num_iterations/(num_rows*num_cols)
37     start_time = time.time()       # 計時
38     global_start = time.time()
39
```

第 9 章 生成式 AI（Generative AI）

```
40    # RGB 三色中心值，輸入圖像以中心值為 0 作轉換
41    norm_means = np.array([103.939, 116.779, 123.68])
42    min_vals = -norm_means
43    max_vals = 255 - norm_means
44
45    # 開始訓練
46    imgs = []
47    for i in range(num_iterations):
48        grads, all_loss = compute_grads(cfg)
49        loss, style_score, content_score = all_loss
50        opt.apply_gradients([(grads, init_image)])
51        clipped = tf.clip_by_value(init_image, min_vals, max_vals)
52        init_image.assign(clipped)
53        end_time = time.time()
55        # 記錄最小損失時的圖像
56        if loss < best_loss:
57            best_loss = loss    # 記錄最小損失
58            best_img = deprocess_img(init_image.numpy())  # 生成圖像
59
60        # 每N個週期數生成圖像
61        if i % display_interval== 0:
62            start_time = time.time()
63
64            # 生成圖像
65            plot_img = init_image.numpy()
66            plot_img = deprocess_img(plot_img)
67            imgs.append(plot_img)
68
69            # IPython.display.clear_output(wait=True)   # 可清除之前的顯示
70            print(f'週期數: {i}')
71            elapsed_time = time.time() - start_time
72            print(f'總損失: {loss:.2e}, 風格損失: {style_score:.2e},' +
73                  f'內容損失: {content_score:.2e}, 耗時: {elapsed_time:.2f}s')
74            IPython.display.display_png(Image.fromarray(plot_img))
75
76    print(f'總耗時: {(time.time() - global_start):.2f}s')
77    # IPython.display.clear_output(wait=True)   # 可清除之前的顯示
78    # 顯示生成的圖像
79    plt.figure(figsize=(14,4))
80    for i,img in enumerate(imgs):
81        plt.subplot(num_rows,num_cols,i+1)
82        plt.imshow(img)
83        plt.axis('off')
84
85    return best_img, best_loss
```

14. 呼叫上述函數，執行模型訓練。

```
1    # 執行訓練
2    model = get_model()
3    best, best_loss = run_style_transfer(content_path,
4                                          style_path, num_iterations=500)
```

- 執行結果：執行 500 個週期，剛開始畫面變化很大，之後轉換逐漸減少，表示損失函數逐步收斂。

9-34

9-6 風格轉換（Style Transfer）-- 人人都可以是畢卡索

第 9 章　生成式 AI（Generative AI）

15. 顯示最佳的圖像。

```
1  # 顯示最佳的圖像
2  Image.fromarray(best)
```

- 執行結果：

9-6 風格轉換（Style Transfer）-- 人人都可以是畢卡索

16. 比較內容圖與生成圖。

```
# 定義顯示函數，比較原圖與生成圖像
def show_results(best_img, content_path, style_path, show_large_final=True):
    plt.figure(figsize=(10, 5))
    content = load_img(content_path)
    style = load_img(style_path)

    plt.subplot(1, 2, 1)
    plt.axis('off')
    plt.imshow(content)
    plt.title('原圖')


    plt.subplot(1, 2, 2)
    plt.axis('off')
    plt.imshow(style)
    plt.title('風格圖')

    if show_large_final:
        plt.figure(figsize=(10, 10))
        plt.axis('off')

        plt.imshow(best_img)
        plt.title('Output Image')
        plt.show()

# 原圖與生成圖像的比較
show_results(best, content_path, style_path)
```

17. 拿另一張內容圖來測試。

```
# 以另一張圖測試
content_path = './style_transfer/Green_Sea_Turtle_grazing_seagrass.jpg'
style_path = './style_transfer/wave.jpg'

best_starry_night, best_loss = run_style_transfer(content_path, style_path)
show_results(best_starry_night, content_path, style_path)
```

9-37

第 9 章　生成式 AI（Generative AI）

- 執行結果：

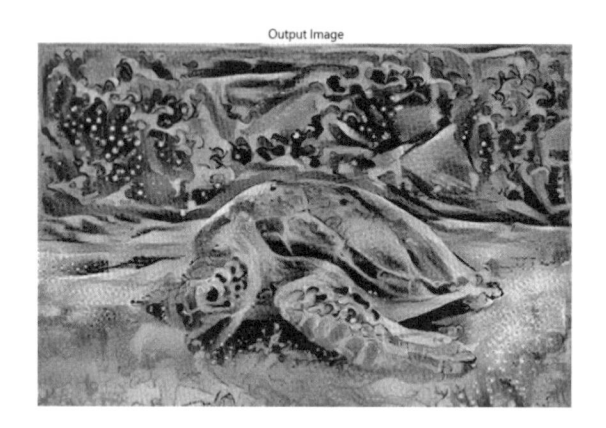

9-7 快速風格轉換（Fast Style Transfer）

TensorFlow 官網提供的範例『Neural Style Transfer』[15]，有提及一些改善的措施，譬如，基本的風格轉換演算法所生成的圖像經常會偏重高頻（high frequency），而高頻會造成邊緣特別明顯，可使用正則化（Regularization）修正損失函數，改善此一現象。

上述程式生成一張圖像需要花 290 秒，在如今的社群媒體時代，就算這酷炫的效果能抓住了大眾的眼球，執行時間過長，使用者也會失去耐心，因此後續有許多學者研究如何提升演算法速度，有興趣的讀者可搜尋『Fast Style Transfer』。另外，也有同學問到，如果用同一張風格圖，對另一張新的內容圖進行風格轉換，也要重新訓練嗎？我們可以仿效一般神經網路將模型儲存，之後輸入新圖片，只需推論，不需重新訓練，如何儲存模型也是一個值得研究的課題。開發美圖秀秀的美圖公司砸了近兩億人民幣，才將速度縮短至 3 秒，可見技術難度頗高，所以，速度絕對是商業模式重要的考量因素。

9-7 快速風格轉換（Fast Style Transfer）

TensorFlow Hub 裡面有許多預先訓練好的模型可以直接套用，包括風格轉換模型，以下範例使用 Fast style transfer 演算法預先訓練好的模型，可快速產生新圖像。

範例. 使用 TensorFlow Hub 的 Fast style transfer 預先訓練模型完成風格轉換。

程式：**09_06_Fast_Style_Transfer.ipynb**，程式修改自『Fast Style Transfer for Arbitrary Styles』[17]。

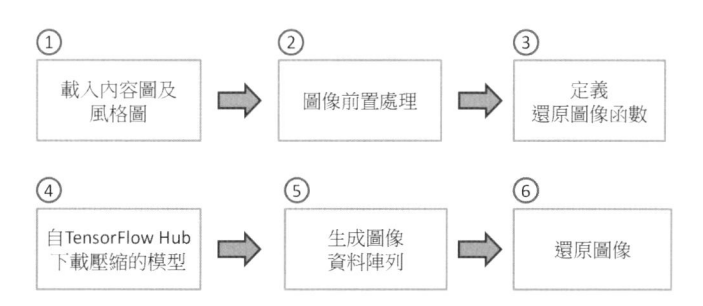

```
①載入內容圖及風格圖 → ②圖像前置處理 → ③定義還原圖像函數
④自TensorFlow Hub下載壓縮的模型 → ⑤生成圖像資料陣列 → ⑥還原圖像
```

1. 載入套件。

```
1  import functools
2  import os
3  from matplotlib import gridspec
4  import matplotlib.pylab as plt
5  import numpy as np
6  import tensorflow as tf
7  import tensorflow_hub as hub
```

2. 定義圖像裁切、載入、前置處理及顯示的函數

```
1  # 定義裁切圖像的函數
2  def crop_center(image):
3    """Returns a cropped square image."""
4    shape = image.shape
5    new_shape = min(shape[1], shape[2])
6    offset_y = max(shape[1] - shape[2], 0) // 2
7    offset_x = max(shape[2] - shape[1], 0) // 2
8    image = tf.image.crop_to_bounding_box(
9        image, offset_y, offset_x, new_shape, new_shape)
10   return image
11
12 # 定義載入圖像並進行前置處理的函數
13 @functools.lru_cache(maxsize=None)
14 def load_image(image_url, image_size=(256, 256), preserve_aspect_ratio=True):
```

第 9 章　生成式 AI（Generative AI）

```
15    """Loads and preprocesses images."""
16    # Cache image file locally.
17    if image_url.startswith('http'):
18        image_path = tf.keras.utils.get_file(os.path.basename(image_url)[-128:], image_url)
19    else:
20        image_path = image_url
21    # Load and convert to float32 numpy array, add batch dimension, and normalize to range [0, 1].
22    img = tf.io.decode_image(
23        tf.io.read_file(image_path),
24        channels=3, dtype=tf.float32)[tf.newaxis, ...]
25    img = crop_center(img)
26    img = tf.image.resize(img, image_size, preserve_aspect_ratio=True)
27    return img
```

```
29  # 定義顯示圖像的函數
30  def show_n(images, titles=('',)):
31      n = len(images)
32      image_sizes = [image.shape[1] for image in images]
33      w = (image_sizes[0] * 6) // 320
34      plt.figure(figsize=(w * n, w))
35      gs = gridspec.GridSpec(1, n, width_ratios=image_sizes)
36      for i in range(n):
37          plt.subplot(gs[i])
38          plt.imshow(images[i][0], aspect='equal')
39          plt.axis('off')
40          plt.title(titles[i] if len(titles) > i else '')
41      plt.show()
```

3. 下載圖像：可使用本機圖像或從網路下載。

```
1  content_img_size = (384, 384)
2  style_img_size = (256, 256)  # Recommended to keep it at 256.
3  content_path = "../style_transfer/chicago.jpg"
4  style_path = "../style_transfer/wave.jpg"
5  content_image = load_image(content_path, content_img_size)
6  style_image = load_image(style_path, style_img_size)
7  style_image = tf.nn.avg_pool(style_image, ksize=[3,3], strides=[1,1], padding='SAME')
8  show_n([content_image, style_image], ['Content image', 'Style image'])
```

4. 自 TensorFlow Hub 下載壓縮的模型。

```
1  # 自 TensorFlow Hub 下載壓縮的模型
2  hub_handle = 'https://tfhub.dev/google/magenta/arbitrary-image-stylization-v1-256/2'
3  hub_module = hub.load(hub_handle)
```

5. 生成新的圖像。

9-7 快速風格轉換（Fast Style Transfer）

```
1  outputs = hub_module(content_image, style_image)
2  stylized_image = outputs[0]
3  show_n([content_image, style_image, stylized_image],
4          titles=['Original content image', 'Style image', 'Stylized image'])
```

- 執行結果：與自行訓練的模型效果相同，但速度較快。

6. 以另一張圖測試：從網路下載圖像。

```
1  storage_url = 'https://storage.googleapis.com/download.tensorflow.org/'
2  content_url = storage_url + 'example_images/YellowLabradorLooking_new.jpg'
3  style_url = storage_url + 'example_images/Vassily_Kandinsky%2C_1913_-_Composition_7.jpg'
4  content_image = load_image(content_url, content_img_size)
5  style_image = load_image(style_url, style_img_size)
6  style_image = tf.nn.avg_pool(style_image, ksize=[3,3], strides=[1,1], padding='SAME')
7  show_n([content_image, style_image], ['Content image', 'Style image'])
```

7. 生成圖像：與前面指令相同。

```
1  outputs = hub_module(content_image, style_image)
2  stylized_image = outputs[0]
3  show_n([content_image, style_image, stylized_image],
4          titles=['Original content image', 'Style image', 'Stylized image'])
```

- 執行結果：

第 9 章　生成式 AI（Generative AI）

風格轉換是一個非常有趣的應用，除了轉換成名畫風格之外，也可將照片卡通化，或是針對臉部美肌，凡此種種應用都值得研究。當然，不只有風格轉換演算法可以這樣玩，其他像 GAN 或 OpenCV 影像處理也都能做到類似的功能，大家一起天馬行空，胡思亂想吧！

9-8　本章小結

編碼器與解碼器（Encoder-decoder）模型是一個非常重要的架構，目前生成式 AI 都是以此架構為基礎，包括 ChatGPT 及 Stable Diffusion，ChatGPT 的基礎演算法 Transformer，如下圖，是一個典型的 Encoder-decoder 架構。

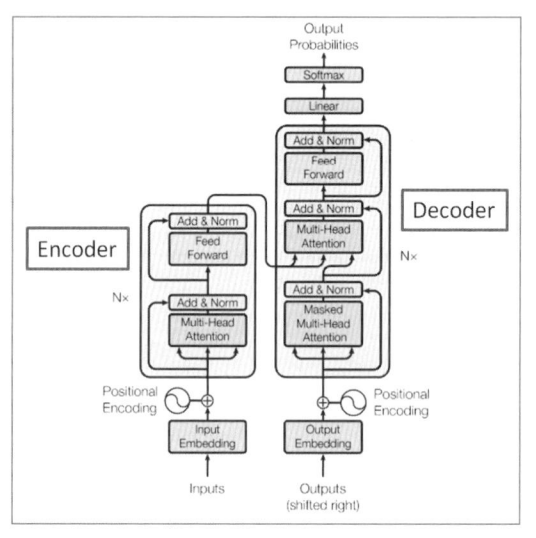

▲ 圖 9.10 Transformer，圖片來源：Attention Is All You Need [18]

而 Stable Diffusion 的基礎演算法是擴散模型（Diffusion Model），如下圖，也是一個典型的 Encoder-decoder 架構，有人形容它是多層的 VAE。

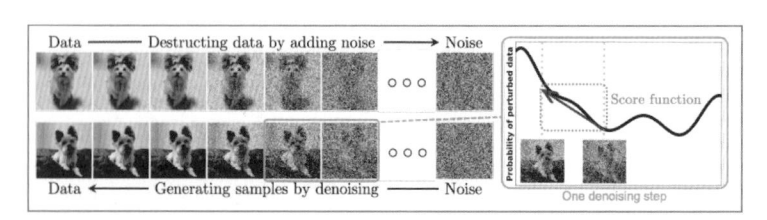

▲ 圖 9.11 Diffusion Model，上層是 Encoder，下層是 Decoder，圖片來源：Diffusion Models: A Comprehensive Survey of Methods and Applications [19]

9-42

Encoder-decoder 模型需要編碼再解碼，並且需要大量的訓練資料，因此訓練時間通常需要較久的時間，必須投資大量的硬體設備才能在短時間訓練好模型，因而引發 IT 大廠競相購買 GPU 顯卡，例如馬斯克一次購買 10 萬張 GPU 顯卡，其他公司也不惶多讓，ChatGPT 模型訓練一次就要花上百萬美金，令人咋舌。

參考資料（References）

[1] An AI-Generated Painting Wins Art Contest, Annoys Artists
(https://shakiroslann.medium.com/an-ai-generated-painting-wins-art-contest-annoys-artists-f3f7c98e3088)

[2] Sony 世界攝影大獎優勝作品「竟是 AI 畫的」
(https://news.ltn.com.tw/news/world/breakingnews/4271589)

[3] Vijay Badrinarayanan、Alex Kendall、Roberto Cipolla,《SegNet: A Deep Convolutional Encoder-Decoder Architecture for Image Segmentation》, 2015
(https://arxiv.org/abs/1511.00561)

[4] PyImageSearch, Introduction to Autoencoders
(https://pyimagesearch.com/2023/07/10/introduction-to-autoencoders/)

[5] Convolutional Variational Autoencoder(CAE)
(https://www.tensorflow.org/tutorials/generative/cvae)

[6] mingukkang, CVAE
(https://github.com/mingukkang/CVAE)

[7] gozsoy, conditional-vae
(https://github.com/mingukkang/CVAE)

[8] PyImageSearch, U-Net Image Segmentation in Keras
(https://pyimagesearch.com/2022/02/21/u-net-image-segmentation-in-keras/)

[9] TensorFlow 官網提供的範例『Image segmentation』
(https://www.tensorflow.org/tutorials/images/segmentation)

[10] Oxford 大學資訊工程學系網址
(https://www.robots.ox.ac.uk/)

第 9 章　生成式 AI（Generative AI）

[11] Nahian Siddique et al., U-Net and its variants for medical image segmentation: theory and applications
(https://arxiv.org/pdf/2011.01118)

[12] Walid Ehab a Yongmin Li a, Performance Analysis of UNet and Variants for Medical Image Segmentation
(https://arxiv.org/pdf/2309.13013)

[13] fast-style-transfer GitHub
(https://github.com/lengstrom/fast-style-transfer)

[14] 翁書婷,《催生全球首位 AI 繪師 Andy，美圖搶攻人工智慧卻面臨一大挑戰》
(https://www.bnext.com.tw/article/47330/ai-andy-meitu)

[15] Leon A. Gatys、Alexander S. Ecker、Matthias Bethge,《A Neural Algorithm of Artistic Style》
(https://arxiv.org/abs/1508.06576)

[16] TensorFlow 官網提供的範例『Neural Style Transfer』
(https://www.tensorflow.org/tutorials/generative/style_transfer)

[17] TensorFlow 官網提供的範例『Fast Style Transfer for Arbitrary Styles』
(https://www.tensorflow.org/hub/tutorials/tf2_arbitrary_image_stylization)

[18] Attention Is All You Need
(https://arxiv.org/pdf/1706.03762)

[19] Diffusion Models: A Comprehensive Survey of Methods and Applications
(https://arxiv.org/pdf/2209.00796)

第 10 章

生成對抗網路（GAN）

話說水能載舟，亦能覆舟，生成式 AI 雖然給人類帶來了許多便利，但也造成不小的危害。近幾年氾濫的深度偽造（Deepfake）就是一例，它利用 AI 技術偽造政治人物與名人的視訊，能夠做到真假難辨，一旦在網路上散播開來，就形成莫大的災難，根據統計，名人色情片 8 成都是偽造的。深度偽造的基礎演算法就是『生成對抗網路』（Generative Adversarial Network, GAN），本章就來探討此一演算法。

Facebook 人工智慧研究院 Yann LeCun 在接受 Quora 專訪時提到：『GAN 及其變形是近十年最有趣的想法』（This, and the variations that are now being pro-posed is the most interesting idea in the last 10 years in ML, in my opinion.），一句話造成 GAN 一炮而紅，其作者 Ian Goodfellow（真的是好傢伙！）也成為各界競相邀請演講的對象。

另外，2018 年 10 月紐約佳士得藝術拍賣會，也賣出第一幅以 GAN 演算法所繪製的肖像畫，最後得標價為 $432,500 美金。有趣的是，畫作右下角還列出 GAN 的損失函數，相關報導可參見『全球首次！AI 創作肖像畫 10 月佳士得拍賣』[1] 及『Is artificial intelligence set to become art's next medium?』[2]。

第 10 章　生成對抗網路（GAN）

▲ 圖 10.1　Edmond de Belamy 肖像畫，圖片來源：佳士得網站 [3]

此後有人統計每 28 分鐘就有一篇與 GAN 相關的論文發表。

10-1　生成對抗網路介紹

關於生成對抗網路有一個很生動的比喻：它是由兩個神經網路所組成，一個網路扮演偽鈔製造者（Counterfeiter），不斷製造假鈔，另一個網路則扮演警察，從偽造者那邊拿到假鈔，並判斷真假，然後，偽造者就根據警察判斷結果不停的改良，最後假鈔變成真假難辨（天啊！這是什麼電影情節！），這就是 GAN 的概念。

偽鈔製造者稱為『生成模型』（Generative model），警察則是『判別模型』（Discriminative model），簡單的架構如下圖：

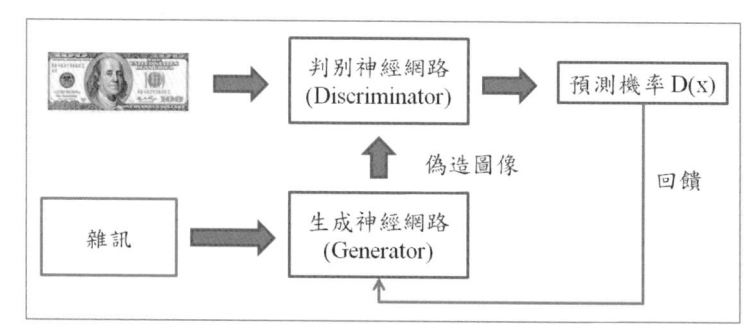

▲ 圖 10.2　生成對抗網路（Generative Adversarial Network, GAN）的架構

處理流程如下：

1. 先訓練判別神經網路：從訓練資料中抽取樣本 (x)，餵入判別神經網路，期望預測機率 D(x) ≒ 1，相反的，判斷來自生成網路的偽造圖片（z），期望預測機率 D（G（z））≒ 0。D：判別模型，G：生成模型。

2. 訓練生成神經網路：剛開始以常態分配或均勻分配產生雜訊（z），餵入生成神經網路，生成偽造圖片，一開始很容易被判定是偽造的圖片。

3. 透過判別網路的反向傳導（Backpropagation），回饋給生成網路，再透過訓練不斷更新權重，改良偽造圖片的準確度（技術），反覆訓練，直到產生精準的圖片，訓練才算完成。

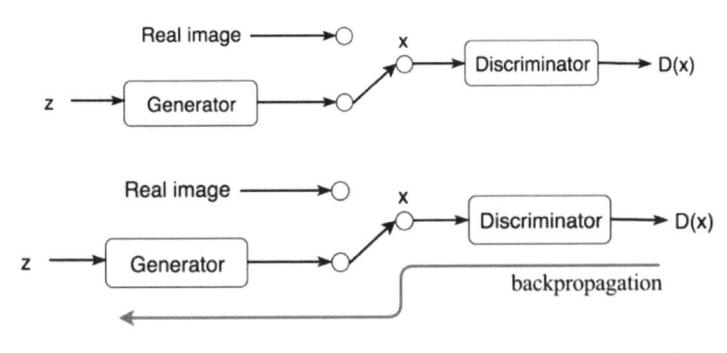

▲ 圖 10.3 判別神經網路的反向傳導（Backpropagation）

GAN 根據以上流程重新定義損失函數。

1. 判別神經網路的損失函數：前半段為真實資料的判別，後半段為偽造資料的判別。

$$\max_{D} V(D) = \mathbb{E}_{\boldsymbol{x} \sim p_{\text{data}}(\boldsymbol{x})}[\log D(\boldsymbol{x})] + \mathbb{E}_{\boldsymbol{z} \sim p_{\boldsymbol{z}}(\boldsymbol{z})}[\log(1 - D(G(\boldsymbol{z})))]$$

recognize real images better   recognize generated images better

- x：真實的訓練資料，故預測機率 D(x) 愈大愈好。
- E：為期望值，因為訓練資料並不完全相同，故預測機率有高有低。
- z：偽造資料，預測機率 D（G（z））愈小愈好，調整為 1- D（G（z）），變成愈大愈好。
- D（x）與 1-D（G（z））兩者相加當然是愈大愈好。

- 取 Log：並不會影響最大化求解，通常機率相乘會造成多次方，不容易求解，故取 Log，使損失函數轉變成一次方函數。

2. 生成神經網路的損失函數：即判別神經網路損失函數的右邊多項式，生成神經網路期望偽造資料被分類為真的機率愈大愈好，故差距愈小愈好。

$$\min_G V(G) = \mathbb{E}_{z \sim p_z(z)}[\log(1 - D(G(z)))]$$

3. 兩個網路損失函數合而為一的表達式如下：

$$\min_G \max_D V(D, G) = \mathbb{E}_{x \sim p_{data}}[\log D(x)] + \mathbb{E}_{z \sim p_z}[\log(1 - D(G(z)))]$$

整個演算法的虛擬碼如下，使用小批量梯度下降法，最小化損失函數。

Algorithm 1 Minibatch stochastic gradient descent training of generative adversarial nets. The number of steps to apply to the discriminator, k, is a hyperparameter. We used $k = 1$, the least expensive option, in our experiments.

---
for number of training iterations **do**
　for k steps **do**
　　• Sample minibatch of m noise samples $\{z^{(1)}, \ldots, z^{(m)}\}$ from noise prior $p_g(z)$.
　　• Sample minibatch of m examples $\{x^{(1)}, \ldots, x^{(m)}\}$ from data generating distribution $p_{data}(x)$.
　　• Update the discriminator by ascending its stochastic gradient:
$$\nabla_{\theta_d} \frac{1}{m} \sum_{i=1}^m \left[\log D\left(x^{(i)}\right) + \log\left(1 - D\left(G\left(z^{(i)}\right)\right)\right) \right].$$
　end for
　　• Sample minibatch of m noise samples $\{z^{(1)}, \ldots, z^{(m)}\}$ from noise prior $p_g(z)$.
　　• Update the generator by descending its stochastic gradient:
$$\nabla_{\theta_g} \frac{1}{m} \sum_{i=1}^m \log\left(1 - D\left(G\left(z^{(i)}\right)\right)\right).$$
end for
The gradient-based updates can use any standard gradient-based learning rule. We used momentum in our experiments.

▲ 圖 10.4 GAN 演算法的虛擬碼

生成網路希望製造出來的圖片越來越逼真，能通過判別網路的檢驗，而判別網路則希望將生成網路所製造的圖片都判定為假資料，兩者目標相反，互相對抗，故稱為『生成對抗網路』。

10-2 生成對抗網路種類

GAN 不是只有一種演算法，其變形非常多，可以參閱『The GAN Zoo』[4]，有上百種演算法，其功能各有不同，譬如：

1. CGAN：參閱『Pose Guided Person Image Generation』[5]，可生成不同的姿勢。

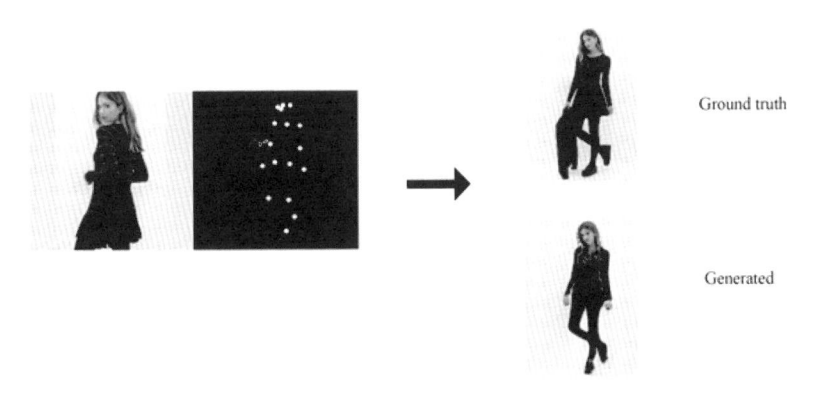

▲ 圖 10.5 CGAN 演算法的姿勢生成

2. ACGAN：參閱『Towards the Automatic Anime Characters Creation with Gen-erative Adversarial Networks』[6]，可生成不同的動漫人物。

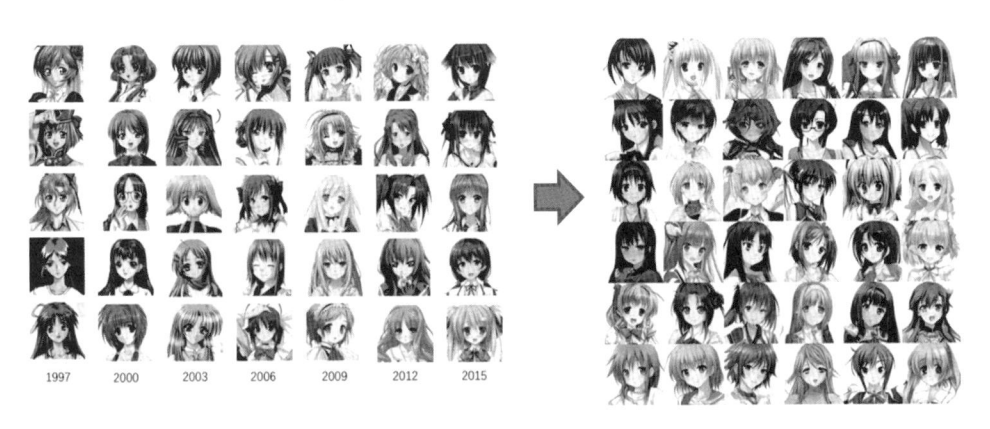

▲ 圖 10.6 ACGAN 演算法，從左邊的動漫角色，生成為右邊的新角色

第 10 章　生成對抗網路（GAN）

MakeGirlsMoe 網站[7] 有一個展示，可利用不同的模型與參數，生成各種動漫人物。

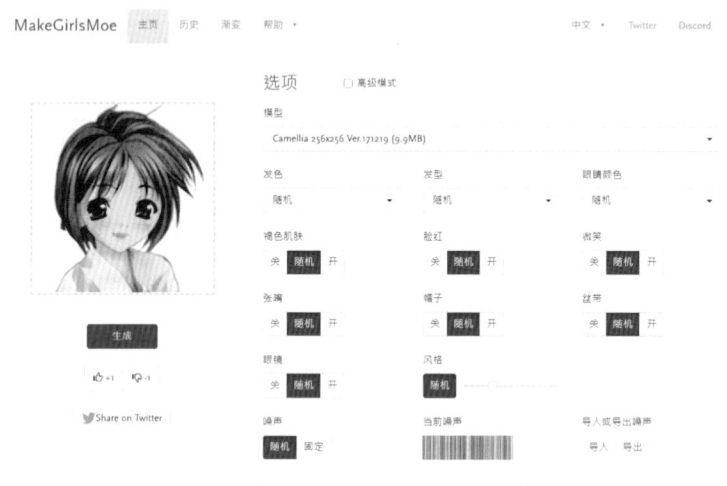

▲ 圖 10.7　ACGAN 展示網站

3. CycleGAN：風格轉換，MIL WebDNN 網站[8] 有一個展示，可選擇不同的風格圖，生成各式風格的照片或視訊。

▲ 圖 10.8　CycleGAN 的展示網站

4. StarGAN：參閱『StarGAN: Unified Generative Adversarial Networks for Multi-Domain Image-to-Image Translation』[9]，生成不同的臉部表情、轉換膚色、髮色或是性別，原始程式碼可參閱 StarGAN GitHub[10]。

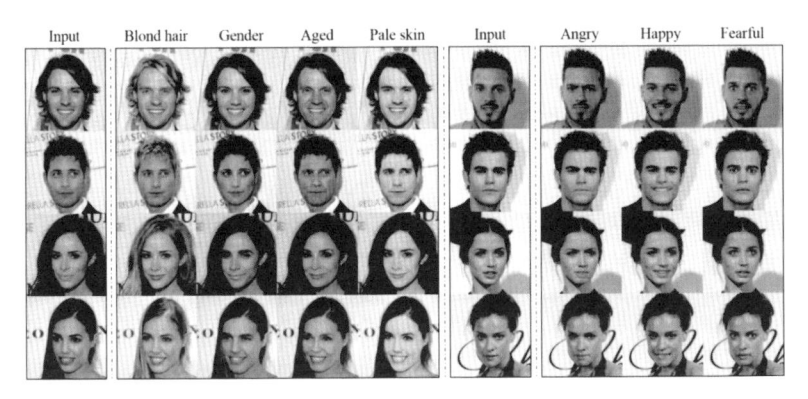

▲ 圖 10.9 StarGAN 展示，將左邊的臉轉換膚色、髮色、性別或表情

5. SRGAN：能生成高解析度的圖像，可參閱『Photo-Realistic Single Image Super-Resolution Using a Generative Adversarial Network』[11]。

▲ 圖 10.10 SRGAN 展示，由左而右從低解析度的圖像生成為高解析度的圖像

6. StyleGAN2：功能與語義分割（Image Segmentation）相反，它是從語義分割圖渲染成實景圖，參閱『Analyzing and Improving the Image Quality of Style-GAN』[12]，原始程式碼可參閱 StyleGAN2 GitHub [13]。

▲ 圖 10.11 StyleGAN2 展示，從右邊的圖像生成為左邊的圖像

第 10 章　生成對抗網路（GAN）

限於篇幅，僅介紹一小部分的演算法，讀者如有興趣可參閱『GAN — Some cool applications of GAN』[14] 一文有更多種演算法的介紹。

大部份的演算法只是修改原創者 GAN 的損失函數，即可產生不同的神奇效果，所以，根據『The GAN Zoo』[4] 的統計，GAN 相關的論文數量呈現爆炸性的成長。

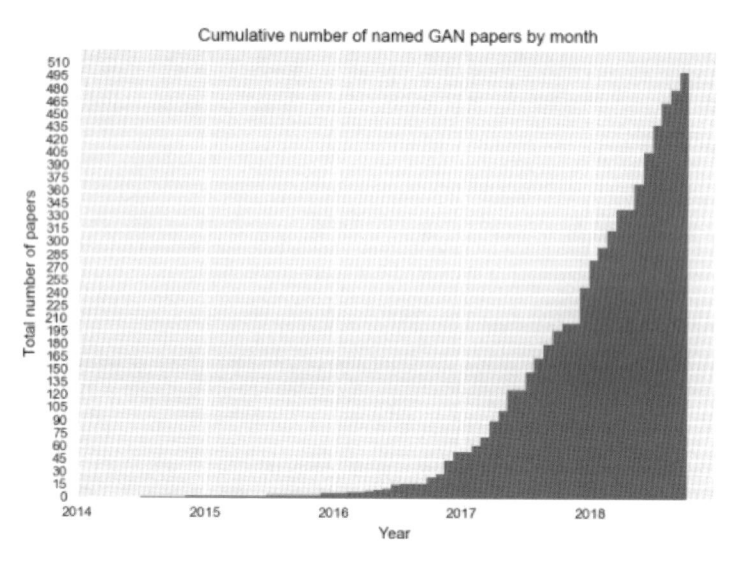

▲ 圖 10.12　與 GAN 有關的論文數量呈現爆炸性的成長

10-3　DCGAN

DCGAN（Deep Convolutional Generative Adversarial Network）演算法是一個較簡單的變形 GAN。

範例 1. 以 MNIST 資料實作 DCGAN 演算法，產生手寫阿拉伯數字。

程式：**10_01_DCGAN_MNIST.ipynb**，修改自『Deep Convolutional Generative Adversarial Network』[15]，筆者以 Colaboratory 測試，較為快速。

10-3 DCGAN

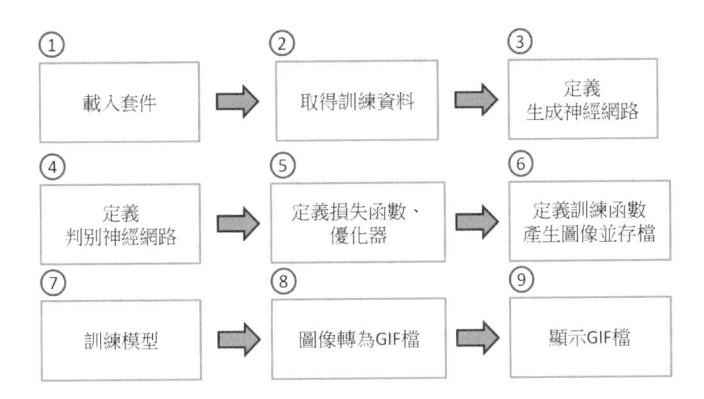

```
①載入套件 → ②取得訓練資料 → ③定義生成神經網路
④定義判別神經網路 → ⑤定義損失函數、優化器 → ⑥定義訓練函數產生圖像並存檔
⑦訓練模型 → ⑧圖像轉為GIF檔 → ⑨顯示GIF檔
```

1. 安裝 imageio 套件,用於產生 GIF 動畫:

    pip install imageio

2. 載入套件。

```python
1  # 載入相關套件
2  import glob
3  import imageio
4  import matplotlib.pyplot as plt
5  import numpy as np
6  import os
7  import PIL
8  import tensorflow as tf
9  from tensorflow.keras import layers
10 import time
11 from IPython import display
```

3. 取得訓練資料,轉為 TensorFlow Dataset。

```python
1  # 取得 MNIST 訓練資料
2  (train_images, train_labels), (_, _) = tf.keras.datasets.mnist.load_data()
3  # 像素標準化
4  train_images = train_images.reshape(train_images.shape[0], 28, 28, 1).astype('float32')
5  train_images = (train_images - 127.5) / 127.5  # 使像素值介於 [-1, 1]
6
7  # 參數設定
8  BUFFER_SIZE = 60000 # 緩衝區大小
9  BATCH_SIZE = 256    # 訓練批量
10
11 # 轉為 Dataset
12 train_dataset = tf.data.Dataset.from_tensor_slices(train_images) \
13         .shuffle(BUFFER_SIZE).batch(BATCH_SIZE).prefetch(BUFFER_SIZE).cache()
```

4. 定義生成神經網路:

    - use_bias=False:訓練不產生偏差項,因為要生成的影像盡量是像素所構成的。

- 採用 LeakyReLU Activation Function：可避免產生 0，以免生成的影像有太多空白。
- Conv2DTranspose：反卷積層，進行上採樣（Up Sampling），由小圖插補為大圖，strides=（2, 2）表示寬高各增大 2 倍。
- 最後產生寬高為（28, 28）的單色向量。

```python
def make_generator_model():
    model = tf.keras.Sequential()
    model.add(layers.Input((100,)))
    model.add(layers.Dense(7*7*256, use_bias=False))
    model.add(layers.BatchNormalization())
    model.add(layers.LeakyReLU())

    model.add(layers.Reshape((7, 7, 256)))
    assert model.output_shape == (None, 7, 7, 256)  # None 代表批量不檢查

    model.add(layers.Conv2DTranspose(128, (5, 5), strides=(1, 1),
                                     padding='same', use_bias=False))
    assert model.output_shape == (None, 7, 7, 128)
    model.add(layers.BatchNormalization())
    model.add(layers.LeakyReLU())

    model.add(layers.Conv2DTranspose(64, (5, 5), strides=(2, 2),
                                     padding='same', use_bias=False))
    assert model.output_shape == (None, 14, 14, 64)
    model.add(layers.BatchNormalization())
    model.add(layers.LeakyReLU())

    model.add(layers.Conv2DTranspose(1, (5, 5), strides=(2, 2),
                    padding='same', use_bias=False, activation='tanh'))
    assert model.output_shape == (None, 28, 28, 1)

    return model
```

5. 建立並測試生成神經網路。

```python
# 產生生成神經網路
generator = make_generator_model()

# 測試產生的雜訊
noise = tf.random.normal([1, 100])
generated_image = generator(noise, training=False)
```

10-3 DCGAN

```
7
8   # 顯示雜訊生成的圖像
9   plt.imshow(generated_image[0, :, :, 0], cmap='gray')
```

- 執行結果：

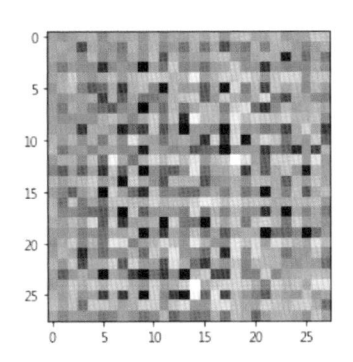

6. 定義判別神經網路：類似一般的 CNN 判別模型，但要去除池化層，避免資訊遺失。

```
1   def make_discriminator_model():
2       model = tf.keras.Sequential()
3       model.add(layers.Input([28, 28, 1]))
4       model.add(layers.Conv2D(64, (5, 5), strides=(2, 2), padding='same'))
5       model.add(layers.LeakyReLU())
6       model.add(layers.Dropout(0.3))
7
8       model.add(layers.Conv2D(128, (5, 5), strides=(2, 2), padding='same'))
9       model.add(layers.LeakyReLU())
10      model.add(layers.Dropout(0.3))
11
12      model.add(layers.Flatten())
13      model.add(layers.Dense(1))
14
15      return model
```

7. 建立並測試判別神經網路：真實影像的預測機率會是較大的值，生成影像的預測機率則會是較小的值，因為無明顯的線條特徵。

```
1   discriminator = make_discriminator_model()
2
3   # 真實的影像預測值會是較大的值，生成的影像預測值會是較小的值
4   decision = discriminator(generated_image)
5   print (f'預測值={decision[0][0]}')
```

- 執行結果：-0.00208456，為很小的值。

8. 測試真實的影像。

```
1  decision = discriminator(train_images[0:5])
2  print (f'預測值={decision}')
```

- 執行結果：為絕對值較大的值。

  [[0.12165608]

  [0.14320901]

  [-0.00153188]

  [0.13788301]

  [0.11824991]]

9. 定義損失函數為二分類交叉熵（BinaryCrossentropy）、優化器為 Adam。
   - 判別神經網路的損失函數為『真實影像』加上『生成影像』的損失函數和，因為判別神經網路會同時接收真實影像和生成影像。

```
1  # 定義損失函數為二分類交叉熵
2  cross_entropy = tf.keras.losses.BinaryCrossentropy(from_logits=True)
3
4  # 定義判別神經網路損失函數為 真實影像 + 生成影像 的損失函數
5  def discriminator_loss(real_output, fake_output):
6      real_loss = cross_entropy(tf.ones_like(real_output), real_output)
7      fake_loss = cross_entropy(tf.zeros_like(fake_output), fake_output)
8      total_loss = real_loss + fake_loss
9      return total_loss
10
11 # 定義生成神經網路損失函數為 生成影像 的損失函數
12 def generator_loss(fake_output):
13     return cross_entropy(tf.ones_like(fake_output), fake_output)
14
15 # 優化器均為 Adam
16 generator_optimizer = tf.keras.optimizers.Adam(1e-4)
17 discriminator_optimizer = tf.keras.optimizers.Adam(1e-4)
```

10. 設定檢查點：因為訓練要花很長的時間，可以設定檢查點定時將模型存檔，萬一中途當掉或者要再加強訓練，還可以從上次的中斷點繼續訓練。

```
1  # 在檢查點模型存檔
2  checkpoint_dir = './dcgan_training_checkpoints'
3  checkpoint_prefix = os.path.join(checkpoint_dir, "ckpt")
4  checkpoint = tf.train.Checkpoint(generator_optimizer=generator_optimizer,
5                                   discriminator_optimizer=discriminator_optimizer,
```

```
6                             generator=generator,
7                             discriminator=discriminator)
```

11. 參數設定。

```
1  # 參數設定
2  EPOCHS = 50                      # 訓練執行週期
3  noise_dim = 100                  # 雜訊向量大小
4  num_examples_to_generate = 16    # 生成筆數
5
6  # 產生亂數(雜訊)
7  seed = tf.random.normal([num_examples_to_generate, noise_dim])
```

12. 定義梯度下降函數：同時對生成神經網路和判別神經網路進行訓練，雜訊與真實圖像也一起餵入判別神經網路，計算損失及梯度，並更新權重。

- @tf.function：會產生運算圖，使函數運算速度加快。

```
1  # 定義梯度下降，分別對判別神經網路、生成神經網路進行訓練
2  @tf.function   # 產生運算圖
3  def train_step(images):
4      noise = tf.random.normal([BATCH_SIZE, noise_dim])
5
6      with tf.GradientTape() as gen_tape, tf.GradientTape() as disc_tape:
7          # 生成神經網路進行訓練
8          generated_images = generator(noise, training=True)
9
10         # 判別神經網路進行訓練
11         real_output = discriminator(images, training=True)              # 真實影像
12         fake_output = discriminator(generated_images, training=True)    # 生成影像
13
14         # 計算損失
15         gen_loss = generator_loss(fake_output)
16         disc_loss = discriminator_loss(real_output, fake_output)
17
18     # 梯度下降
19     gradients_of_generator = gen_tape.gradient(gen_loss, generator.trainable_variables)
20     gradients_of_discriminator = disc_tape.gradient(disc_loss,
21                                                    discriminator.trainable_variables)
22
23     # 更新權重
24     generator_optimizer.apply_gradients(zip(gradients_of_generator,
25                                             generator.trainable_variables))
26     discriminator_optimizer.apply_gradients(zip(gradients_of_discriminator,
27                                                  discriminator.trainable_variables))
```

13. 定義訓練函數：同時間會產生圖像並存檔。

```
1  # 定義訓練函數
2  def train(dataset, epochs):
3      for epoch in range(epochs):
4          start = time.time()
5
```

```
 6            for image_batch in dataset:
 7                train_step(image_batch)
 8
 9            # 產生圖像
10            display.clear_output(wait=True)
11            generate_and_save_images(generator, epoch + 1, seed)
12
13            # 每 10 個執行週期存檔一次
14            if (epoch + 1) % 10 == 0:
15                checkpoint.save(file_prefix = checkpoint_prefix)
16
17            print ('epoch {} 花費 {} 秒'.format(epoch + 1, time.time()-start))
18
19        # 顯示最後結果
20        display.clear_output(wait=True)
21        generate_and_save_images(generator, epochs, seed)
```

14. 定義產生圖像的函數

```
 1  # 產生圖像並存檔
 2  def generate_and_save_images(model, epoch, test_input):
 3      # 預測
 4      predictions = model(test_input, training=False)
 5
 6      # 顯示 4x4 的格子
 7      fig = plt.figure(figsize=(4, 4))
 8      for i in range(predictions.shape[0]):
 9          plt.subplot(4, 4, i+1)
10          plt.imshow(predictions[i, :, :, 0] * 127.5 + 127.5, cmap='gray')
11          plt.axis('off')
12
13      # 存檔
14      plt.savefig('./GAN_result/image_at_epoch_{:04d}.png'.format(epoch))
15      plt.show()
```

15. 訓練模型：訓練過程會產生動畫效果。

```
 1  train(train_dataset, EPOCHS)
```

- 執行結果：訓練了 50 個週期，數字已隱約成形。

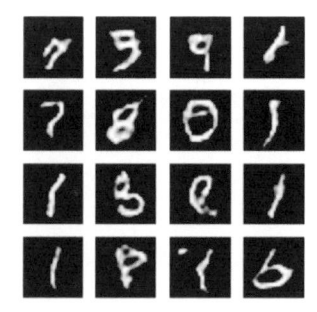

16. 將訓練過程中的存檔圖像轉為 GIF 檔，並顯示 GIF 檔。

```
1   # 產生 GIF 檔
2   anim_file = './GAN_result/dcgan.gif'
3   with imageio.get_writer(anim_file, mode='I') as writer:
4       filenames = glob.glob('./GAN_result/image*.png')
5       filenames = sorted(filenames)
6       for filename in filenames:
7           # print(filename)
8           image = imageio.imread(filename)
9           writer.append_data(image)
10
11  # 顯示 GIF 檔
12  import tensorflow_docs.vis.embed as embed
13  embed.embed_file(anim_file)
```

- 執行結果：注意，GIF 檔會不斷循環播放，所以也可以打開檔案總管點選大圖顯示來檢視。

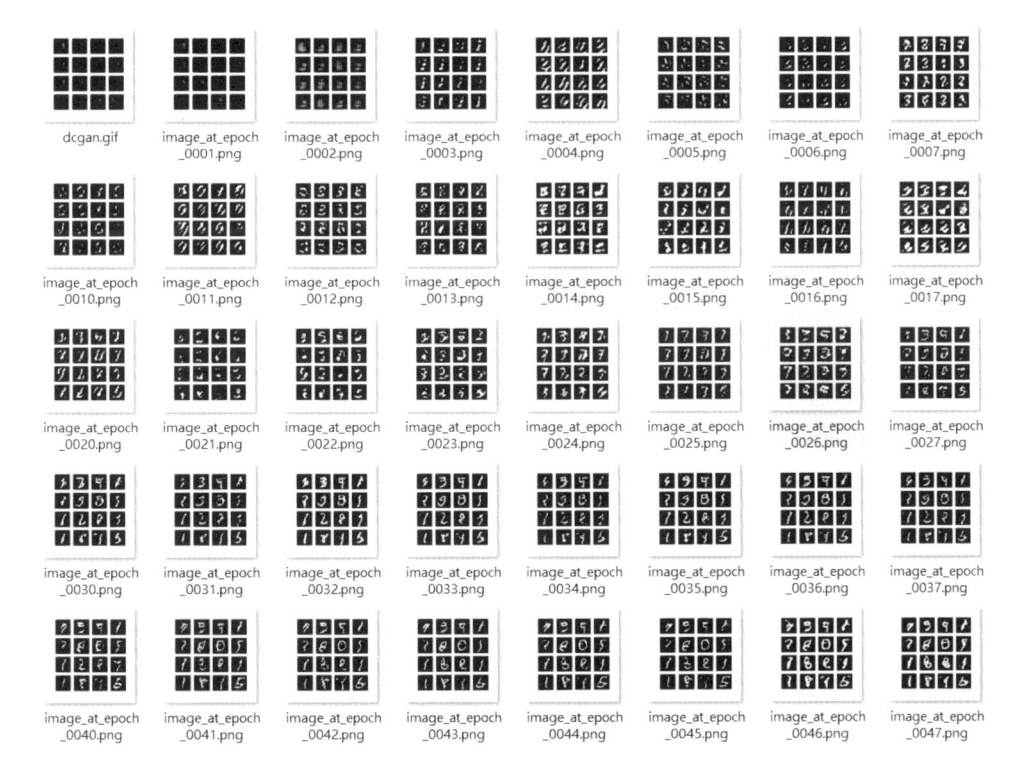

17. 安裝 tensorflow-docs 套件，以播放 GIF 檔。

    pip install tensorflow-docs

第 10 章　生成對抗網路（GAN）

18. 播放 GIF 檔：可以看到完整的訓練過程，由一開始的雜訊到最後愈來愈清晰。

```
1  import tensorflow_docs.vis.embed as embed
2  embed.embed_file(anim_file)
```

19. 還原（restore）模型：測試還原後是否仍可產生訓練後的成果。

    checkpoint.restore（tf.train.latest_checkpoint（checkpoint_dir））

20. 輸入雜訊，請生成模型產生圖像。

```
1  # 產生圖像
2  def generate_images(model, test_input):
3      # 預測
4      predictions = model(test_input, training=False)
5  
6      # 顯示 4x4 的格子
7      fig = plt.figure(figsize=(4, 4))
8      for i in range(predictions.shape[0]):
9              plt.subplot(4, 4, i+1)
10             plt.imshow(predictions[i, :, :, 0] * 127.5 + 127.5, cmap='gray')
11             plt.axis('off')
12 
13     plt.show()
14 generate_images(generator, seed)
```

- 執行結果：可產生相同效果的圖片。

21. 再訓練 5 週期。

```
1  train(train_dataset, 5)
```

- 執行結果：比之前更清晰，表示還原模型成功，可接續上次模型繼續訓練（Continuous learning）。

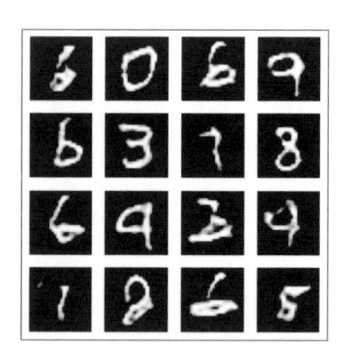

10-3 DCGAN

再以名人臉部資料集，生成近似真實的圖像，程式修改自 Keras 官網『DCGAN to generate face images』[16]。

範例 2. 以名人臉部資料集實作 DCGAN 演算法，相關邏輯幾乎沒有改變，只是資料集置換，從檔案中讀取訓練資料。

程式：**10_02_DCGAN_Face.ipynb**。

1. 載入相關套件。

```
1  # 載入相關套件
2  import tensorflow as tf
3  from tensorflow import keras
4  from tensorflow.keras import layers
5  import numpy as np
6  import matplotlib.pyplot as plt
7  import os
```

2. 訓練資料集為名人臉部：約 1.3GB，下載 img_align_celeba.zip，解壓縮至 celeba_gan 資料夾，gdown 是專門自 Google 雲端硬碟下載檔案的套件，可下載大尺寸的檔案，且比 wget 指令快速。

```
1  zip_dir = './celeba_gan'
2  dataset = keras.preprocessing.image_dataset_from_directory(
3      zip_dir, label_mode=None, image_size=(64, 64), batch_size=32
4  )
5  os.makedirs(zip_dir)
6
7  url = "https://drive.google.com/uc?id=1O7m1010EJjLE5QxLZiM9Fpjs7Oj6e684"
8  output = "./data.zip"
9  gdown.download(url, output, quiet=True)
10
11 with ZipFile("./data.zip", "r") as zipobj:
12     zipobj.extractall("./celeba_gan")
```

3. 產生資料集（dataset），並縮放圖像寬度及高度為（64, 64），使用資料集（Dataset）讀取較具效率，逐批載入記憶體，不需一次載入所有檔案。

```
1  # 從celeba_gan目錄產生資料集
2  dataset = keras.preprocessing.image_dataset_from_directory(
3      "celeba_gan", label_mode=None, image_size=(64, 64), batch_size=32
4  )
5  # 像素標準化
6  dataset = dataset.map(lambda x: x / 255.0)
```

- 執行結果：共有 202599 個圖檔，同屬一個類別（以子資料夾為類別名稱）。

10-17

4. 載入並顯示第一個圖檔。

```
1  # 載入並顯示第一個圖檔
2  image = next(iter(dataset))
3  plt.axis("off")
4  plt.imshow((image.numpy() * 255).astype("int32")[0])
```

- 執行結果：

5. 定義判別神經網路：使用一般的 CNN 結構，包含卷積層與完全連接層，用來辨識輸入的影像，但為了避免影像資訊減損，故不包含池化層。

```
1   # 判別神經網路
2   discriminator = keras.Sequential(
3       [
4           keras.Input(shape=(64, 64, 3)),
5           layers.Conv2D(64, kernel_size=4, strides=2, padding="same"),
6           layers.LeakyReLU(alpha=0.2),
7           layers.Conv2D(128, kernel_size=4, strides=2, padding="same"),
8           layers.LeakyReLU(alpha=0.2),
9           layers.Conv2D(128, kernel_size=4, strides=2, padding="same"),
10          layers.LeakyReLU(alpha=0.2),
11          layers.Flatten(),
12          layers.Dropout(0.2),
13          layers.Dense(1, activation="sigmoid"),
14      ],
15      name="discriminator",
16  )
17  discriminator.summary()
```

- 執行結果：輸入圖像維度為（64, 64, 3），寬高各為 64，RGB 3 顏色。

6. 定義生成神經網路：使用一般的解碼器結構，含完全連接層、反卷積層（Conv2DTranspose）與卷積層，可輸出特徵向量。

10-3 DCGAN

```
1   # 生成神經網路
2   latent_dim = 128
3
4   generator = keras.Sequential(
5       [
6           keras.Input(shape=(latent_dim,)),
7           layers.Dense(8 * 8 * 128),
8           layers.Reshape((8, 8, 128)),
9           layers.Conv2DTranspose(128, kernel_size=4, strides=2, padding="same"),
10          layers.LeakyReLU(alpha=0.2),
11          layers.Conv2DTranspose(256, kernel_size=4, strides=2, padding="same"),
12          layers.LeakyReLU(alpha=0.2),
13          layers.Conv2DTranspose(512, kernel_size=4, strides=2, padding="same"),
14          layers.LeakyReLU(alpha=0.2),
15          layers.Conv2D(3, kernel_size=5, padding="same", activation="sigmoid"),
16      ],
17      name="generator",
18  )
19  generator.summary()
```

- 執行結果：輸入隨機向量大小為（8 * 8 * 128），最後網路輸出為（64, 64, 3），與判別神經網路輸入的維度一致。

```
Model: "generator"
_____
Layer (type)                 Output Shape              Param #
=================================================================
dense_4 (Dense)              (None, 8192)              1056768

reshape_1 (Reshape)          (None, 8, 8, 128)         0

conv2d_transpose_3 (Conv2DTr (None, 16, 16, 128)       262272

leaky_re_lu_12 (LeakyReLU)   (None, 16, 16, 128)       0

conv2d_transpose_4 (Conv2DTr (None, 32, 32, 256)       524544

leaky_re_lu_13 (LeakyReLU)   (None, 32, 32, 256)       0

conv2d_transpose_5 (Conv2DTr (None, 64, 64, 512)       2097664

leaky_re_lu_14 (LeakyReLU)   (None, 64, 64, 512)       0

conv2d_10 (Conv2D)           (None, 64, 64, 3)         38403
=================================================================
Total params: 3,979,651
Trainable params: 3,979,651
Non-trainable params: 0
_____
```

第 10 章　生成對抗網路（GAN）

7. 定義 GAN：組合兩個網路。

```
1   # 定義GAN，組合兩個網路
2   class GAN(keras.Model):
3       def __init__(self, discriminator, generator, latent_dim):
4           super(GAN, self).__init__()
5           self.discriminator = discriminator
6           self.generator = generator
7           self.latent_dim = latent_dim
8
9       # 編譯：定義損失函數
10      def compile(self, d_optimizer, g_optimizer, loss_fn):
11          super(GAN, self).compile()
12          self.d_optimizer = d_optimizer
13          self.g_optimizer = g_optimizer
14          self.loss_fn = loss_fn
15          self.d_loss_metric = keras.metrics.Mean(name="d_loss")
16          self.g_loss_metric = keras.metrics.Mean(name="g_loss")
17
18      # 效能指標：判別神經網路、生成神經網路
19      @property
20      def metrics(self):
21          return [self.d_loss_metric, self.g_loss_metric]
22
23      # 訓練
24      def train_step(self, real_images):
25          # 隨機抽樣 batch_size 筆，維度大小：latent_dim(128)
26          batch_size = tf.shape(real_images)[0]
27          random_latent_vectors = tf.random.normal(shape=(batch_size, self.latent_dim))
28
29          # 生成圖像
30          generated_images = self.generator(random_latent_vectors)
31
32          # 與訓練資料結合
33          combined_images = tf.concat([generated_images, real_images], axis=0)
34
35          # 訓練資料的標籤設為 1，生成圖像的標籤設為 0
36          labels = tf.concat(
37              [tf.ones((batch_size, 1)), tf.zeros((batch_size, 1))], axis=0
38          )
39
40          # 將標籤加入雜訊，此步驟非常重要
41          labels += 0.05 * tf.random.uniform(tf.shape(labels))
42
43          # 訓練判別神經網路
44          with tf.GradientTape() as tape:
45              predictions = self.discriminator(combined_images)
46              d_loss = self.loss_fn(labels, predictions)
47          grads = tape.gradient(d_loss, self.discriminator.trainable_weights)
48          self.d_optimizer.apply_gradients(
49              zip(grads, self.discriminator.trainable_weights)
50          )
51
52          # 隨機抽樣 batch_size 筆，維度大小：latent_dim(128)
53          random_latent_vectors = tf.random.normal(shape=(batch_size, self.latent_dim))
54
55          # 生成圖像的標籤設為 0
56          misleading_labels = tf.zeros((batch_size, 1))
```

```
57
58          # 訓練生成神經網路，注意，不可更新判別神經網路的權重，只更新生成神經網路的權重
59          with tf.GradientTape() as tape:
60              predictions = self.discriminator(self.generator(random_latent_vectors))
61              g_loss = self.loss_fn(misleading_labels, predictions)
62          grads = tape.gradient(g_loss, self.generator.trainable_weights)
63          self.g_optimizer.apply_gradients(zip(grads, self.generator.trainable_weights))
64
65          # 計算效能指標
66          self.d_loss_metric.update_state(d_loss)
67          self.g_loss_metric.update_state(g_loss)
68          return {
69              "d_loss": self.d_loss_metric.result(),
70              "g_loss": self.g_loss_metric.result(),
71          }
```

8. 自訂 Callback：在訓練過程中儲存圖像。

```
1  # 建立自訂的 Callback，在訓練過程中儲存圖像
2  class GANMonitor(keras.callbacks.Callback):
3      def __init__(self, num_img=3, latent_dim=128):
4          self.num_img = num_img
5          self.latent_dim = latent_dim
6
7      # 在每一執行週期結束時產生圖像
8      def on_epoch_end(self, epoch, logs=None):
9          random_latent_vectors = tf.random.normal(shape=(self.num_img, self.latent_dim))
10         generated_images = self.model.generator(random_latent_vectors)
11         generated_images *= 255
12         generated_images.numpy()
13         for i in range(self.num_img):
14             # 儲存圖像
15             img = keras.preprocessing.image.array_to_img(generated_images[i])
16             img.save("./GAN_generated/img_%03d_%d.png" % (epoch, i))
```

9. 訓練模型：在每個執行週期結束時產生 10 張圖像，依據原程式的註解，訓練週期正常需要 100 次，才能產生令人驚豔的圖片。

```
1  # 訓練模型
2  epochs = 1   # 訓練週期正常需要100次
3
4  gan = GAN(discriminator=discriminator, generator=generator, latent_dim=latent_dim)
5  gan.compile(
6      d_optimizer=keras.optimizers.Adam(learning_rate=0.0001),
7      g_optimizer=keras.optimizers.Adam(learning_rate=0.0001),
8      loss_fn=keras.losses.BinaryCrossentropy(),
9  )
10
11 # 產生10張圖像
12 gan.fit(
13     dataset, epochs=epochs, callbacks=[GANMonitor(num_img=10, latent_dim=latent_dim)]
14 )
```

第 10 章　生成對抗網路（GAN）

- 執行結果：筆者的 PC 執行 1 週期就需花 2~3 個小時，如果單純使用 CPU，那可以先去睡個覺，隔天再來看結果了，這時候 GPU 卡就顯得格外重要。

- 執行 1 週期結果：還非常模糊。

img_000_0.png　img_000_1.png　img_000_2.png　img_000_3.png　img_000_4.png　img_000_5.png　img_000_6.png　img_000_7.png　img_000_8.png　img_000_9.png

- 執行 30 個週期的結果：如下圖，雖然有改善，但仍然不符預期，依據官網說明至少要執行 200 個週期。

10-4 Progressive GAN

Progressive GAN 也稱為 Progressive Growing GAN 或 PGGAN，源自於 NVIDIA 2017 年發表的文章『Progressive Growing of GANs for Improved Quality, Stability, and Variation』[17]，可以生成高畫質且穩定的圖像，小圖像透過層層的神經層不斷擴大，直到所要求的尺寸為止，大部份是針對人臉的生成。

10-4 Progressive GAN

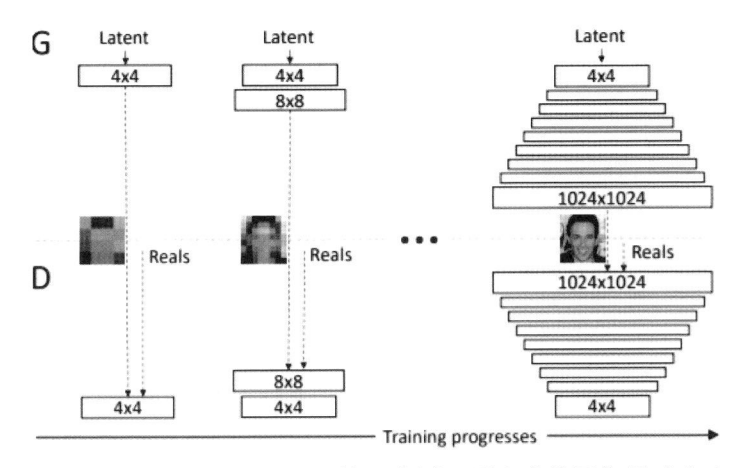

▲ 圖 10.13 Progressive GAN 的示意圖，要生成的圖像尺寸愈大，神經層就增加愈多，圖片來源：『Progressive Growing of GANs for Improved Quality, Stability, and Variation』[12]

Progressive GAN 厲害的地方是演算法生成的尺寸可以大於訓練資料集的任何圖像，稱為『超解析度』（Super Resolution）。網路架構如下：

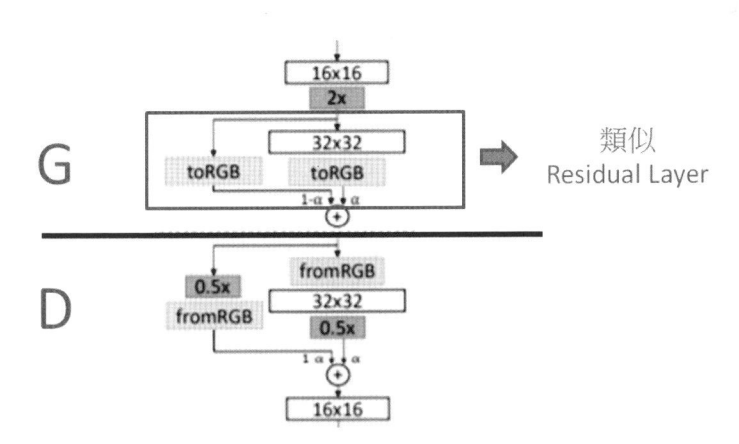

▲ 圖 10.14 Progressive GAN 的網路架構

生成網路（G）使用類似像 Residual 神經層，一邊輸入原圖像，另一邊輸入為反卷積層，使用權重 α，訂定兩個輸入層的比例。判別網路（D）與生成網路（G）做反向操作，進行辨識。使用名人臉部資料集進行模型訓練，根據論文估計，使用 8 顆 Tesla V100 GPU，大約要訓練 4 天左右，才可以得到不錯的效果。讀到這裡，對我們平民百姓來

第 10 章　生成對抗網路（GAN）

說，簡直是晴天霹靂，根本不用玩了。還好，TensorFlow Hub 提供預先訓練好的模型，我們馬上來測試一下吧。

範例 3. 再拿名人臉部資料集來實作 Progressive GAN 演算法。

程式：**10_03_ Progressive_GAN_Face.ipynb**。

訓練資料集：名人臉部資料集。

- 使用隨機向量，流程如下：

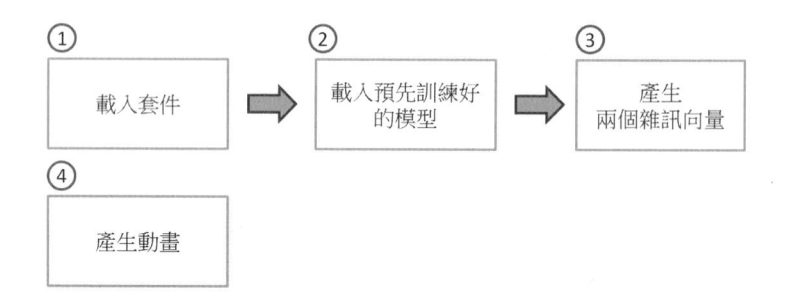

①載入套件 ➡ ②載入預先訓練好的模型 ➡ ③產生兩個雜訊向量

④產生動畫

- 若使用自訂的圖像，流程如下：

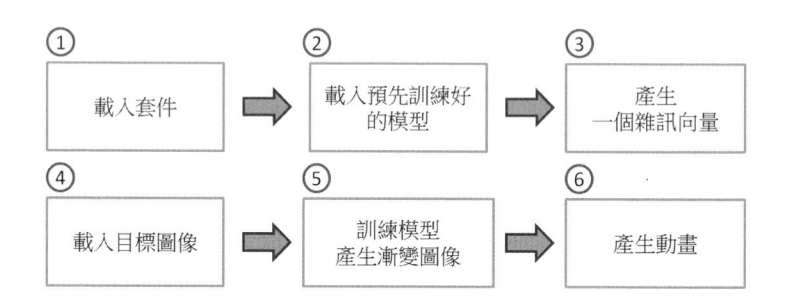

①載入套件 ➡ ②載入預先訓練好的模型 ➡ ③產生一個雜訊向量

④載入目標圖像 ➡ ⑤訓練模型產生漸變圖像 ➡ ⑥產生動畫

1. 安裝 scikit-image、imageio 套件：之前範例已安裝過。

    pip install scikit-image imageio

2. 載入套件。

```
1  # 載入相關套件
2  from absl import logging
3  import imageio
4  import PIL.Image
5  import matplotlib.pyplot as plt
6  import numpy as np
7  import tensorflow as tf
```

10-4 Progressive GAN

```
 8  import tensorflow_hub as hub
 9  import time
10  from IPython import display
11  from skimage import transform
```

3. 定義顯示圖像的函數、轉為動畫的函數。

```
 1  # 生成網路的輸入向量尺寸
 2  latent_dim = 512
 3
 4  # 定義顯示圖像的函數
 5  def display_image(image):
 6      image = tf.constant(image)
 7      image = tf.image.convert_image_dtype(image, tf.uint8)
 8      return PIL.Image.fromarray(image.numpy())
 9
10  # 定義顯示一序列圖像轉為動畫的函數
11  from tensorflow_docs.vis import embed
12  def animate(images):
13      images = np.array(images)
14      converted_images = np.clip(images * 255, 0, 255).astype(np.uint8)
15      imageio.mimsave('./animation.gif', converted_images)
16      return embed.embed_file('./animation.gif')
17
18  # 工作記錄檔只記錄錯誤等級以上的訊息
19  logging.set_verbosity(logging.ERROR)
```

4. 插補兩個向量，產生動畫：採非線性插補，在超球面上插補兩個向量，這樣才能有較平滑的轉換，隨機抽取兩張圖像，由第一張圖像開始，漸變成第二張圖像。

```
 1  # 在超球面上插補兩向量
 2  def interpolate_hypersphere(v1, v2, num_steps):
 3      v1_norm = tf.norm(v1)
 4      v2_norm = tf.norm(v2)
 5      v2_normalized = v2 * (v1_norm / v2_norm)
 6
 7      vectors = []
 8      for step in range(num_steps):
 9          interpolated = v1 + (v2_normalized - v1) * step / (num_steps - 1)
10          interpolated_norm = tf.norm(interpolated)
11          interpolated_normalized = interpolated * (v1_norm / interpolated_norm)
12          vectors.append(interpolated_normalized)
13      return tf.stack(vectors)
14
15  # 插補兩向量
16  def interpolate_between_vectors():
17      # 產生兩個雜訊向量
18      v1 = tf.random.normal([latent_dim])
19      v2 = tf.random.normal([latent_dim])
20
21      # 產生 25 個步驟的插補
22      vectors = interpolate_hypersphere(v1, v2, 50)
23
```

第 10 章　生成對抗網路（GAN）

```
24      # 產生一序列的圖像
25      interpolated_images = progan(vectors)['default']
26      return interpolated_images
27
28  # 插補兩向量，產生動畫
29  interpolated_images = interpolate_between_vectors()
30  animate(interpolated_images)
```

- 執行結果：由左圖漸變為右圖。

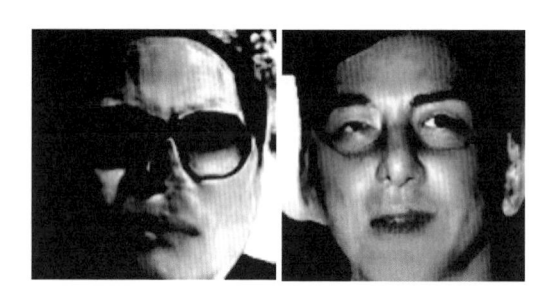

5. 使用隨機亂數或自訂的圖像：由第 5 行的 image_from_module_space 變數來調控，設為 True 表示使用隨機亂數；False 表示使用自訂的圖像。

```
1   # 使用自訂的圖像
2   image_path='./images_face/face_test.png'
3
4   # True：使用隨機亂數，False：使用自訂的圖像
5   image_from_module_space = True
6
7   tf.random.set_seed(0)
8
9   def get_module_space_image():
10      vector = tf.random.normal([1, latent_dim])
11      images = progan(vector)['default'][0]
12      return images
13
14  # 使用隨機亂數
15  if image_from_module_space:
16      target_image = get_module_space_image()
17  else: # 使用自訂的圖像
18      image = imageio.imread(image_path)
19      target_image = transform.resize(image, [128, 128])
20
21  # 顯示圖像
22  display_image(target_image)
```

- 執行結果：目標圖像。

10-4 Progressive GAN

6. 以 PGGAN 處理的圖像作為起始圖像。

```
1  # 以PGGAN 處理的圖像作為起始圖像
2  initial_vector = tf.random.normal([1, latent_dim])
3
4  # 顯示圖像
5  display_image(progan(initial_vector)['default'][0])
```

- 執行結果：起始圖像。

7. 找到最接近的特徵向量，漸變成第二張圖像：模型訓練的過程中，每 5 步驟就會產生一個圖像。

```
1  # 找到最接近的特徵向量
2  def find_closest_latent_vector(initial_vector, num_optimization_steps,
3      images = []
4      losses = []
5
6      vector = tf.Variable(initial_vector)
7      optimizer = tf.optimizers.Adam(learning_rate=0.01)
8      # 以 MAE 為損失函數
9      loss_fn = tf.losses.MeanAbsoluteError(reduction="sum")
10
11     # 訓練
12     for step in range(num_optimization_steps):
13         if (step % 100)==0:
14             print()
15         print('.', end='')
16
17         # 梯度下降
```

```
18            with tf.GradientTape() as tape:
19                image = progan(vector.read_value())['default'][0]
20
21            # 每 5 步驟產生一個圖像
22            if (step % steps_per_image) == 0:
23                img = tf.keras.preprocessing.image.array_to_img(image.numpy())
24                img.save(f"./PGGAN_generated/img_{step:03d}.png")
25                images.append(image.numpy())
26
27            # 計算損失
28            target_image_difference = loss_fn(image, target_image[:,:,:3])
29            # 正則化
30            regularizer = tf.abs(tf.norm(vector) - np.sqrt(latent_dim))
31            loss = tf.cast(target_image_difference, tf.double) + tf.cast(regularizer, tf.double)
32            losses.append(loss.numpy())
33
34            # 根據梯度,更新權重
35            grads = tape.gradient(loss, [vector])
36            optimizer.apply_gradients(zip(grads, [vector]))
37
38        return images, losses
39
40
41 num_optimization_steps=200
42 steps_per_image=5
43 images, loss = find_closest_latent_vector(initial_vector, num_optimization_steps, steps_per_image)
```

8. 顯示訓練過程的損失函數。

```
1 # 顯示訓練過程的損失函數
2 plt.plot(loss)
3 plt.ylim([0,max(plt.ylim())])
```

9. 顯示動畫。

```
1 animate(np.stack(images))
```

- 執行結果:由左圖漸變為右圖。

- 修改步驟 4 第 5 行的 image_from_module_space 變數控制,更改為 False,使用自訂的圖像,臉部特寫必須一致,效果才會比較好。

10-4 Progressive GAN

- 起始圖像如下：

- 自訂的圖像如下，為目標圖像。

- 執行結果：

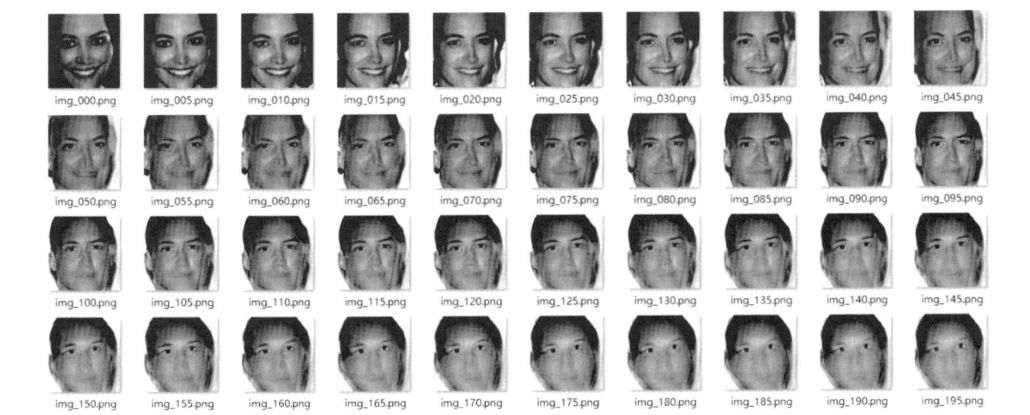

第 10 章　生成對抗網路（GAN）

10-5 Conditional GAN

DCGAN 生成的圖片是隨機的，以 MNIST 資料集而言，生成圖像的確會是數字，但無法控制要生成哪一個數字。Conditional GAN 增加了一個條件（Condition），即目標變數 Y，用來控制生成的數字，與之前 CVAE 一樣。

Conditional GAN 也稱為 cGAN，它是 Mehdi Mirz 等學者於 2014 年發表的一篇文章『Conditional Generative Adversarial Nets』[18] 中所提出的演算法，它修改 GAN 損失函數如下：

$$\min_G \max_D V(D, G) = \mathbb{E}_{\boldsymbol{x} \sim p_{\text{data}}(\boldsymbol{x})}[\log D(\boldsymbol{x}|\boldsymbol{y})] + \mathbb{E}_{\boldsymbol{z} \sim p_z(\boldsymbol{z})}[\log(1 - D(G(\boldsymbol{z}|\boldsymbol{y})))]$$

將單純的 D(x) 改為條件機率 D（x|y）。

範例. 以 Fashion MNIST 資料集實作 Conditional GAN 演算法。

程式：**10_04_CGAN_FashionMNIST.ipynb**。

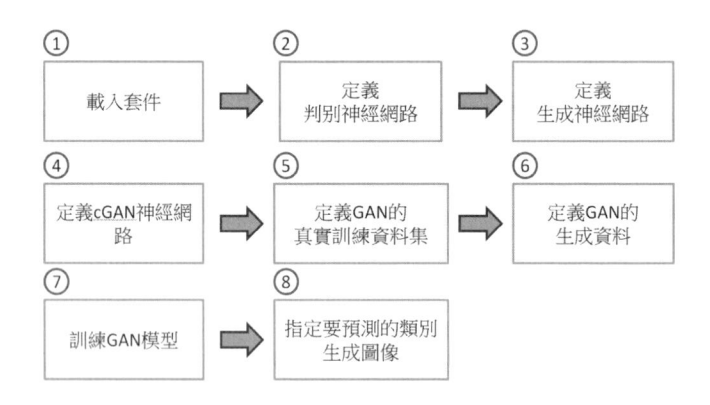

① 載入套件 → ② 定義判別神經網路 → ③ 定義生成神經網路

④ 定義cGAN神經網路 → ⑤ 定義GAN的真實訓練資料集 → ⑥ 定義GAN的生成資料

⑦ 訓練GAN模型 → ⑧ 指定要預測的類別生成圖像

1. 載入套件。

```
1  # 載入相關套件
2  from numpy import zeros, ones, expand_dims
3  from numpy.random import randn
4  from numpy.random import randint
5  from tensorflow.keras.datasets.fashion_mnist import load_data
6  from tensorflow.keras.optimizers import Adam
7  from tensorflow.keras.models import Model
8  from tensorflow.keras.layers import Input, Dense, Reshape, Flatten
9  from tensorflow.keras.layers import Dropout, Embedding, Concatenate, LeakyReLU
10 from tensorflow.keras.layers import Conv2D, Conv2DTranspose
```

10-5 Conditional GAN

2. 定義判別神經網路：把 Fashion MNIST 資料集的標籤（Label）也視為特徵變數（x），因為我們要指定生成圖像的類別，透過嵌入層（Embedding）轉成 50 個向量。注意，第 28 行 Adam 的參數 lr 在新版的 TensorFlow 應改為 learning_rate。

```python
# 定義判別神經網路
def define_discriminator(in_shape=(28,28,1), n_classes=10):
    # 輸入 Y
    in_label = Input(shape=(1,))
    li = Embedding(n_classes, 50)(in_label)
    li = Dense(in_shape[0] * in_shape[1])(li)
    li = Reshape((in_shape[0], in_shape[1], 1))(li)
    
    # 輸入圖像
    in_image = Input(shape=in_shape)
    
    # 結合 Y 及圖像
    merge = Concatenate()([in_image, li])
    
    # 抽樣(downsampling)
    fe = Conv2D(128, (3,3), strides=(2,2), padding='same')(merge)
    fe = LeakyReLU(alpha=0.2)(fe)
    fe = Conv2D(128, (3,3), strides=(2,2), padding='same')(fe)
    fe = LeakyReLU(alpha=0.2)(fe)
    fe = Flatten()(fe)
    fe = Dropout(0.4)(fe)
    out_layer = Dense(1, activation='sigmoid')(fe)
    
    # 定義模型的輸入及輸出
    model = Model([in_image, in_label], out_layer)
    
    # 編譯
    opt = Adam(lr=0.0002, beta_1=0.5)
    model.compile(loss='binary_crossentropy', optimizer=opt, metrics=['accuracy'])
    return model
```

3. 定義生成神經網路：加入 Y。

```python
# 定義生成神經網路
def define_generator(latent_dim, n_classes=10):
    # 輸入 Y
    in_label = Input(shape=(1,))
    # embedding for categorical input
    li = Embedding(n_classes, 50)(in_label)
    li = Dense(7 * 7)(li)
    li = Reshape((7, 7, 1))(li)
    
    # 輸入圖像
    in_lat = Input(shape=(latent_dim,))
    gen = Dense(128 * 7 * 7)(in_lat)
    gen = LeakyReLU(alpha=0.2)(gen)
    gen = Reshape((7, 7, 128))(gen)
    
    # 結合 Y 及圖像
    merge = Concatenate()([gen, li])
```

第 10 章　生成對抗網路（GAN）

```python
18
19      # 上採樣(upsampling)
20      gen = Conv2DTranspose(128, (4,4), strides=(2,2), padding='same')(merge)
21      gen = LeakyReLU(alpha=0.2)(gen)
22      gen = Conv2DTranspose(128, (4,4), strides=(2,2), padding='same')(gen)
23      gen = LeakyReLU(alpha=0.2)(gen)
24      out_layer = Conv2D(1, (7,7), activation='tanh', padding='same')(gen)
25
26      # 定義模型的輸入及輸出
27      model = Model([in_lat, in_label], out_layer)
28      return model
```

4. 定義 cGAN 神經網路：結合判別神經網路和生成神經網路。

```python
1   # 定義cGAN神經網路
2   def define_gan(g_model, d_model):
3       d_model.trainable = False           # 判別神經網路不重新訓練
4       gen_noise, gen_label = g_model.input # 取得生成神經網路的輸入
5       gen_output = g_model.output          # 取得生成神經網路的輸出
6
7       # 取得判別神經網路的輸出
8       gan_output = d_model([gen_output, gen_label])
9
10      # 定義模型的輸入及輸出
11      model = Model([gen_noise, gen_label], gan_output)
12
13      # 編譯
14      opt = Adam(lr=0.0002, beta_1=0.5)
15      model.compile(loss='binary_crossentropy', optimizer=opt)
16      return model
```

5. 定義 GAN 的真實訓練資料集：真實資料的標籤（Y）均為 1。

```python
1   # 載入 fashion mnist 資料集
2   def load_real_samples():
3       (trainX, trainy), (_, _) = load_data()
4
5       # 增加一維作為色彩通道
6       X = expand_dims(trainX, axis=-1)
7       X = X.astype('float32')
8       # 標準化，使像素值介於 [-1,1]
9       X = (X - 127.5) / 127.5
10      return [X, trainy]
11
12  # 定義模型的輸入 n 筆
13  def generate_real_samples(dataset, n_samples):
14      images, labels = dataset
15      # 隨機抽樣 n 筆
16      ix = randint(0, images.shape[0], n_samples)
17      # fashion mnist 資料集的 X、Y 均作為 GAN 的輸入
18      X, labels = images[ix], labels[ix]
19      # GAN 的 Y 均為 1
20      y = ones((n_samples, 1))
21      return [X, labels], y
```

10-5 Conditional GAN

6. 定義 GAN 的生成資料。

```
1   # 生成隨機向量
2   def generate_latent_points(latent_dim, n_samples, n_classes=10):
3       # 隨機向量
4       x_input = randn(latent_dim * n_samples)
5       z_input = x_input.reshape(n_samples, latent_dim)
6       # 產生 n 個類別的 Y
7       labels = randint(0, n_classes, n_samples)
8       return [z_input, labels]
9
10  # 定義模型的生成資料 n 筆
11  def generate_fake_samples(generator, latent_dim, n_samples):
12      # 生成隨機向量
13      z_input, labels_input = generate_latent_points(latent_dim, n_samples)
14      # 生成圖像
15      images = generator.predict([z_input, labels_input])
16      # 產生均為 0 的 Y
17      y = zeros((n_samples, 1))
18      return [images, labels_input], y
```

7. 訓練模型：訓練 GAN 模型，內含判別神經網路、生成神經網路。

```
1   # 訓練模型
2   def train(g_model, d_model, gan_model, dataset, latent_dim, n_epochs=100, n_batch=128):
3       bat_per_epo = int(dataset[0].shape[0] / n_batch)
4       half_batch = int(n_batch / 2)
5
6       # 訓練
7       for i in range(n_epochs):
8           for j in range(bat_per_epo):
9               # 隨機抽樣 n 筆
10              [X_real, labels_real], y_real = generate_real_samples(dataset, half_batch)
11              # 更新判別神經網路權重
12              d_loss1, _ = d_model.train_on_batch([X_real, labels_real], y_real)
13
14              # 生成隨機向量
15              [X_fake, labels], y_fake = generate_fake_samples(g_model, latent_dim, half_batch)
16              # 更新判別神經網路權重
17              d_loss2, _ = d_model.train_on_batch([X_fake, labels], y_fake)
18
19              # 生成一批的隨機向量
20              [z_input, labels_input] = generate_latent_points(latent_dim, n_batch)
21              # 產生均為 1 的 Y
22              y_gan = ones((n_batch, 1))
23
24              # 訓練模型
25              g_loss = gan_model.train_on_batch([z_input, labels_input], y_gan)
26
27              # 顯示損失
28              print('>%d, %d/%d, d1=%.3f, d2=%.3f g=%.3f' %
29                  (i+1, j+1, bat_per_epo, d_loss1, d_loss2, g_loss))
30
31      # 模型存檔
32      g_model.save('cgan_generator.h5')
33
34  # 參數設定
35  latent_dim = 100                        # 隨機向量尺寸
36
```

第 10 章　生成對抗網路（GAN）

```
37  d_model = define_discriminator()              # 建立判別神經網路
38  g_model = define_generator(latent_dim)        # 建立生成神經網路
39  gan_model = define_gan(g_model, d_model)      # 建立 GAN 神經網路
40
41  dataset = load_real_samples()                 # 讀取訓練資料
42  # 訓練
43  train(g_model, d_model, gan_model, dataset, latent_dim)
```

- 執行結果：模型訓練需要幾個小時的時間，先去做其他事，再回來看結果。

8. 定義雜訊生成、顯示圖像的函數。

```
1   # 載入相關套件
2   from numpy import asarray
3   from tensorflow.keras.models import load_model
4   from matplotlib import pyplot
5
6   # 生成隨機資料
7   def generate_latent_points(latent_dim, n_samples, n_classes=10):
8       # generate points in the latent space
9       x_input = randn(latent_dim * n_samples)
10      # reshape into a batch of inputs for the network
11      z_input = x_input.reshape(n_samples, latent_dim)
12      # generate labels
13      labels = randint(0, n_classes, n_samples)
14      return [z_input, labels]
15
16  # 顯示圖像
17  def save_plot(examples, n):
18      # 繪製 n x n 個圖像
19      for i in range(n * n):
20          pyplot.subplot(n, n, 1 + i)
21          pyplot.axis('off')
22          # cmap='gray_r'：反轉黑白，因 MNIST 像素與 RGB 相反
23          pyplot.imshow(examples[i, :, :, 0], cmap='gray_r')
24      pyplot.show()
```

9. 預測並顯示結果。

```
1   # 載入模型
2   model = load_model('cgan_generator.h5')
3   # 生成 100 筆資料
4   latent_points, labels = generate_latent_points(100, 100)
5   # 標記 0~9
6   labels = asarray([x for _ in range(10) for x in range(10)])
7
8   # 預測並顯示結果
9   X = model.predict([latent_points, labels])
10  # 將像素範圍由 [-1,1] 轉換為 [0,1]
11  X = (X + 1) / 2.0
12  # 繪圖
13  save_plot(X, 10)
```

- 執行結果：結果非常理想，我們指定標籤（Label）就可以生成該類別的圖像。

上例是運用 Conditional GAN 很簡單的例子，只是把標籤（Label）一併當作 X，輸入模型中訓練，作為條件（Condition）或限制條件（Constraint）。另外還有很多延伸的作法，例如 ColorGAN，它把前置處理的輪廓圖作為條件，與雜訊一併當作 X，就可以生成與原圖相似的圖像，並且可以為灰階圖上色，相關細節可參閱『Colorization Using ConvNet and GAN』[19] 或『End-to-End Conditional GAN-based Architectures for Image Colourisation』[20]。

Grayscale Image　　Edge Image　　Ground Truth Image

▲ 圖 10.15 ColorGAN，圖片來源：『Colorization Using ConvNet and GAN』[14]

10-6 Pix2Pix

Pix2Pix 為 Conditional GAN 演算法的應用，出自於 Phillip Isola 等學者在 2016 年發表的『Image-to-Image Translation with Conditional Adversarial Networks』[21]，它能夠將影像進行像素的轉換，故稱為 Pix2Pix，可應用於：

第 10 章　生成對抗網路（GAN）

1. 將語義分割的街景圖轉換為真實圖像
2. 將語義分割的建築外觀轉換為真實圖像
3. 將衛星照轉換為地圖
4. 將白天圖像轉換為夜晚圖像
5. 將輪廓圖轉為實物圖像

▲ 圖 10.16 將語義分割的街景圖轉換為真實圖像，以下圖片來源均來自『Image-to-Image Translation with Conditional Adversarial Networks』[16]

▲ 圖 10.17 將衛星照轉換為地圖，反之亦可

▲ 圖 10.18 將白天圖像轉換為夜晚圖像

10-6 Pix2Pix

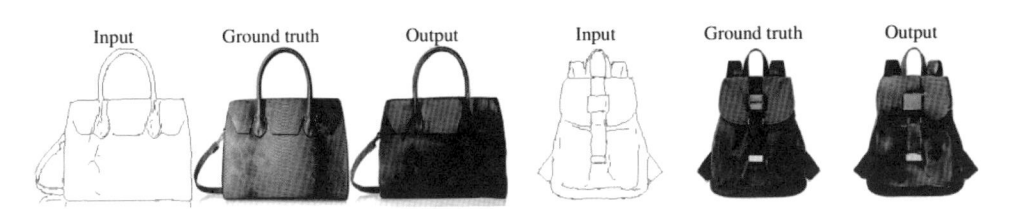

▲ 圖 10.19 將輪廓圖轉為實物圖像

生成網路採用的 U-net 結構，引進了『Skip-connect』的技巧，即每一層反卷積層的輸入都是『前一層的輸出』加『與該層對稱的卷積層的輸出』，解碼時可從對稱的編碼器得到對應的資訊，使得生成的圖像保有原圖像的特徵。

判別網路額外考慮輸入圖像的判別，將真實圖像、生成圖像與輸入圖像合而為一，作為判別網路的輸入，進行辨識。原生的 GAN 在預測像素時，是以真實資料對應的『單一像素』進行辨識，然而 Pix2Pix 則引用 PatchGAN 的思維，利用卷積將圖像切成多個較小的區域，每個像素與對應的『區域』進行辨識（Softmax），計算最大可能的輸出。PatchGAN 可參見『Image-to-Image Translation with Conditional Adversarial Networks』[16] 一文。

範 例．以 CMP Facade Database 資料集實作 Pix2Pix GAN 演算法。CMP Facade Database[22] 共有 12 類的建築物局部外型，如外觀（façade）、造型（molding）、屋簷（cornice）、柱子（pillar）、窗戶（window）、門（door）等。資料集自 pix2pix datasets[23] 下載。

程式：**10_05_Pix2Pix.ipynb**，注意，執行此範例，請去睡個覺，再來看結果:)，原文估計使用單片 V100 GPU，訓練一個週期約需 15 秒。

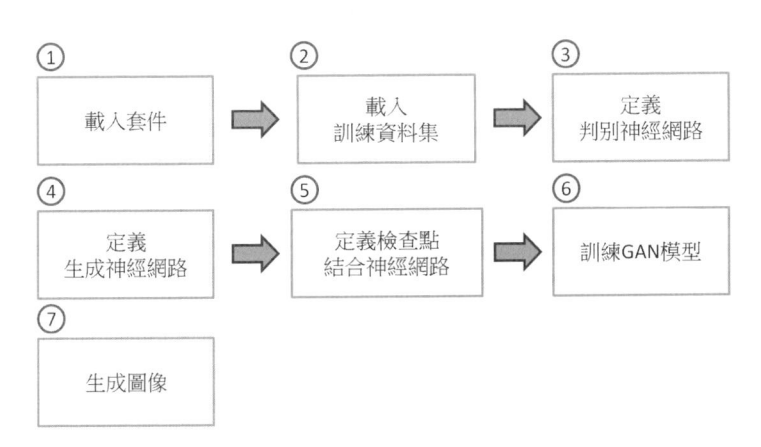

① 載入套件 → ② 載入訓練資料集 → ③ 定義判別神經網路

④ 定義生成神經網路 → ⑤ 定義檢查點結合神經網路 → ⑥ 訓練GAN模型

⑦ 生成圖像

10-37

第 10 章　生成對抗網路（GAN）

1. 載入套件。

```python
# 載入相關套件
import tensorflow as tf
import os
import time
from matplotlib import pyplot as plt
from IPython import display
```

2. 參數設定並定義圖像處理的函數。

```python
# 參數設定
PATH = './CMP Facade Database/facades/'
BUFFER_SIZE = 400   # 緩衝區大小
BATCH_SIZE = 1      # 批量
IMG_WIDTH = 256     # 圖像寬度
IMG_HEIGHT = 256    # 圖像高度

# 載入圖像
def load(image_file):
    # 讀取圖檔
    image = tf.io.read_file(image_file)
    image = tf.image.decode_jpeg(image)

    w = tf.shape(image)[1]

    # 高為寬的一半
    w = w // 2
    real_image = image[:, :w, :]
    input_image = image[:, w:, :]

    # 轉為浮點數
    input_image = tf.cast(input_image, tf.float32)
    real_image = tf.cast(real_image, tf.float32)

    return input_image, real_image

# 縮放圖像
def resize(input_image, real_image, height, width):
    input_image = tf.image.resize(input_image, [height, width],
                                  method=tf.image.ResizeMethod.NEAREST_NEIGHBOR)
    real_image = tf.image.resize(real_image, [height, width],
                                 method=tf.image.ResizeMethod.NEAREST_NEIGHBOR)

    return input_image, real_image

# 隨機裁切圖像
def random_crop(input_image, real_image):
    stacked_image = tf.stack([input_image, real_image], axis=0)
    cropped_image = tf.image.random_crop(
        stacked_image, size=[2, IMG_HEIGHT, IMG_WIDTH, 3])

    return cropped_image[0], cropped_image[1]
```

10-38

10-6 Pix2Pix

```python
44  # 標準化,使像素值介於 [-1, 1]
45  def normalize(input_image, real_image):
46      input_image = (input_image / 127.5) - 1
47      real_image = (real_image / 127.5) - 1
48
49      return input_image, real_image
50
51  # 隨機轉換
52  @tf.function()
53  def random_jitter(input_image, real_image):
54      # 縮放圖像為 286 x 286 x 3
55      input_image, real_image = resize(input_image, real_image, 286, 286)
56
57      # 隨機裁切至 256 x 256 x 3
58      input_image, real_image = random_crop(input_image, real_image)
59
60      if tf.random.uniform(()) > 0.5:
61          # 水平翻轉
62          input_image = tf.image.flip_left_right(input_image)
63          real_image = tf.image.flip_left_right(real_image)
64
65      return input_image, real_image
```

3. 顯示任一張訓練圖片。

```python
1  # 隨意顯示一張訓練圖片
2  inp, re = load(PATH+'train/100.jpg')
3  # 顯示圖像
4  plt.figure()
5  plt.imshow(inp/255.0)
6  plt.figure()
7  plt.imshow(re/255.0)
```

- 執行結果:左為語義分割圖,右為實景圖。

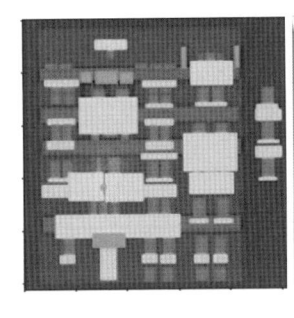

4. 隨機轉換測試:也可作為資料增補。

```python
1  # 隨機轉換測試
2  plt.figure(figsize=(6, 6))
```

10-39

```
3  for i in range(4):
4      rj_inp, rj_re = random_jitter(inp, re)
5      plt.subplot(2, 2, i+1)
6      plt.imshow(rj_inp/255.0)
7      plt.axis('off')
8  plt.show()
```

- 執行結果：

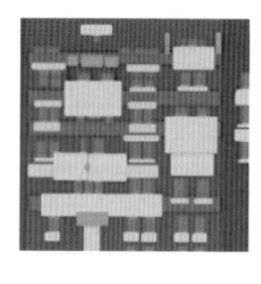

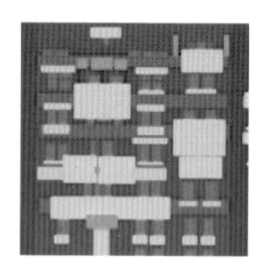

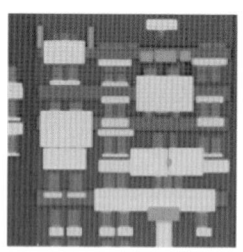

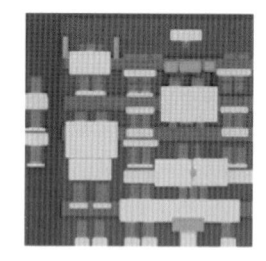

5. 載入訓練資料，轉換為 TensorFlow Dataset。

```
1  def load_image_train(image_file):
2      input_image, real_image = load(image_file)
3      input_image, real_image = random_jitter(input_image, real_image)
4      input_image, real_image = normalize(input_image, real_image)
5
6      return input_image, real_image
7
8  def load_image_test(image_file):
9      input_image, real_image = load(image_file)
10     input_image, real_image = resize(input_image, real_image,
11                                      IMG_HEIGHT, IMG_WIDTH)
12     input_image, real_image = normalize(input_image, real_image)
13
14     return input_image, real_image
15
16 train_dataset = tf.data.Dataset.list_files(PATH+'train/*.jpg')
17 train_dataset = train_dataset.map(load_image_train,
18                                  num_parallel_calls=tf.data.AUTOTUNE)
19 train_dataset = train_dataset.shuffle(BUFFER_SIZE)
20 train_dataset = train_dataset.batch(BATCH_SIZE)
21
```

10-6 Pix2Pix

```
21
22 test_dataset = tf.data.Dataset.list_files(PATH+'test/*.jpg')
23 test_dataset = test_dataset.map(load_image_test)
24 test_dataset = test_dataset.batch(BATCH_SIZE)
```

6. 定義採樣（down sampling）、上採樣（up sampling）函數，並做簡單測試。

```
1  OUTPUT_CHANNELS = 3
2
3  # 定義採樣函數
4  def downsample(filters, size, apply_batchnorm=True):
5      initializer = tf.random_normal_initializer(0., 0.02)
6
7      result = tf.keras.Sequential()
8      result.add(
9          tf.keras.layers.Conv2D(filters, size, strides=2, padding='same',
10                         kernel_initializer=initializer, use_bias=False))
11
12     if apply_batchnorm:
13         result.add(tf.keras.layers.BatchNormalization())
14
15     result.add(tf.keras.layers.LeakyReLU())
16
17     return result
18
19 down_model = downsample(3, 4)
20 down_result = down_model(tf.expand_dims(inp, 0))
21 print(down_result.shape)
```

- 執行結果：（1, 128, 128, 3）。

```
1  # 定義上採樣函數
2  def upsample(filters, size, apply_dropout=False):
3      initializer = tf.random_normal_initializer(0., 0.02)
4
5      result = tf.keras.Sequential()
6      result.add(
7          tf.keras.layers.Conv2DTranspose(filters, size, strides=2,
8                                  padding='same',
9                                  kernel_initializer=initializer,
10                                 use_bias=False))
11
12     result.add(tf.keras.layers.BatchNormalization())
13
14     if apply_dropout:
15         result.add(tf.keras.layers.Dropout(0.5))
16
17     result.add(tf.keras.layers.ReLU())
18
19     return result
20
21 up_model = upsample(3, 4)
22 up_result = up_model(down_result)
23 print(up_result.shape)
```

第 10 章　生成對抗網路（GAN）

- 執行結果：（1, 256, 256, 3）。

7. 定義生成神經網路：U-Net 結構。

```
1   def Generator():
2       inputs = tf.keras.layers.Input(shape=[256, 256, 3])
3   
4       down_stack = [
5           downsample(64, 4, apply_batchnorm=False),    # (bs, 128, 128, 64)
6           downsample(128, 4),    # (bs, 64, 64, 128)
7           downsample(256, 4),    # (bs, 32, 32, 256)
8           downsample(512, 4),    # (bs, 16, 16, 512)
9           downsample(512, 4),    # (bs, 8, 8, 512)
10          downsample(512, 4),    # (bs, 4, 4, 512)
11          downsample(512, 4),    # (bs, 2, 2, 512)
12          downsample(512, 4),    # (bs, 1, 1, 512)
13      ]
14  
15      up_stack = [
16          upsample(512, 4, apply_dropout=True),    # (bs, 2, 2, 1024)
17          upsample(512, 4, apply_dropout=True),    # (bs, 4, 4, 1024)
18          upsample(512, 4, apply_dropout=True),    # (bs, 8, 8, 1024)
19          upsample(512, 4),    # (bs, 16, 16, 1024)
20          upsample(256, 4),    # (bs, 32, 32, 512)
21          upsample(128, 4),    # (bs, 64, 64, 256)
22          upsample(64, 4),     # (bs, 128, 128, 128)
23      ]
24  
25      initializer = tf.random_normal_initializer(0., 0.02)
26      # (bs, 256, 256, 3)
27      last = tf.keras.layers.Conv2DTranspose(OUTPUT_CHANNELS, 4, strides=2,
28              padding='same', kernel_initializer=initializer, activation='tanh')
29  
30      x = inputs
31  
32      # Downsampling through the model
33      skips = []
34      for down in down_stack:
35          x = down(x)
36          skips.append(x)
37  
38      skips = reversed(skips[:-1])
39  
40      # Upsampling and establishing the skip connections
41      for up, skip in zip(up_stack, skips):
42          x = up(x)
43          x = tf.keras.layers.Concatenate()([x, skip])
44  
45      x = last(x)
46  
47      return tf.keras.Model(inputs=inputs, outputs=x)
```

10-6 Pix2Pix

8. 繪製生成神經網路模型。

```
1  # 建立生成神經網路
2  generator = Generator()
3
4  # 繪製生成神經網路模型
5  tf.keras.utils.plot_model(generator, show_shapes=True, dpi=64)
```

- 執行結果：為 U 型結構。

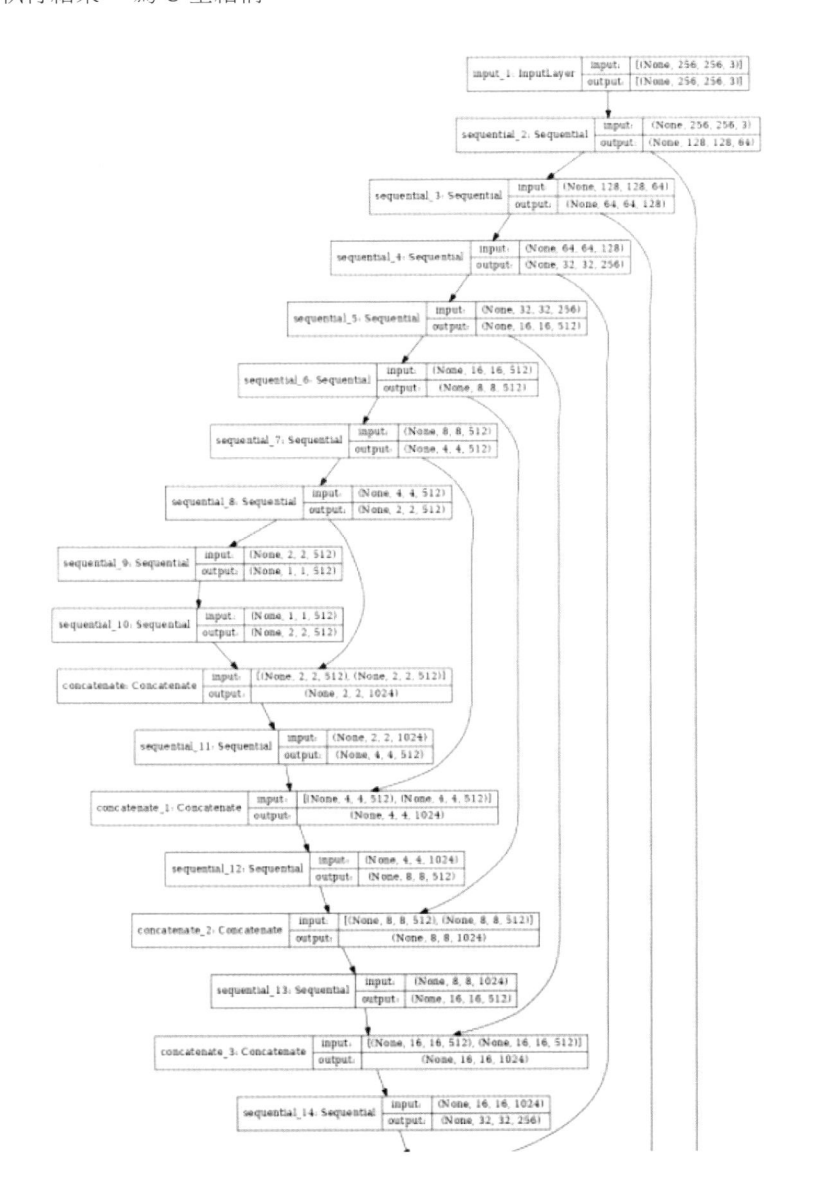

10-43

第 10 章 生成對抗網路（GAN）

9. 測試生成神經網路：取第一筆資料。

```
1  # 測試生成神經網路
2  gen_output = generator(inp[tf.newaxis, ...], training=False)
3  plt.imshow(gen_output[0, ...])
```

- 執行結果：只有隱約的形狀。

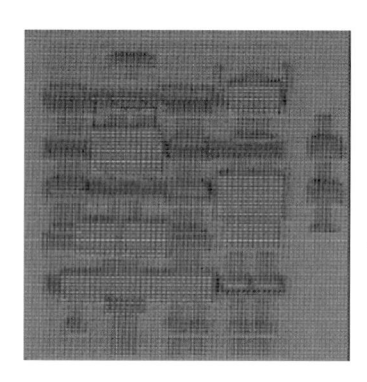

10. 定義判別神經網路。

```
1   # 定義判別神經網路
2   def Discriminator():
3       initializer = tf.random_normal_initializer(0., 0.02)
4   
5       inp = tf.keras.layers.Input(shape=[256, 256, 3], name='input_image')
6       tar = tf.keras.layers.Input(shape=[256, 256, 3], name='target_image')
7   
8       x = tf.keras.layers.concatenate([inp, tar])  # (bs, 256, 256, channels*2)
9   
10      down1 = downsample(64, 4, False)(x)   # (bs, 128, 128, 64)
11      down2 = downsample(128, 4)(down1)     # (bs, 64, 64, 128)
12      down3 = downsample(256, 4)(down2)     # (bs, 32, 32, 256)
13  
14      # (bs, 34, 34, 256)
15      zero_pad1 = tf.keras.layers.ZeroPadding2D()(down3)
16      # (bs, 31, 31, 512)
17      conv = tf.keras.layers.Conv2D(512, 4, strides=1,
18                                    kernel_initializer=initializer,
19                                    use_bias=False)(zero_pad1)
20  
21      batchnorm1 = tf.keras.layers.BatchNormalization()(conv)
22  
23      leaky_relu = tf.keras.layers.LeakyReLU()(batchnorm1)
24  
25      # (bs, 33, 33, 512)
26      zero_pad2 = tf.keras.layers.ZeroPadding2D()(leaky_relu)
27  
28      # (bs, 30, 30, 1)
29      last = tf.keras.layers.Conv2D(1, 4, strides=1,
```

10-6 Pix2Pix

```
30                                            kernel_initializer=initializer)(zero_pad2)
31
32      return tf.keras.Model(inputs=[inp, tar], outputs=last)
```

11. 建立判別神經網路,並繪製模型。

```
1  # 建立判別神經網路
2  discriminator = Discriminator()
3
4  # 繪製生成神經網路模型
5  tf.keras.utils.plot_model(discriminator, show_shapes=True, dpi=64)
```

- 執行結果: 主要為卷積層。

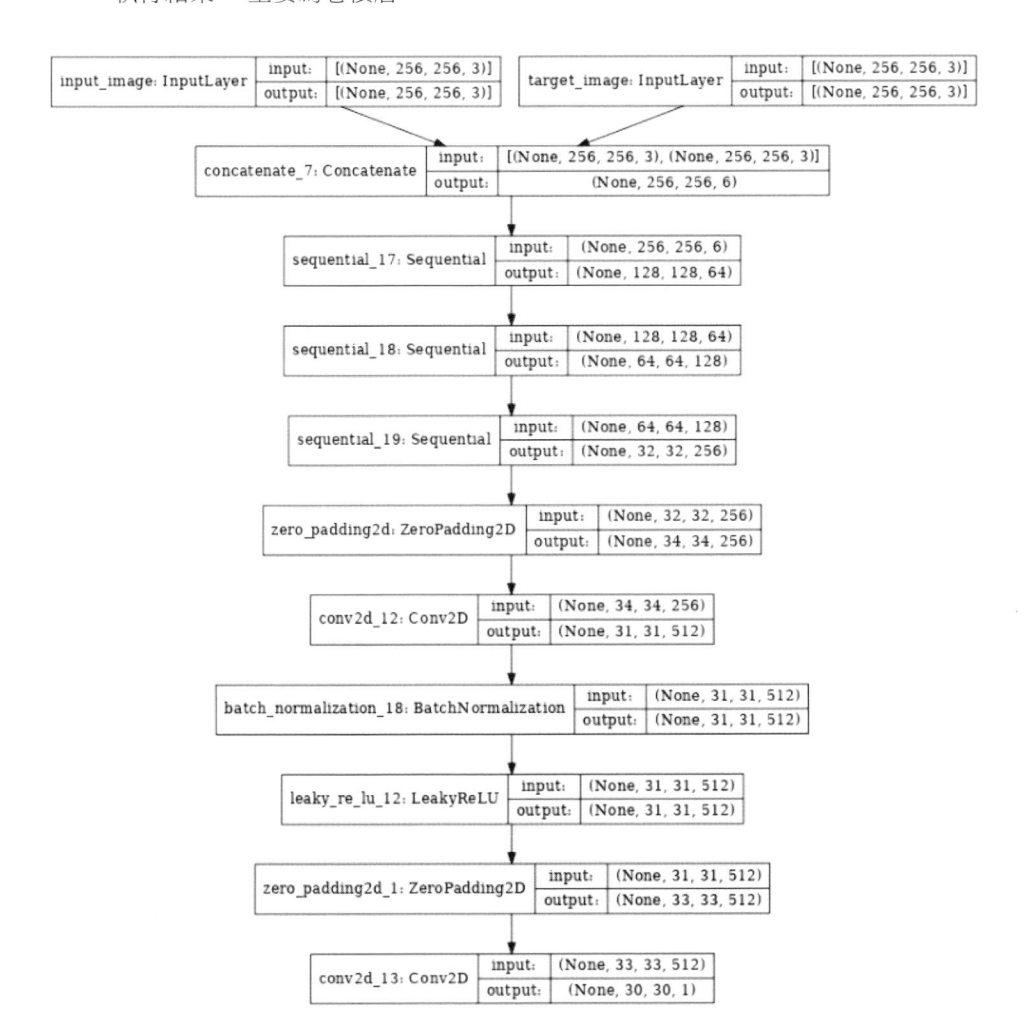

10-45

第 10 章 生成對抗網路（GAN）

12. 測試判別神經網路：取第一筆資料。

```
1  # 測試判別神經網路
2  disc_out = discriminator([inp[tf.newaxis, ...], gen_output], training=False)
3  plt.imshow(disc_out[0, ..., -1], vmin=-20, vmax=20, cmap='RdBu_r')
4  plt.colorbar()
```

- 執行結果：

13. 定義損失函數。

```
1   # 定義損失函數為二分類交叉熵
2   LAMBDA = 100
3   loss_obj = tf.keras.losses.BinaryCrossentropy(from_logits=True)
4
5   # 定義判別網路損失函數
6   def discriminator_loss(disc_real_output, disc_generated_output):
7       real_loss = loss_object(tf.ones_like(disc_real_output), disc_real_output)
8
9       generated_loss = loss_object(tf.zeros_like(disc_generated_output), disc_generated_output)
10
11      total_disc_loss = real_loss + generated_loss
12
13      return total_disc_loss
14
15  # 定義生成網路損失函數
16  def generator_loss(disc_generated_output, gen_output, target):
17      gan_loss = loss_object(tf.ones_like(disc_generated_output), disc_generated_output)
18
19      # mean absolute error
20      l1_loss = tf.reduce_mean(tf.abs(target - gen_output))
21
22      total_gen_loss = gan_loss + (LAMBDA * l1_loss)
23
24      return total_gen_loss, gan_loss, l1_loss
```

14. 定義優化器。

```
1  # 定義優化器
2  generator_optimizer = tf.keras.optimizers.Adam(2e-4, beta_1=0.5)
3  discriminator_optimizer = tf.keras.optimizers.Adam(2e-4, beta_1=0.5)
```

15. 定義檢查點：可設定以下函數，直接生成圖像。

```
1  # 定義檢查點
2  checkpoint_dir = './Pix2Pix_training_checkpoints'
3  checkpoint_prefix = os.path.join(checkpoint_dir, "ckpt")
4  checkpoint = tf.train.Checkpoint(generator_optimizer=generator_optimizer,
5                                   discriminator_optimizer=discriminator_optimizer,
6                                   generator=generator,
7                                   discriminator=discriminator)
```

16. 生成圖像，並顯示。

```
1  # 生成圖像，並顯示
2  def generate_images(model, test_input, tar):
3      prediction = model(test_input, training=True)
4      plt.figure(figsize=(15, 15))
5
6      display_list = [test_input[0], tar[0], prediction[0]]
7      title = ['Input Image', 'Ground Truth', 'Predicted Image']
8
9      # 顯示 輸入圖像、真實圖像、與生成圖像
10     for i in range(3):
11         plt.subplot(1, 3, i+1)
12         plt.title(title[i])
13         # 轉換像素質介於 [0, 1]
14         plt.imshow(display_list[i] * 0.5 + 0.5)
15         plt.axis('off')
16     plt.show()
17
18 # 取一筆資料測試
19 for example_input, example_target in test_dataset.take(1):
20     generate_images(generator, example_input, example_target)
```

- 執行結果：還未訓練的結果。

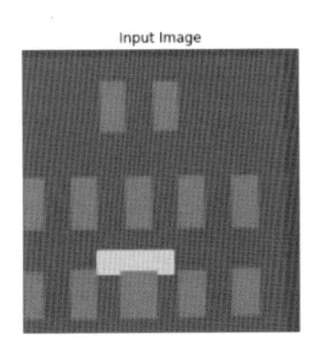

第 10 章　生成對抗網路（GAN）

17. 定義訓練模型的函數。

```python
import datetime
# 參數設定
EPOCHS = 150
log_dir="logs/"

summary_writer = tf.summary.create_file_writer(
    log_dir + "fit/" + datetime.datetime.now().strftime("%Y%m%d-%H%M%S"))

# 定義訓練模型的函數
@tf.function
def train_step(input_image, target, epoch):
    with tf.GradientTape() as gen_tape, tf.GradientTape() as disc_tape:
        gen_output = generator(input_image, training=True)

        disc_real_output = discriminator([input_image, target], training=True)
        disc_generated_output = discriminator([input_image, gen_output], training=True)

        gen_total_loss, gen_gan_loss, gen_l1_loss = generator_loss(disc_generated_output, gen_output, target)
        disc_loss = discriminator_loss(disc_real_output, disc_generated_output)

    generator_gradients = gen_tape.gradient(gen_total_loss, generator.trainable_variables)
    discriminator_gradients = disc_tape.gradient(disc_loss, discriminator.trainable_variables)

    generator_optimizer.apply_gradients(zip(generator_gradients, generator.trainable_variables))
    discriminator_optimizer.apply_gradients(zip(discriminator_gradients, discriminator.trainable_variables))

    with summary_writer.as_default():
        tf.summary.scalar('gen_total_loss', gen_total_loss, step=epoch)
        tf.summary.scalar('gen_gan_loss', gen_gan_loss, step=epoch)
        tf.summary.scalar('gen_l1_loss', gen_l1_loss, step=epoch)
        tf.summary.scalar('disc_loss', disc_loss, step=epoch)
```

```python
def fit(train_ds, epochs, test_ds):
    for epoch in range(epochs):
        start = time.time()

        display.clear_output(wait=True)

        for example_input, example_target in test_ds.take(1):
            generate_images(generator, example_input, example_target)
        print("Epoch: ", epoch)

        # Train
        for n, (input_image, target) in train_ds.enumerate():
            print('.', end='')
            if (n+1) % 100 == 0:
                print()
            train_step(input_image, target, epoch)
        print()

        # saving (checkpoint) the model every 20 epochs
        if (epoch + 1) % 20 == 0:
            checkpoint.save(file_prefix=checkpoint_prefix)

        print ('Time taken for epoch {} is {} sec\n'.format(epoch + 1, time.time()-start))

    checkpoint.save(file_prefix=checkpoint_prefix)
```

18. 訓練模型：可先啟動 Tensorboard 監看工作記錄，在 Notebook 中啟動如下。

```python
# 啟動 Tensorboard
%load_ext tensorboard
%tensorboard --logdir {log_dir}
```

```
1  # 訓練模型
2  fit(train_dataset, EPOCHS, test_dataset)
```

19. 取 5 筆資料測試。

```
1  # 取 5 筆資料測試
2  for inp, tar in test_dataset.take(5):
3      generate_images(generator, inp, tar)
```

- 執行結果：以下僅只截圖 2 筆資料，經過訓練後，預測的圖像已沒有樹木或欄杆了。

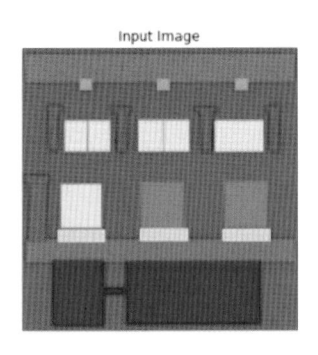

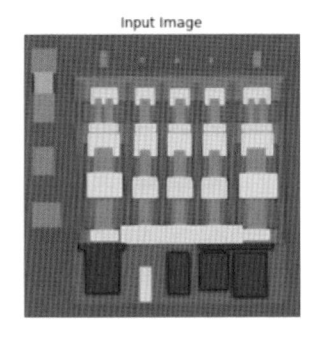

10-7 CycleGAN

前面 GAN 演算法處理的都是成對轉換資料，CycleGAN 則是針對非成對的資料生成圖像。成對的意思是一張原始圖像對應一張目標圖像，下圖右方表示多對多的資料，也就是給予不同的場域（Domain），原始圖像就可以合成指定場景的圖像。

第 10 章　生成對抗網路（GAN）

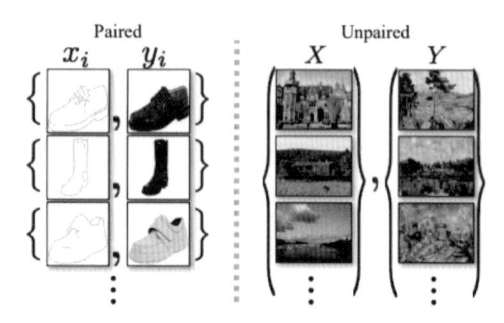

▲ 圖 10.20 成對的資料（左方）vs. 非成對的資料（右方），以下圖片來源均來自『Unpaired Image-to-Image Translation using Cycle-Consistent Adversarial Networks』[24]

CycleGAN 或稱 Cycle-Consistent GAN，是 Jun-Yan Zhu 等學者於 2017 年發表的一篇文章『Unpaired Image-to-Image Translation using Cycle-Consistent Adver-sarial Networks』[18] 中提出的演算法，概念如下圖：

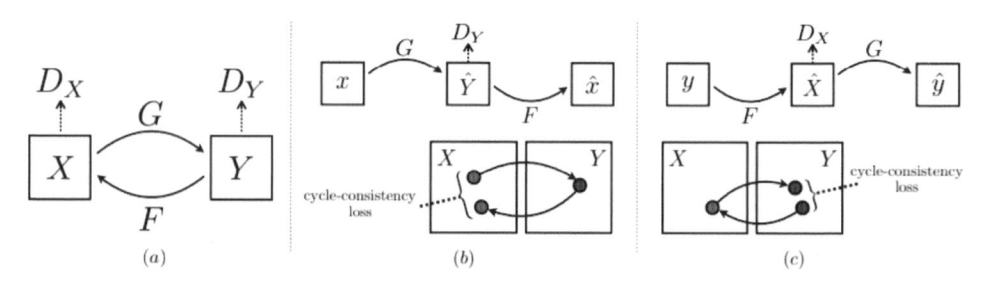

▲ 圖 10.21 CycleGAN 網路結構

- 上圖（a）：有兩個生成網路，G 將圖像由 X 場域（Domain）生成 Y 場域的圖像，F 網路則是相反功能，由 Y 場域（Domain）生成 X 場域的圖像。

- 上圖（b）：引進 cycle consistency losses 概念，可以做到 x → G(x) → F（G(x)）≈ x，即 x 經過 G、F 轉換，可得到近似 x 的圖像，稱為 Forward cycle-consistency loss。

- 上圖（c）：從另一場域 y 開始，也可以做到 y → F（y）→ G（F（y））≈ y，稱為 Backward cycle-consistency loss。

- 整個模型類似兩個 GAN 網路的組合，具備循環機制，因此，損失函數如下：

L（G, F, DX, DY）= LGAN（G, D_Y, X, Y）+ LGAN（F, D_X, Y, X）+λLcyc（G, F）

10-7 CycleGAN

其中：

$$\mathcal{L}_{\text{cyc}}(G, F) = \mathbb{E}_{x \sim p_{\text{data}}(x)}[\|F(G(x)) - x\|_1] + \mathbb{E}_{y \sim p_{\text{data}}(y)}[\|G(F(y)) - y\|_1]$$

λ 控制 G、F 損失函數的相對重要性。

這種機制可應用到影像增強（Photo Enhancement）、影像彩色化（Image Colorization）、風格轉換（Style Transfer）等功能，如下圖，幫一般的馬匹塗上斑馬紋。

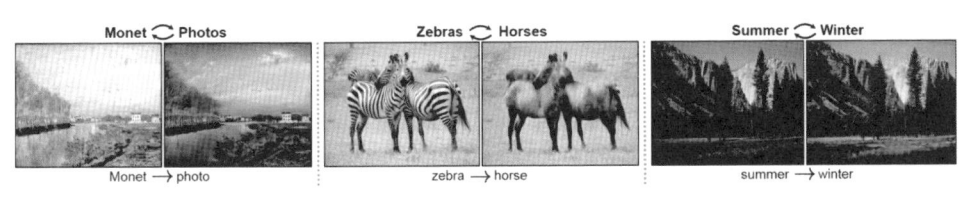

▲ 圖 10.22 CycleGAN 的功能展示

範例. 以 horse2zebra 資料集實作 CycleGAN 演算法。流程與上一節大致相同，所以不再複製 / 貼上，節省篇幅。

程式：**10_06_ CycleGAN.ipynb**。

1. 載入套件。

```
1  # 載入相關套件
2  import tensorflow as tf
3  import tensorflow_datasets as tfds
4  from tensorflow_examples.models.pix2pix import pix2pix
5
6  import os
7  import time
8  import matplotlib.pyplot as plt
9  from IPython.display import clear_output
10
11 # 類似 prefetch()，可以提升資料集存取效能
12 AUTOTUNE = tf.data.AUTOTUNE
```

2. 載入訓練資料：tfds.load 會載入內建資料集，相關資料可參考 Tensorflow 官網有關 CycleGAN 的描述 [25]。

```
1  # 載入訓練資料
2  dataset, metadata = tfds.load('cycle_gan/horse2zebra',
3                                with_info=True, as_supervised=True)
4
5  train_horses, train_zebras = dataset['trainA'], dataset['trainB']
6  test_horses, test_zebras = dataset['testA'], dataset['testB']
```

10-51

3. 定義圖像處理的函數：與上一節相同。

```python
# 參數設定
BUFFER_SIZE = 1000  # 緩衝區大小
BATCH_SIZE = 1      # 批量
IMG_WIDTH = 256     # 圖像寬度
IMG_HEIGHT = 256    # 圖像高度

# 隨機裁切圖像
def random_crop(image):
    cropped_image = tf.image.random_crop(
            image, size=[IMG_HEIGHT, IMG_WIDTH, 3])

    return cropped_image

# 標準化，使像素值介於 [-1, 1]
def normalize(image):
    image = tf.cast(image, tf.float32)
    image = (image / 127.5) - 1
    return image

# 隨機轉換


def random_jitter(image):
    # 縮放圖像為 286 x 286 x 3
    image = tf.image.resize(image, [286, 286],
            method=tf.image.ResizeMethod.NEAREST_NEIGHBOR)

    # 隨機裁切至 256 x 256 x 3
    image = random_crop(image)

    # 水平翻轉
    image = tf.image.random_flip_left_right(image)
    return image
```

4. 定義資料前置處理的函數，設定資料集屬性。

```python
# 訓練資料前置處理
def preprocess_image_train(image, label):
    image = random_jitter(image)
    image = normalize(image)
    return image

# 測試資料前置處理
def preprocess_image_test(image, label):
    image = normalize(image)
    return image

# 設定資料集屬性
train_horses = train_horses.map(
        preprocess_image_train, num_parallel_calls=AUTOTUNE).cache().shuffle(
        BUFFER_SIZE).batch(1)
```

10-7 CycleGAN

5. 資料測試。

```
1   # 各取一筆資料測試
2   sample_horse = next(iter(train_horses))
3   sample_zebra = next(iter(train_zebras))
4
5   plt.subplot(121)
6   plt.title('Horse')
7   # 轉換為 [0, 1]
8   plt.imshow(sample_horse[0] * 0.5 + 0.5)
9
10  plt.subplot(122)
11  plt.title('Horse with random jitter')
12  # 將像素值由 [-1, 1] 轉換為 [0, 1]，才能顯示
13  plt.imshow(random_jitter(sample_horse[0]) * 0.5 + 0.5)
```

- 執行結果：左為原圖，右為隨機轉換的圖。

6. 定義 CycleGAN 神經網路：借用上一節的 Pix2Pix 網路結構。

```
1   # 定義生成神經網路
2   OUTPUT_CHANNELS = 3
3
4   # 以 Pix2Pix 的模型建立 CycleGAN
5   generator_g = pix2pix.unet_generator(OUTPUT_CHANNELS, norm_type='instancenorm')
6   generator_f = pix2pix.unet_generator(OUTPUT_CHANNELS, norm_type='instancenorm')
7
8   discriminator_x = pix2pix.discriminator(norm_type='instancenorm', target=False)
9   discriminator_y = pix2pix.discriminator(norm_type='instancenorm', target=False)
10
11  # 生成圖像
12  to_zebra = generator_g(sample_horse)
13  to_horse = generator_f(sample_zebra)
14  plt.figure(figsize=(8, 8))
15  contrast = 8
16
17  # 顯示圖片
18  imgs = [sample_horse, to_zebra, sample_zebra, to_horse]
19  title = ['Horse', 'To Zebra', 'Zebra', 'To Horse']
20
21  for i in range(len(imgs)):
22      plt.subplot(2, 2, i+1)
23      plt.title(title[i])
```

10-53

第 10 章 生成對抗網路（GAN）

```
24      if i % 2 == 0:
25          plt.imshow(imgs[i][0] * 0.5 + 0.5)
26      else:
27          plt.imshow(imgs[i][0] * 0.5 * contrast + 0.5)
28  plt.show()
```

- 執行結果：左為原圖，右為未訓練前的結果。

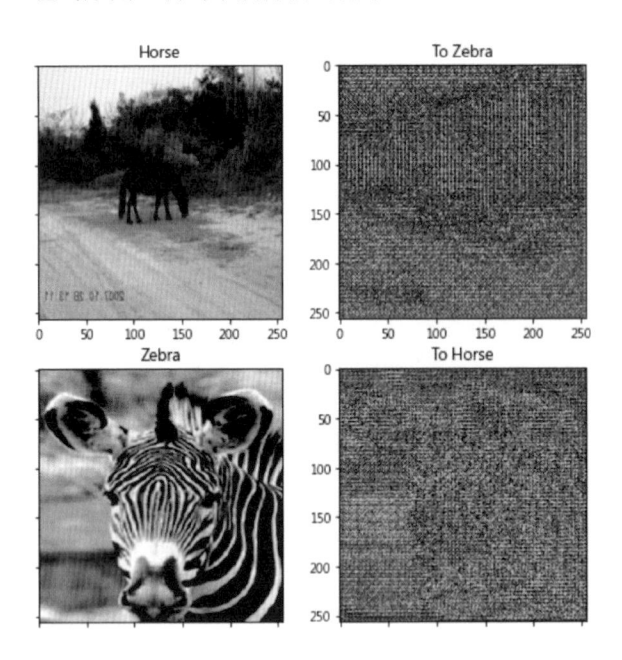

7. 判別網路圖像測試。

```
1  # 判別網路圖像測試
2  plt.figure(figsize=(8, 8))
3
4  plt.subplot(121)
5  plt.title('Is a real zebra?')
6  # 使用判別網路辨識圖像
7  plt.imshow(discriminator_y(sample_zebra)[0, ..., -1], cmap='RdBu_r')
8
9  plt.subplot(122)
10 plt.title('Is a real horse?')
11 # 使用判別網路辨識圖像
12 plt.imshow(discriminator_x(sample_horse)[0, ..., -1], cmap='RdBu_r')
13
14 plt.show()
```

- 執行結果：未訓練前的結果，左圖為 X 判別網路的結果，右圖為 Y 判別網路的結果。

10-7 CycleGAN

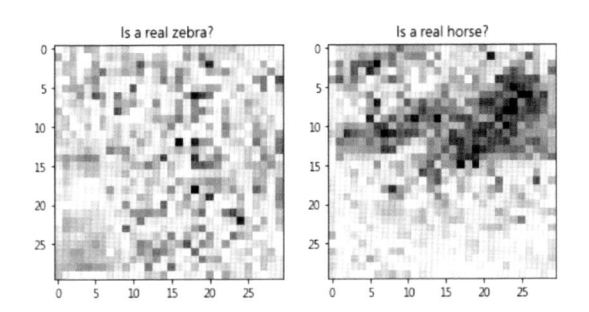

8. 定義損失函數為二分類交叉熵：包括判別網路、生成網路、循環損失與 Identity 損失函數。Identity 損失函數為輸入 Y 與 Y 生成網路的差異，X 亦同，正常的話差異應為 0。

```
1   # 定義損失函數為二分類交叉熵
2   LAMBDA = 10
3   loss_obj = tf.keras.losses.BinaryCrossentropy(from_logits=True)
4
5   # 定義判別網路損失函數
6   def discriminator_loss(real, generated):
7       # 真實資料的損失
8       real_loss = loss_obj(tf.ones_like(real), real)
9
10      # 生成資料的損失
11      generated_loss = loss_obj(tf.zeros_like(generated), generated)
12
13      # 總損失
14      total_disc_loss = real_loss + generated_loss
15
16      return total_disc_loss * 0.5
17
18  # 定義生成網路損失函數
19  def generator_loss(generated):
20      return loss_obj(tf.ones_like(generated), generated)
21
22  # 定義循環損失函數，參見圖 10.21 CycleGAN 網路結構
23  def calc_cycle_loss(real_image, cycled_image):
24      loss1 = tf.reduce_mean(tf.abs(real_image - cycled_image))
25      return LAMBDA * loss1
26
27  # 定義 Identity 損失函數，|G(Y)-Y| + |f(X)-X|
28  def identity_loss(real_image, same_image):
29      loss = tf.reduce_mean(tf.abs(real_image - same_image))
30      return LAMBDA * 0.5 * loss
```

9. 定義優化器：採用 Adam。

```
1   # 定義優化器
2   generator_g_optimizer = tf.keras.optimizers.Adam(2e-4, beta_1=0.5)
3   generator_f_optimizer = tf.keras.optimizers.Adam(2e-4, beta_1=0.5)
4
```

第 10 章　生成對抗網路（GAN）

```
5  discriminator_x_optimizer = tf.keras.optimizers.Adam(2e-4, beta_1=0.5)
6  discriminator_y_optimizer = tf.keras.optimizers.Adam(2e-4, beta_1=0.5)
```

10. 定義檢查點：將以上網路函數放入。

```
1   # 定義檢查點
2   checkpoint_path = "./CycleGAN_checkpoints/train"
3
4   ckpt = tf.train.Checkpoint(generator_g=generator_g,
5                              generator_f=generator_f,
6                              discriminator_x=discriminator_x,
7                              discriminator_y=discriminator_y,
8                              generator_g_optimizer=generator_g_optimizer,
9                              generator_f_optimizer=generator_f_optimizer,
10                             discriminator_x_optimizer=discriminator_x_optimizer,
11                             discriminator_y_optimizer=discriminator_y_optimizer)
12
13  ckpt_manager = tf.train.CheckpointManager(ckpt, checkpoint_path, max_to_keep=5)
14
15  # 如果檢查點存在，回復至最後一個檢查點
16  if ckpt_manager.latest_checkpoint:
17      ckpt.restore(ckpt_manager.latest_checkpoint)
18      print ('Latest checkpoint restored!!')
```

11. 定義生成圖像與顯示的函數。

```
1   # 定義生成圖像及顯示的函數
2   def generate_images(model, test_input):
3       prediction = model(test_input)
4
5       plt.figure(figsize=(12, 12))
6
7       display_list = [test_input[0], prediction[0]]
8       title = ['Input Image', 'Predicted Image']
9
10      for i in range(2):
11          plt.subplot(1, 2, i+1)
12          plt.title(title[i])
13          # getting the pixel values between [0, 1] to plot it.
14          plt.imshow(display_list[i] * 0.5 + 0.5)
15          plt.axis('off')
16      plt.show()
```

12. 定義訓練模型的函數。

```
1   # 參數設定
2   EPOCHS = 40
3
4   # 定義訓練模型的函數
5   @tf.function
6   def train_step(real_x, real_y):
7       # persistent=True：表示 tf.GradientTape 會重複使用
8       with tf.GradientTape(persistent=True) as tape:
```

10-7 CycleGAN

```
 9          # Generator G translates X -> Y
10          # Generator F translates Y -> X
11          fake_y = generator_g(real_x, training=True)
12          cycled_x = generator_f(fake_y, training=True)
13
14          fake_x = generator_f(real_y, training=True)
15          cycled_y = generator_g(fake_x, training=True)
16
17          # same_x and same_y are used for identity loss.
18          same_x = generator_f(real_x, training=True)
19          same_y = generator_g(real_y, training=True)
20
21          disc_real_x = discriminator_x(real_x, training=True)
22          disc_real_y = discriminator_y(real_y, training=True)
23
24          disc_fake_x = discriminator_x(fake_x, training=True)
25          disc_fake_y = discriminator_y(fake_y, training=True)
26
27          # 計算生成網路損失
28          gen_g_loss = generator_loss(disc_fake_y)
29          gen_f_loss = generator_loss(disc_fake_x)
30
31          # 計算循環損失
32          total_cycle_loss = calc_cycle_loss(real_x, cycled_x) + calc_cycle_loss(real_y, cycled_y)
33
34          # 計算總損失 Total generator loss = adversarial loss + cycle loss
35          total_gen_g_loss = gen_g_loss + total_cycle_loss + identity_loss(real_y, same_y)
36          total_gen_f_loss = gen_f_loss + total_cycle_loss + identity_loss(real_x, same_x)
37
38          disc_x_loss = discriminator_loss(disc_real_x, disc_fake_x)
39          disc_y_loss = discriminator_loss(disc_real_y, disc_fake_y)
40
41      # 計算生成網路梯度
42      generator_g_gradients = tape.gradient(total_gen_g_loss,
43                                            generator_g.trainable_variables)
44      generator_f_gradients = tape.gradient(total_gen_f_loss,
45                                            generator_f.trainable_variables)
46
47      # 計算判別網路梯度
48      discriminator_x_gradients = tape.gradient(disc_x_loss,
49                                                discriminator_x.trainable_variables)
50      discriminator_y_gradients = tape.gradient(disc_y_loss,
51                                                discriminator_y.trainable_variables)
52
53      # 更新權重
54      generator_g_optimizer.apply_gradients(zip(generator_g_gradients,
55                                                generator_g.trainable_variables))
56
57      generator_f_optimizer.apply_gradients(zip(generator_f_gradients,
58                                                generator_f.trainable_variables))
59
60      discriminator_x_optimizer.apply_gradients(zip(discriminator_x_gradients,
61                                                    discriminator_x.trainable_variables))
62
63      discriminator_y_optimizer.apply_gradients(zip(discriminator_y_gradients,
64                                                    discriminator_y.trainable_variables))
```

第 10 章　生成對抗網路（GAN）

13. 訓練模型。

```
# 訓練模型
for epoch in range(EPOCHS):
    start = time.time()

    n = 0
    for image_x, image_y in tf.data.Dataset.zip((train_horses, train_zebras)):
        train_step(image_x, image_y)
        if n % 10 == 0:
            print ('.', end='')
        n += 1

    clear_output(wait=True)
    # 產生圖像
    generate_images(generator_g, sample_horse)

    # 檢查點存檔
    if (epoch + 1) % 5 == 0:
        ckpt_save_path = ckpt_manager.save()
        print ('Saving checkpoint for epoch {} at {}'.format(epoch+1, ckpt_save_path))

    # 計時
    print ('Time taken for epoch {} is {} sec\n'.format(epoch + 1, time.time()-start))
```

- 執行結果：訓練中的結果如下，有逐步的轉變，第一排右圖為第 9 週期的結果，第二排右圖為第 10 週期的結果，圖像已有明顯改善，第三排右圖為第 12 週期的結果，已集中在馬匹的處理上。

10-7 CycleGAN

Time taken for epoch 12 is 1216.7267162799835 sec

- 訓練的最終結果：效果很好，馬匹已加上了斑馬紋。

- 每個訓練週期均執行約 1200 秒，即 20 分鐘，全部 40 個週期，筆者大概執行了 15 個小時。

14. 取 5 筆資料測試。

```
1  # 取 5 筆資料測試
2  for inp in test_horses.take(5):
3      generate_images(generator_g, inp)
```

- 執行結果：以下僅只截圖 2 筆資料，左側為原圖，右側為預測的圖像，效果比訓練樣本差，應該是因為訓練的執行週期不足，原文作者執行了 200 個週期，如果真的照做，這次就不用睡覺了，乾脆去日本玩個三天兩夜，再回來看結果。

10-59

10-8 CartoonGAN

最後介紹一個有趣的 CartoonGAN 演算法，可以把圖像轉換為日本漫畫家的風格，類似上一章的風格轉換，相關資訊如下：

1. 論文：『CartoonGAN: Generative Adversarial Networks for Photo Cartoonization』[26]。
2. 簡介：『用 CartoonGAN 及 TensorFlow 2 生成新海誠與宮崎駿動畫』[27]。
3. 模型訓練及推論的使用：mnicnc404 CartoonGan-tensorflow GitHub[28]。

CartoonGAN 演算法最重要的是重新定義損失函數：

1. 總損失 = 生成對抗損失（adversarial loss）+ 內容損失（content loss）

$$\mathcal{L}(G,D) = \mathcal{L}_{adv}(G,D) + \omega\mathcal{L}_{con}(G,D)$$

2. 其中生成對抗損失有拆解成 3 部份：

$$\begin{aligned}\mathcal{L}_{adv}(G,D) = &\mathbb{E}_{c_i \sim S_{data}(c)}[\log D(c_i)] \\ &+ \mathbb{E}_{e_j \sim S_{data}(e)}[\log(1 - D(e_j))] \\ &+ \mathbb{E}_{p_k \sim S_{data}(p)}[\log(1 - D(G(p_k)))]\end{aligned}$$

10-8 CartoonGAN

詳細說明請參閱論文,以下僅介紹使用方法:

1. 訓練:GitHub 內有 train.py,可訓練模型,指令含參數如下:

```
python train.py \
    --batch_size 8 \
    --pretrain_epochs 1 \
    --content_lambda .4 \
    --pretrain_learning_rate 2e-4 \
    --g_adv_lambda 8. \
    --generator_lr 8e-5 \
    --discriminator_lr 3e-5 \
    --style_lambda 25. \
    --light \
    --dataset_name {your dataset name}
```

2. 資料集內容如下:

```
datasets
└── YourDataset [your dataset name]
    ├── testA [(must) 8 real-world images for evaluation]
    ├── trainA [(must) (source) real-world images]
    ├── trainB [(must) (target) cartoon images]
    └── trainB_smooth [(must, but can be generated by running scripts/smooth.py)]
```

3. 模型推論:範例包括宮崎駿(Hayao)、新海誠(Shinkai)及細田守(Hosoda)風格,可一次指定多個風格,例如:

- python cartoonize.py --styles shinkai hayao hosoda
- 預設輸入資料夾:input_images。
- 預設輸出資料夾:output_images。
- 執行結果如下:

    原圖:

轉換結果：

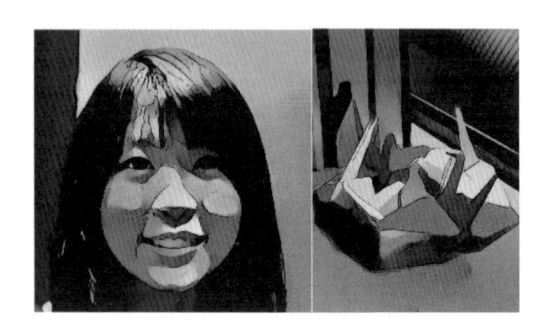

另外有一個專案，針對臉部進行卡通化的轉換也非常棒，執行速度很快，請參閱『SystemErrorWang FacialCartoonization』[29]。執行指令如下：

python inference.py

- 程式以 PyTorch 撰寫，需先安裝 PyTorch 套件。
- weight.pth 為模型權重檔。
- 預設輸入資料夾：images。
- 預設輸出資料夾：results。
- 執行結果如下：左圖為原圖，右圖為轉換結果。

10-9 GAN 挑戰

這一章我們認識了許多種不同的 GAN 演算法，由於大部分是由同一組學者發表的，因此可以看到演化的脈絡。原生 GAN 加上條件後，變成 Conditional GAN，再將生成網路改成對稱型的 U-Net 後，就變成 Pix2Pix GAN，接著再設定兩個 Pix2Pix 循環的網路，就衍生出 CycleGAN。除此之外，許多的演算法也會修改損失函數的定義，來產生各種意想不到的效果。本書介紹的演算法只是滄海一粟，更多的內容可參考李宏毅老師的 PPT『Introduction of Generative Adversarial Network（GAN）』[30]。

另一方面，GAN 不光是應用在圖像上，還可以結合自然語言處理（NLP）、強化學習（RL）等技術，擴大應用範圍，像是高解析圖像生成、虛擬人物的生成、資料壓縮、文字轉語音（Text To Speech, TTS）、醫療、天文、物理、遊戲…等，可以參閱『Tutorial on Deep Generative Models』[31] 一文。

縱使 GAN 應用廣泛，但仍然存在一些挑戰：

1. 生成的圖像模糊：因為神經網路是根據訓練資料求取迴歸，類似求取每個樣本在不同範圍的平均值，所以生成的圖像會是相似點的平均，導致圖像模糊。必須有非常大量訓練資料，加上相當多的訓練週期，才能產生畫質較佳的圖像。另外，GAN 對超參數特別敏感，包括學習率、濾波器（Filter）尺寸，初始值設定得不好，造成生成的資料過差時，判別網路就會都判定為偽，到最後生成網路只能一直產生少數類別的資料了。

2. 梯度消失（Vanishing Gradient）：當生成的資料過差時，判別網路判定為真的機率接近 0，梯度會變得非常小，因此就無法提供良好的梯度來改善生成器，造成生成器梯度消失。發生這種情形時可以多使用 leaky ReLU activation function、簡化判別網路結構、或增加訓練週期加以改善。

3. 模式崩潰（Mode Collapse）：是指生成器生成的內容過於雷同，缺少變化。如果訓練資料的類別不只一種，生成網路則會為了讓判別網路辨識的準確率提高，而專注在比較擅長的類別，導致生成的類別缺乏多樣性。以製造偽鈔來舉例，假設鈔票分別有 100 元、500 元與 1000 元，若偽鈔製造者比較善於製作 500 元紙鈔，模型可能就會全部都製作 500 元的偽鈔。

4. 執行訓練時間過久：這是最大的問題了吧，反覆實驗的時候，假如沒有相當的硬體支援，每次調整個參數都要折磨好幾天，再多的耐心也會消磨殆盡。

10-10 深度偽造（Deepfake）

深度偽造（Deepfake）是目前很夯的技術，也是一個 AI 危害人類社會的明顯證據。BuzzFeed.com 在 2018 年放上一段影片，名叫『You Won't Believe What Obama Says

第 10 章　生成對抗網路（GAN）

In This Video!』[32]，影片中歐巴馬總統的演說全是偽造的，嘴型和聲音都十分逼真，震驚世人，自此以後，各界瘋狂研究深度偽造影片的製作，使得網路上的充斥真假難辨的影片，造成非常嚴重的假新聞災難。

深度偽造大部分是在視訊中換臉，由於人在說話時頭部會自然轉動，有各種角度的特寫，因此，必須要收集特定人 360 度的臉部圖像，才能讓演算法成功置換。從網路媒體中收集名人的各種影像是最容易的方式，所以，網路上流傳最多的大部分是偽造名人的影片，如政治人物、明星等。

深度偽造的技術基礎來自 GAN，類似於前面介紹的 CycleGAN，架構如下圖，也能結合臉部辨識的功能，在抓到臉部特徵點（Landmark）後，就可以進行原始臉部與要置換臉部的互換。

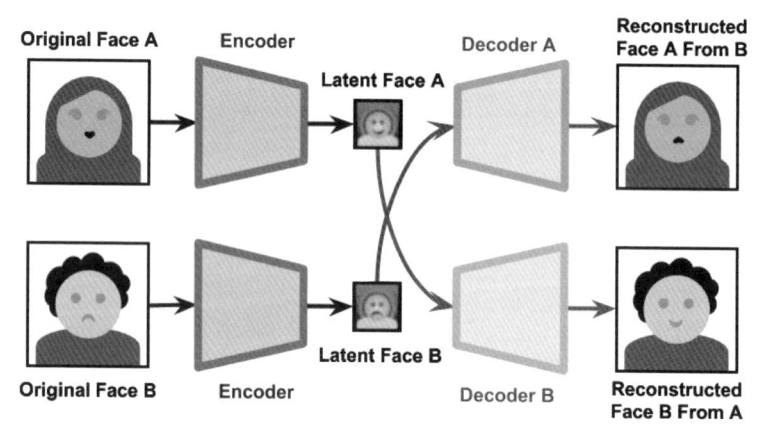

▲ 圖 10.23 深度偽造的架構示意圖，
圖片來源：『Understanding the Technology Behind DeepFakes』[33]

Aayush Bansal 等學者在 2018 年發表『Recycle-GAN: Unsupervised Video Retargeting』[34]，RecycleGAN 是擴充 CycleGAN 的演算法，它的損失函數額外加上了時間同步的相關性（Temporal Coherence），如下：

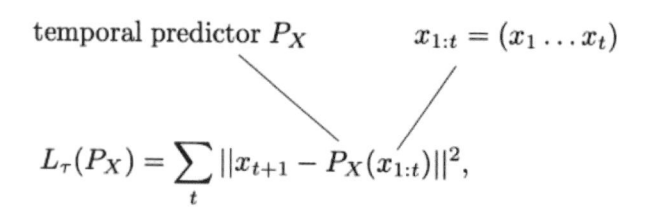

$$L_\tau(P_X) = \sum_t \|x_{t+1} - P_X(x_{1:t})\|^2,$$

其中 temporal predictor P_X，$x_{1:t} = (x_1 \ldots x_t)$

10-10 深度偽造（Deepfake）

類似時間序列（Time Series），t+1 時間點的圖像應該是 1 至 t 時間點的圖像的延續。因此，生成網路的損失函數如下：

$$L_r(G_X, G_Y, P_Y) = \sum_t \|x_{t+1} - G_X(P_Y(G_Y(x_{1:t})))\|^2$$

其中：$G_y(x_i)$ 是將 x_i 轉成 y_i 的生成網路

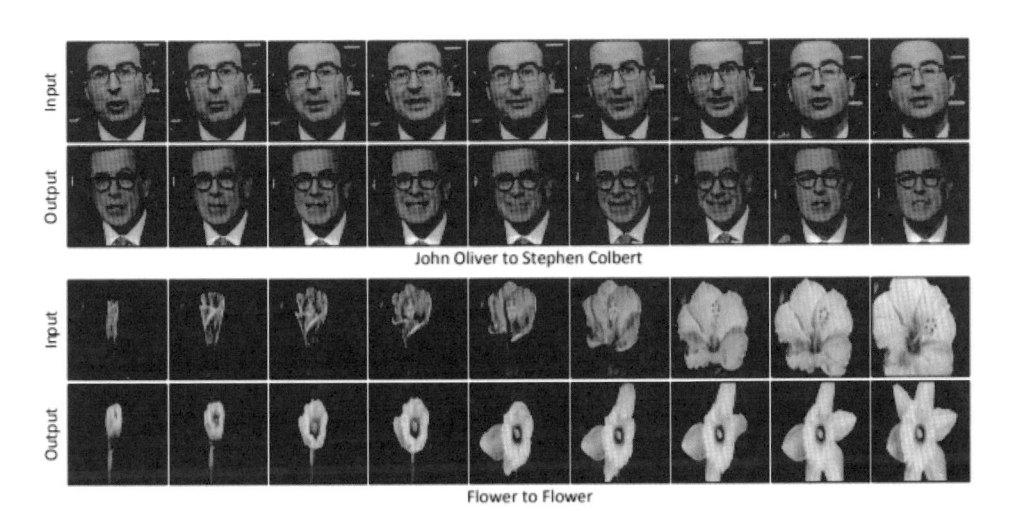

▲ 圖 10.24 視訊是連續的變化，因此 t+1 時間點的圖像應該是 1 至 t 時間點的圖像的延續，圖片來源：『Recycle-GAN: Unsupervised Video Retargeting』[21]

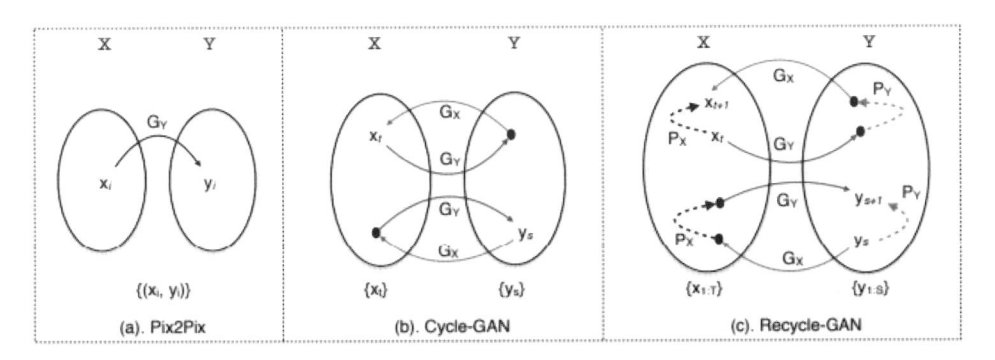

▲ 圖 10.25 RecycleGAN 演算法的演進，
圖片來源：『Recycle-GAN: Unsupervised Video Retargeting』[21]

第 10 章　生成對抗網路（GAN）

（a）Pix2Pix 是成對（Paired data）轉換。

（b）CycleGAN 是循環轉換，使用成對的網路架構。

（c）RecycleGAN 加上時間同步的相關性（Px、Py）。

因此整體的 RecycleGAN 的損失函數，包括以下部份：

$$\min_{G,P} \max_{D} L_{rg}(G,P,D) = \underbrace{L_g(G_X,D_X) + L_g(G_Y,D_Y)}_{\text{GAN objective}} + \underbrace{\lambda_{rx}L_r(G_X,G_Y,P_Y) + \lambda_{ry}L_r(G_Y,G_X,P_X)}_{\text{Recycle Loss}} + \underbrace{\lambda_{\tau x}L_\tau(P_X) + \lambda_{\tau y}L_\tau(P_Y)}_{\text{Recurrent loss}}$$

另外，還有 Face2Face、嘴型同步技術（Lip-syncing technology）等演算法，有興趣的讀者可以參閱 Jonathan Hui 的『Detect AI-generated Images & Deep-fakes』[35]系列文章，裡面有大量的圖片展示，十分有趣。

Deepfake 的實作可參閱『DeepFakes in 5 minutes』[36]一文，它介紹如何利用 DeepFaceLab 套件，在很短的時間內製作出深度偽造的影片，原始程式碼放在 DeepFaceLab GitHub[37]，網頁附有一個視訊『Mini tutorial』說明，只要按步驟執行腳本（Scripts），就可以順利完成影片，不過，它需要更強的硬體設備，筆者就不敢測試了。

Deepfake 造成了嚴重的假新聞災難，許多學者及企業提出反制的方法來辨識真假，簡單的像是『Detect AI-generated Images & Deepfakes（Part 1）』[38]一文所述，可以從臉部邊緣是否模糊、是否有隨機的雜訊、臉部是否對稱等細節來辨別，當然也有大公司推出可辨識影片真假的工具，例如微軟的『Microsoft Video Authenticator』，可參閱 ITHome 相關的報導[39]。

不管是 Deepfake 還是反制的演算法，未來發展都值得關注，這起事件也讓科學家留意到科學的發展必須兼顧倫理與道德，否則，好萊塢科幻片的劇情就不再只是幻想，人類有可能走向自我毀滅的道路。

參考資料（References）

[1] 自由時報,《全球首次！AI 創作肖像畫 10 月佳士得拍賣》, 2018
(https://news.ltn.com.tw/news/world/breakingnews/2529174)

[2] 佳士得官網《Is artificial intelligence set to become art's next medium?》
(https://www.christies.com/features/A-collaboration-between-two-artists-one-human-one-a-machine-9332-1.aspx)

[3] 佳士得官網關於 Edmond de Belamy 肖像畫的介紹
(https://www.christies.com/lot/lot-edmond-de-belamy-from-la-famille-de-6166184)

[4] the-gan-zoo GitHub
(https://github.com/hindupuravinash/the-gan-zoo)

[5] Liqian Ma、Xu Jia、Qianru Sun 等人,《Pose Guided Person Image Genera-tion》, 2018
(https://arxiv.org/pdf/1705.09368.pdf)

[6] Yanghua Jin、Jiakai Zhang 等 人,《Towards the Automatic Anime Characters Creation with Generative Adversarial Networks》, 2017
(https://arxiv.org/pdf/1708.05509.pdf)

[7] MakeGirlsMoe 網站
(https://make.girls.moe/#/)

[8] MIL WebDNN 網站
(https://mil-tokyo.github.io/webdnn/)

[9] Yanghua Jin、Jiakai Zhang 等 人,《Towards the Automatic Anime Characters Creation with Generative Adversarial Networks》, 2017
(https://arxiv.org/pdf/1708.05509.pdf)

[10] StarGAN GitHub
(https://github.com/yunjey/stargan)

[11] Christian Ledig、Lucas Theis、Ferenc Huszár 等人,《Photo-Realistic Single Image Super-Resolution Using a Generative Adversarial Network》, 2017
(https://arxiv.org/pdf/1609.04802.pdf)

10-67

[12] Tero Karras、Samuli Laine、Miika Aittala 等人,《Analyzing and Improving the Image Quality of StyleGAN》, 2020
(https://arxiv.org/pdf/1912.04958.pdf)

[13] StyleGAN2 GitHub
(https://github.com/NVlabs/stylegan2)

[14] Jonathan Hui,《GAN — Some cool applications of GAN》, 2018
(https://jonathan-hui.medium.com/gan-some-cool-applications-of-gans-4c9ecca35900)

[15] Deep Convolutional Generative Adversarial Network
(https://www.tensorflow.org/tutorials/generative/dcgan)

[16] Keras 官網範例『DCGAN to generate face images』
(https://keras.io/examples/generative/dcgan_overriding_train_step/)

[17] Tero Karras、Timo Aila、Samuli Laine 等人,《Progressive Growing of GANs for Improved Quality, Stability, and Variation》, 2017
(https://arxiv.org/abs/1710.10196)

[18] Mehdi Mirza、Simon Osindero,《Conditional Generative Adversarial Nets》, 2014
(https://arxiv.org/abs/1411.1784)

[19] Qiwen Fu、Wei-Ting Hsu、Mu-Heng Yang,《Colorization Using ConvNet and GAN》, 2017
(http://cs231n.stanford.edu/reports/2017/pdfs/302.pdf)

[20] ColorGAN GitHub
(https://github.com/bbc/ColorGAN#end-to-end-conditional-gan-based-architectures-for-image-colourisation)

[21] Phillip Isola、Jun-Yan Zhu、Tinghui Zhou,《Image-to-Image Translation with Conditional Adversarial Networks》, 2016
(https://arxiv.org/abs/1611.07004)

[22] CMP Facade Database
(https://cmp.felk.cvut.cz/~tylecr1/facade/)

[23] pix2pix datasets
(https://efrosgans.eecs.berkeley.edu/pix2pix/datasets/)

參考資料（References）

[24] Jun-Yan Zhu、Taesung Park、Phillip Isola、Alexei A. Efros,《Unpaired Image-to-Image Translation using Cycle-Consistent Adversarial Networks》, 2017
(https://arxiv.org/abs/1703.10593)

[25] Tensorflow 官網有關 CycleGAN 的說明
(https://www.tensorflow.org/datasets/catalog/cycle_gan)

[26] CartoonGAN: Generative Adversarial Networks for Photo Cartoonization
(https://openaccess.thecvf.com/content_cvpr_2018/papers/Chen_CartoonGAN_Generative_Adversarial_CVPR_2018_paper.pdf)

[27] 用 CartoonGAN 及 TensorFlow 2 生成新海誠與宮崎駿動畫
(https://leemeng.tw/generate-anime-using-cartoongan-and-tensorflow2.html)

[28] mnicnc404 CartoonGan-tensorflow GitHub
(https://github.com/mnicnc404/CartoonGan-tensorflow)

[29] SystemErrorWang FacialCartoonization
(https://github.com/SystemErrorWang/FacialCartoonization)

[30] 李宏毅老師的 PPT『Introduction of Generative Adversarial Network (GAN)』
(https://speech.ee.ntu.edu.tw/~tlkagk/slide/Tutorial_HYLee_GAN.pdf)

[31] Shakir Mohamed、Danilo Rezende,《Tutorial on Deep Generative Models》, 2017
(http://www.shakirm.com/slides/DeepGenModelsTutorial.pdf)

[32] You Won't Believe What Obama Says In This Video!
(https://www.youtube.com/watch?v=cQ54GDm1eL0)

[33] Alan Zucconi,《Understanding the Technology Behind DeepFakes》, 2018
(https://www.alanzucconi.com/2018/03/14/understanding-the-technology-behind-deepfakes/)

[34] Aayush Bansal、Shugao Ma、Deva Ramanan、Yaser Sheikh,《Recycle-GAN: Unsupervised Video Retargeting》, 2018
(https://arxiv.org/abs/1808.05174)

[35] Jonathan Hui,《Detect AI-generated Images & Deepfakes》, 2020
(https://jonathan-hui.medium.com/detect-ai-generated-images-deepfakes-part-1-b518ed5075f4)

[36] Louis (What's AI) Bouchard,《DeepFakes in 5 minutes》, 2020

(https://pub.towardsai.net/deepfakes-in-5-minutes-155c13d48fa3)

[37] DeepFaceLab GitHub

(https://github.com/iperov/DeepFaceLab)

[38] Jonathan Hui,《Detect AI-generated Images & Deepfakes (Part 1)》, 2020
(https://jonathan-hui.medium.com/detect-ai-generated-images-deepfakes-part-1-b518ed5075f4)

[39] 林妍溱,《微軟開發能判別 Deepfake 影像及內容變造的技術》, 2020

(https://www.ithome.com.tw/news/139740)

第 11 章
擴散模型
(Diffusion Model)

擴散模型（Diffusion Model）是目前最夯的影像生成演算法，包括 MidJourney、Stable Diffusion、DALLE…等影像生成工具底層都是使用擴散模型，它結合了 VAE 及 GAN 的特點，藉由編碼（Encode）掌握特徵，再透過解碼（Decode）生成圖像，過程中引進 GAN 概念，透過不斷學習，訓練出自雜訊生成圖像的模型。

本章討論以下內容：

1. 擴散模型（Diffusion Model）的原理。
2. MidJourney 簡介。
3. Stable Diffusion 的簡介與本機安裝。
4. OpenAI DALL·E 使用。

第 11 章　擴散模型（Diffusion Model）

11-1 擴散模型（Diffusion Model）原理

擴散（Diffusion）的原意是『液體從濃度較高的區域向濃度較低的區域移動』，如下圖，加一滴藍色墨水至裝有清水的瓶子內，墨水會逐漸擴散，直至全部清水被渲染成藍色為止，這個過程稱為擴散，而擴散模型（Diffusion Model）原理就是利用這個概念發展而成的演算法。

▲ 圖 11.1 擴散（Diffusion），圖片來源：『What Is Diffusion in Chemistry?』[1]

擴散模型（Diffusion Model）與 Encoder-Decoder 類似，分為兩個子網路：

1. 正向擴散（Forward Diffusion）：將一張原來清晰的圖像不斷地加入雜訊，直到整張圖像完全變成雜訊為止，類似擴散的原意，加墨水至裝有清水的瓶子內，直至全部清水被渲染成藍色。

2. 反向擴散（Reverse Diffusion）：從雜訊的圖像還原成清晰的圖像，是一個去噪（Denosing）的過程。

與 Encoder-Decoder 模型不同的是，它是多層的 VAE，是逐層加入雜訊，而非像 GAN 的 Generator 子網路只在源頭輸入雜訊，如下圖。

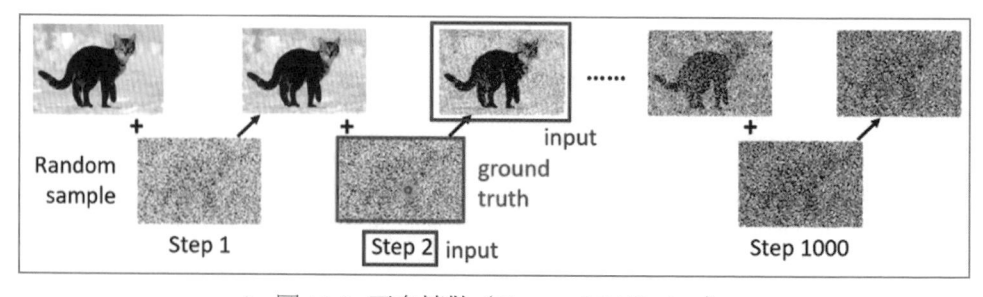

▲ 圖 11.2 正向擴散（Forward Diffusion），
圖片來源：『實作理解 Diffusion Model: 來自 DDPM 的簡化概念』[2]

11-1 擴散模型（Diffusion Model）原理

透過加入雜訊的過程，讓網路來學習如何 Reverse Diffusion，生成圖像，如下圖。

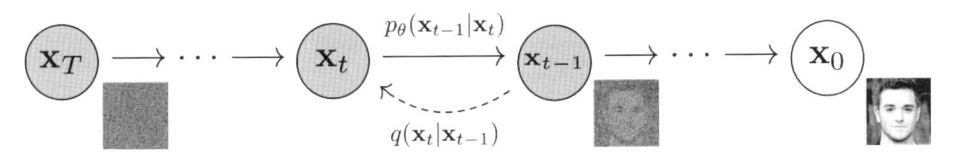

▲ 圖 11.3 正向擴散（Forward Diffusion）與反向擴散（Reverse Diffusion）示意圖，圖片來源：『Denoising Diffusion Probabilistic Models』[3]

詳細步驟說明如下：

1. 上圖下方是 Forward Diffusion，上方是 Reverse Diffusion。
2. Forward Diffusion 引進馬可夫鏈（Markov chain），依上一張圖的機率分佈加入雜訊（Noise），機率分佈公式如下：

$$q(\mathbf{x}_t \mid \mathbf{x}_{t-1}) = \mathcal{N}(\mathbf{x}_t; \mu_t = \sqrt{1-\beta_t}\mathbf{x}_{t-1}, \Sigma_t = \beta_t \mathbb{I})$$

- q 是平均數 μ_t，變異數 Σ_t 的常態分配函數。
- β_t：加入雜訊的比例（權重）。
- $\mathbb{I}$：單位矩陣（Identity matrix）。
- β_t：加入雜訊的比例（權重）。

- t：時序（time step），即步驟。
- 第 t 張與第 0 張圖雜訊的關係如下，為 0~t-1 張累計的總和：

$$q(\mathbf{x}_{1:T}|\mathbf{x}_0) := \prod_{t=1}^{T} q(\mathbf{x}_t|\mathbf{x}_{t-1})$$

3. Reverse Diffusion 是上圖的上方，逐步去噪，還原成清晰的圖像，公式如下：

$$p_\theta(\mathbf{x}_{t-1} \mid \mathbf{x}_t) = \mathcal{N}(\mathbf{x}_{t-1}; \mu_\theta(\mathbf{x}_t, t), \Sigma_\theta(\mathbf{x}_t, t))$$

11-3

第 11 章　擴散模型（Diffusion Model）

- p 是平均數 μ_θ，變異數 Σ_θ 的常態分配函數，μ_θ 是經由訓練得到的，而 Σ_θ 可以是固定值或經由訓練得到。
- Reverse Diffusion 就是訓練模型求最佳解 θ（類似權重）。
- 損失函數就是最小化生成圖像 $p_\theta(x_{t-1} \mid x_t)$ 與真實圖像 $q(x_{t-1}\, x_t)$ 之差，即兩個機率分配的 KL 散度（Kullback-Leibler Divergence）。

$$L = \mathbb{E}_{x_0,t,\epsilon}\left[\|\epsilon - \epsilon_\theta(x_t, t)\|^2\right]$$

ε 是 Forward Diffusion 加入的雜訊。

ε_θ 是模型的雜訊預測值。

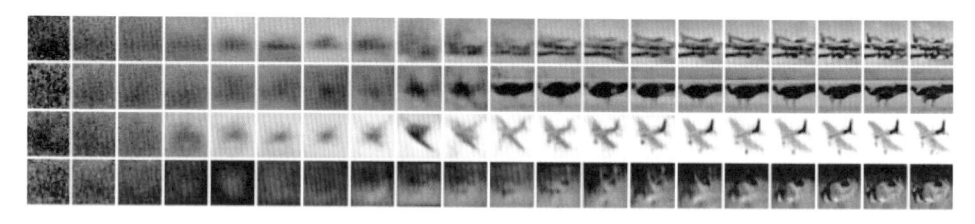

▲ 圖 11.4 反向擴散（Reverse Diffusion）示意圖，
圖片來源：『Denoising Diffusion Probabilistic Models』

4. 模型生成過程是輸入單純的雜訊，不斷的反向擴散（Reverse Diffusion），直到 X_0，即清晰的圖像為止，與 GAN 類似。

『Step by Step visual introduction to Diffusion Models』[4] 一文有很棒的動畫展示，顯現 Forward Diffusion 及 Reverse Diffusion 的過程，可透過滑鼠操作滑桿觀察圖像變化，有助於 Diffusion Model 的理解，讀者可實際體驗一下。

11-1 擴散模型（Diffusion Model）原理

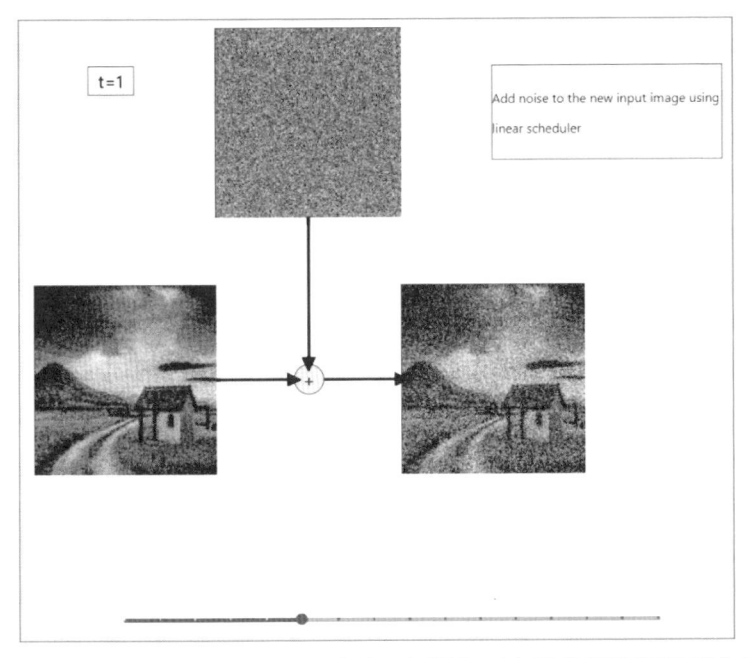

▲ 圖 11.5 Forward Diffusion，可透過滑鼠操作下方滑桿觀察圖像變化及說明

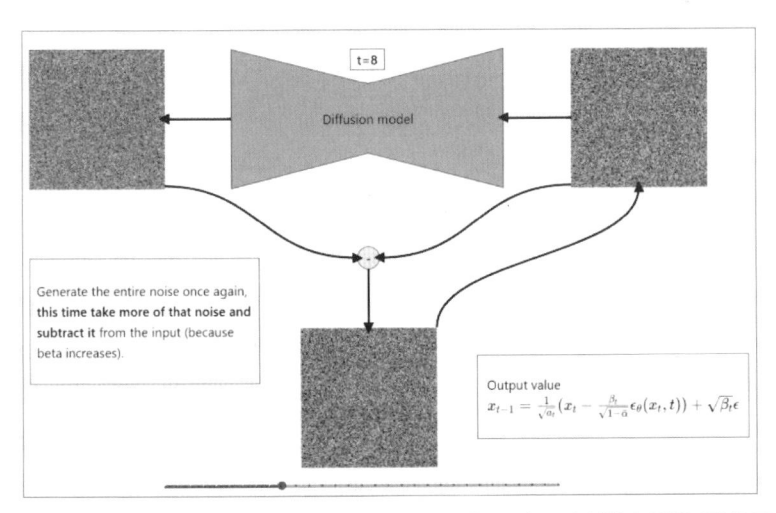

▲ 圖 11.6 Reverse Diffusion，可透過滑鼠操作下方滑桿觀察圖像變化及說明

第 11 章　擴散模型（Diffusion Model）

而 MidJourney、Stable Diffusion、DALLE…等影像生成工具均是輸入一段提示（Prompt），生成符合文字描述的圖像，那是因為他們將圖像及物件描述一併輸入模型訓練，演算法簡稱 CLIP（Contrastive Language-Image Pre-training），如下圖。

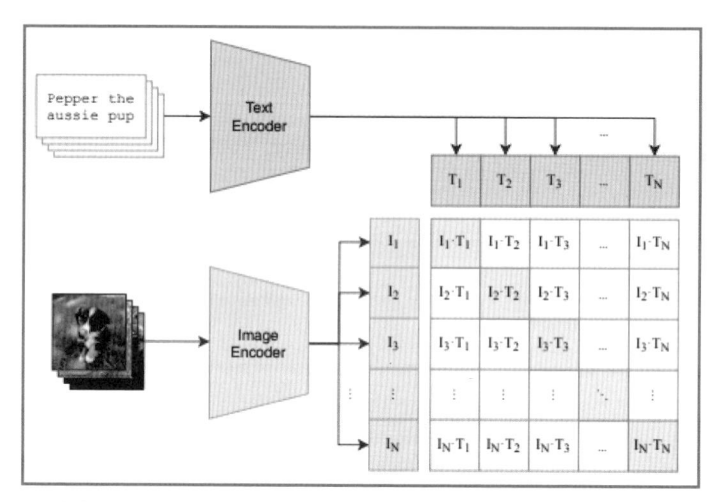

▲ 圖 11.7 圖像及物件描述一併輸入模型訓練，圖片來源：
『Learning Transferable Visual Models From Natural Language Supervision』[5]

Stable Diffusion 採用 Latent Diffusion Model 模型架構[6]，如下圖，可參閱『What are Stable Diffusion Models and Why are they a Step Forward for Image Generation?』[7]，針對每一個模塊（Block）有很詳細的說明。

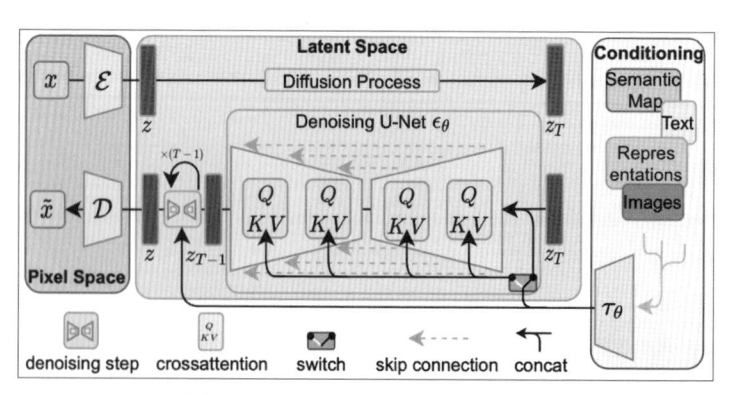

▲ 圖 11.8 Stable Diffusion 模型架構，
圖片來源：『High-Resolution Image Synthesis with Latent Diffusion Models』[8]

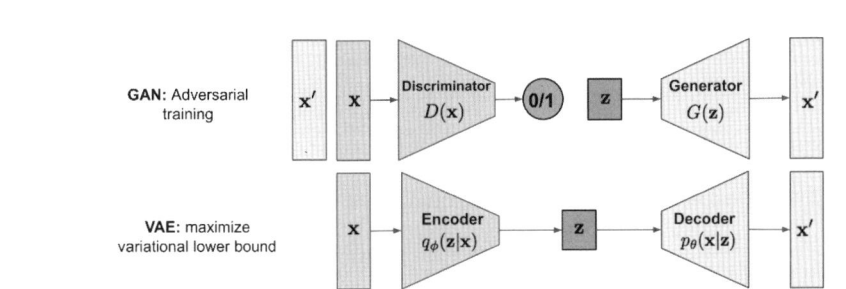

最後以一張圖比較 GAN、VAE 與 Diffusion Model。

▲ 圖 11.9 GAN、VAE 與 Diffusion Model 比較，圖片來源：『What are Diffusion Models？』[9]

- GAN：以一些圖像 (x) 及雜訊（z）訓練模型，生成圖像（X'）。
- VAE：以一些圖像 (x) 訓練模型，萃取特徵的機率分配（z），再利用機率分配生成圖像（X'）。
- Diffusion Model：以一些圖像 (x) 及逐層加入的雜訊（z）訓練模型，學習生成圖像的機制。

11-2 擴散模型（Diffusion Model）實作

- 由於擴散模型演算法非常複雜且需巨量的資料才能訓練出可用的模型，一般人沒辦法實作模型訓練，因此，我們只針對核心的演算法進行實驗，最初演算法是 2020/06/19 由美國加州大學柏克萊分校研發的，稱為 Denoising Diffusion Probabilistic Models（DDPM）[10]，上一節的 Forward/Reverse Diffusion 公式，即 DDPM 的內容，之後又提出 Denoising Diffusion Implicit Models（DDIM）[11]，OpenAI 也在 2021/02/18 提出 Improved DDPM[12]，都是在改善 Reverse Diffusion 生成圖像的速度。以下我就來看看 DDPM 的程式碼及執行結果。

第 11 章　擴散模型（Diffusion Model）

範例1. DDPM實作，程式來自『Diffusion Models. What are they, how do they work』[13]，複製放在 ddpm 資料夾。

1. 安裝額外的套件：

    pip install fire pydantic tqdm

2. 讀取隨機產生瑞士卷形狀的資料：使用 Scikit-learn 套件生成亂數資料。

```
from sklearn.datasets import make_swiss_roll

N = 1 << 10 # 1024
X = make_swiss_roll(n_samples=N, noise=1e-1)[0][:, [0, 2]] / 10.0
X.shape
```

- 執行結果：（1024, 2），表 1024 筆資料，2 個特徵。

3. 資料繪圖。

```
import matplotlib.pyplot as plt

plt.scatter(X[:, 0], X[:, 1])
```

- 執行結果：

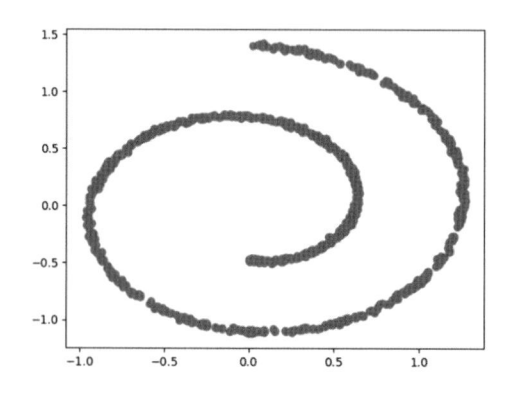

4. DDPM 訓練，並產生動畫。

```
from diffusers import DDPM
from models import BasicDiscreteTimeModel
from main import train, animate

```

11-2 擴散模型（Diffusion Model）實作

```
5  # 參數設定
6  d_model = 128
7  n_layers = 2
8  batch_size = 128
9  n_epochs = 400
10 sample_size = 512
11 steps_between_sampling = 20
12 seed = 42
13 n_steps = 100
14
15 print("Creating model")
16 model = BasicDiscreteTimeModel(d_model=d_model, n_layers=n_layers)
17 ddpm = DDPM(n_steps=n_steps)
18
19 print("Training")
20 result = train(
21     model=model,
22     ddpm=ddpm,
23     batch_size=batch_size,
24     n_epochs=n_epochs,
25     sample_size=sample_size,
26     steps_between_sampling=steps_between_sampling,
27     seed=seed,
28 )
29 # 產生動畫
30 anim = animate(result.samples, save=False)
```

- 執行結果：Forward Diffusion 將瑞士卷形狀的資料變成雜訊。

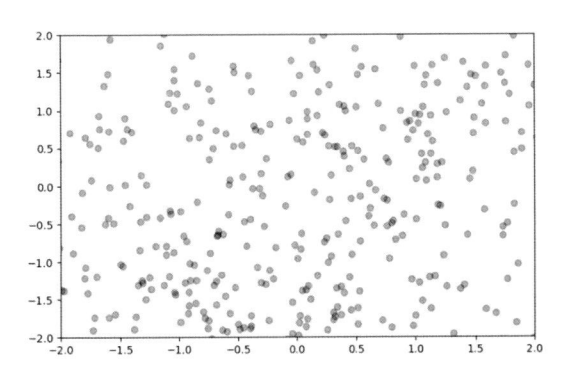

第 11 章　擴散模型（Diffusion Model）

5. 顯示訓練過程的動畫。

```
1  from IPython.display import HTML
2  HTML(anim.to_jshtml())
```

- 執行結果：可按下方播放鍵，觀察 Reverse Diffusion 過程，從亂數還原為瑞士卷形狀的資料。

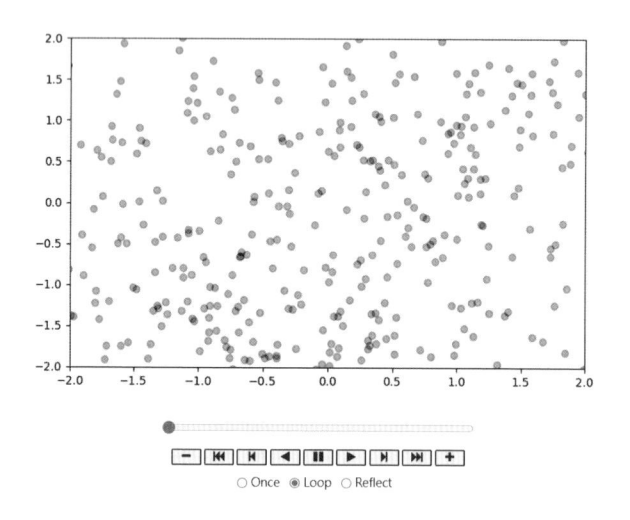

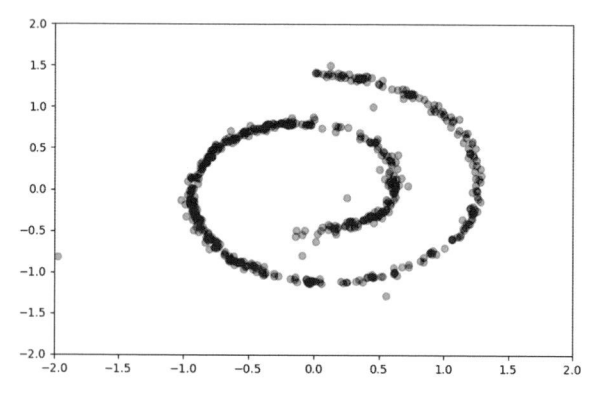

上述範例主要呼叫 main.py 中的 train 函數，該函數讀入瑞士卷形狀的資料進行 DDPM 模型訓練，再以模型進行生成圖像。DDPM 演算法定義在 diffusers.py，包括生成圖像的 sample 函數。

範例 2. 使用 DDPM 訓練 MNIST 模型，並生成影像，程式來自『SingleZombie/DL-Demos』[14]，複製放在 ddpm_mnist 資料夾。

1. 自 GitHub 下載程式：

   git clone https://github.com/SingleZombie/DL-Demos.git

2. 安裝套件：

   python setup.py develop

3. 切換至本書的範例資料夾 ddpm_mnist。

4. 執行主程式：程式使用 PyTorch 開發，預設使用 GPU，要以 CPU 執行可修改第 86 行為 cpu。訓練 100 執行週期，要花費較長時間，如要簡單測試，可修改第 17 行。

5. 執行結果：

   - 會建立模型檔：model_unet_res.pth。
   - 生成圖像 work_dirs\diffusion.jpg，如下，效果比 AutoEncoder 來的好。

11-3 MidJourney 簡介

MidJourney 是影像生成工具的始祖，在 2022 年 7 月推出，造成一股熱潮，只要輸入簡單的文字描述，稱為提示（Prompt），就可以產生非常美觀且符合文字描述的圖片，以下簡單說明使用程序。

第 11 章　擴散模型（Diffusion Model）

1. MidJourney 目前必須付費訂閱才能使用，無試用期間，訂閱方案如下，主要差異在使用的時間長短。

	Basic Plan	Standard Plan	Pro Plan	Mega Plan
Monthly Subscription Cost	$10	$30	$60	$120
Annual Subscription Cost	$96 ($8 / month)	$288 ($24 / month)	$576 ($48 / month)	$1152 ($96 / month)
Fast GPU Time	3.3 hr/month	15 hr/month	30 hr/month	60 hr/month
Relax GPU Time	-	Unlimited	Unlimited	Unlimited
Purchase Extra GPU Time	$4/hr	$4/hr	$4/hr	$4/hr
Work Solo In Your Direct Messages	✓	✓	✓	✓
Stealth Mode	-	-	✓	✓

▲ 圖 11.10 MidJourney 訂閱方案，圖片來源：『MidJourney Plan Comparison』[15]

2. 使用介面有兩種：
   - 網頁。
   - Discord 社群軟體。

3. 以下僅介紹網頁使用方式，Discord 操作方式也很類似。

4. 訂閱後，瀏覽 https://www.midjourney.com/ 。

5. 在網頁上方輸入提示，點選『create』選單，會生成四張低解析度的圖片。

6. 將滑鼠停留在圖片上，會出現按鈕，點擊後可選擇生成高解析度或模糊的圖片，或指定風格（Style）…等，也可以依據選擇的圖片重新產生四張相類型的圖片。

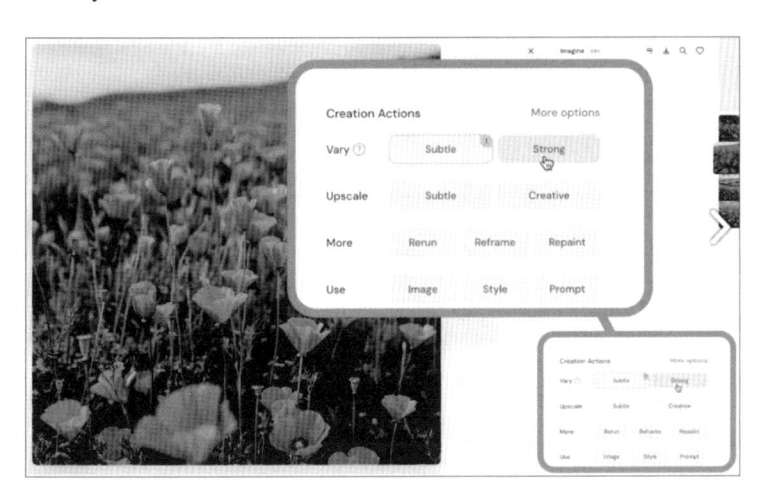

7. 生成高解析度之後可以下載圖像。

8. 在輸入提示的同時也可以輸入參數，直接產生期望的輸出格式，如下圖右方以『--』開頭的參數名稱，參數說明可參閱『MidJourney Parameter List』[16]。

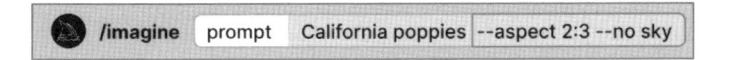

11-4 Stable Diffusion 簡介

Stable Diffusion 是 Stability AI 公司在 2022 年發佈的產品，目前已發展至 3.5 版[17]，功能包括：

1. 以文生圖（Text to image）：輸入提示，生成圖像。

2. 以圖生圖（Image modification）：輸入圖像，生成相關的圖像。

3. 圖像修復（Inpainting）：輸入圖像，修復有缺陷或被遮罩的部份。

4. 圖像延伸（Outpainting）：輸入圖像，可依圖像延伸相關的內容，如下，左圖為名畫『Girl with a Pearl Earring』，右圖為圖像延伸的結果，工具會依照提示及原圖生成合理的內容，『DALL·E: Introducing outpainting』[18] 一文還展示圖像延伸的動畫，如果讀者有興趣可以參閱。

第 11 章　擴散模型（Diffusion Model）

▲ 圖 11.11　MidJourney 訂閱方案，圖片來源：『DALL·E: Introducing outpainting』

5. ControlNet：以文生圖時可以加入條件，更精準的指定生成的內容，例如指定風格，包括素描、油畫…等。

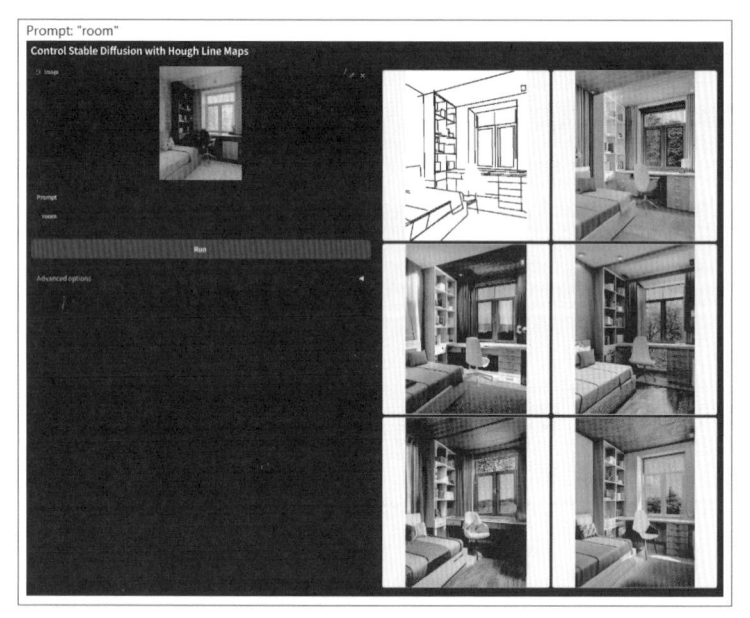

▲ 圖 11.12　ControlNet 展示，圖片來源：『ControlNet GitHub』[19]

Stable Diffusion 可以在雲端網站操作，也可以在本機安裝。

1. 雲端網站的服務稱為 DreamStudio，也有開源稱為 StableStudio。
2. 本機安裝以『AUTOMATIC1111 Stable Diffusion Web UI』提供的方式最為盛行，提供網頁介面，可安裝多種模型，另外還有 ComfyUI，提供節點是的管道（Pipeline），可設定各種加工效果。

11-5 DreamStudio 使用

進入 DreamStudio 首頁,選擇試用,出現以下畫面:

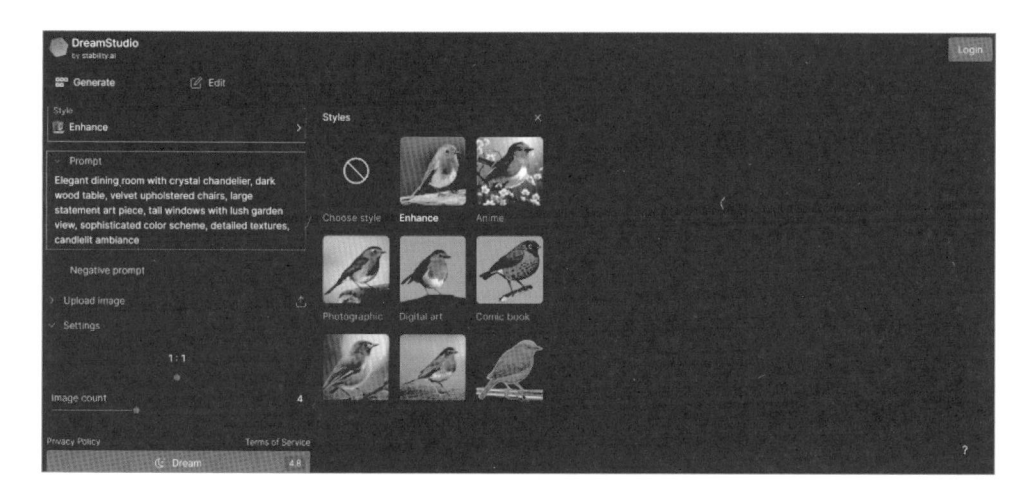

1. 在右上角註冊或登入。
2. 在左方指定輸出風格(Style)。
3. 輸入提示(Prompt):頁面已經有範例,除了一般提示(Prompt),還有負面提示(Negative Prompt),聲明哪些內容不要產生。
4. 點擊下方的『Dream』:生成圖像。

實驗:

1. 筆者先嘗試輸入中文,系統出現錯誤訊息。
2. 改輸入『Totoro held an umbrella for the girl. It was raining lightly.』,生成下圖:

11-15

3. 上圖 girl 生成小龍貓，將提示加上 little，即『Totoro held an umbrella for the little girl. It was raining lightly.』，生成下圖，已置換為小女孩：

4. 提示如果能鉅細靡遺，就能產生心中期望的圖像，如何撰寫提示的指引稱為『提示工程』（Prompt Engineering），可參閱 Stable Diffusion 官方指引『Stable Diffusion prompt: a definitive guide』[20]，主要分為 8 項關鍵字：

- 主題（Subject）。
- 輸出材質（Medium）：例如插圖、油畫、3D 渲染或攝影⋯。
- 風格（Style）：例如印象派（Impressionist）、超現實主義（Surrealist）、普普藝術（Pop art）⋯。
- 引用知名圖庫網站（Art-sharing website）：引用知名圖庫網站如 Artsta-tion、Deviant Art 等，引導輸出影像偏向這些網站風格。
- 解析度（Resolution）。
- 其他細節（Additional details）。
- 顏色（Color）。
- 燈光（Lighting）。

官方指引有針對每一項關鍵字舉例說明，非常實用，值得一讀。

11-6 Stable Diffusion 本機安裝

Stable Diffusion 模型提供免費下載，可以在本機安裝，最簡單的方式是使用 Hugging Face API，其中 Hugging Face 管道（Pipeline）是最簡單的模型推論用法，但是使用時要記住以下幾點：

11-6 Stable Diffusion 本機安裝

1. 先註冊 Hugging Face 帳號[21]。
2. 至個人帳號設定產生 Access Token[22]。

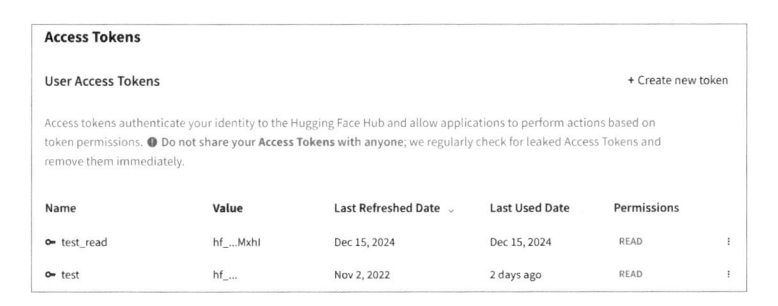

3. 新增環境變數,記錄 Access Token。

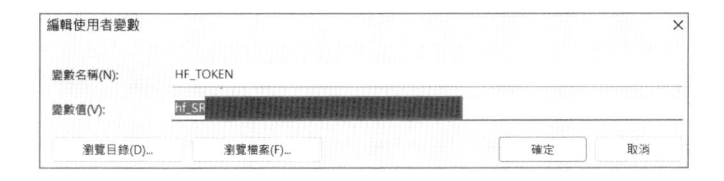

4. 下載某些模型前,要先線上簽授權協議。

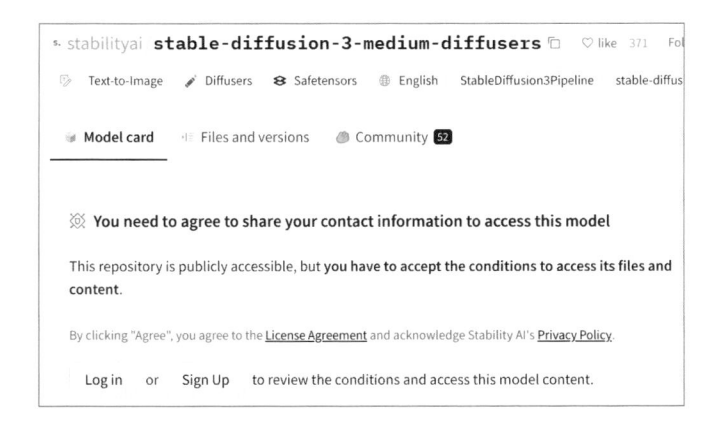

5. 執行前要登入帳號:需先安裝 transformers。
   - pip install transformers
   - huggingface-cli login

範例. 使用 Hugging Face API,載入 Stable Diffusion 模型生成圖像,程式為 01_HG_Stable_Diffusion_API_test.py,修改自 Hugging Face Stable Diffusion 3 Pipeline [23],需

第 11 章　擴散模型（Diffusion Model）

下載 10 多 GB 的模型，如本機硬碟容量不足或無較大記憶體的 GPU 卡，可在 Colab 上執行。

1. 安裝套件：除了 PyTorch 外還需安裝以下套件。

    - pip install diffusers accelerate transformers -U

2. 載入套件。

```
2  import torch
3  from diffusers import StableDiffusion3Pipeline
```

3. 使用管道（Pipeline）載入模型：管道是預先安排好的一系列工作，以下是下載 Stable Diffusion v3 模型，也可以下載其他版本模型，較早的版本檔案較小，生成圖像速度較快，但品質較差。

```
6  pipe = StableDiffusion3Pipeline.from_pretrained("stabilityai/stable-diffusion-3-medium-diffusers"
7                                                  , torch_dtype=torch.float16)
8  pipe.to("cuda")  # 複製模型到GPU記憶體
```

4. 進行模型推論，生成圖像：在第 12 行輸入提示，需要執行一段時間，視 GPU 卡效能而定。

```
11  image = pipe(
12      prompt="A snake holding a sign that says Happy Lunar Year",
13      negative_prompt="",
14      num_inference_steps=28,
15      height=1024,
16      width=1024,
17      guidance_scale=7.0,
18  ).images[0]
```

- 執行結果：得到一個可愛的圖像。

5. 圖像存檔。

11-18

11-6 Stable Diffusion 本機安裝

```
21  image.save("sd3_hello_world.png")
```

目前網路上有一些工具平台方便安裝，特點如下：

Feature	ComfyUI	Automatic1111	Fooocus/RuinedFooocus	SD Next	SwarmUI
Interface	Advanced	Friendly	Simplified	Simplified	Friendly
Models	SD + Flux	Stable Diffusion only	SD XL only	Various txt2img/img2img	SD + Flux
Performance	Efficient	Optimizable	Solid	Optimized	High
Features	Graph workflows, memory	Plugins, upscaling	Styles, Img2Img	Multi-platform, models	Modular, AI video
License	Open-source (GPL)	Open-source (Free)	Open-source (Free)	Open-source (Free)	Open-source (Free)

▲ 圖 13. 11 Stable Diffusion 工具平台，
圖片來源：『Awesome LLM Resources List 』[24]

其中『AUTOMATIC1111 Stable Diffusion Web UI 』[25] 較為簡易，它使用 Gradio 套件實作 Stable Diffusion 的網頁介面，以下說明在 Windows 作業系統下的安裝程序：

1. 最好具備 GPU 顯卡，且最好要有 12GB 記憶體以上，因為模型大小動不動就 5GB 以上。

2. 最好建立 Python 虛擬環境，因為某些套件並不是安裝最新版，可能會蓋掉既有較新的版本。

   conda create --name stable python=3.10

3. 切換虛擬環境：

   conda activate stable

4. 自發行網址 [26] 下載安裝檔 sd.webui.zip，並解壓縮。

5. 自解壓縮的資料夾下，依序執行 update.bat、run.bat，會帶出網頁介面，如下圖。

第 11 章　擴散模型（Diffusion Model）

- 在最上方選擇模型，目前只有一個『v1-5-pruned-emaonly.safetensors』。
- 接著輸入提示及負面提示：筆者輸入『generate cartoon, which has Totoro hold an umbrella with the little girl in a raining night』。
- 下方還有一些選項可以調整。

Sampling method	Schedule type
DPM++ 2M	Automatic

Sampling steps　20

Hires. fix　　Refiner

Width　512

Height　512

Batch count　1

Batch size　1

CFG Scale　7

- 點擊『Generate』按鈕。

6. 執行結果：不太理想，如下圖示筆者執行將近 10 次的結果，與 MidJourney 比較品質真的差太多了。下方有一些按鈕可點擊，如開啟歷程記錄、存檔、…等。

11-20

11-6 Stable Diffusion 本機安裝

7. 除了以文生圖外，還有以圖生圖等其他功能。

| txt2img | img2img | Extras | PNG Info | Checkpoint Merger | Train | Settings | Extensions |

8. 另外還有圖像修復、模型微調（Lora）…等功能。

| Generation | Textual Inversion | Hypernetworks | Checkpoints | Lora |
| img2img | Sketch | Inpaint | Inpaint sketch | Inpaint upload | Batch |

9. 圖像修復（Inpainting）：可選擇『Inpaint sketch』，依下列步驟測試：
   - 先上載一個圖檔。
   - 點擊『sketch』，然後使用『刷子』（brush）在圖像上劃上一些線條。
   - 再點擊『inpaint sketch』，可以看到線條被修復，模型會以周遭的顏色修補線條。
   - 點擊『inpaint』，可回復線條。

10. 加入其他模型：Stable Diffusion Web UI 允許載入其他模型，依據『AUTOMATIC1111 Stable Diffusion Web UI wiki』[27] 的『Feature』頁面提供的資訊，目前有 2 個模型可加入，其中 sd_xl_base_1.0_0.9vae 載入有錯誤訊息，我們嘗試另一個模型 sd_xl_refiner_1.0_0.9vae，點超連結 [28]，再點擊『download』下載至 sd.webui\webui\models\Stable-diffusion 資料夾內。

11. 再回到 Stable Diffusion Web UI，點擊『Stable Diffusion checkpoint』右下方的 refresh 圖示，選擇『sd_xl_refiner_1.0_0.9vae.safetensors』，即會將下載的 sd_xl_refiner_1.0_0.9vae.safetensors 轉換為 model.pt 存入 VAE-approx 資料夾中，之後就可以操作上述功能了。

第 11 章　擴散模型（Diffusion Model）

```
Stable Diffusion checkpoint
sd_xl_refiner_1.0_0.9vae.safetensors [8d0ce6c0 ▼
```

12. 再從新輸入提示『Totoro hold an umbrella with the little girl human in a raining night』，生成下列圖像，品質似乎好一些。

11-7 Stable Diffusion API

Stable Diffusion Web UI 也可以變成 API Server，修改 sd.webui\webui\webui-user.bat，將 COMMANDLINE_ARGS 加上【--api】，如下：

- set COMMANDLINE_ARGS=--api

重新執行 sd.webui 資料夾下的 run.bat，它就變成 API Server 了。

範例. 撰寫用戶端程式呼叫 API Server

程式：02_Stable_Diffusion_API_test.ipynb，修改自 w-e-w GitHub[29]。

1. 載入套件。

```
1  import urllib.request
2  import base64
3  import json
4  import time
5  import os
```

11-7 Stable Diffusion API

2. 定義圖像編碼與解碼函數。

```python
def encode_file_to_base64(path):
    with open(path, 'rb') as file:
        return base64.b64encode(file.read()).decode('utf-8')

def decode_and_save_base64(base64_str, save_path):
    with open(save_path, "wb") as file:
        file.write(base64.b64decode(base64_str))
```

3. 以文生圖（Text to image）：

- Payload：定義以文生圖的參數，定義 prompt 為『1 golden-hair girl』。
- 以 urllib.request.Request 呼叫 API Server，URL 為 http://127.0.0.1:7860/sdapi/v1/txt2img。
- 傳回來的資料以 JSON 格式解析。
- 呼叫 decode_and_save_base64 將 base 64 字串解碼為圖像並存檔。

```python
payload = {
    "prompt": "masterpiece, (best quality:1.1), 1 golden-hair girl <lora:lora_model:1>",
    "negative_prompt": "",
    "seed": 1,
    "steps": 20,
    "width": 512,
    "height": 512,
    "cfg_scale": 7,
    "sampler_name": "DPM++ 2M",
    "n_iter": 1,
    "batch_size": 1,
}

webui_server_url = 'http://127.0.0.1:7860'
api_endpoint = f'{webui_server_url}/sdapi/v1/txt2img'
data = json.dumps(payload).encode('utf-8')
request = urllib.request.Request(
    api_endpoint,
    headers={'Content-Type': 'application/json'},
    data=data,
)
response = urllib.request.urlopen(request)
response = json.loads(response.read().decode('utf-8'))
image = response.get('images')[0]
decode_and_save_base64(image, './out.png')
```

第 11 章　擴散模型（Diffusion Model）

4. 顯示輸出圖檔。

```
1  import matplotlib.pyplot as plt
2
3  def show_image(image_path):
4      image = plt.imread(image_path)
5      fig, ax = plt.subplots()
6      im = ax.imshow(image)
7      ax.axis('off')
8      plt.show()
9
10 show_image('./out.png')
```

5. 執行結果：得到一個金髮女子圖像。

6. 以圖生圖（Image modification）。

- batch_size：定義一次產生幾張圖。
- Payload：需輸入一張圖及提示，模型會依照提示修改圖像。
- 以 urllib.request.Request 呼叫 API Server，URL 為 http://127.0.0.1:7860/sdapi/v1/img2img。
- 傳回來的資料以 JSON 格式解析。
- 傳回 2 張圖，以迴圈呼叫 decode_and_save_base64 將 base 64 字串解碼為圖像並存檔。

```
1  out_dir = 'img2img'
2  os.makedirs(out_dir, exist_ok=True)
3
4  batch_size = 2
5  init_images = [
6      encode_file_to_base64("./images/in.png"),
7  ]
8  payload = {
```

11-24

11-7 Stable Diffusion API

```
9        "prompt": "blue hair",
10       "seed": 1,
11       "steps": 20,
12       "width": 512,
13       "height": 512,
14       "denoising_strength": 0.5,
15       "n_iter": 1,
16       "init_images": init_images,
17       "batch_size": batch_size if len(init_images) == 1 else len(init_images),
18       # "mask": encode_file_to_base64(r"B:\path\to\mask.png")
19   }
```

```
21   webui_server_url = 'http://127.0.0.1:7860'
22   api_endpoint = f'{webui_server_url}/sdapi/v1/img2img'
23   data = json.dumps(payload).encode('utf-8')
24   request = urllib.request.Request(
25       api_endpoint,
26       headers={'Content-Type': 'application/json'},
27       data=data,
28   )
29   response = urllib.request.urlopen(request)
30   response = json.loads(response.read().decode('utf-8'))
31   for index, image in enumerate(response.get('images')):
32       save_path = f'{out_dir}/{index+1:02d}.png'
33       decode_and_save_base64(image, save_path)
34       show_image(save_path)
```

7. 執行結果：

- 輸入圖像：

- 輸出圖像：把小孩的頭髮變成藍色，而非大人。

第 11 章　擴散模型（Diffusion Model）

11-8 Stable Diffusion Extension

Stable Diffusion Web UI 可以安裝擴充模型（Extension），以下以 ControlNet 為例，它是『Adding Conditional Control to Text-to-Image Diffusion Models』[30] 實作，類似 Conditional VAE，在既有的 Stable Diffusion 模型加上各種風格的指定。

首先在【Extensions】頁籤內點擊【Install from URL】子頁籤，輸入 URL（https://github.com/Mikubill/sd-webui-controlnet.git），如下，再點擊【Install】：

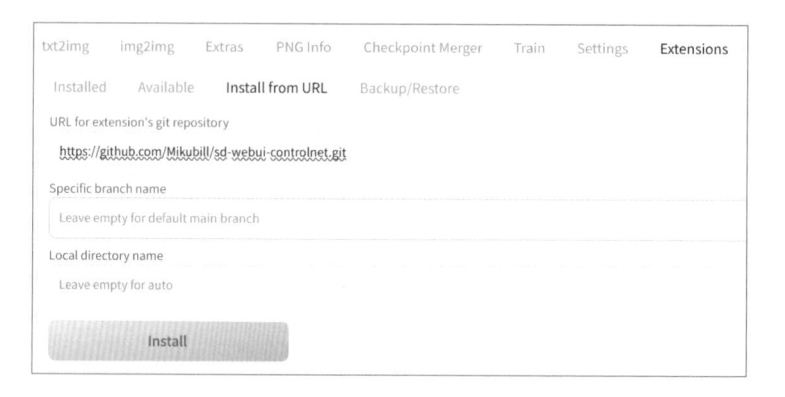

- 以上的 URL 係 ControlNet 擴充模型，請參閱 sd-webui-controlnet GitHub[31]。
- 點擊【Install】後，擴充模型會下載至 sd.webui/webui/extensions/sd-webui-controlnet 資料夾。
- 點擊【Installed】可以看到 ControlNet 擴充模型，點擊 ControlNet 對應的 URL。

- 該網頁【Download Models】段落內有超連結，點擊後的網頁中有許多模型，筆者選擇【Stable Diffusion 1.5 / 2.0 中型模型】[32]，再點擊右上方的【⋮】，選擇【Clone repository】，出現 Git 指令，在 sd.webui\webui\models 資料夾下執行：

    git lfs install

    git clone https://huggingface.co/comfyanonymous/ControlNet-v1-1_fp16_safetensors

- 點擊【Check for updates】及【Apply and restart UI】。
- 重新啟動 Stable Diffusion Web UI。

11-8 Stable Diffusion Extension

測試：

1. 點選【img2img】。
2. 輸入提示【a man with black hair】。
3. 拖曳 images\lena.jpg 至提示下方的框中。
4. 點選【ControlNet v1.1.455】右方的下拉圖示，勾選【enable】，下方有許多效果，點選【SoftEdge】，強調線條。

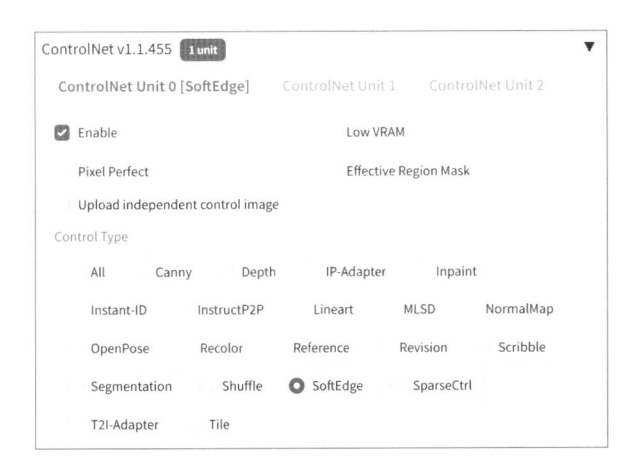

5. 【Control Mode】選擇【ControlNet is more important】，表示較注重圖像的風格，若選【My prompt is more important】，表示較注重提示。
6. 點擊【Generate】按鈕，執行結果如下，左方為原圖，右方為生成圖像，可以看出右圖承襲左圖的風格，又依提示生成男子圖像，以下結果是筆者點擊【Generate】很多次，才得到比較好的結果。

更多的說明請參閱『sd-webui-controlnet GitHub』。

11-27

第 11 章 擴散模型（Diffusion Model）

也可以直接使用 Python 執行 ControlNet 生成圖像的功能，不需要 Stable Di-ffusion Web UI，請參閱『lllyasviel/ControlNet GitHub』[33]，筆者安裝不順利，改用 v1.1[34]，也未成功，仍將程序記錄如下：

1. 下載專案後解壓縮。

2. 切換至專案資料夾，直接執行以下指令安裝：

    conda env create -f environment.yaml

3. 上一步驟會安裝 CPU103 的 PyTorch，如果要使用 GPU，須再執行以下指令：

    conda install pytorch==1.12.1 torchvision==0.13.1 torchaudio==0.12.1 cudatoolkit=11.6 -c pytorch -c conda-forge

4. 自 HuggingFace lllyasviel/ControlNet-v1-1[35] 下載模型，複製 *.pth 至 Control-Net-v1-1-nightly-main\models 資料夾內。

5. 自 HuggingFace stable-diffusion-v1-5 [36] 下載模型，複製 v1-5-pruned.ckpt 至 ControlNet-v1-1-nightly-main\models 資料夾內。

6. 切換至虛擬環境：Conda activate control-v11

7. 至專案資料夾下執行範例程式：

    python gradio_canny.py

8. 開啟瀏覽器，輸入 http://localhost:7860/。

9. 拖曳 ControlNet-v1-1-nightly-main\test_imgs\bird.png 至瀏覽器的 Image 欄位，並在下方輸入提示【bird】，點擊【Run】。

10. 執行一段時間後，結果如下，不過筆者等了 1000 多秒仍未執行成功。

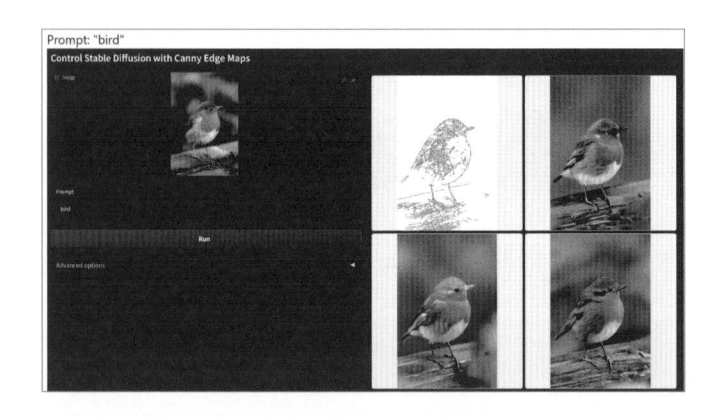

11-9 ControlNet in Diffusers

Diffusers 是很簡單好用的預先訓練擴散模型套件,用於產生影像、音訊甚至分子 3D 結構。無論是簡單的推論還是訓練自己的擴散模型都可以使用此套件。搭配 ControlNet 也比『lllyasviel/ControlNet GitHub』官網來的簡單,直接使用 HuggingFace 的預先訓練模型載入標準語法,程式簡單易懂。

範例. ControlNet in Diffusers 測試,程式 03_ControlNet_test.ipynb 修改自『Ultra fast ControlNet with Diffusers』[37]。

1. 除了 PyTorch 外還需安裝許多套件,有些套件必須要支援長檔名,若是在 Windows 作業系統,請滑鼠雙擊 enable-long-paths.reg,再執行 ControlNet_install.bat,內含許多套件。

2. 下載測試圖片『戴著珍珠耳環的女孩』。

```
1 from diffusers import StableDiffusionControlNetPipeline
2 from diffusers.utils import load_image
3
4 image = load_image(
5     "https://hf.co/datasets/huggingface/documentation-images/resolve/main/diffusers/input_image_vermeer.png"
6 )
7 image
```

- 執行結果:

3. 使用 Canny 演算法取得圖像輪廓。

```
1 import cv2
2 from PIL import Image
3 import numpy as np
4
5 image = np.array(image)
6
7 low_threshold = 100
8 high_threshold = 200
```

第 11 章　擴散模型（Diffusion Model）

```
 9
10  image = cv2.Canny(image, low_threshold, high_threshold)
11  image = image[:, :, None]
12  image = np.concatenate([image, image, image], axis=2)
13  canny_image = Image.fromarray(image)
14  canny_image
```

- 執行結果：

4. 下載 controlnet-canny 及 stable diffusion 模型，合併兩者。

```
 1  from diffusers import StableDiffusionControlNetPipeline, ControlNetModel
 2  from diffusers import UniPCMultistepScheduler
 3  import torch
 4
 5  # download controlnet-canny、stable diffusion models, combine both.
 6  controlnet = ControlNetModel.from_pretrained("lllyasviel/sd-controlnet-canny", torch_dtype=torch.float16)
 7  pipe = StableDiffusionControlNetPipeline.from_pretrained(
 8      "runwayml/stable-diffusion-v1-5", controlnet=controlnet, torch_dtype=torch.float16
 9  )
10  pipe.scheduler = UniPCMultistepScheduler.from_config(pipe.scheduler.config)
11  # enable GPU
12  pipe.enable_model_cpu_offload()
13  # xFormers wasn't build with CUDA support
14  # pipe.enable_xformers_memory_efficient_attention()
```

- pipe.enable_model_cpu_offload：模型會自動載入至 GPU。

- pipe.enable_xformers_memory_efficient_attention()：使用 xFormers 套件可節省記憶體的使用，可惜它不支援 GPU，故與上一行只能擇一使用。

5. 定義多格圖像顯示的函數，以便顯示模型生成的圖像。

```
 1  # 定義多格圖像顯示的函數
 2  def image_grid(imgs, rows, cols):
 3      assert len(imgs) == rows * cols
 4
 5      w, h = imgs[0].size
 6      grid = Image.new("RGB", size=(cols * w, rows * h))
 7      grid_w, grid_h = grid.size
```

11-9 ControlNet in Diffusers

```
 8
 9     for i, img in enumerate(imgs):
10         grid.paste(img, box=(i % cols * w, i // cols * h))
11     return grid
```

6. 影像生成。

```
14  prompt = ", best quality, extremely detailed"
15  prompt = [t + prompt for t in ["Sandra Oh", "Kim Kardashian", "rihanna", "taylor swift"]]
16  generator = [torch.Generator(device="cpu").manual_seed(2) for i in range(len(prompt))]
17
18  output = pipe(
19      prompt,
20      canny_image,
21      negative_prompt=["monochrome, lowres, bad anatomy, worst quality, low quality"] * len(prompt),
22      generator=generator,
23      num_inference_steps=20,
24  )
25  # 圖像顯示
26  image_grid(output.images, 2, 2)
```

- 執行結果：原圖換臉成 4 個女星，她們的名字均輸入至提示中。

7. 再測試另一組人名，均為歷任美國總統。

```
 1  prompt = ", best quality, extremely detailed"
 2  prompt = [t + prompt for t in ["Donald Trump", "Barack Obama",
 3                                 "Joseph Robinette Biden", "Abraham Lincoln"]]
 4  generator = [torch.Generator(device="cpu").manual_seed(2) for i in range(len(prompt))]
 5
 6  output = pipe(
 7      prompt,
 8      canny_image,
 9      negative_prompt=["monochrome, lowres, bad anatomy, worst quality, low quality"] * len(prompt),
10      generator=generator,
11      num_inference_steps=20,
12  )
```

第 11 章 擴散模型（Diffusion Model）

```
13
14 image_grid(output.images, 2, 2)
```

- 執行結果：

8. 使用『馬鈴薯頭先生』（Mr Potato Head）風格的 ControlNet。

```
1  model_id = "sd-dreambooth-library/mr-potato-head"
2  pipe = StableDiffusionControlNetPipeline.from_pretrained(
3      model_id,
4      controlnet=controlnet,
5      torch_dtype=torch.float16,
6  )
7  pipe.scheduler = UniPCMultistepScheduler.from_config(pipe.scheduler.config)
8  pipe.enable_model_cpu_offload()
9  # pipe.enable_xformers_memory_efficient_attention()
```

9. 輸入提示『馬鈴薯頭先生的照片，最佳品質，必須非常詳細』。

```
1  generator = torch.manual_seed(2)
2  prompt = "a photo of sks mr potato head, best quality, extremely detailed"
3  output = pipe(
4      prompt,
5      canny_image,
6      negative_prompt="monochrome, lowres, bad anatomy, worst quality, low quality",
7      generator=generator,
8      num_inference_steps=20,
9  )
10 output.images[0]
```

- 執行結果：

10. 使用具有深度（depth）風格的模型。

```
1  controlnet = ControlNetModel.from_pretrained(
2      "lllyasviel/sd-controlnet-depth", torch_dtype=torch.float16
3  )
4  pipe = StableDiffusionControlNetPipeline.from_pretrained(
5      "runwayml/stable-diffusion-v1-5", controlnet=controlnet, safety_checker=None, torch_dtype=torch.float16
6  )
```

```
7  pipe.scheduler = UniPCMultistepScheduler.from_config(pipe.scheduler.config)
8  pipe.enable_model_cpu_offload()
9  # pipe.enable_xformers_memory_efficient_attention()
```

11. 使用另一張圖像測試。

```
1  from transformers import pipeline
2
3  image = load_image("https://huggingface.co/lllyasviel/sd-controlnet-depth/resolve/main/images/stormtrooper.png")
4  image
```

- 執行結果：

第 11 章　擴散模型（Diffusion Model）

12. 生成圖像：提示為『衝鋒隊的講座』（Stormtrooper's lecture）。

```
1  generator = torch.manual_seed(2)
2  depth_estimator = pipeline('depth-estimation')
3  image = depth_estimator(image)['depth']
4  image = np.array(image)
5  image = image[:, :, None]
6  image = np.concatenate([image, image, image], axis=2)
7  image = Image.fromarray(image)
8
9  output = pipe("Stormtrooper's lecture", image, num_inference_steps=20)
10 image = output.images[0]
11 # image.save('./stormtrooper_depth_out.png')
12 image
```

- 執行結果：

13. 姿勢偵測（Pose detection）：模型可抓到關鍵點，使用 4 張瑜珈人像。

```
1  urls = "yoga1.jpeg", "yoga2.jpeg", "yoga3.jpeg", "yoga4.jpeg"
2  imgs = [
3      load_image("https://hf.co/datasets/YiYiXu/controlnet-testing/resolve/main/" + url)
4      for url in urls
5  ]
6
7  image_grid(imgs, 2, 2)
```

- 執行結果：

11-9 ControlNet in Diffusers

14. 抓取關鍵點。

```
1  from controlnet_aux import OpenposeDetector
2
3  model = OpenposeDetector.from_pretrained("lllyasviel/ControlNet")
4
5  poses = [model(img) for img in imgs]
6  image_grid(poses, 2, 2)
```

- 執行結果：

15. 依據關鍵點生成人像。

第 11 章　擴散模型（Diffusion Model）

```
1  controlnet = ControlNetModel.from_pretrained(
2      "fusing/stable-diffusion-v1-5-controlnet-openpose", torch_dtype=torch.float16
3  )
4
5  model_id = "runwayml/stable-diffusion-v1-5"
6  pipe = StableDiffusionControlNetPipeline.from_pretrained(
7      model_id,
8      controlnet=controlnet,
9      torch_dtype=torch.float16,
10 )
11 pipe.scheduler = UniPCMultistepScheduler.from_config(pipe.scheduler.config)
12 pipe.enable_model_cpu_offload()
13 # pipe.enable_xformers_memory_efficient_attention()
```

16. 生成超級英雄圖像。

```
1  generator = [torch.Generator(device="cpu").manual_seed(2) for i in range(4)]
2  prompt = "super-hero character, best quality, extremely detailed"
3  output = pipe(
4      [prompt] * 4,
5      poses,

6      negative_prompt=["monochrome, lowres, bad anatomy, worst quality, low quality"] * 4,
7      generator=generator,
8      num_inference_steps=20,
9  )
10 image_grid(output.images, 2, 2)
```

- 執行結果：

在範例檔最下面還有更多的風格說明。

11-10 NitroFusion

在撰寫本書的同時（2024/12），英國薩里（Surrey）大學發表一個新的模型 NitroFusion[38]，可以生成精美的圖像，筆者迫不及待的測試一番，比之前的 Stable Diffusion 模型輸出品質好太多了，根據論文『NitroFusion: High-Fidelity Single-Step Diffusion through Dynamic Adversarial Training』[39] 說明，NitroFusion 採用 GAN 架構，不同的是，生成子網路使用與擴散模型一樣的多步（Multi-step）教師模型，再轉化單步（Single-step）的學生模型，配合動態的判別網路池（Dynamic Discriminator Head Pool），如下圖：

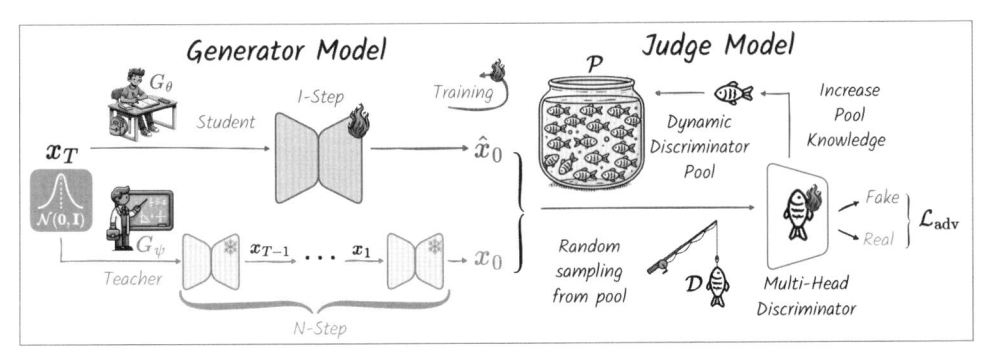

▲ 圖 11.14 NitroFusion 架構，圖片來源：『NitroFusion: High-Fidelity Single-Step Diffusion through Dynamic Adversarial Training』

由於設備限制不能進行訓練，因此，我們就著重在模型的推論，使用 API 生成圖像，觀察輸出的品質。

範例. NitroFusion API 測試，程式：**04_NitroFusion_test.ipynb**。

1. 安裝 PyTorch、diffusers，見之前範例說明。

2. 建立模型。

```
from diffusers import LCMScheduler
class TimestepShiftLCMScheduler(LCMScheduler):
    def __init__(self, *args, shifted_timestep=250, **kwargs):
        super().__init__(*args, **kwargs)
        self.register_to_config(shifted_timestep=shifted_timestep)
    def set_timesteps(self, *args, **kwargs):
        super().set_timesteps(*args, **kwargs)
        self.origin_timesteps = self.timesteps.clone()
        self.shifted_timesteps = (self.timesteps * self.config.shifted_timestep
```

第 11 章　擴散模型（Diffusion Model）

```
10                                / self.config.num_train_timesteps).long()
11         self.timesteps = self.shifted_timesteps
12     def step(self, model_output, timestep, sample, generator=None, return_dict=True):
13         if self.step_index is None:
14             self._init_step_index(timestep)
15         self.timesteps = self.origin_timesteps
16         output = super().step(model_output, timestep, sample, generator, return_dict)
17         self.timesteps = self.shifted_timesteps
18         return output
```

3. 載入模型：基礎模型仍為 Stable Diffusion，擴充模型為 NitroFusion（nitrosd-realism_unet.safetensors）。

```
1  import torch
2  from diffusers import DiffusionPipeline, UNet2DConditionModel
3  from huggingface_hub import hf_hub_download
4  from safetensors.torch import load_file
5  import matplotlib.pyplot as plt
6
7  # 載入模型
8  base_model_id = "stabilityai/stable-diffusion-xl-base-1.0"
9  repo = "ChenDY/NitroFusion"
10 ckpt = "nitrosd-realism_unet.safetensors"
11 unet = UNet2DConditionModel.from_config(base_model_id, subfolder="unet").to("cuda", torch.float16)
12 unet.load_state_dict(load_file(hf_hub_download(repo, ckpt), device="cuda"))
13 scheduler = TimestepShiftLCMScheduler.from_pretrained(base_model_id, subfolder="scheduler", shifted_timestep=250)
14 scheduler.config.original_inference_steps = 4
15 # 另一擴充模型(NitroSD-Vibrant)
16 # ckpt = "nitrosd-vibrant_unet.safetensors"
17 # unet = UNet2DConditionModel.from_config(base_model_id, subfolder="unet").to("cuda", torch.float16)
18 # unet.load_state_dict(load_file(hf_hub_download(repo, ckpt), device="cuda"))
19 # scheduler = TimestepShiftLCMScheduler.from_pretrained(base_model_id, subfolder="scheduler", shifted_timestep=500)
20 # scheduler.config.original_inference_steps = 4
21 pipe = DiffusionPipeline.from_pretrained(
22     base_model_id,
23     unet=unet,
24     scheduler=scheduler,
25     torch_dtype=torch.float16,
26     variant="fp16",
27 ).to("cuda")
```

4. 生成圖像：相關參數筆者並未深入研究。

```
1  prompt = "An illustration of Totoro hold an umbrella with the little girl human in a raining night"
2  image = pipe(
3      prompt=prompt,
4      num_inference_steps=1,    # support 1 - 4 inference steps
5      guidance_scale=0,
6  ).images[0]
7
8  plt.imshow(image)
9  plt.axis('off')
10 plt.show()
```

- 執行結果：圖像輸出品質比基礎模型好很多，但仍會產生荒誕內容，必須重複執行，才會得到較佳的圖像，還是值得期待未來的發展。

11-11 OpenAI DALL·E

上述本機安裝的模型都佔很大的記憶體，如下圖工作管理員所示，差點就把筆者的 GPU 記憶體佔滿了。

```
使用率              專屬 GPU 記憶體
2%                 15.7/16.0 GB
GPU 記憶體          共用 GPU 記憶體
18.0/31.9 GB       2.3/15.9 GB
```

因此，還是花點小錢訂閱雲端工具，可能 Total Owner Cost（TOC）較低，接著就來看看 OpenAI 的影像生成工具 DALL·E，目前已發展至 DALL·E 3，它整合至 ChatGPT 使用者介面內，故可以瀏覽 https://chatgpt.com/，註冊並登入後即可使用 DALL·E，如下圖：

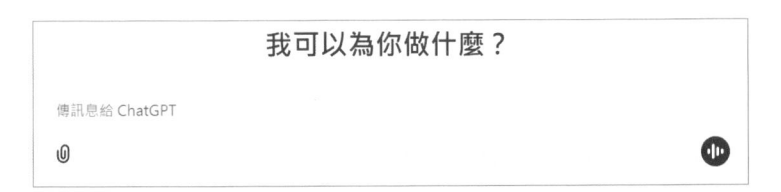

1. 輸入提示【An illustration of Totoro hold an umbrella with the little girl human in a raining night】，按 Enter 鍵。
2. 執行結果：效果太棒了，右上角還有下載的圖示。

第 11 章　擴散模型（Diffusion Model）

3. 也可以以圖生圖，先點選提示框左下角的【📎】，上傳圖檔（lena.jpg），並輸入提示【generate a man with black hair with attached image style】，按 Enter 鍵。

4. 執行結果：如下圖，並顯示說明【Here is the generated image of a man with black hair in a vintage style, inspired by the provided image. Let me know if you'd like any further refinements!】，出現試用限制訊息。

5. 也可以使用程式碼生成圖像，參閱『OpenAI Image generation』文件 [40]，程式碼如下，程式檔案為 05_DALLE_test.ipynb：

```python
from openai import OpenAI
client = OpenAI()

response = client.images.generate(
    model="dall-e-3",
    prompt="a pretty golden-hair mature girl",
    size="1024x1024",
    quality="standard",
    n=1,
)

image_url = response.data[0].url
image_url
```

11-40

11-11 OpenAI DALL·E

6. 執行結果：會出現一長串的超連結（URL），如下。

    'https://oaidalleapiprodscus.blob.core.windows.net/private/org-8b5wOBmK82BeorSFOMxwOTtc/user-pIUWnt8WnVXuiz1Dp90n2CVi/img-cuF1PNYpQAclUqMGv8ZiRNmT.png?st=2024-12-07T12%3A59%3A09Z&se=2024-12-07T14%3A59%3A09Z&sp=r&sv=2024-08-04&sr=b&rscd=inline&rsct=image/png&skoid=d505667d-d6c1-4a0a-bac7-5c84a87759f8&sktid=a48cca56-e6da-484e-a814-9c849652bcb3&skt=2024-12-06T22%3A50%3A44Z&ske=2024-12-07T22%3A50%3A44Z&sks=b&skv=2024-08-04&sig=SWmwluMPs8S/ca3fVQFkVCfkcN7DPlasJO9mUjS3OII%3D'

7. 點選上述超連結（URL），得到圖片如下：

8. Inpainting：準備一個 PNG 格式的檔案，使用繪圖軟體將要修復的部份刪除，存成另一個檔，例如 mask.png。

```
1  from openai import OpenAI
2  client = OpenAI()
3
4  response = client.images.edit((
5      model="dall-e-2",
6      image=open("sunlit_lounge.png", "rb"),
7      mask=open("mask.png", "rb"),
8      prompt="A sunlit indoor lounge area with a pool containing a flamingo",
9      n=1,
10     size="1024x1024"
11 )
12 image_url = response.data[0].url
13 image_url
```

第 11 章　擴散模型（Diffusion Model）

- 執行結果：會產生超連結，點選上述超連結，可取得圖片。左圖為原圖，中間為遮照圖，右圖為修復結果。

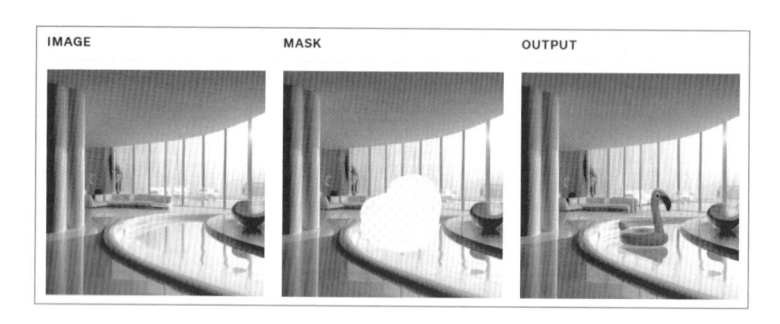

11-12　本章小結

上述生成的圖像均為繪圖風格，目前各家工具已開始發展出真人照片的生成，例如 Flux、MidJourney，除此之外，許多模型還可以生成 3D 模型、視訊（Text to Video）、3D 遊戲、音樂…等，例如 OpenAI Sora、Claude 3.5 Sonnet，視訊品質令人驚豔。影像生成已經廣被設計產業及電玩開發業者使用，已嚴重影響傳統圖庫業者及初階設計師的生計，一般企業需要插圖，花點小錢訂閱工具軟體，就可以省掉相關人員的薪資費用，如果要廣告影片，也可以將靜態圖片，加上提示，就可以產生符合情境描述的影片了 [41]。

▲ 圖 11.15　OpenAI Sora 展示『漫步東京』，圖片來源：『OpenAI Sora』[42]

另外，要注意生成的影像版權，一般而言，生成的影像只有使用權，沒有著作權或所有權，各工具軟體的授權內容有所差異，使用時必須注意授權條約，不要勿觸法律。

參考資料（References）

[1] What Is Diffusion in Chemistry?
(https://www.thoughtco.com/definition-of-diffusion-604430)

[2] 實作理解 Diffusion Model: 來自 DDPM 的簡化概念
(https://medium.com/ai-blog-tw/ 邊實作邊學習 diffusion-model- 從 ddpm 的簡化概念理解 -4c565a1c09c)

[3] Denoising Diffusion Probabilistic Models
(https://arxiv.org/pdf/2006.11239)

[4] Lilian Weng, What are Diffusion Models?
(https://arxiv.org/pdf/2006.11239)

[5] Alec Radford et al., Learning Transferable Visual Models From Natural Language Supervision
(https://arxiv.org/pdf/2006.11239)

[6] Stable Diffusion Wikiperia
(https://en.wikipedia.org/wiki/Stable_Diffusion)

[7] J. Rafid Siddiqui, What are Stable Diffusion Models and Why are they a Step Forward for Image Generation?
(https://towardsdatascience.com/what-are-stable-diffusion-models-and-why-are-they-a-step-forward-for-image-generation-aa1182801d46)

[8] Robin Rombach et al., High-Resolution Image Synthesis with Latent Diffusion Models
(https://arxiv.org/pdf/2112.10752)

[9] Lilian Weng, What are Diffusion Models?
(https://arxiv.org/pdf/2006.11239)

[10] Jonathan Ho et al., Denoising Diffusion Probabilistic Models (DDPM)
(https://arxiv.org/pdf/2006.11239)

[11] Jiaming Song et al., Denoising Diffusion Implicit Models (DDIM)
(https://arxiv.org/pdf/2010.02502)

[12] Alex Nichol/Prafulla Dhariwal, Improved Denoising Diffusion Probabilistic Models
(https://arxiv.org/pdf/2102.09672)

[13] Jonathan Kernes, Diffusion Models
(https://towardsdatascience.com/diffusion-models-91b75430ec2)

[14] SingleZombie/DL-Demo
(https://github.com/SingleZombie/DL-Demos/tree/master)

[15] MidJourney Plan Comparison
(https://docs.midjourney.com/docs/plans)

[16] MidJourney Parameter List
(https://docs.midjourney.com/docs/parameter-list)

[17] Stable Diffusion 維基百科
(https://docs.midjourney.com/docs/parameter-list)

[18] DALL·E: Introducing outpainting
(https://openai.com/index/dall-e-introducing-outpainting/)

[19] ControlNet GitHub
(https://github.com/lllyasviel/ControlNet)

[20] Stable Diffusion prompt: a definitive guide
(https://stable-diffusion-art.com/prompt-guide/)

[21] Hugging Face Sign Up
(https://huggingface.co/)

[22] Hugging Face Sign Up
(https://huggingface.co/)

[23] Hugging Face Stable Diffusion 3 Pipeline
(https://huggingface.co/docs/diffusers/main/en/api/pipelines/stable_diffusion/stable_diffusion_3)

參考資料（References）

[24] Awesome LLM Resources List
(https://github.com/ilsilfverskiold/Awesome-LLM-Resources-List/blob/main/img-gen/README.md#generative-image-tools)

[25] AUTOMATIC1111 Stable Diffusion Web UI
(https://github.com/AUTOMATIC1111/stable-diffusion-webui)

[26] AUTOMATIC1111 Stable Diffusion Web UI 發行網址
(https://github.com/AUTOMATIC1111/stable-diffusion-webui/releases/tag/v1.0.0-pre)

[27] AUTOMATIC1111 Stable Diffusion Web UI wiki
(https://github.com/AUTOMATIC1111/stable-diffusion-webui/wiki)

[28] sd_xl_refiner_1.0_0.9vae 超連結
(https://huggingface.co/stabilityai/stable-diffusion-xl-refiner-1.0/blob/main/sd_xl_refiner_1.0_0.9vae.safetensors)

[29] w-e-w GitHub
(https://gist.github.com/w-e-w/0f37c04c18e14e4ee1482df5c4eb9f53)

[30] Adding Conditional Control to Text-to-Image Diffusion Models
(https://arxiv.org/abs/2302.05543)

[31] sd-webui-controlnet GitHub
(https://github.com/Mikubill/sd-webui-controlnet)

[32] Stable Diffusion 1.5 / 2.0 中型模型
(https://huggingface.co/comfyanonymous/ControlNet-v1-1_fp16_safetensors/tree/main)

[33] lllyasviel/ControlNet GitHub
(https://github.com/lllyasviel/ControlNet)

[34] lllyasviel/ControlNet v1.1 GitHub
(https://github.com/lllyasviel/ControlNet-v1-1-nightly)

[35] HuggingFace lllyasviel/ControlNet-v1-1
(https://huggingface.co/lllyasviel/ControlNet-v1-1/tree/main)

第 11 章　擴散模型（Diffusion Model）

[36] HuggingFace stable-diffusion-v1-5/stable-diffusion-v1-5
(https://huggingface.co/stable-diffusion-v1-5/stable-diffusion-v1-5/blob/main/v1-5-pruned.ckpt)

[37] Ultra fast ControlNet with Diffusers
(https://huggingface.co/blog/controlnet)

[38] NitroFusion, Hugging Face
(https://huggingface.co/ChenDY/NitroFusion)

[39] Dar-Yen Chen et al., NitroFusion: High-Fidelity Single-Step Diffusion through Dynamic Adversarial Training
(https://arxiv.org/pdf/2412.02030)

[40] OpenAI Image generation
(https://platform.openai.com/docs/guides/images)

[41] OpenAI's New Sora Model Is Amazing
(https://medium.com/coding-beauty/new-openai-sora-model-56e399cd259f)

[42] OpenAI Sora
(https://openai.com/index/sora/)

第12章 其他影像應用

除了上述提到的影像應用,還有許多的應用已被企業採用,包括:

1. 臉部辨識(Facial Recognition)。
2. 光學文字辨識(OCR)。
3. 車牌辨識(Automatic Number Plate Recognition, ANPR)。
4. 圖像去背(Background Removing)。

以下我們就逐一來討論這些課題。

12-1 臉部辨識(Facial Recognition)

臉部辨識(Facial Recognition)的應用面向非常廣泛,國內廠商不論是系統廠商、PC廠商、NAS廠商,甚至是電信業者,都已涉獵此一領域,推出各種五花八門的相關產品,已經有以下這些應用類型:

第 12 章　其他影像應用

1. 智慧保全：結合門禁系統，運用在家庭、學校、員工宿舍、飯店、機場登機檢查、出入境比對、黑名單/罪犯/失蹤人口比對…等方面。
2. 考勤系統：上下班臉部刷卡取代卡片。
3. 商店即時監控：即時辨識 VIP 和黑名單客戶的進出，進行客戶關懷、發送折扣碼、或記錄停留時間，作為商品陳列與改善經營效能的參考依據。
4. 快速結帳：以臉部辨識取代刷卡付帳。
5. 人流統計：針對有人數容量限制的公共場所，如百貨公司、遊樂園、體育場館，透過臉部辨識，進行人數控管。
6. 情緒分析：辨識臉部情緒，發生意外時能迅速通報救援，或進行滿意度調查。
7. 社群軟體上傳照片的辨識：標註朋友姓名等。

依據技術類別可細分為：

1. 臉部偵測（Face Detection）：與物件偵測類似，因此運用物件偵測技術即可做到此功能，偵測圖像中有那些臉部和其位置。
2. 臉部特徵點檢測（Facial Landmarks Detection）：偵測臉部的特徵點，用來比對兩張臉是否屬於同一人。
3. 臉部追蹤（Face Tracking）：在視訊中追蹤移動中的臉部，可辨識人移動的軌跡。
4. 臉部辨識（Face Recognition），分為兩種：
   4.1 臉部識別（Face Identification）：從 N 個人中找出最相似的人。
   4.2 臉部驗證（Face Verification）：驗證臉部是否相符，例如，出入境檢查旅客是否與其護照上的大頭照相符合。

各項臉部辨識技術及支援的套件，如下圖：

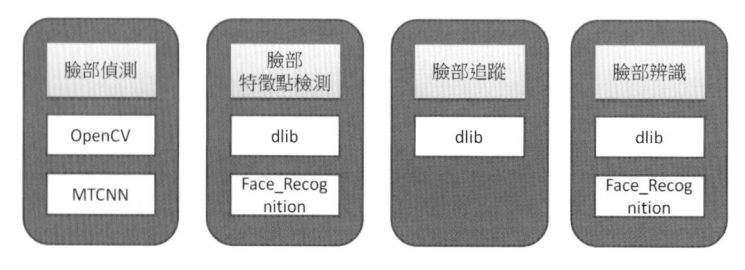

臉部偵測	臉部特徵點檢測	臉部追蹤	臉部辨識
OpenCV	dlib	dlib	dlib
MTCNN	Face_Recognition		Face_Recognition

▲ 圖 12.1　臉部辨識的技術類別與相關套件支援

12-2 臉部偵測（Face Detection）

OpenCV 使用 Haar Cascades 演算法來進行各種物件的偵測，它會將各種物件的特徵記錄在 XML 檔案，稱為『級聯分類器』（Cascade File），可在 OpenCV 安裝資料夾內找到（haarcascade_*.xml），筆者已把相關檔案複製到範例程式資料夾下。

Haar Cascades 技術發展較早，辨識速度快，能夠作到即時偵測，缺點則是準確度較差，容易造成偽陽性（False Positive），即將非臉部的圖像誤認為臉部，它的原理類似卷積，如下圖所示，以各種濾波器（Filters）掃描圖像，像是眼部比臉頰暗，鼻樑比臉頰亮等。

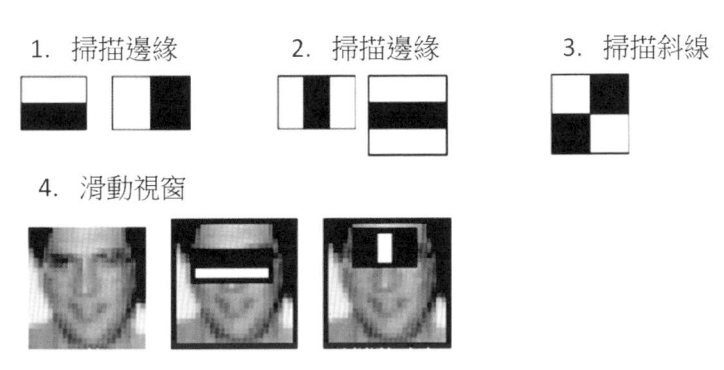

1. 掃描邊緣
2. 掃描邊緣
3. 掃描斜線
4. 滑動視窗

▲ 圖 12.2 Haar Cascades 以濾波器（Filters） 掃描圖像

範例. 使用 OpenCV 進行臉部偵測（Face Detection）。

程式：**12_01_ 臉部偵測 _opencv.ipynb**。

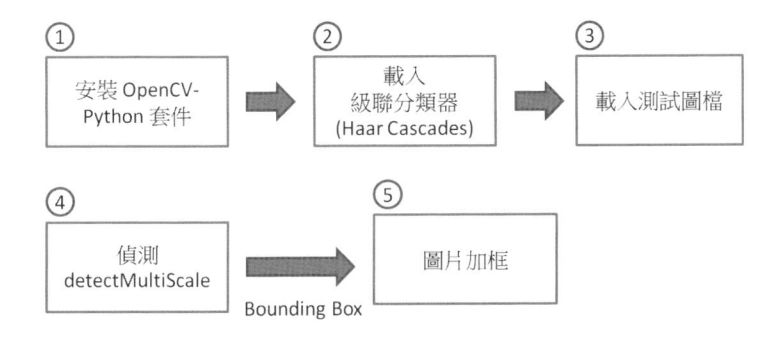

① 安裝 OpenCV-Python 套件 → ② 載入級聯分類器 (Haar Cascades) → ③ 載入測試圖檔

④ 偵測 detectMultiScale → Bounding Box → ⑤ 圖片加框

1. 安裝 OpenCV 套件：

    pip install opencv-python

第 12 章　其他影像應用

2. 載入相關套件。

```
1  # 載入相關套件
2  import cv2
3  from cv2 import CascadeClassifier
4  from cv2 import rectangle
5  import matplotlib.pyplot as plt
6  from cv2 import imread
```

3. 載入臉部的級聯分類器（face cascade file）。

```
1  # 載入臉部級聯分類器(face cascade file)
2  face_cascade = './cascade_files/haarcascade_frontalface_alt.xml'
3  classifier = cv2.CascadeClassifier(face_cascade)
```

4. 載入測試圖檔。

```
1   # 載入圖檔
2   image_file = "./images_face/teammates.jpg"
3   image = imread(image_file)
4
5   # OpenCV 預設為 BGR 色系，轉為 RGB 色系
6   im_rgb = cv2.cvtColor(image, cv2.COLOR_BGR2RGB)
7
8   # 顯示圖像
9   plt.imshow(im_rgb)
10  plt.axis('off')
11  plt.show()
```

- 執行結果：

5. 偵測臉部並顯示圖像。

```
1  # 偵測臉部
2  bboxes = classifier.detectMultiScale(image)
3  # 臉部加框
```

12-4

12-2 臉部偵測（Face Detection）

```
4  for box in bboxes:
5      # 取得框的座標及寬高
6      x, y, width, height = box
7      x2, y2 = x + width, y + height
8      # 加白色框
9      rectangle(im_rgb, (x, y), (x2, y2), (255,255,255), 2)
10
11 # 顯示圖像
12 plt.imshow(im_rgb)
13 plt.axis('off')
14 plt.show()
```

- 執行結果：全部人的臉都有被正確偵測到。

6. 載入另一圖檔。

```
1  # 載入圖檔
2  image_file = "./images_face/classmates.jpg"
3  image = imread(image_file)
4
5  # OpenCV 預設為 BGR 色系，轉為 RGB 色系
6  im_rgb = cv2.cvtColor(image, cv2.COLOR_BGR2RGB)
7
8  # 顯示圖像
9  plt.imshow(im_rgb)
10 plt.axis('off')
11 plt.show()
```

- 執行結果：臉部特寫。

12-5

7. 偵測臉部並顯示圖像。

```
1  # 偵測臉部
2  bboxes = classifier.detectMultiScale(image)
3  # 臉部加框
4  for box in bboxes:
5      # 取得框的座標及寬高
6      x, y, width, height = box
7      x2, y2 = x + width, y + height
8      # 加紅色框
9      rectangle(im_rgb, (x, y), (x2, y2), (255,0,0), 5)
10
11 # 顯示圖像
12 plt.imshow(im_rgb)
13 plt.axis('off')
14 plt.show()
```

- 執行結果：就算圖像中的臉部占據畫面較大，偵測亦可正確偵測。

8. 同時載入眼睛與微笑的級聯分類器。

```
1  # 載入眼睛級聯分類器(eye cascade file)
2  eye_cascade = './cascade_files/haarcascade_eye_tree_eyeglasses.xml'
3  classifier = cv2.CascadeClassifier(eye_cascade)
4
5  # 載入微笑級聯分類器(smile cascade file)
6  smile_cascade = './cascade_files/haarcascade_smile.xml'
7  smile_classifier = cv2.CascadeClassifier(smile_cascade)
```

9. 偵測臉部並顯示圖像。

```
1  # 偵測臉部
2  bboxes = classifier.detectMultiScale(image)
3  # 臉部加框
4  for box in bboxes:
5      # 取得框的座標及寬高
6      x, y, width, height = box
7      x2, y2 = x + width, y + height
```

12-2 臉部偵測（Face Detection）

```
 8        # 加白色框
 9        rectangle(im_rgb, (x, y), (x2, y2), (255,0,0), 5)
10
11  # 偵測微笑
12  # scaleFactor=2.5：掃描時每次縮減掃描視窗的尺寸比例。
13  # minNeighbors=20：每一個被選中的視窗至少要有鄰近且合格的視窗數
14  bboxes = smile_classifier.detectMultiScale(image, 2.5, 20)
15  #微笑加框
16  for box in bboxes:
17      # 取得框的座標及寬高
18      x, y, width, height = box
19      x2, y2 = x + width, y + height
20      # 加白色框
21      rectangle(im_rgb, (x, y), (x2, y2), (255,0,0), 5)
22  #     break
23
24  # 顯示圖像
25  plt.imshow(im_rgb)
26  plt.axis('off')
27  plt.show()
```

- 執行結果：左邊人臉的眼睛少抓了一個，嘴巴誤抓好幾個。這是筆者調整 detectMultiScale 參數多次後，所能得到的較佳結果。

detectMultiScale 相關參數的介紹如下：

1. scaleFactor：設定每次掃描視窗縮小的尺寸比例，設定較小值，會偵測到較多合格的視窗。

2. minNeighbors：每一個被選中的視窗至少要有鄰近且合格的視窗數，設定較大值，會讓偽陽性降低，但會使偽陰性提高。

3. minSize：小於這個設定值，會被過濾掉，格式為（w, h）。

4. maxSize：大於這個設定值，會被過濾掉，格式為（w, h）。

12-3 MTCNN 演算法

Haar Cascades 技術發展較早，使用很簡單，但是要能準確偵測，必須因應圖像的色澤、光線、物件大小來調整參數，並不容易。因此，近幾年發展改用深度學習演算法進行臉部偵測，較知名的演算法 MTCNN 係由 Kaipeng Zhang 等學者於 2016 年『Joint Face Detection and Alignment using Multi-task Cascaded Convolutional Networks』發表[1]。

MTCNN 的架構是運用影像金字塔加上三個神經網路，如下圖所示，四個部分的功能各為：

1. 影像金字塔（Image Pyramid）：擷取不同尺寸的臉部。
2. 建議網路（Proposal Network or P-Net）：類似區域推薦，找出候選的區域。
3. 強化網路（Refine Network or R-Net）：找出合格框（bounding boxes）。
4. 輸出網路（Output Network or O-Net）：找出臉部特徵點（Landmarks）。

乍看下來，會不會覺得有些熟悉？其實 MTCNN 的作法與物件偵測演算法 Faster R-CNN 類似。

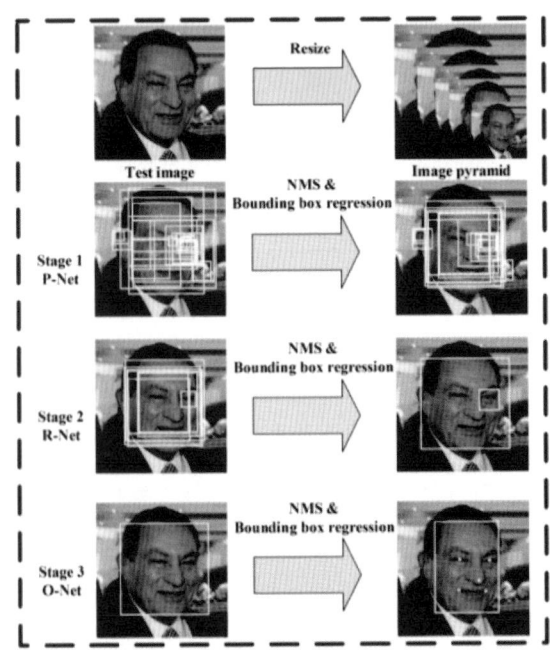

▲ 圖 12.3 MTCNN 使用影像金字塔加上三個神經網路

12-3 MTCNN 演算法

範例. 使用 MTCNN 進行臉部偵測。

程式：**12_02_ 臉部偵測 _mtcnn.ipynb**。

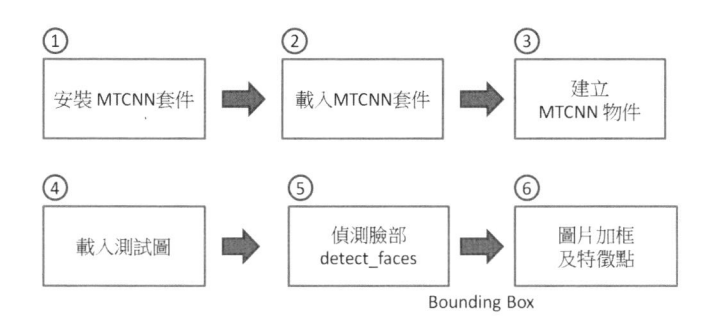

1. 安裝 MTCNN 套件：

   pip install mtcnn

2. 載入相關套件，包含 MTCNN。

```
1  # 安裝套件：pip install mtcnn
2  # 載入相關套件
3  import matplotlib.pyplot as plt
4  from matplotlib.patches import Rectangle, Circle
5  from mtcnn.mtcnn import MTCNN
```

3. 載入並顯示圖檔。

```
1  # 載入圖檔
2  image_file = "./images_face/classmates.jpg"
3  image = plt.imread(image_file)
4
5  # 顯示圖像
6  plt.imshow(image)
7  plt.axis('off')
8  plt.show()
```

- 執行結果：

第 12 章 其他影像應用

4. 建立 MTCNN 物件,偵測臉部。

```
1  # 建立 MTCNN 物件
2  detector = MTCNN()
3
4  # 偵測臉部
5  faces = detector.detect_faces(image)
```

5. 臉部增加框與特徵點,並顯示圖像。

```
1   # 臉部加框
2   ax = plt.gca()
3   for result in faces:
4       # 取得框的座標及寬高
5       x, y, width, height = result['box']
6       # 加紅色框
7       rect = Rectangle((x, y), width, height, fill=False, color='red')
8       ax.add_patch(rect)
9
10      # 特徵點
11      for key, value in result['keypoints'].items():
12          # create and draw dot
13          dot = Circle(value, radius=5, color='green')
14          ax.add_patch(dot)
15
16  # 顯示圖像
17  plt.imshow(image)
18  plt.axis('off')
19  plt.show()
```

- 執行結果:特徵點包括眼睛、鼻子、嘴角,詳細內容請參閱程式。

6. 玩個小技巧,將每張臉個別顯示出來。

```
1  # 臉部加框
2  plt.figure(figsize=(8,6))
3  ax = plt.gca()
4
5  for i, result in enumerate(faces):
```

12-4 臉部追蹤（Face Tracking）

```
 6      # 取得框的座標及寬高
 7      x1, y1, width, height = result['box']
 8      x2, y2 = x1 + width, y1 + height
 9
10      # 顯示圖像
11      plt.subplot(1, len(faces), i+1)
12      plt.axis('off')
13      plt.imshow(image[y1:y2, x1:x2])
14  plt.show()
```

- 執行結果：

12-4 臉部追蹤（Face Tracking）

臉部追蹤（Face Tracking）可在影片中追蹤特定人的臉部，本文使用 Face Recognition 套件，它是以 dlib 為基礎的套件，所以須先安裝 dlib 套件，安裝 dlib 會從 C/C++ 原始碼建置，在 Windows 作業環境下，必須備妥下列工具：

1. Microsoft Visual Studio 2022 或 2019。
2. CMake for Windows：目前 Microsoft Visual Studio 2022 已內含 CMake，安裝時可勾選『CMake』，下圖最後一項。

開啟 Microsoft Visual Studio 選單中的『Developer command prompt』，執行安裝套件的指令，因為他已經將 CMake 加入 Path 路徑中。

第 12 章　其他影像應用

依序安裝 dlib、Face Recognition 套件：

- pip install dlib

- pip install face-recognition

範例 1. 使用 Face Recognition 套件進行臉部偵測。

程式：**12_03_ 臉部偵測 _Face_Recognition.ipynb**。

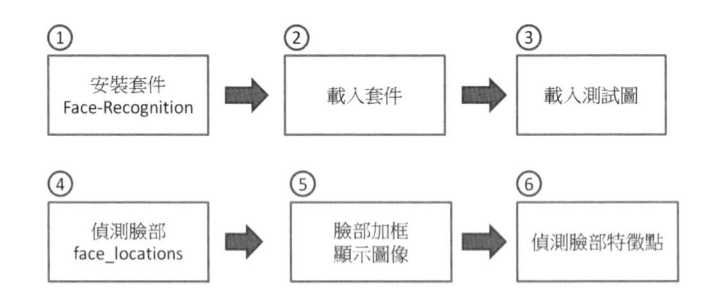

```
①安裝套件 Face-Recognition → ②載入套件 → ③載入測試圖
④偵測臉部 face_locations → ⑤臉部加框 顯示圖像 → ⑥偵測臉部特徵點
```

1. 載入相關套件。

```
1  # 安裝套件: pip install face-recognition
2  # 載入相關套件
3  import matplotlib.pyplot as plt
4  from matplotlib.patches import Rectangle, Circle
5  import face_recognition
```

2. 載入並顯示圖檔。

```
1  # 載入圖檔
2  image_file = "./images_face/classmates.jpg"
3  image = plt.imread(image_file)
4
5  # 顯示圖像
6  plt.imshow(image)
7  plt.axis('off')
8  plt.show()
```

- 執行結果：

12-4 臉部追蹤（Face Tracking）

3. 呼叫 face_locations 函數偵測臉部。

```
1  # 偵測臉部
2  faces = face_recognition.face_locations(image)
```

4. 臉部加框，顯示圖像，注意，框的座標所代表的方向依序為上／左／下／右（逆時鐘）。

```
1   # 臉部加框
2   ax = plt.gca()
3   for result in faces:
4       # 取得框的座標
5       y1, x1, y2, x2 = result
6       width, height = x2 - x1, y2 - y1
7       # 加紅色框
8       rect = Rectangle((x1, y1), width, height, fill=False, color='red')
9       ax.add_patch(rect)
10
11  # 顯示圖像
12  plt.imshow(image)
13  plt.axis('off')
14  plt.show()
```

- 執行結果：

5. 偵測臉部特徵點並顯示。

```
1   # 偵測臉部特徵點並顯示
2   from PIL import Image, ImageDraw
3
4   # 載入圖檔
5   image = face_recognition.load_image_file(image_file)
6
7   # 轉為 Pillow 圖像格式
8   pil_image = Image.fromarray(image)
9
10  # 取得圖像繪圖物件
11  d = ImageDraw.Draw(pil_image)
12
13  # 偵測臉部特徵點
```

12-13

第 12 章　其他影像應用

```
14  face_landmarks_list = face_recognition.face_landmarks(image)
15
16  for face_landmarks in face_landmarks_list:
17      # 顯示五官特徵點
18      for facial_feature in face_landmarks.keys():
19          print(f"{facial_feature} 特徵點: {face_landmarks[facial_feature]}\n")
20
21      # 繪製特徵點
22      for facial_feature in face_landmarks.keys():
23          d.line(face_landmarks[facial_feature], width=5, fill='green')
24
25  # 顯示圖像
26  plt.imshow(pil_image)
27  plt.axis('off')
28  plt.show()
```

- 執行結果如下：

- 五官特徵點的座標。

```
chin 特徵點: [(958, 485), (968, 525), (982, 562), (999, 598), (1022, 630), (1054, 657), (1092, 677), (1135, 693), (1179, 689), (1220, 670), (1249, 639), (1274, 606), (1291, 567), (1298, 524), (1296, 478), (1291, 433), (1283, 387)]

left_eyebrow 特徵點: [(969, 464), (978, 434), (1002, 417), (1032, 413), (1061, 415)]

right_eyebrow 特徵點: [(1119, 397), (1142, 373), (1172, 361), (1204, 364), (1228, 382)]

nose_bridge 特徵點: [(1098, 440), (1107, 477), (1115, 512), (1124, 548)]

nose_tip 特徵點: [(1092, 557), (1112, 562), (1133, 565), (1151, 552), (1167, 538)]

left_eye 特徵點: [(1006, 473), (1019, 458), (1038, 454), (1058, 461), (1042, 467), (1024, 472)]

right_eye 特徵點: [(1147, 436), (1160, 417), (1179, 409), (1201, 414), (1186, 423), (1167, 430)]

top_lip 特徵點: [(1079, 606), (1100, 595), (1121, 586), (1142, 585), (1160, 576), (1186, 570), (1215, 567), (1207, 571), (1164, 585), (1145, 593), (1125, 596), (1088, 605)]

bottom_lip 特徵點: [(1215, 567), (1197, 598), (1176, 619), (1155, 628), (1134, 631), (1109, 626), (1079, 606), (1088, 605), (1128, 612), (1149, 610), (1168, 601), (1207, 571)]
```

- 臉部輪廓畫線。

範例 2. 使用 Face-Recognition 套件進行視訊臉部追蹤。

程式：**12_04_ 臉部追蹤 _Face_Recognition.ipynb**，修改自 face-recognition GitHub 的範例 [2]。

12-4 臉部追蹤（Face Tracking）

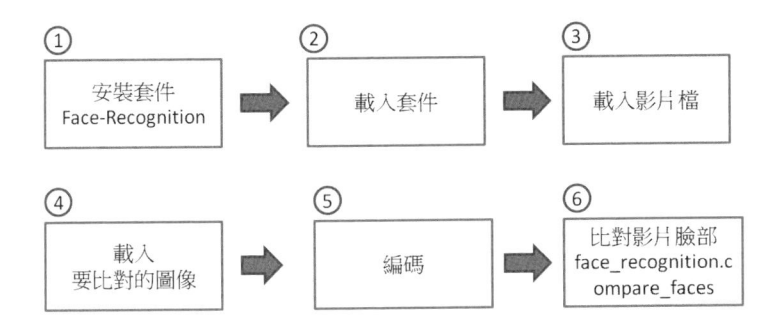

```
①安裝套件 Face-Recognition ➡ ②載入套件 ➡ ③載入影片檔
④載入要比對的圖像 ➡ ⑤編碼 ➡ ⑥比對影片臉部 face_recognition.compare_faces
```

1. 之前曾安裝 opencv-contrib-python，目前只需一般版，先解除安裝 OpenCV 套件，再安裝一般版：

   pip uninstall opencv-python opencv-contrib-python -y

   pip install opencv-python

2. 載入相關套件。

```python
1  # 安裝套件：pip install face-recognition
2  # 載入相關套件
3  import matplotlib.pyplot as plt
4  from matplotlib.patches import Rectangle, Circle
5  import face_recognition
6  import cv2
```

3. 載入影片檔。

```python
1  # 載入影片檔
2  input_movie = cv2.VideoCapture("./images_face/short_hamilton_clip.mp4")
3  length = int(input_movie.get(cv2.CAP_PROP_FRAME_COUNT))
4  print(f'影片幀數：{length}')
```

- 執行結果：影片總幀數為 275。

4. 指定輸出檔名，注意，影片解析度設為（640, 360），故輸入的影片不得低於此解析度，否則輸出檔將會無法播放。

```python
1  # 指定輸出檔名
2  fourcc = cv2.VideoWriter_fourcc(*'XVID')
3  # 每秒幀數(fps):29.97，影片解析度(Frame Size):(640, 360)
4  output_movie = cv2.VideoWriter('./images_face/output.avi',
5                                  fourcc, 29.97, (640, 360))
```

第 12 章　其他影像應用

5. 載入要辨識的圖像,範例設定這兩個人:Lin-Manuel Miranda(美國歌手)與 Barack Obama(美國總統),需先編碼(Encode)為向量,以利臉部比對。

```
1  # 載入要辨識的圖像
2  image_file = 'lin-manuel-miranda.png' # 美國歌手
3  lmm_image = face_recognition.load_image_file("./images_face/"+image_file)
4  # 取得圖像編碼
5  lmm_face_encoding = face_recognition.face_encodings(lmm_image)[0]
6
7  # obama
8  image_file = 'obama.jpg' # 美國總統
9  obama_image = face_recognition.load_image_file("./images_face/"+image_file)
10 # 取得圖像編碼
11 obama_face_encoding = face_recognition.face_encodings(obama_image)[0]
12
13 # 設定陣列
14 known_faces = [
15     lmm_face_encoding,
16     obama_face_encoding
17 ]
18
19 # 目標名稱
20 target_face_names = ['lin-manuel-miranda', 'obama']
```

6. 變數初始化。

```
1  # 變數初始化
2  face_locations = []    # 臉部位置
3  face_encodings = []    # 臉部編碼
4  frame_number = 0       # 幀數
```

7. 比對臉部並存檔。

```
1  # 偵測臉部並寫入輸出檔
2  while(input_movie.isOpened()):
3      # 讀取一幀影像
4      ret, frame = input_movie.read()
5      frame_number += 1
6
7      # 影片播放結束,即跳出迴圈
8      if not ret:
9          break
10         # 會造成記憶體不足
11         # print("Cannot receive frame")   # 如果讀取錯誤,印出訊息
12         # continue
13
14     # 將 BGR 色系轉為 RGB 色系
```

12-16

12-4 臉部追蹤（Face Tracking）

```
15      rgb_frame = cv2.cvtColor(frame, cv2.COLOR_BGR2RGB)
16      # rgb_frame = np.ascontiguousarray(frame[:, :, ::-1])
17
18      # 找出臉部位置
19      face_locations = face_recognition.face_locations(rgb_frame)
20      # 編碼
21      face_encodings = face_recognition.face_encodings(rgb_frame, face_locations)
23      # 比對臉部
24      face_names = []
25      for face_encoding in face_encodings:
26          # 比對臉部編碼是否與圖檔符合
27          match = face_recognition.compare_faces(known_faces, face_encoding,
28                                                 tolerance=0.50)
29
30          # 找出符合臉部的名稱
31          # print(len(match))
32          name = None
33          for i in range(len(match)):
34              if match[i] and i < len(target_face_names):
35                  name = target_face_names[i]
36                  break
37
38          face_names.append(name)
40      # 輸出影片標記臉部位置及名稱
41      for (top, right, bottom, left), name in zip(face_locations, face_names):
42          if not name:
43              continue
44
45          # 加框
46          # print(top, right, bottom, left)
47          cv2.rectangle(frame, (left, top), (right, bottom), (0, 0, 255), 2)
48
49          # 標記名稱
50          cv2.rectangle(frame, (left, bottom - 25), (right, bottom), (0, 0, 255)
51                        , cv2.FILLED)
52          font = cv2.FONT_HERSHEY_DUPLEX
53          cv2.putText(frame, name, (left + 6, bottom - 6), font, 0.5,
54                      (255, 255, 255), 1)
55
56      # 將每一幀影像存檔
57      # print("Writing frame {} / {}".format(frame_number, length))
58      output_movie.write(frame)
60  # 關閉輸入檔
61  input_movie.release()
62  output_movie.release()
63  print('Save ./images_face/output.avi OK')
64  # 關閉所有視窗
65  cv2.destroyAllWindows()
```

第 12 章　其他影像應用

- 執行結果：

    Save ./images_face/output.avi OK

8. 輸出的影片為 images_face/output.avi 檔案：觀看影片後發現，偵測速度較慢，Obama 並未偵測到，因圖片檔是正面照，而影像檔則是側面的畫面。但瑕不掩瑜，大致上仍追蹤得到主要影像的動態。

範例 3. 改用 WebCam 進行臉部即時追蹤。

程式：**12_05_臉部追蹤_webcam.ipynb**，修改自 face-recognition GitHub 的範例[18]。由於步驟重複的部分很多，所以只說明與**範例 2** 有差異的程式碼。

1. 以讀取 WebCam 取代載入影片檔。

```
1  # 指定第一台 webcam
2  video_capture = cv2.VideoCapture(0)
```

2. 讀取 WebCam 一幀影像：第 4 行。

```
1  # 偵測臉部並即時顯示
2  while True:
3      # 讀取一幀影像
4      ret, frame = video_capture.read()
```

3. 偵測臉部的處理均相同，但存檔改成即時顯示。

```
46      # 顯示每一幀影像
47      cv2.imshow('Video', frame)
```

4. 按 q 即可跳出迴圈。

```
49      # 按 q 即跳出迴圈
50      if cv2.waitKey(1) & 0xFF == ord('q'):
51          break
```

原作者還示範一個例子，可在 Raspberry pi 執行，即時進行臉部追蹤。

▌12-5 臉部特徵點偵測

偵測臉部特徵點可使用 Face-Recognition、dlib 或者 OpenCV 套件，以上這三種都可偵測到 68 個特徵點，如下圖：

12-18

12-5 臉部特徵點偵測

▲ 圖 12.4 臉部 68 個特徵點的位置

範例 1. 使用 dlib 實作臉部特徵點的偵測，dlib 函數庫包含機器學習、數值分析、計算機視覺、影像處理…等功能。

程式：**12_06_ 臉部特徵點偵測 .ipynb**。

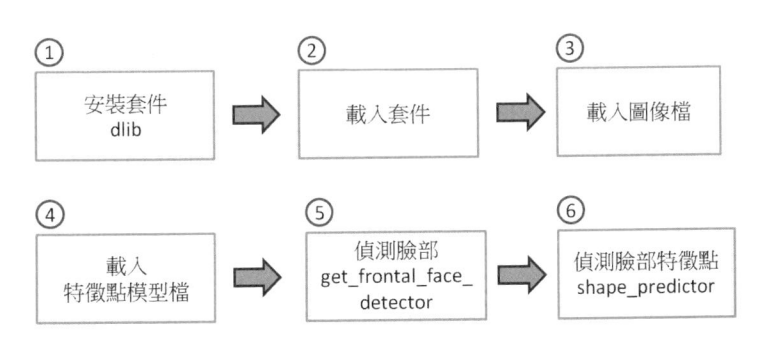

① 安裝套件 dlib → ② 載入套件 → ③ 載入圖像檔

④ 載入特徵點模型檔 → ⑤ 偵測臉部 get_frontal_face_detector → ⑥ 偵測臉部特徵點 shape_predictor

1. 載入套件。

```
1  # 載入相關套件
2  import dlib
3  import cv2
4  import matplotlib.pyplot as plt
5  from matplotlib.patches import Rectangle, Circle
6  from imutils import face_utils
```

第 12 章　其他影像應用

2. 載入並顯示圖檔。

```
1  # 載入圖檔
2  image_file = "./images_face/classmates.jpg"
3  image = plt.imread(image_file)
4
5  # 顯示圖像
6  plt.imshow(image)
7  plt.axis('off')
8  plt.show()
```

- 執行結果：

3. 偵測臉部特徵點並顯示：

- dlib 特徵點模型檔為 shape_predictor_68_face_landmarks.dat，可偵測 68 個點，如果只需要偵測 5 個點就好，可載入 shape_predictor_5_face_landmarks.dat。

- dlib.get_frontal_face_detector：偵測臉部。

- dlib.shape_predictor：偵測臉部特徵點。

```
1   # 載入 dlib 以 HOG 基礎的臉部偵測模型
2   model_file = "shape_predictor_68_face_landmarks.dat"
3   detector = dlib.get_frontal_face_detector()
4   predictor = dlib.shape_predictor(model_file)
5
6   # 偵測圖像的臉部
7   rects = detector(image)
8
9   print(f'偵測到{len(rects)}張臉部.')
10  # 偵測每張臉的特徵點
11  for (i, rect) in enumerate(rects):
12      # 偵測特徵點
13      shape = predictor(image, rect)
14
15      # 轉為 NumPy 陣列
16      shape = face_utils.shape_to_np(shape)
17
18      # 標示特徵點
19      for (x, y) in shape:
```

12-20

```
20        cv2.circle(image, (x, y), 10, (0, 255, 0), -1)
21
22 # 顯示圖像
23 plt.imshow(image)
24 plt.axis('off')
25 plt.show()
```

- 執行結果：

偵測到2張臉部.

4. 偵測視訊檔也沒問題，按 Esc 鍵即可提前結束。

```
1  # 讀取視訊檔
2  cap = cv2.VideoCapture('./images_face/hamilton_clip.mp4')
3  while True:
4      # 讀取一幀影像
5      _, image = cap.read()
6
7      # 偵測圖像的臉部
8      rects = detector(image)
9      for (i, rect) in enumerate(rects):
10         # 偵測特徵點
11         shape = predictor(image, rect)
12         shape = face_utils.shape_to_np(shape)
13
14         # 標示特徵點
15         for (x, y) in shape:
16             cv2.circle(image, (x, y), 2, (0, 255, 0), -1)
17
18     # 顯示影像
19     cv2.imshow("Output", image)
20
21     k = cv2.waitKey(5) & 0xFF    # 按 Esc 跳離迴圈
22     if k == 27:
23         break
24
25 # 關閉輸入檔
26 cap.release()
27 # 關閉所有視窗
28 cv2.destroyAllWindows()
```

第 12 章　其他影像應用

OpenCV 針對臉部特徵點的偵測，提供三種演算法：

1. FacemarkLBF：Shaoqing Ren 等學者於 2014 年發表『Face Alignment at 3000 FPS via Regressing Local Binary Features』所提出的 [3]。

2. FacemarkAAM：Georgios Tzimiropoulos 等學者於 2013 年發表『Optimization problems for fast AAM fitting in-the-wild』所提出的 [4]。

3. FacemarkKamezi：V.Kazemi 和 J. Sullivan 於 2014 年發表『One Millisecond Face Alignment with an Ensemble of Regression Trees』所提出的 [5]。

我們分別實驗一下，看看有什麼差異。

範例 2. 使用 OpenCV 套件進行臉部特徵點偵測，程式為 **12_07_ 臉部特徵點偵測 _ OpenCV.ipynb**。

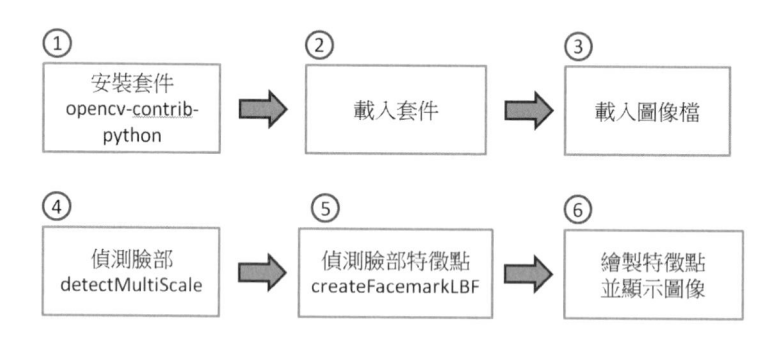

1. 須安裝 OpenCV 擴充版：
    - 解除安裝：pip uninstall opencv-python opencv-contrib-python
    - 安裝套件：pip install opencv-contrib-python

2. 載入套件。

```
1  # 解除安裝套件: pip uninstall opencv-python opencv-contrib-python
2  # 安裝套件:    pip install opencv-contrib-python
3  # 載入相關套件
4  import cv2
5  import numpy as np
6  from matplotlib import pyplot as plt
```

3. 載入並顯示圖檔：使用 Lena.jpg 圖像測試。

```
1  # 載入圖檔
2  image_file = "./images_Object_Detection/lena.jpg"
3  image = cv2.imread(image_file)
```

```
 4
 5  # 顯示圖像
 6  image_RGB = cv2.cvtColor(image, cv2.COLOR_BGR2RGB)
 7  plt.imshow(image_RGB)
 8  plt.axis('off')
 9  plt.show()
```

- 執行結果：

4. 使用 FacemarkLBF 偵測臉部特徵點。

```
 1  # 偵測臉部
 2  cascade = cv2.CascadeClassifier("./cascade_files/haarcascade_frontalface_alt2.xml")
 3  faces = cascade.detectMultiScale(image , 1.5, 5)
 4  print("faces", faces)
 5
 6  # 建立臉部特徵點偵測的物件
 7  facemark = cv2.cv2.face.createFacemarkLBF()
 8  # 訓練模型 lbfmodel.yaml 下載自：
 9  # https://raw.githubusercontent.com/kurnianggoro/GSOC2017/master/data/lbfmodel.yaml
10  facemark .loadModel("lbfmodel.yaml")
11  # 偵測臉部特徵點
12  ok, landmarks1 = facemark.fit(image , faces)
13  print ("landmarks LBF", ok, landmarks1)
```

- 執行結果：顯示臉部和特徵點的座標。

```
faces [[225 205 152 152]]
landmarks LBF True [array([[[201.31314, 268.08807],
        [201.5153 , 293.1106 ],
        [204.91422, 317.07196],
        [210.71988, 340.4278 ],
        [222.97098, 360.37122],
        [240.34521, 375.51422],
        [260.10678, 386.35587],
        [280.64197, 392.04227],
        [298.6573 , 390.89835],
        [311.434  , 384.88406],
        [318.37827, 371.23538],
        [324.82266, 357.113  ],
        [331.87363, 342.1786 ],
        [339.7072 , 327.1501 ],
        [346.04462, 311.9719 ],
        [349.2847 , 296.59448],
        [348.95883, 280.12585],
        [236.43172, 252.06743],
```

5. 繪製特徵點並顯示圖像。

```
1  # 繪製特徵點
2  for p in landmarks1[0][0]:
3      cv2.circle(image, tuple(p), 5, (0, 255, 0), -1)
4
5  # 顯示圖像
6  image_RGB = cv2.cvtColor(image, cv2.COLOR_BGR2RGB)
7  plt.imshow(image_RGB)
8  plt.axis('off')
9  plt.show()
```

- 執行結果:很準確,可惜無法偵測到左上角被帽子遮蔽的部分。

6. 改用 FacemarkAAM 來偵測臉部特徵點。

```
1  # 建立臉部特徵點偵測的物件
2  facemark = cv2.face.createFacemarkAAM()
3  # 訓練模型 aam.xml 下載自:
4  # https://github.com/berak/tt/blob/master/aam.xml
5  facemark.loadModel("aam.xml")
6  # 偵測臉部特徵點
7  ok, landmarks2 = facemark.fit(image , faces)
8  print ("Landmarks AAM", ok, landmarks2)
```

7. 繪製特徵點並顯示圖像:過程與前面的程式碼相同。

```
1  # 繪製特徵點
2  for p in landmarks2[0][0]:
3      cv2.circle(image, tuple(p), 5, (0, 255, 0), -1)
4
5  # 顯示圖像
6  image_RGB = cv2.cvtColor(image, cv2.COLOR_BGR2RGB)
7  plt.imshow(image_RGB)
8  plt.axis('off')
9  plt.show()
```

12-5 臉部特徵點偵測

- 執行結果：左上角反而多出一些錯誤的特徵點。

8. 換用 FacemarkKamezi 偵測臉部特徵點。

```
1  # 建立臉部特徵點偵測的物件
2  facemark = cv2.face.createFacemarkKazemi()
3  # 訓練模型 face_landmark_model.dat 下載自：
4  # https://github.com/opencv/opencv_3rdparty/tree/contrib_face_alignment_20170818
5  facemark.loadModel("face_landmark_model.dat")
6  # 偵測臉部特徵點
7  ok, landmarks2 = facemark.fit(image , faces)
8  print ("Landmarks Kazemi", ok, landmarks2)
```

9. 繪製特徵點並顯示圖像：過程與前面的程式碼相同。

```
1  # 繪製特徵點
2  for p in landmarks2[0][0]:
3      cv2.circle(image, tuple(p), 5, (0, 255, 0), -1)
4
5  # 顯示圖像
6  image_RGB = cv2.cvtColor(image, cv2.COLOR_BGR2RGB)
7  plt.imshow(image_RGB)
8  plt.axis('off')
9  plt.show()
```

- 執行結果：左上角也是多出一些錯誤的特徵點。

第 12 章　其他影像應用

12-6　臉部驗證（Face Verification）

偵測完臉部特徵點後，利用線性代數的法向量比較多張臉的特徵點，就能找出哪一張臉最相似，使用 Face-Recognition 或 dlib 套件都可以。

範例. 使用 Face-Recognition 或 dlib 套件，比對哪一張臉最相似。

程式：**12_08_ 臉部驗證 .ipynb**。

1. 載入相關套件。

```
1  # 載入相關套件
2  import face_recognition
3  import numpy as np
4  from matplotlib import pyplot as plt
```

2. 載入所有要比對的圖檔。

```
1   # 載入圖檔
2   known_image_1 = face_recognition.load_image_file("./images_face/jared_1.jpg")
3   known_image_2 = face_recognition.load_image_file("./images_face/jared_2.jpg")
4   known_image_3 = face_recognition.load_image_file("./images_face/jared_3.jpg")
5   known_image_4 = face_recognition.load_image_file("./images_face/obama.jpg")
6
7   # 標記圖檔名稱
8   names = ["jared_1.jpg", "jared_2.jpg", "jared_3.jpg", "obama.jpg"]
9
10  # 顯示圖像
```

12-26

12-6 臉部驗證（Face Verification）

```
11  unknown_image = face_recognition.load_image_file("./images_face/jared_4.jpg")
12  plt.imshow(unknown_image)
13  plt.axis('off')
14  plt.show()
```

- 執行結果：

3. 圖像編碼：使用 face_recognition.face_encodings 函數編碼。

```
1  # 圖像編碼
2  known_image_1_encoding = face_recognition.face_encodings(known_image_1)[0]
3  known_image_2_encoding = face_recognition.face_encodings(known_image_2)[0]
4  known_image_3_encoding = face_recognition.face_encodings(known_image_3)[0]
5  known_image_4_encoding = face_recognition.face_encodings(known_image_4)[0]
6  known_encodings = [known_image_1_encoding, known_image_2_encoding,
7                     known_image_3_encoding, known_image_4_encoding]
8  unknown_encoding = face_recognition.face_encodings(unknown_image)[0]
```

4. 使用 face_recognition.compare_faces 進行比對。

```
1  # 比對
2  results = face_recognition.compare_faces(known_encodings, unknown_encoding)
3  print(results)
```

- 執行結果：[True, True, True, False]，前三筆符合，完全正確。

5. 載入相關套件：改用 dlib 套件。

```
1  # 載入相關套件
2  import dlib
3  import cv2
4  import numpy as np
5  from matplotlib import pyplot as plt
```

第 12 章　其他影像應用

6. 載入模型：包括特徵點偵測、編碼、臉部偵測。

```
1  # 載入模型
2  pose_predictor_5_point = dlib.shape_predictor("shape_predictor_5_face_landmarks.dat")
3  face_encoder = dlib.face_recognition_model_v1("dlib_face_recognition_resnet_model_v1.dat")
4  detector = dlib.get_frontal_face_detector()
```

7. 定義臉部編碼與比對的函數：由於 dlib 無相關現成的函數，必須自行撰寫。

```
1   # 找出哪一張臉最相似
2   def compare_faces_ordered(encodings, face_names, encoding_to_check):
3       distances = list(np.linalg.norm(encodings - encoding_to_check, axis=1))
4       return zip(*sorted(zip(distances, face_names)))
5
6
7   # 利用線性代數的法向量比較兩張臉的特徵點
8   def compare_faces(encodings, encoding_to_check):
9       return list(np.linalg.norm(encodings - encoding_to_check, axis=1))
10
11  # 圖像編碼
12  def face_encodings(face_image, number_of_times_to_upsample=1, num_jitters=1):
13      # 偵測臉部
14      face_locations = detector(face_image, number_of_times_to_upsample)
15      # 偵測臉部特徵點
16      raw_landmarks = [pose_predictor_5_point(face_image, face_location)
17                       for face_location in face_locations]
18      # 編碼
19      return [np.array(face_encoder.compute_face_descriptor(face_image,
20                       raw_landmark_set, num_jitters)) for
21                       raw_landmark_set in raw_landmarks]
```

8. 載入圖檔並顯示。

```
1   # 載入圖檔
2   known_image_1 = cv2.imread("./images_face/jared_1.jpg")
3   known_image_2 = cv2.imread("./images_face/jared_2.jpg")
4   known_image_3 = cv2.imread("./images_face/jared_3.jpg")
5   known_image_4 = cv2.imread("./images_face/obama.jpg")
6   unknown_image = cv2.imread("./images_face/jared_4.jpg")
7   names = ["jared_1.jpg", "jared_2.jpg", "jared_3.jpg", "obama.jpg"]
8
9   # 轉換 BGR 為 RGB
10  known_image_1 = known_image_1[:, :, ::-1]
11  known_image_2 = known_image_2[:, :, ::-1]
12  known_image_3 = known_image_3[:, :, ::-1]
13  known_image_4 = known_image_4[:, :, ::-1]
14  unknown_image = unknown_image[:, :, ::-1]
```

9. 圖像編碼。

```
1  # 圖像編碼
2  known_image_1_encoding = face_encodings(known_image_1)[0]
3  known_image_2_encoding = face_encodings(known_image_2)[0]
4  known_image_3_encoding = face_encodings(known_image_3)[0]
```

```
5  known_image_4_encoding = face_encodings(known_image_4)[0]
6  known_encodings = [known_image_1_encoding, known_image_2_encoding,
7                     known_image_3_encoding, known_image_4_encoding]
8  unknown_encoding = face_encodings(unknown_image)[0]
```

10. 比對。

```
1  # 比對
2  computed_distances = compare_faces(known_encodings, unknown_encoding)
3  computed_distances_ordered, ordered_names = compare_faces_ordered(known_encodings,
4                                                      names, unknown_encoding)
5  print('比較兩張臉的法向量距離：', computed_distances)
6  print('排序：', computed_distances_ordered)
7  print('依相似度排序：', ordered_names)
```

- 執行結果：顯示兩張臉的法向量距離，數字愈小表示愈相似。

```
比較兩張臉的法向量距離： [0.3998327850880958, 0.4104153798439364, 0.3913189516694114, 0.9053701677487068]
排序： (0.3913189516694114, 0.3998327850880958, 0.4104153798439364, 0.9053701677487068)
依相似度排序： ('jared_3.jpg', 'jared_1.jpg', 'jared_2.jpg', 'obama.jpg')
```

12-7 光學文字辨識（OCR）

光學文字辨識（Optical Character Recognition, OCR）是辨識圖像或文件中的文字，以節省大量的輸入時間或抄寫錯誤，可應用於支票號碼/金額辨識、車牌辨識（Automatic Number Plate Recognition, ANPR）、證件、保單…等，但也有人拿來破解登入用的圖形碼驗證（Captcha），例如搶票工具，我們就來看看如何實作 OCR 辨識。

Tesseract OCR 是目前很盛行的 OCR 軟體，HP 公司於 2005 年開放原始程式碼（Open Source），以 C++ 開發而成的，可由原始程式碼建置，或直接安裝已建置好的程式，在這裡我們採取後者，自 Tesseract OCR GitHub[6] 下載最新版 exe 檔，直接執行即可，安裝時可勾選各國語言資料。安裝完成後，將安裝路徑（例如 C:\Program Files\Tesseract-OCR）放入環境變數 path 內。若要以 Python 呼叫 Tesseract OCR，需額外安裝 pytesseract 套件，指令如下：

- pip install pytesseract

最簡單的測試指令如下：

- tesseract ＜圖檔＞ ＜辨識結果檔＞

例如辨識一張發票檔（./images_ocr/receipt.png）的指令為：

- tesseract ../images_ocr/receipt.png ../images_ocr/result.txt -l eng --psm 6

其中 -l eng：為辨識英文，--psm 6：指單一區塊（a single uniform block of text）。相關的參數請參考 Tesseract OCR 使用說明 [7]。

```
            Welcome to Mel's

Check #: 0001                  12/20/11
Server: Josh F                  4:38 PM
Table: 7/1                    Guests: 2
---------------------------------------
2 Beef Burgr (@9.95/ea)           19.90
   SIDE: Fries
1 Bud Light                        3.79
1 Bud                              4.50
---------------------------------------
Sub-total                         28.19
Sales Tax                          2.50
TOTAL                             30.69

---------------------------------------
Balance Due                       30.69

       Thank you for your patronage!
```

▲ 圖 12.5 發票，

圖片來源：A comprehensive guide to OCR with Tesseract, OpenCV and Python [8]

- 執行結果：幾乎全對，只有特殊符號誤判。

```
Welcome to Mel's
Check #: 0001 12/20/11
Server: Jesh F 4:38 PM
Table: 7/1 Guests: 2
2 Beef Burgr (€9.95/ea) 19,90
SIDE: Fries

1 Bud Light 3.79
1 Bud 4.50
Sub-total 28.19
Sales Tax 2.50
TOTAL 30.69
Balance Due 30.69

Thank you for your patronage! :
```

範例. 以 Python 呼叫 Tesseract OCR API，辨識中、英文。

程式：**12_09_Tesseract_OCR.ipynb**。

1. 載入套件。

```python
# 載入相關套件
import cv2
import pytesseract
import matplotlib.pyplot as plt
```

12-7 光學文字辨識（OCR）

2. 載入並顯示圖檔。

```
1  # 載入圖檔
2  image = cv2.imread('./images_ocr/receipt.png')
3
4  # 顯示圖檔
5  image_RGB = cv2.cvtColor(image, cv2.COLOR_BGR2RGB)
6  plt.figure(figsize=(10,6))
7  plt.imshow(image_RGB)
8  plt.axis('off')
9  plt.show()
```

3. OCR 辨識：呼叫 image_to_string 函數。

```
1  # 參數設定
2  custom_config = r'--psm 6'
3  # OCR 辨識
4  print(pytesseract.image_to_string(image, config=custom_config))
```

- 執行結果：與直接下指令的辨識結果大致相同。

4. 只辨識數字：參數設定加『outputbase digits』。

```
1  # 參數設定，只辨識數字
2  custom_config = r'--psm 6 outputbase digits'
3  # OCR 辨識
4  print(pytesseract.image_to_string(image, config=custom_config))
```

- 執行結果：

```
0001 122011
4338-
71 2
29.95 19.90
1 3.79
1 4.50
- 28.19
2.50
0 30.69
30.69
```

5. 只辨識有限字元。

```
1  # 參數設定白名單，只辨識有限字元
2  custom_config = r'-c tessedit_char_whitelist=abcdefghijklmnopqrstuvwxyz --psm 6'
3  # OCR 辨識
4  print(pytesseract.image_to_string(image, config=custom_config))
```

12-31

第 12 章　其他影像應用

- 執行結果：

```
elcometoels
heck
erverdeshf
able uests
eefurgrea
efries

udight
ud
ubtotal
alesfax
a
alanceue

hankyouforyourpatronage a
```

6. 設定黑名單：只辨識有限字元。

```
1  # 參數設定黑名單，只辨識有限字元
2  custom_config = r'-c tessedit_char_blacklist=abcdefghijklmnopqrstuvwxyz --psm 6'
3  # OCR 辨識
4  print(pytesseract.image_to_string(image, config=custom_config))
```

- 執行結果：

```
W   M]'
C #: 0001 12/20/11
S: J F 4:38 PM
T: 7/1 G: 2
2 B B (€9.95/) 19,90
SIDE: F

1 B L 3.79
1 B 4.50
S-] 28.19
S T 2.50
TOTAL 30.69
BI D 30.69

T    é! '
```

7. 辨識多國文字：先載入並顯示圖檔。

```
1  # 載入圖檔
2  image = cv2.imread('./images_ocr/chinese.png')
3
4  # 顯示圖檔
5  image_RGB = cv2.cvtColor(image, cv2.COLOR_BGR2RGB)
6  plt.figure(figsize=(10,6))
7  plt.imshow(image_RGB)
8  plt.axis('off')
9  plt.show()
```

- 圖檔如下：

  > Tesseract OCR 是目前很盛行的 OCR 軟體，HP 公司於 2005 年開放原始程式碼
  > (Open Source)，以 C++開發而成的，可由原始程式碼建置安裝，或直接安裝已建
  > 置好的程式，我們採取後者，先至 https://github.com/UB-Mannheim/tesseract/wiki
  > 下載最新版 exe 檔，直接執行即可，安裝完成後，將安裝路徑下 bin 子目錄放入
  > 環境變數 Path 內。若要以 Python 呼叫 Tesseract OCR，需額外安裝 pytesseract 套
  > 件，指令如下：。
  > pip install pytesseract。

8. 辨識多國文字。

```
1  # 辨識多國文字，中文繁體、日文及英文
2  custom_config = r'-l chi_tra+jpn+eng --psm 6'
3  # OCR 辨識
4  print(pytesseract.image_to_string(image, config=custom_config))
```

- 執行結果：中文/英文夾雜的辨識效果也不錯，改採新細明體字型，效果也不錯。

- Tesseract 支援的語言如下：

afr (Afrikaans), amh (Amharic), ara (Arabic), asm (Assamese), aze (Azerbaijani), aze_cyrl (Azerbaijani - Cyrilic), bel (Belarusian), ben (Bengali), bod (Tibetan), bos (Bosnian), bre (Breton), bul (Bulgarian), cat (Catalan; Valencian), ceb (Cebuano), ces (Czech), chi_sim (Chinese simplified), chi_tra (Chinese traditional), chr (Cherokee), cos (Corsican), cym (Welsh), dan (Danish), deu (German), div (Dhivehi), dzo (Dzongkha), ell (Greek, Modern, 1453-), eng (English), enm (English, Middle, 1100-1500), epo (Esperanto), equ (Math / equation detection module), est (Estonian), eus (Basque), fas (Persian), fao (Faroese), fil (Filipino), fin (Finnish), fra (French), frk (Frankish), frm (French, Middle, ca.1400-1600), fry (West Frisian), gla (Scottish Gaelic), gle (Irish), glg (Galician), grc (Greek, Ancient, to 1453), guj (Gujarati), hat (Haitian; Haitian Creole), heb (Hebrew), hin (Hindi), hrv (Croatian), hun (Hungarian), hye (Armenian), iku (Inuktitut), ind (Indonesian), isl (Icelandic), ita (Italian), ita_old (Italian - Old), jav (Javanese), jpn (Japanese), kan (Kannada), kat (Georgian), kat_old (Georgian - Old), kaz (Kazakh), khm (Central Khmer), kir (Kirghiz; Kyrgyz), kmr (Kurdish Kurmanji), kor (Korean), kor_vert (Korean vertical), lao (Lao), lat (Latin), lav (Latvian), lit (Lithuanian), ltz (Luxembourgish), mal (Malayalam), mar (Marathi), mkd (Macedonian), mlt (Maltese), mon (Mongolian), mri (Maori), msa (Malay), mya (Burmese), nep (Nepali), nld (Dutch; Flemish), nor (Norwegian), oci (Occitan post 1500), ori (Oriya), osd (Orientation and script detection module), pan (Panjabi; Punjabi), pol (Polish), por (Portuguese), pus (Pushto; Pashto), que (Quechua), ron (Romanian; Moldavian; Moldovan), rus (Russian), san (Sanskrit), sin (Sinhala; Sinhalese), slk (Slovak), slv (Slovenian), snd (Sindhi), spa (Spanish; Castilian), spa_old (Spanish; Castilian - Old), sqi (Albanian), srp (Serbian), srp_latn (Serbian - Latin), sun (Sundanese), swa (Swahili), swe (Swedish), syr (Syriac), tam (Tamil), tat (Tatar), tel (Telugu), tgk (Tajik), tha (Thai), tir (Tigrinya), ton (Tonga), tur (Turkish), uig (Uighur; Uyghur), ukr (Ukrainian), urd (Urdu), uzb (Uzbek), uzb_cyrl (Uzbek - Cyrilic), vie (Vietnamese), yid (Yiddish), yor (Yoruba)

▲ 圖 12.6 Tesseract 支援的語言，圖片來源：Tesseract 官網的語言列表 [9]

12-8 EasyOCR

Tesseract OCR 辨識效果雖然不錯，但是對於多段文字區隔的內容會有缺失，例如表格、證件、廣告…等，每一區域的辨識結果應獨立顯示，另外，字體大小不一、多種色彩也是問題，以下為 Tesseract OCR 辨識的結果，圖像均來自 EasyOCR GitHub[10]。

範例 1. Tesseract OCR 辨識測試。

程式：**12_09_Tesseract_OCR.ipynb** 下半部。

第 12 章　其他影像應用

1. 辨識路牌，含簡體中文及英文。

```
1  # 載入圖檔
2  image = cv2.imread('../images_ocr/sim_chinese.jpg')
3  # 顯示圖檔
4  image_RGB = cv2.cvtColor(image, cv2.COLOR_BGR2RGB)
5  plt.figure(figsize=(10,6))
6  plt.imshow(image_RGB)
7  plt.axis('off')
8  plt.show()
9
10 # 辨識多國文字，中文、日文及英文
11 custom_config = r'-l chi_sim+eng --psm 6'
12 # OCR 辨識
13 print(pytesseract.image_to_string(image, config=custom_config))
```

- 執行結果：左方為原圖，右方為辨識結果，效果不佳，而且【西 315】、【東 309】應合併顯示。

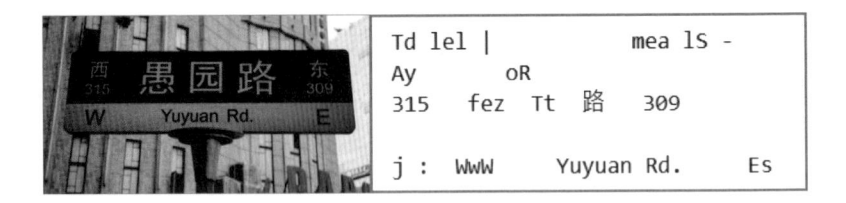

```
Td lel |                mea ls -
Ay           oR
315      fez    Tt    路    309

j :    WwW       Yuyuan Rd.       Es
```

2. 辨識日文告示牌，含日文及英文。

```
1  # 載入圖檔
2  image = cv2.imread('../images_ocr/japanese.jpg')
3  # 顯示圖檔
4  image_RGB = cv2.cvtColor(image, cv2.COLOR_BGR2RGB)
5  plt.figure(figsize=(10,6))
6  plt.imshow(image_RGB)
7  plt.axis('off')
8  plt.show()
9
10 # 辨識日文
11 custom_config = r'-l chi_sim+jpn+eng'
12 # OCR 辨識
13 print(pytesseract.image_to_string(image, config=custom_config))
```

- 執行結果：下方為原圖，辨識結果為【清潔できれいな港区を】，只辨識出一種顏色的文字。

12-34

12-8 EasyOCR

EasyOCR 使用神經網路訓練出來的模型可解決上述的問題,並提供額外的資訊,包括位置、可信度,讓使用者自行決定要如何拼湊辨識結果。

範例 2. EasyOCR 辨識測試。

程式:**12_10_EasyOCR_test.ipynb**。

1. 需安裝 PyTorch、EasyOCR 套件,後者指令如下:

    pip install easyocr

2. 載入套件。

```
1  import easyocr
2  import cv2
3  import matplotlib.pyplot as plt
```

3. 定義顯示圖片函數。

```
2  def show_image(image_path):
3      # 載入圖檔
4      image = cv2.imread(image_path)
5      # 顯示圖檔
6      image_RGB = cv2.cvtColor(image, cv2.COLOR_BGR2RGB)
7      plt.figure(figsize=(10,6))
8      plt.imshow(image_RGB)
9      plt.axis('off')
10     plt.show()
```

4. 辨識路牌,含簡體中文及英文。

```
1  # 載入語系
2  reader = easyocr.Reader(['ch_sim','en'])
3
4  # 顯示圖片
5  image_path = '../images_ocr/sim_chinese.jpg'
```

12-35

第 12 章　其他影像應用

```
6  show_image(image_path)
7  # 辨識
8  result = reader.readtext(image_path)
9  result
```

- 執行結果：下方分別為原圖及辨識結果，每一區段含 3 項資訊，位置（上下左右的座標）、辨識文字及可信度。

```
[([[224, 55], [461, 55], [461, 122], [224, 122]], '高鐵左營站', 0.8849726511608671),
 ([[228, 107], [317, 107], [317, 147], [228, 147]], 'HSR', 0.998867080380785),
 ([[330, 112], [465, 112], [465, 152], [330, 152]],
  'Station',
  0.9995088549573461),
 ([[154, 151], [471, 151], [471, 225], [154, 225]],
  '汽車臨停接送區',
  0.9991856250451403),
 ([[163, 209], [229, 209], [229, 245], [163, 245]], 'Car', 0.9998338570549938),
 ([[240, 210], [387, 210], [387, 249], [240, 249]],
  'Kiss and',
  0.8644983962391248),
 ([[397, 215], [473, 215], [473, 251], [397, 251]],
  'Ride',
  0.8671890373592814)]
```

5. 若只要顯示辨識文字，可加參數。

```
2  reader.readtext(image_path, detail = 0)
```

- 執行結果：[' 西 ', ' 愚園路 ', ' 東 ', '315', '309', 'W', 'Yuyuan Rd.', 'E']。

6. 辨識路牌，含簡體中文及英文。

```
1  # 載入語系
2  reader = easyocr.Reader(['ch_sim','en'])
3
4  # 顯示圖片
5  image_path = '../images_ocr/sim_chinese.jpg'
```

12-36

12-8 EasyOCR

```
6  show_image(image_path)
7  # 辨識
8  result = reader.readtext(image_path)
9  result
```

- 執行結果：下方分別為原圖及辨識結果。

```
[([[86, 80], [134, 80], [134, 128], [86, 128]], '西', 0.8112799833398299),
 ([[189, 75], [469, 75], [469, 165], [189, 165]], '愚园路', 0.9578828861756024),
 ([[517, 81], [565, 81], [565, 123], [517, 123]], '东', 0.994182132573517),
 ([[78, 126], [136, 126], [136, 156], [78, 156]], '315', 0.9999933927534448),
 ([[514, 126], [574, 126], [574, 156], [514, 156]], '309', 0.9999630407897285),
 ([[79, 173], [125, 173], [125, 213], [79, 213]], 'W', 0.3244718485863274),
 ([[226, 170], [414, 170], [414, 220], [226, 220]],
  'Yuyuan Rd。',
  0.859484846374734),
 ([[529, 173], [569, 173], [569, 213], [529, 213]], 'E', 0.5590796767109083)]
```

7. 辨識日文告示牌，含日文及英文。

```
1  # 載入語系
2  reader = easyocr.Reader(['ja','en'])
3
4  # 顯示圖片
5  image_path = '../images_ocr/japanese.jpg'
6  show_image(image_path)
7  # 辨識
8  result = reader.readtext(image_path)
9  result
```

- 執行結果：下方分別為原圖及辨識結果。

第 12 章　其他影像應用

```
[([[71, 49], [489, 49], [489, 159], [71, 159]], 'ポイ橘て禁止』', 0.4121011430607613),
 ([[95, 149], [461, 149], [461, 235], [95, 235]],
  'NOLTTB',
  0.47329141767693006),
 ([[80, 232], [475, 232], [475, 288], [80, 288]],
  '清潔できれいな港区を',
  0.6718399295133427),
 ([[109, 289], [437, 289], [437, 333], [109, 333]],
  '潜 区 MNATOCITY',
  0.7224698964668003)]
```

8. 身份證正面辨識：使用自訂的圖像測試，身份證來自內政部的樣本。

```
1  reader = easyocr.Reader(['ch_tra','en'])
2  image_path = '../images_ocr/ID_1.jpg'
3  show_image(image_path)
4  result = reader.readtext(image_path)
5  result
```

- 執行結果：下方分別為原圖及辨識結果，效果似乎不佳，但筆者使用正式的身份證，辨識結果非常棒。

```
[([[56, 24], [128, 24], [128, 50], [56, 50]], '中輒國民', 0.000599920516833663),
 ([[52, 58], [88, 58], [88, 90], [52, 90]], '樣', 0.999994277962287),
 ([[122, 58], [162, 58], [162, 90], [122, 90]], '本', 0.9993911716451862),
 ([[45, 81], [71, 81], [71, 101], [45, 101]], '陳', 0.993427620869852),
 ([[80, 78], [142, 78], [142, 102], [80, 102]], '筱 玲', 0.4553695580166064),
 ([[13, 115], [165, 115], [165, 135], [13, 135]],
  '年舶民圍 57年6月5日',
  0.14864740190147943),
 ([[199, 117], [241, 117], [241, 131], [199, 131]],
  '仕%!',
  0.005414885728442887),
 ([[17, 141], [175, 141], [175, 157], [17, 157]],
  'f_15 S4年-74 1日titi#5',
  0.009994214551325908),
 ([[176, 135], [262, 135], [262, 166], [176, 166]],
  'A234567890',
  0.7338360591182215)]
```

9. 身份證反面辨識。

```
1  reader = easyocr.Reader(['ch_tra','en'])
2  image_path = '../images_ocr/ID_2.jpg'
3  show_image(image_path)
4  result = reader.readtext(image_path)
5  result
```

- 執行結果：使用正式的身份證效果會非常好。

EasyOCR 支援 80 多種語言，也可以結合 Gradio 製作展示網頁，模型架構如下：

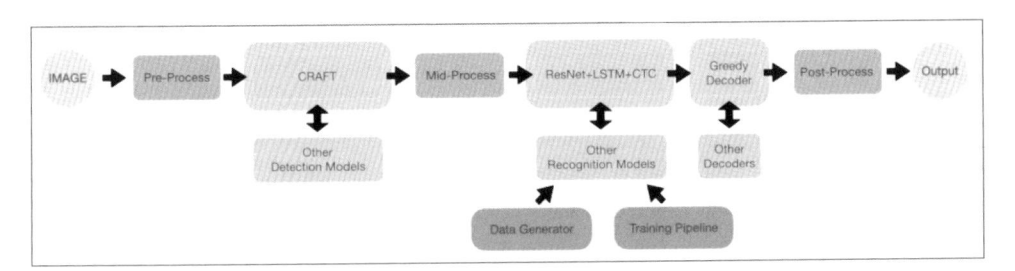

主要的文字辨識演算法為 CRAFT（Character Region Awareness for Text Detection），可參閱論文[11]。另外還有 keras-ocr 套件[12]，採用 TensorFlow 開發，也值得一試。最近有大語言模型 Llama 3.2-Vision 也支援 OCR 功能，可參閱 imanoop7/Ollama-OCR GitHub[13]。

12-9 車牌辨識（ANPR）

車牌辨識（Automatic Number Plate Recognition, ANPR）系統已行之有年了，早期用像素逐點辨識，或將數字細線化後，再比對線條，但最近幾年已改採深度學習進行辨識，可應用到許多場域：

1. 機車檢驗：檢驗單位的電腦會安裝攝影鏡頭進行車牌辨識。
2. 停車場：當車輛進場時，系統會先辨識車牌並儲存，出場時會再辨識車牌比對，檢查是否已繳款。

範例. 以 OpenCV 及 Tesseract OCR 進行車牌辨識。

程式：**12_11_ 車牌辨識 .ipynb**，修改自『Car License Plate Recognition using Raspberry Pi and OpenCV』[14]。

第 12 章　其他影像應用

```
①                        ②                        ③
安裝套件                  載入套件                  載入圖像檔
Tesseract OCR
pytesseract

④                        ⑤                        ⑥
轉為灰階                  取得等高線區域            轉為近似多邊形
萃取輪廓                                          尋找四邊形

⑦                        ⑧                        ⑨
找出車牌的上下            仿射                      車牌號碼
左右的座標                                          OCR 辨識
```

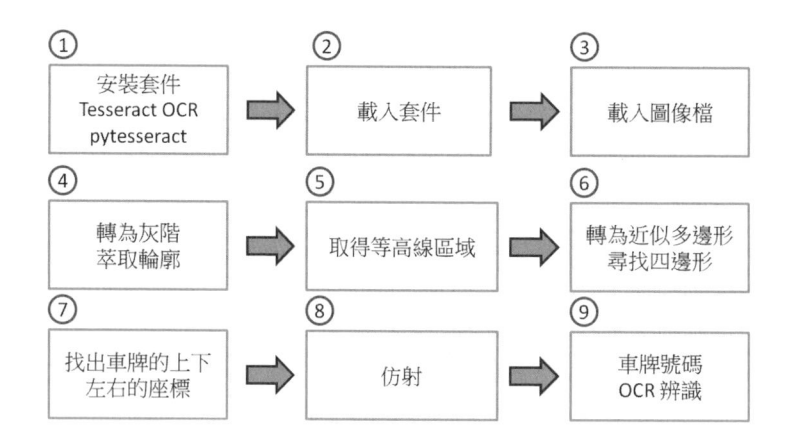

1. 載入套件。

```
1  # 載入相關套件
2  import cv2
3  import imutils
4  import numpy as np
5  import matplotlib.pyplot as plt
6  import pytesseract
7  from PIL import Image
```

2. 載入並顯示圖檔。

```
1  # 載入圖檔
2  image = cv2.imread('./images_ocr/2.jpg',cv2.IMREAD_COLOR)
3
4  # 顯示圖檔
5  image_RGB = cv2.cvtColor(image, cv2.COLOR_BGR2RGB)
6  plt.imshow(image_RGB)
7  plt.axis('off')
8  plt.show()
```

- 執行結果：此測試圖來自原程式。

12-9 車牌辨識（ANPR）

3. 先轉為灰階，會比較容易辨識，再萃取輪廓。

```python
# 萃取輪廓
gray = cv2.cvtColor(image, cv2.COLOR_BGR2GRAY)  # 轉為灰階
gray = cv2.bilateralFilter(gray, 11, 17, 17)    # 模糊化，去除雜訊
edged = cv2.Canny(gray, 30, 200)                # 萃取輪廓

# 顯示圖檔
plt.imshow(edged, cmap='gray')
plt.axis('off')
plt.show()
```

- 執行結果：

4. 取得等高線區域，並排序，取前 10 個區域。

```python
# 取得等高線區域，並排序，取前10個區域
cnts = cv2.findContours(edged.copy(), cv2.RETR_TREE, cv2.CHAIN_APPROX_SIMPLE)
cnts = imutils.grab_contours(cnts)
cnts = sorted(cnts, key = cv2.contourArea, reverse = True)[:10]
```

5. 找第一個含四個點的等高線區域：將等高線區域轉為近似多邊形，接著尋找四邊形的等高線區域。

```python
# 找第一個含四個點的等高線區域
screenCnt = None
for i, c in enumerate(cnts):
    # 計算等高線區域周長
    peri = cv2.arcLength(c, True)
    # 轉為近似多邊形
    approx = cv2.approxPolyDP(c, 0.018 * peri, True)
    # 等高線區域維度
    print(c.shape)
    
    # 找第一個含四個點的多邊形
    if len(approx) == 4:
        screenCnt = approx
        print(i)
        break
```

6. 在原圖上繪製多邊形，框住車牌。

```
1  # 在原圖上繪製多邊形，即車牌
2  if screenCnt is None:
3      detected = 0
4      print("No contour detected")
5  else:
6      detected = 1
7
8  if detected == 1:
9      cv2.drawContours(image, [screenCnt], -1, (0, 255, 0), 3)
10     print(f'車牌座標=\n{screenCnt}')
```

7. 去除車牌以外的圖像，找出車牌的上下左右的座標，計算車牌寬高。

```
1  # 去除車牌以外的圖像
2  mask = np.zeros(gray.shape,np.uint8)
3  new_image = cv2.drawContours(mask,[screenCnt],0,255,-1,)
4  new_image = cv2.bitwise_and(image, image, mask=mask)
5
6  # 轉為浮點數
7  src_pts = np.array(screenCnt, dtype=np.float32)
8
9  # 找出車牌的上下左右的座標
10 left = min([x[0][0] for x in src_pts])
11 right = max([x[0][0] for x in src_pts])
12 top = min([x[0][1] for x in src_pts])
13 bottom = max([x[0][1] for x in src_pts])
14
15 # 計算車牌寬高
16 width = right - left
17 height = bottom - top
18 print(f'寬度={width}, 高度={height}')
```

8. 仿射（affine transformation），將車牌轉為矩形：仿射可將偏斜的梯形轉為矩形，筆者發現等高線區域的各點座標都是以**逆時針排列**，因此，當要找出第一點座標在哪個方向時，通常它會位在上方或左方，所以不需考慮右下角。

```
1  # 計算仿射(affine transformation)的目標區域座標，須與擷取的等高線區域座標順序相同
2  if src_pts[0][0][0] > src_pts[1][0][0] and src_pts[0][0][1] < src_pts[3][0][1]:
3      print('起始點為右上角')
4      dst_pts = np.array([[width, 0], [0, 0], [0, height], [width, height]], dtype=np.float32)
5  elif src_pts[0][0][0] < src_pts[1][0][0] and src_pts[0][0][1] > src_pts[3][0][1]:
6      print('起始點為左下角')
7      dst_pts = np.array([[0, height], [width, height], [width, 0], [0, 0]], dtype=np.float32)
8  else:
9      print('起始點為左上角')
10     dst_pts = np.array([[0, 0], [0, height], [width, height], [width, 0]], dtype=np.float32)
11
12 # 仿射
13 M = cv2.getPerspectiveTransform(src_pts, dst_pts)
14 Cropped = cv2.warpPerspective(gray, M, (int(width), int(height)))
```

12-9 車牌辨識（ANPR）

9. 車牌號碼 OCR 辨識。

```
1  # 車牌號碼 OCR 辨識
2  text = pytesseract.image_to_string(Cropped, config='--psm 11')
3  print("車牌號碼：",text)
```

- 執行結果： /HR.26 BR 9044，有誤認一些非字母或數字的符號，可直接去除，這樣車牌辨識的結果就完全正確。

10. 顯示原圖和車牌。

```
1   # 顯示原圖及車牌
2   cv2.imshow('Orignal image',image)
3   cv2.imshow('Cropped image',Cropped)
4
5   # 車牌存檔
6   cv2.imwrite('Cropped.jpg', Cropped)
7
8   # 按 Enter 鍵結束
9   cv2.waitKey(0)
10
11  # 關閉所有視窗
12  cv2.destroyAllWindows()
```

- 執行結果：

- 車牌：

11. 再使用 images_ocr/1.jpg 測試，車牌為 NAX-6683，辨識為 NAY-6683，X 誤認為 Y，有可能是台灣車牌的字型不同，可使用台灣車牌字型供 Tesseract OCR 使用。

第 12 章　其他影像應用

- 車牌：

另外，筆者實驗發現，若鏡頭拉遠或拉近，而造成車牌過大或過小的話，都有可能辨識錯誤，所以，實際進行時，鏡頭最好與車牌距離能固定，會比較容易辨識。假如圖像的畫面太雜亂，取到的車牌區域也有可能是錯的，而這問題相對容易處理，當 OCR 辨識不到字或者字數不足時，就再找其他的等高線區域，即可解決。

從這個範例可以得知，通常一個實際的案例，並不會像內建的資料集一樣，可以直接套用，常常都需要進行前置處理，如灰階化、萃取輪廓、找等高線區域、仿射等等，資料清理（Data Clean）完才可餵入演算法加以訓練，而且為了適應環境變化，這些工作還必須反覆進行。所以有人統計，光收集資料、整理資料、特徵工程等工作就佔專案 85% 的時間，只有把最 boring、dirty 的工作處理好，才是專案成功的關鍵因素，這與參加 Kaggle 競賽是截然不同的感受，魔鬼總是藏在細節裡。

12-10 影像去背（Background Removing）

影像去背（Background Removing）是 PhotoShop、Canva…等影像處理軟體必備的功能，用途包括：

1. 產品資料：通常我們會希望擁有一張乾淨背景的產品照片可合成到任何背景上。
2. 個人照片：個人簡介或證照附上的照片。
3. 網頁設計、明信片。
4. 廣告素材。

12-10 影像去背（Background Removing）

5. 遊戲及 App 的人物造型。
6. 教育訓練素材。

本節要介紹 Rembg 工具[15]，它是以深度學習訓練的模型，主要的演算法是 U^2-Net，屬於物件偵測演算法 U-Net 變形，可詳閱『U^2-Net: Going Deeper with Nested U-Structure for Salient Object Detection』[16] 及 GitHub[17]。

CPU 版本安裝程序：

1. 安裝 ONNX Runtime：ONNX 是模型的共通格式，TensorFlow、PyTorch 模型均可匯出此一格式。

   pip install onnxruntime

2. 安裝 Rembg：

   pip install "rembg[cli]"

GPU 版本安裝程序：

1. 安裝 CUDA 及 TensorRT：TensorRT 是 Nvidia 高效能的深度學習推論平台（a platform for high performance deep learning inference），必須先安裝 CUDA toolkit，可參考『CUDA Toolkit Downloads』[18]，如使用深度學習，通常也要安裝 CuDNN，可參考『CuDNN Downloads』[19]，之後再下載 TensorRT 壓縮檔，並解壓縮，之後執行 python 資料夾下的 whl 檔案。

   pip install tensorrt-10.7.0-cp312-none-win_amd64.whl

   ** 注意，須隨環境不同選用適合的 whl 檔名。

2. 設定 Path 路徑：包括 CUDA toolkit 的 bin 資料夾、CuDNN 的 bin\ 版本資料夾及 Tensor-RT 的 lib 資料夾，如下圖。

3. 安裝 GPU 版的 ONNX Runtime：

   pip install onnxruntime-gpu

4. 安裝 GPU 版的 Rembg：

   pip install "rembg[gpu,cli]"

CPU 及 GPU 版的差別在於去背的速度，不過，GPU 安裝程序非常複雜，除了上述的安裝程序，筆者還嘗試多種組合：

1. 只安裝 CuDNN、CUDA，不安裝 TensorRT。
2. 安裝 CuDNN、CUDA、TensorRT，但安裝 CPU 版的 ONNX Runtime、Rembg。

結果發現最後一種去背速度最快，但無法證明是否使用了 GPU，這部分要請讀者針對本身的硬體進行測試。測試的過程不禁讓筆者要埋怨 Nvidia 的文件要寫得如此複雜，為什麼不寫個簡單程式判別 GPU 顯卡型號，自動安裝最適合的版本，目前使用者還要手動選擇版本，並設定 Path 路徑，一步錯則萬劫不復。

使用程序有 2 種：

1. 使用指令去背。
2. 撰寫 python 程式。

最簡單的指令：

- rembg i <input file name> <output file name>

1. 測試 Rembg GitHub 範例檔：在 src\12 資料夾下執行以下指令。

   rembg i ../images_rembg/animal-1.jpg 1.jpg

   下方左圖為原圖，右圖為去背結果，幾近完美，images_rembg 資料夾下還有其他檔案，讀者可以試試看。

12-10 影像去背（Background Removing）

2. 測試其他檔案：

   rembg i ..\images_face\jared_1.jpg 2.jpg

   下方左圖為原圖，右圖為去背結果，衣服處理有瑕疵，Rembg 應該是針對主物件進行處理，其他背景效果會較不理想。

3. 測試兩個人的檔案：

   rembg i ..\images_face\classmates.jpg 3.jpg

   下方左圖為原圖，右圖為去背結果，也很理想。

4. 也可以一次處理多個檔案：

   rembg i <input folder> <output folder>

範例．撰寫 python 程式，呼叫 Rembg API。

程式：12_12_Rembg_test.ipynb。

1. 載入套件。

```
1  from rembg import remove
2  from PIL import Image
```

第 12 章　其他影像應用

2. 載入並顯示圖檔。

```
# 載入圖檔
input_path = r'..\images_face\jared_1.jpg'
input = Image.open(input_path)
# 顯示圖檔
input.show()
```

3. 去背：呼叫 remove 即可。

```
output_path = 'output.png'
output = remove(input)
output.show()
output.save(output_path)
```

另外，Rembg 底層支援各種的 U^2-Net 模型[20]，可以生成各種效果，例如下圖的素描、模糊、透明、遮罩…等效果，API 使用可參考 U^2-Net GitHub[21] 的範例程式。

12-11　本章小結

這一章我們介紹許多影像的應用，傳統影像處理的演算法已逐步被深度學習演算法取代，目前後者效果可能還不如前者，但是，它代表的意義是我們可以使用大量的資料提升準確率，若加上推理（Reasoning），終有一日深度學習演算法會超越傳統演算法。

影像應用不勝枚舉，更多深度學習的範例可參見『Keras Code Examples』[22]。

參考資料（References）

[1] Kaipeng Zhang、Zhanpeng Zhang、Zhifeng Li、Yu Qiao,《Joint Face Detection and Alignment using Multi-task Cascaded Convolutional Networks》, 2016
(https://arxiv.org/abs/1604.02878)

[2] Face Recognition GitHub 的範例
(https://github.com/ageitgey/face_recognition)

[3] Shaoqing Ren、Xudong Cao、Yichen Wei 等人,《Face Alignment at 3000 FPS via Regressing Local Binary Features》, 2014
(http://www.jiansun.org/papers/CVPR14_FaceAlignment.pdf)

[4] Georgios Tzimiropoulos、Maja Pantic,《Optimization problems for fast AAM fitting in-the-wild》, 2013
(https://ibug.doc.ic.ac.uk/media/uploads/documents/tzimiro_pantic_iccv2013.pdf)

[5] V.Kazemi、J. Sullivan,《One Millisecond Face Alignment with an Ensemble of Regression Trees》, 2014
(http://www.csc.kth.se/~vahidk/face_ert.html)

[6] Tesseract OCR GitHub
(https://github.com/UB-Mannheim/tesseract/wiki)

[7] Tesseract OCR 使用說明
(https://github.com/tesseract-ocr/tesseract/blob/master/doc/tesseract.1.asc)

[8] Filip Zelic、Anuj Sable,《A comprehensive guide to OCR with Tesseract, OpenCV and Python》, 2021
(https://nanonets.com/blog/ocr-with-tesseract/)

[9] Tesseract 官網的語言列表
(https://github.com/tesseract-ocr/tesseract/blob/master/doc/tesseract.1.asc#LANGUAGES)

[10] EasyOCR GitHub
(https://github.com/iwstkhr/python-ocr-sample)

[11] Youngmin Baek et al., Character Region Awareness for Text Detection
(https://arxiv.org/pdf/1904.01941)

[12] keras-ocr 套件
(https://github.com/faustomorales/keras-ocr)

[13] imanoop7/Ollama-OCR GitHub
(https://github.com/imanoop7/Ollama-OCR)

[14] Aswinth Raj,《Car License Plate Recognition using Raspberry Pi and OpenCV》, 2019
(https://circuitdigest.com/microcontroller-projects/license-plate-recognition-using-raspberry-pi-and-opencv)

[15] RembgGitHub
(https://github.com/danielgatis/rembg)

[16] Xuebin Qin et al., U 2 -Net: Going Deeper with Nested U-Structure for Salient Object Detection
(https://arxiv.org/pdf/2005.09007)

[17] Xuebin Qin et al., U 2 -Net GitHub
(https://github.com/xuebinqin/U-2-Net)

[18] CUDA Toolkit Downloads
(https://developer.nvidia.com/cuda-downloads)

[19] CuDNN Downloads
(https://developer.nvidia.com/cudnn)

[20] Rembg 底層支援各種的 U2-Net 模型
(https://github.com/danielgatis/rembg?tab=readme-ov-file#models)

[21] U2-Net GitHub
(https://github.com/xuebinqin/U-2-Net)

[22] Keras Code Examples
(https://keras.io/examples/)

MEMO

MEMO

深智數位
股份有限公司

深智數位
股份有限公司